Current Topics in Microbiology and Immunology

Volume 421

The review series *Current Topics in Microbiology and Immunology* provides a synthesis of the latest research findings in the areas of molecular immunology, bacteriology and virology. Each timely volume contains a wealth of information on the featured subject. This review series is designed to provide access to up-to-date, often previously unpublished information.
2017 Impact Factor: 5.829, 5-Year Impact Factor: 4.427

More information about this series at http://www.springer.com/series/82

Steffen Backert
Editor

Molecular Mechanisms of Inflammation: Induction, Resolution and Escape by *Helicobacter pylori*

Responsible series editor: Klaus Aktories

Springer

Editor
Steffen Backert
Division of Microbiology,
Department of Biology
University of Erlangen-Nuremberg
Erlangen, Germany

ISSN 0070-217X ISSN 2196-9965 (electronic)
Current Topics in Microbiology and Immunology
ISBN 978-3-030-15137-9 ISBN 978-3-030-15138-6 (eBook)
https://doi.org/10.1007/978-3-030-15138-6

Library of Congress Control Number: 2019933904

This Springer imprint is published by the registered company Springer Nature Switzerland AG.
The registered company address is: Gewerbestrasse 11, 6330 Cham, Switzerland

Foreword: New Directions for Investigating Detrimental and Enabling Interactions Between *Helicobacter pylori* and Its Host

Grant support: NIH CA-116087, DK-58404, DK-58587, and CA-77955

Address correspondence to: Richard M. Peek Jr.
Division of Gastroenterology
Vanderbilt University
1030C MRB IV
2215B Garland Avenue
Nashville, TN 37232-2279
Phone: 615/322-5200
Fax: 615/343-6229
email: richard.peek@vanderbilt.edu

Helicobacter pylori infects the stomachs of approximately 4.4 billion persons, over half of the global population, justifying its stature as the most common bacterial infection worldwide. Virtually, all persons infected by this organism develop co-existing gastritis, a signature feature of which is the capacity to persist for decades. Microbial persistence implies a relationship in which the signals of the colonizing organism affect the signals of the host, allowing host and bacteria to participate in a dynamic equilibrium, the intensity of which is dependent upon environmental influences. Unfortunately, there are biological costs for these long-term relationships. For example, chronic infection with *H. pylori* incurs the highest known risk for gastric cancer, and *H. pylori* is the only bacterial species designated by the World Health Organization as a class I carcinogen. With an estimated 1 million new cases per year, gastric adenocarcinoma claims >780,000 lives annually and approximately 80% of the gastric cancer burden and 5.5% of all malignancies worldwide are attributable to *H. pylori*-induced inflammation and injury. Eradication of *H. pylori* significantly decreases the risk of developing cancer in infected individuals; however, universal test and treat strategies for *H. pylori* are not feasible due to the high prevalence of infection, the fact that only a subset of colonized persons ever develop neoplasia, and the expense and side effects of antibiotic therapy. These observations, in conjunction with evidence that carriage of certain strains is inversely related to esophagal adenocarcinoma, atopic diseases, and inflammatory bowel disease, underscore the importance of identifying mechanisms that regulate pathologic interactions of *H. pylori* with its host which promote the disease. All of the chapters contained in this unique book broadly and comprehensively describe the latest insights about such mechanisms.

The *cag* pathogenicity island is a fascinating *H. pylori* strain-specific gene locus that encodes a type IV secretion system (T4SS), and *cag*$^+$ strains markedly augment the risk for disease compared to *cag*$^-$ strains. The protein product of the *cagA* gene (CagA) is translocated by the T4SS into epithelial cells, undergoes tyrosine phosphorylation, and activates a eukaryotic phosphatase (SHP-2) and various other signaling factors, leading to pathogenic cellular responses. Non-phosphorylated CagA also exerts pathologic effects via induction of pro-inflammatory signaling, activation of β-catenin, and disruption of apical-junctional complexes. Recently, elegant studies have demonstrated that gastric stem and progenitor cells can be specifically targeted and activated by *cag*$^+$ *H. pylori*, which likely contributes to the enhanced carcinogenic risk incurred by these strains.

In addition to the *cag* T4SS, the *H. pylori* genome contains an unusually high proportion of open reading frames that encode outer membrane proteins (OMPs). Consistent with sequence data, *H. pylori* strains express multiple paralogous OMPs, several of which bind to the defined receptors on gastric epithelial cells. The first defined *H. pylori* host receptor partnership involved BabA, a highly conserved OMP encoded by the strain-specific gene *babA2*, which binds the Lewis histo-blood-group antigen Le^b. A second ligand–receptor coupling involved SabA, an *H. pylori* adhesin that binds the sialyl-Lewis^x (Le^x) antigen, an established tumor antigen and marker of gastric dysplasia that is upregulated by chronic gastric inflammation. Host integrin receptors also represent an entry point for CagA injection, and an important role is played by CagL, a T4SS-pilus-localized protein. CagL bridges the T4SS to $\alpha_5\beta_1$-integrins on target cells and activates host cell focal adhesion kinase (FAK) and Src to ensure that CagA is phosphorylated directly at its site of injection. Additional Cag proteins (CagA, CagI, CagY) also bind β_1 integrin and induce conformational changes of integrin heterodimers, which permits CagA translocation. Recently, a fourth cognate *H. pylori* adhesin-host receptor partnership was identified as an *H. pylori* outer membrane protein, HopQ, which was shown to bind to members of the host carcinoembryonic antigen-related cell adhesion molecule (CEACAM) family of receptors, required for CagA translocation and subsequent signaling within epithelial cells. This remarkable finding has identified an additional target for novel antimicrobial therapeutic strategies as well as providing a framework for new studies of *H. pylori* pathogenesis.

H. pylori has not only evolved mechanisms to aberrantly activate inflammatory signaling but to also evade host recognition and clearance. Toll-like receptors (TLRs) orchestrate immune responses targeting pathogens and bridge innate and adaptive immunity via selective recognition of pathogen-associated molecular patterns (PAMPs). However, *H. pylori* harbors multiple PAMPs that interact differently with TLRs than the respective counterparts in other mucosal pathogens. *H. pylori* FlaA is a non-inflammatory molecule in terms of its ability to activate TLR5. *H. pylori* LPS contains an anergic lipid A core that induces an attenuated TLR4-mediated response. We and others have shown that deacetylation of peptidoglycan allows *H. pylori* to evade host clearance and that pre-activation of Nod1 suppresses *H. pylori*-induced signaling via activation of a Nod1-dependent negative feedback loop. Thus, *H. pylori* has evolved to express an array of diverse phenotypes to subvert obstacles presented by the host, which promotes long-term colonization and, inherently, the development of the disease.

Environmental conditions such as high salt diets or iron deficiency can also augment the risk for disease by enhancing the virulence of *H. pylori*. Exposure of *H. pylori* cag^{+} strains to increasing concentrations of salt markedly enhances the expression of CagA, and the levels of induction are dependent upon a salt-sensitive motif within the CagA promoter. Iron deficiency also accelerates the progression to carcinogenesis within the context of *H. pylori* infection and iron depletion augments assembly of the *cag* T4SS and translocation of CagA into host epithelial cells. Are there other elements that contribute to *H. pylori*-induced diseases? Studies in mice have determined that the gastric microbiota collaborates with *H. pylori* to augment the progression to gastric cancer, and persons with premalignant and malignant gastric lesions have distinct microbial populations in the stomach, which is termed dysbiosis. Restoration of gastric dysbiosis is an exciting area for future research when viewed within the context of data demonstrating that manipulation of the intestinal microbiome improves tumor responses to checkpoint inhibitor therapies. Specifically for *H. pylori*, computational modeling has shown that non-*H. pylori* bacterial lateral gene transfer occurs frequently in gastric cancers and, of interest, 2 of the 5 most targeted host genetic loci for bacterial DNA integration are *CEACAM5* and *CEACAM6*, which encode proteins known to be up-regulated in gastric cancer. Over-expression of CEACAMs in three-dimensional tissue culture systems aberrantly affects tissue organization and architecture and leads to the formation of cancer in mice. Heightened expression of CEACAMs induced by a dysbiotic gastric microbial community may also facilitate *H. pylori* binding, thereby conferring detrimental consequences for the host over prolonged periods of time.

Together, I feel very much delighted to be asked by Steffen Backert to provide this Foreword for the volume on "Molecular Mechanisms of Inflammation: Induction, Resolution and Escape by *Helicobacter pylori*" that he has edited. I was fascinated by the topic list and impressed by the fantastic group of international experts contributing to this book. Ultimately, the book covers crucial investigations that focus on *H. pylori*-induced diseases that will help to construct a paradigm for other cancers that arise from inflammatory foci such as hepatocellular carcinoma and cholangiocarcinoma. Similarly, chronic esophagitis, pancreatitis, and ulcerative colitis each confer a significantly increased risk for the development of cancer within their respective anatomic sites. Thus, a comprehensive understanding of how *H. pylori* initiates gastric cancer will impact our knowledge of how chronic inflammation leads to malignant degeneration in other organ systems. The insight contained in each of these chapters provides such an understanding and is, therefore, a welcome addition to this field.

Nashville, TN, USA

Richard M. Peek, Jr.
Departments of Medicine and Pathology
Microbiology, and Immunology
Vanderbilt University Medical Center

Contents

Abbreviations

~P	Phosphate group
ABCG2	ATP-binding cassette subfamily member G2
Abl	Abelson kinase
ACE	Angiotensin-converting enzyme
ACRG	Asian cancer research group
ADP	Adenosine diphosphate
ADP-Hep	ADP-β-D-manno-heptose
AID	Activation-induced cytidine deaminase
AIEC	Adherent-invasive *Escherichia coli*
AJ	Adherens junction
Aka	Also known as
AKAP	A-kinase-anchoring protein
AKT/PKB	Protein kinase B
ALDH1	Aldehyde dehydrogenase 1
AlpA	Adherence-associated lipoprotein A
AlpB	Adherence-associated lipoprotein B
ALPK1	Alpha protein kinase1
AmiE	Amidase
AmiF	Formidase
AMP	Antimicrobial peptide
AP-1	Activator protein-1
APC	Antigen-presenting cell
APE1	Apurinic/apyrimidinic endonuclease 1
APOBEC	Apolipoprotein B mRNA-editing enzyme, catalytic polypeptide-like
Arg2	Arginase II
ARHGAP6	Gene name for: Rho GTPase-activating protein 6
ARHGAP26	Gene name for: Rho GTPase-activating protein 26
ARID1A	AT-rich interactive domain-containing protein 1A
ASC	Adult stem cell, adaptor protein

ASPP2	Apoptosis-stimulating protein of p53 2
ATF-2	Activating transcription factor 2
ATM	Ataxia telangiectasia mutated kinase
ATP	Adenosine triphosphate
ATR	ATM and Rad3-related kinase
ATRIP	ATR-interacting protein
B2M	Beta-2-microglobulin
BabA	Blood group antigen-binding adhesin A
BabB	Blood group antigen-binding adhesin B
BAC	Bacterial artificial chromosome
BAFF	B-cell-activating factor
Barx1	BarH-like homeobox 1
BCL2	B-cell lymphoma 2 gene
BCL3	B-cell lymphoma 3-encoded protein
BCL10	B-cell lymphoma 10 gene
BCR	B-cell receptor
BE	Barrett's esophagus
BER	Base excision repair
BIR	Baculovirus inhibitor of apoptosis repeat protein domain
BIRC3	Baculoviral IAP repeat containing 3
BMDC	Bone marrow-derived cell
BMM	Bone marrow-derived macrophage
BMP	Bone morphogenic protein
BrdU	Bromodeoxyuridine
C/EBP	CCAAT/enhancer-binding protein
cag	Cytotoxin-associated gene
cag3	Cytotoxin-associated gene 3
cagA	Cytotoxin-associated gene A
CagA	Protein encoded by *cagA*
cagC	Cytotoxin-associated gene C
cagI	Cytotoxin-associated gene I
cagL	Cytotoxin-associated gene L
cagM	Cytotoxin-associated gene M
*cag*PAI	*cag* pathogenicity island
cagT	Cytotoxin-associated gene T
cagX	Cytotoxin-associated gene X
cagY	Cytotoxin-associated gene Y
cAMP	Cyclic adenosine monophosphate
CARD	Caspase recruitment protein domain
CARMA	CARD-containing protein
Cas9	CRISPR-associated protein 9
CASK	Ca/Calmodulin-dependent serine protein kinase
CASP 1	Caspase 1
Cav1	Caveolin 1
CBM	CARD11-BCL10-MALT1 protein complex

CCKBR	Cholecystokinin B receptor (also called CCK_2)
CCL/CXCL	Chemokine ligand
CCND1	Gene name for: Cyclin D1
CCNE1	Gene name for: Cyclin E1
CCR	CC-chemokine receptor
CcrM	*Caulobacter crescentus* DNA methyltransferase
CD	Cluster of differentiation
Cdc6	Cell division control protein 6
CDH1	Gene name for: E-cadherin
CDKN	Cyclin-dependent kinase inhibitor gene
Cdx2	Caudal-type homeobox 2
CEA	Carcinoembryonic antigen
CEACAM	Carcinoembryonic antigen-related cell adhesion molecule
CEA-TCB	Carcinoembryonic antigen-T-cell-specific antibody
CFU	Colony-forming unit
CG	Cholesteryl glucoside
CGD	Chronic granulomatous disease
CGRP	Calcitonin gene-related peptide
CGT	Cholesterol glycosyl-transferase
Chk1	Checkpoint kinase 1
Chk2	Checkpoint kinase 2
CIMP	CpG island methylator phenotype
CIN	Chromosomal instability
CK1	Casein kinase 1
CLD18	Gene name for Claudin 18
CLP	Common lymphoid progenitor
CLR	C-type lectin receptor
CLS	CBF1 suppressor of hairless Lag-1
CM	CagA multimerization sequence
c-Met	Tyrosine kinase (also called hepatocyte growth factor receptor)
comB	*H. pylori* competence gene B
comH	*H. pylori* competence gene H
Cox1	Cyclooxygenase 1
Cox2	Cyclooxygenase 2
CpG	Cytosine-guanine repeats
CRE	cAMP response element
CREB	cAMP responsive element-binding protein
CRISPR	Clustered regularly interspaced short palindromic repeats
CRPIA	Conserved repeat responsible for phosphorylation independent activity
CSC	Cancer stem cells
CSF/MCSF	Colony-stimulating factor
CSFR	Colony-stimulating factor receptor
Csk	Carboxy-terminal Src kinase
CSMD	CUB and sushi multiple domains

c-Src	Cellular sarcoma kinase
CT	Cholera toxin
CTLA4	Cytotoxic T lymphocyte-associated antigen 4
CTNNA1	Gene name for: α-1-Catenin
CTNNB1	Gene name for: β-1-Catenin
CTNND1	Gene name for: δ-1-Catenin
CXCL1	C-X-C motif chemokine ligand 1 (also called keratinocyte chemokine, KC)
CXCL10	C-X-C motif chemokine ligand 10 (also called IL-10)
CXCL8	C-X-C motif chemokine ligand 8 (also called IL-8)
DC	Dendritic cell
DC-SIGN	Dendritic cell-specific HIV-1 receptor (also called CD209)
DDR	DNA damage response
DFMO	Difluoromethylornithine
DGC	Diffuse gastric cancer
DIMT1	DIM 1 dimetyladenosine transferase 1 homolog
DKK	Dickkopf 1
DLBCL	Diffuse large B cell lymphoma
DNA	Deoxyribonucleic acid
DP	Double-positive
DSB	Double-strand DNA breaks
dsDNA	Double-stranded DNA
DU	Duodenal ulcer
DupA	Duodenal ulcer promoting gene A
EAC	Esophageal adenocarcinoma
EBER	EBV encoded small RNA
EBNA	EBV nuclear antigen
EBNA-LP	EBNA-leader protein
EBV	Epstein–Barr virus
ECD	Extracellular domain
ECL	Enterochromaffin-like
ECM	Extracellular matrix
EGF	Epidermal growth factor
EGFR	Epidermal growth factor receptor
EHEC	Enterohemorrhagic *Escherichia coli*
EMT	Epithelial mesenchymal transition
Epi-CAM	Epithelial cell adhesion molecule
EPIYA	Glu-Pro-Ile-Tyr-Ala sequence motif
ER	Endoplasmic reticulum
ERBB2(HER2)	Erb-b2 receptor tyrosine kinase 2
ERCC	Excision repair cross-complementing gene
ERK1/2	Extracellular signal-regulated kinase 1/2
ESC	Embryonic stem cell
FADD	Fas-associated protein with death domain
FAK	Focal adhesion kinase

FAP	Familial adenomatous polyposis
FasL	Fas ligand
FBXO24	F-box protein 24
FDA	Food and Drug Administration
Fe	Ferrum (iron)
FGC	Familial gastric cancer
FGF-10	Fibroblast growth factor-10
FGFR2	Fibroblast growth factor receptor 2
FHIT	Fragile histidine triad
Fic	Filamentation-induced by cAMP
FIGC	Familial intestinal gastric cancer
FISH	Fluorescence in situ hybridization
FlaA	Flagellin A
FlaB	Flagellin B
FliC	Flagellin FliC
Foxal	Forkhead box A1
FoxP1	Forkhead box P1
FoxP3	Forkhead box P3
FtsK	Filamentous temperature-sensitive cell division protein K
FZD7	Frizzled 7
GalNAc	α-N-acetylgalactosamine
GAP	GTPase-activating protein
GAPPS	Gastric adenocarcinoma and proximal polyposis of the stomach
GATA6	GATA-binding protein 6
gB	EBV glycoprotein B
GC	Gastric cancer
GCB	Germinal centre B cell
G-DIF	Diffuse subtype of GC
GEC	Gastric epithelial cell
GEJC	Gastroesophageal junction carcinomas
GERD	Gastro-esophageal reflux disease
GGT	Gamma-glutamyl transpeptidase
gH	EBV glycoprotein H
G-INT	Intestinal subtype of GC
GITR	Glucocorticoid-induced TNFR-related protein
gL	EBV glycoprotein L
GlcNAc	N-acetyloglucosamine
GMDS	GDP-mannose 4,6-dehydratase
gp42	EBV glycoprotein 42
gp130	Glycoprotein 130
gp350	EBV glycoprotein 350
Grb2	Growth factor receptor-bound protein 2
GSH	Glutathione (reduced form)
GSK-3β	Glycogen synthase kinase 3 beta
GSTP1	Glutathione S-transferase P

GWAS	Genome-wide-associated study
GyrA	Subunit A of DNA gyrase
GyrB	Subunit B of DNA gyrase
H2AX	Histone H2A variant X
H_2RA	H_2-receptor antagonist
hBD	Human beta-defensin
HBP	Heptose-1,7-bisphosphate
HBV	Hepatitis B virus
HDGC	Hereditary diffuse gastric cancer
HER2/neu	Human epidermal growth factor receptor 2
HGF	Hepatocyte growth factor
HHI	*H. pylori* HtrA inhibitor
HIF-1	Hypoxia-inducible factor-1
HIV	Human immunodeficiency virus
HKα	H,K-ATPase α subunit
HLA	Human leukocyte antigen class II
HLA-B	Histocompatibility complex class I
HNF4α	Hepatocyte nuclear factor alpha
HNPCC	Hereditary non-polyposis colorectal cancer
Hof	Helicobacter OMP family protein
Hop	Helicobacter Outer membrane protein
HopB	Helicobacter Outer membrane protein B
HopC	Helicobacter Outer membrane protein C
HopD	Helicobacter Outer membrane protein D
HopH	Helicobacter Outer membrane protein H
HopP	Helicobacter Outer membrane protein P
HopQ	Helicobacter Outer membrane protein Q
HopS	Helicobacter Outer membrane protein S
HopT	Helicobacter Outer membrane protein T
HopU	Helicobacter Outer membrane protein U
HopZ	Helicobacter Outer membrane protein Z
Hor	Helicobacter Outer-related protein
HPA	Heparanase
HpaA	Neuraminyllactose-binding hemagglutinin
HSC	Hematopoietic stem cells
Hsp60	Heat shock protein 60
HspB	Heat shock protein B
HtrA	High temperature requirement A
HU	Histone-like protein
Hupki	Human TP53 knock-in
IARC	International agency for research on cancer
IBD	Inflammatory bowel disease
ICOS	Inducible co-stimulator
ICOS-L	Inducible co-stimulator ligand

ICPi	Immune checkpoint inhibitors
IEL	Intra-epithelial lymphocytes
IFN	Interferon
IFNGR	Interferon gamma receptor
Ig	Immunoglobulin
IgA/B	Immunoglobulin A/B
IGFIIR	Insulin-like growth factor II receptor
IGH	Immunoglobulin heavy chain
IGHV	Immunoglobulin heavy chain variable region
IHF	Integration host factor
IKK	IκB kinase
IL	Interleukin
IL-1	Interleukin-1
IL-12	Interleukin-12
IL-17	Interleukin-17
IL-1R	Interleukin-1 receptor
IL-1RA	Interleukin-1 receptor antagonist
IL-6	Interleukin-6
IL-8	Interleukin-8
IL-21	Interleukin-21
IL-23	Interleukin-23
IM	Intestinal metaplasia
IMC	Inner membrane complex
iNKT	Invariant NK T-cells
iNOS	Inducible nitric oxide synthase
INS-GAS	Insulin-gastrin
INSR	Insulin receptor
IP_3	Inositol triphosphate
IPD	Interpulse duration
IPEX	Immunodysregulation polyendocrinopathy enteropathy X-linked
iPSC	Induced pluripotent stem cell
IPSID	Immunoproliferative small intestinal disease
IRAK	IL-1 receptor-associated kinase
IRF	Interferon-regulatory factor
IRG	Interferon-response gene
ISGF3	Interferon-stimulated gene factor 3
ITAM	Immunoreceptor tyrosine-based activation motif
ITIM	Immunoreceptor tyrosine-based inhibition motif
JAK	Janus kinase
JAM	Junctional adhesion molecule
JNK	c-Jun kinase
JUP	Junction plakoglobin
KCNQ1	Potassium voltage-gated channel subfamily Q member 1
KLF	Krüppel-like factor
KO	Knockout

Kras	Kirsten rat sarcoma oncogene
LabA	LacdiNAc-specific adhesin
LacdiNAc	GalNAcβ1–4GlcNAc
LAT	Linker for activation of T cells
LBP	Lipopolysaccharide-binding protein
LCK	Lymphocyte-specific protein tyrosine kinase
LDL-C	Low-density lipoprotein-cholesterol
Le^a	Lewis A antigen
Le^b	Lewis B antigen
LES	Lower esophageal sphincter
Le^x	Lewis X antigen
Le^y	Lewis Y antigen
LFA-1	Lymphocyte function-associated antigen 1
LFS	Li-Fraumeni syndrome
Lgr5	Leucin-rich-repeat-containing G-protein coupled receptor 5
LIMA1	LIM domain and actin-binding protein-1
LL37	37-residue amphipathic α-helical cathelicidin
LMO	LIM domain-only protein fragment
LMP	Latent membrane protein
LOH	Loss of heterozygosity
LPS	Lipopolysaccharide
Lrp	Global regulatory protein
LRR	Leucin-rich repeat
LSP1	Lymphocyte-specific protein 1
LT	Labile toxin
LTβ	Lymphotoxin-β
M1	Type 1 macrophage
M2	Type 2 macrophage
m4C	N4-methylcytosine
m5C	5-methylcytosine
m6A	N6-methyladenine
MAdCAM-1	Mucosal addressin cell adhesion molecule 1
MAGI-1	Membrane-associated guanylate kinase with inverted orientation 1
MAGUK	Membrane-associated guanylate kinase-like protein domain
MALT	Mucosa-associated lymphoid tissue
MALT1	MALT lymphoma translocation protein 1
MAMP	Microorganism-associated molecular pattern
MAP	Mitogen-activated protein
MAPK	Mitogen-activated protein kinase
MARK	Microtubule affinity-regulating kinase
MCL-1	Myeloid cell leukemia 1
MD2	Lymphocyte antigen 96
MDC1	Mediator of DNA damage checkpoint protein 1
MDCK	Madin-Darby canine kidney
MDM2	Mouse double minute 2 homolog

MDSC	Myeloid-derived suppressor cell
MGMT	6-*O*-methylguanine-DNA transferase
MHC	Major histocompatibility complex
Mincle	Macrophage inducible C-type lectin
miR	Non-coding micro-RNA (also called micro-RNA)
miRNA	Non-coding micro-RNA
Mist1	Basic helix-loop-helix family member a15
MKI	MARK kinase inhibitor
MLC	Myosin light chain
MLCK	Myosin light chain kinase
MLH	Human homolog of MMR from *Escherichia coli*
MLN	Mesenteric lymph nodes
MLST	Multilocus sequence typing
MMP	Matrix metalloprotease
MMR	DNA mismatch repair
MOI	Multiplicity of infection
MPF	Mating pair formation
Mre11	Meiotic recombination 11 homolog
MRN	MRE11-RAD50-NBS1 complex
mRNA	Messenger ribonucleic acid
MRR	Middle repeat region (CagY)
MS	Multiple sclerosis
MSI	Microsatellite instability
Msi-1	Musashi RNA-binding protein-1
MSI-H	High microsatellite instability
MSI-L	Low microsatellite instability
MSS	Microsatellite stable
Mtase	Methyltransferase
mtDNA	Mitochondrial DNA
MTHFR	Methylenetetrahydrofolate reductase
mTOR	Mechanistic target of rapamycin
MUC5AC	Mucin 5AC
MUC6	Mucin-6
MukB	SMC homolog
MUPP	Multi-PDZ domain protein
MYC	MYC proto-oncogene
MyD88	Myeloid differentiation primary response gene 88
MZB	Marginal zone B
mβCD	Methyl-β-cyclodextrin
NADPH	Nicotinamide adenine dinucleotide phosphate
NALP1	Nucleotide-binding domain, leucine-rich-containing family, pyrin domain-containing 1 (also called NLRP1)
NapA	Neutrophil-activating protein A (also called HP-NAP or Dps)
Nbs1	Nijmegen breakage syndrome 1
NCI	National Cancer Institute of the United States

NEIL	Nei-like protein
NEMO	NF-κB essential modifier
NER	Nucleotide excision repair
NF-κB	Nuclear factor-kappa B
NFAT	Nuclear factor of activated T cells
NGS	Next-generation sequencing
NHEJ	Non-homologous end joining
NICD	Notch intracellular domain
NIK	NF-κB-inducing kinase
NK	Natural killer
NLR	Nucleotide-binding domain and leucine-rich-repeat-containing-proteins
NLRP	Nucleotide-binding domain, leucine-rich-containing family, pyrin domain-containing
NO	Nitric oxide
NOD	Nucleotide-binding oligomerization domain
NOG	Noggin
NOS2	Nitric oxide synthase 2
NOX	NAPDH oxidase
NOXA1	NADPH oxidase activator 1
NRR	Negative regulatory region
NSAIDs	Non-steroidal anti-inflammatory drugs
NTH1	Endonuclease III homolog 1
nTreg	Natural T regulatory
OCT1	Octamer transcription factor1
ODC	Ornithine decarboxylase
ODN	Oligodeoxynucleotide
OGG	Oxoguanine DNA glycosylase
OGG1	8-oxo-guanine-DNA glycosylase 1
OipA	Outer inflammatory protein A
OMP	Outer membrane protein
OMV	Outer membrane vesicle
$ONOO^{-}$	Peroxinitrite
ORC1	Origin recognition complex subunit 1
ORF	Open reading frame
OSCC	Oesophageal squamous cell carcinoma
p38	Mitogen-activated protein kinase p38
p53	Tumor suppressor protein
p53BP1	p53-binding protein 1
PAI	Pathogenicity island
PAMP	Pathogen-associated molecular pattern
PAR1	Protease-activated receptor 1
Par1b	Partitioning-defective kinase 1b
ParA	Chromosome partitioning protein ParA
ParB	Chromosome partitioning protein ParB

PARK	Parkin gene
PARP1	Poly [ADP-ribose] polymerase 1
parS	Centromere-like sequence
PCR	Polymerase chain reaction
PD-1	Programmed death protein 1
pDC	Plasmacytoid dendritic cell
PDCA-1	Plasmacytoid dendritic cell antigen-1
PDCD1LG2	Programmed cell death 1 ligand 2
PDL1/B7H1	Programmed cell death ligand 1
Pdx1	Pancreatic and duodenal homeobox 1
PG	Peptidoglycan
PGC	Pepsinogen C
PGE2	Prostaglandin E2
PGE-M	Prostaglandin E2 metabolite
PHLPP1	PH domain leucine rich phosphatase 1
PI3K	Phosphatidylinositol 3-kinase
PJS	Peutz-Jeghers syndrome
PKC	Protein kinase C
PKLR	Pyruvate kinase isozymes R/L
PLC	Phosphoinositide phospholipase C
PLCE1 1	Phosphatidylinositol-4,5-bisphosphate phosphodiesterase epsilon-1
PLD1	Phospholipase D1
PMN	Polymorphonuclear neutrophil
PPARγ	Peroxisome proliferator-activated receptor γ
PPI	Proton pump inhibitor
Pre-B	Pre-B cell receptor in precursor B cells
PRK2	Protein kinase C-related kinase 2
PRKAA1	Protein Kinase AMP-Activated Catalytic Subunit Alpha 1
PRR	Pattern recognition receptor
PS	Phosphatidylserine
PSC	Pluripotent stem cells
PSCA	Prostate stem cell antigen
PTEN	Phosphatase and tensin homolog
PTGS2	Prostaglandin endoperoxide synthase 2
PUD	Peptic ulcer disease
Rac1	Ras-related C3 botulinum toxin substrate 1
RAG	Recombination-activating Genes
RegIII	Regenerating islet-derived III
REL	Rel homology domain (RHD) in NF-κB transcription factors
RGD	Arg-Gly-Asp sequence motif
RGDLXXL	Arg-Gly-Asp-Leu/Met-X-X-Leu/Ile sequence motif
RhoA	Ras homolog gene family A
RHS	RGD helper sequence
RIDA	Regulatory inactivation of DnaA activity

RIG1	Retinoic acid inducible gene 1
RIPK2	Receptor-interacting serine/threonine kinase 2
R-M systems	Restriction-modification systems
RNA	Ribonucleic acid
RNF43	Ring finger protein 43
RNS	Reactive nitrogen species
RocF	Urea-producing arginase
ROCK	Rho-Kinase
ROS	Reactive oxygen species
RPTP	Receptor protein tyrosine phosphatase
RSPO	R-spondin1
RTK	Receptor tyrosine kinase
RT-PCR	Reverse transcriptase-polymerase chain reaction
RUNX3	Runt-related transcription factor 3
SabA	Sialic acid-binding adhesin
SCF	Stem cell factor
SCID	Severe combined immune deficiency
SeqA	Sequestration protein A
SFK	Src family kinase
SH2	Src homology 2
SH3	Src homology 3
SHH	Sonic hedgehog
SHP-1/2	Src homology region 2 domain-containing phosphatase-1/2
siRNA	Small interfering RNA
SLB	Single layer antiparallel β-sheet
SLC1A2	Solute carrier family 1 member 2
SLT	Soluble lytic transglycosylase
SMAD2	Mothers against decapentaplegic homolog 2
SMC	Structure maintenance of chromosomes
SMOX	Spermine oxidase
SNP	Single-nucleotide polymorphism
Soj	Sporulation initiation inhibitor Soj
Sox4	Sex-determining region Y (SRY)-box 4
Sox9	Sex-determining region Y (SRY)-box 2
SP1	Specificity protein 1
SP-D	Surfactant-binding protein D
SPEM	Spasmolytic polypeptide-expressing metaplasia
Spo0J	Chromosome partitioning protein
SPR	Surface plasmon resonance
Src	Sarcoma kinase
SSB	Single-strand binding
ssDNA	Single-stranded DNA
STAT	Signal transducer and activator of transcription
STAT1/3/4/6	Signal transducer and activator of transcription factor-1/3/4/6
STMN1	Stathmin

T3SS	Type III secretion system
T4SS	Type IV secretion system
TAK1	Transforming growth factor beta-activated kinase 1
TAKK	Transforming growth factor β-activating kinase
TAM	Tumor-associated macrophage
TBK-1	Serine/threonine protein kinase-1
TCF/LEF	T-cell-specific transcription factor/lymphoid enhancer-binding factor
TCGA	The Cancer Genome Atlas
TCR	T-cell receptor
TdT	Terminal deoxynucleotidyl transferase
TER	Transepithelial electrical resistance
TFF	Trefoil factor family
TFH	T follicular helper
TGCT	Tenosynovial giant cell tumor
TGFR	Transforming growth factor receptor
TGF-β	Transforming growth factor beta
TGF-βRII	Transforming growth factor β receptor II
Th	T helper cell
Th1	T helper 1 cell
Th17	T helper 17 cell
TIFA	TRAF-interacting protein with forkhead-associated domain
TIMP3	Metalloproteinase inhibitor 3
Tipα	TNF-inducing factor α
TIR	Toll/IL-1 receptor domain
TIRAP	TIR domain containing receptor protein
TJ	Tight junction
TLR	Toll-like receptor
TME	Tumor microenvironment
TNF	Tumor necrosis factor
TNFR1	TNF receptor 1
Tnfrsf19	Tumor necrosis factor receptor super family 19
TP53	Tumor protein p53
TRAF	TNF receptor-associated factor
TRD	Target recognition domain
TRD1	Target recognition domain 1
TRD2	Target recognition domain 2
Treg	T regulatory cell
TRIF	TIR-domain-containing adapter-inducing interferon-β
TT	Tetanus toxoid
TWIST1	Twist family BHLH transcription factor 1
UC	Ulcerative colitis
Ure	Urease
UTR	Untranslated region
UV	Ultraviolet

vacA i	Intermediate segment of *vacA* gene
vacA m	Middle segment of *vacA* gene
vacA s	Signal segment of *vacA* gene
VacA	Vacuolating cytotoxin A
Vav	Rac-specific nucleotide exchange factor
VEGFA	Vascular endothelial growth factor A
VEGFR	Vascular endothelial growth factor Receptor
Vil1	Villin-1
VirB10	Inner core protein of the *Agrobacterium tumefaciens* T4SS
VirB4	Cytoplasmic ATPase of the *Agrobacterium tumefaciens* T4SS
VirD4	Channel ATPase of the *Agrobacterium tumefaciens* T4SS
v-Src	Viral sarcoma kinase
WHO	World Health Organization
WWOX	WW domain containing oxidoreductase
XRCC 1/3	X-ray repair cross-complementing group 1/3
αCAG	Cholesteryl-6'-O-tetradecanoyl-α-D-glucopyranoside
αCG	Cholesteryl-α-D-glucopyranoside
αCPG	Cholesteryl-6'-O-phosphatidyl-α-D-glucopyranoside

Inflammation, Immunity, and Vaccine Development for the Gastric Pathogen *Helicobacter pylori*

Tamaki Ikuse, Thomas G. Blanchard and Steven J. Czinn

Contents

Abstract It has been over 30 years since a link was established between *H. pylori* infection of the gastric mucosa and the development of chronic gastric diseases. Research in rodent models supported by data from human tissue demonstrated that the host immune response to *H. pylori* is limited by host regulatory T cells. Immunization has been shown to induce a potent Th1- and Th17-mediated immune response capable of eradicating or at least significantly reducing the bacterial load of *H. pylori* in the stomach in small animal models. These results have not translated well to humans. Clinical trials employing many of the strategies used in rodents for oral immunization including the use of a mucosal adjuvant such as *Escherichia coli* LT or delivery by attenuated enteric bacteria have failed to limit *H. pylori* infection and have highlighted the potential toxicity of exotoxin-based mucosal adjuvants. A recent study, however, utilizing a recombinant fusion protein of *H. pylori* urease and the subunit B of *E. coli* LT, was performed on over 4000 children. Efficacy of over 70% was demonstrated

---

T. Ikuse
Department of Pediatric and Adolescent Medicine, Juntendo University Graduate School of Medicine, Tokyo, Japan
e-mail: taikuse@juntendo.ac.jp

T. G. Blanchard (✉) · S. J. Czinn
Department of Pediatrics, University of Maryland School of Medicine, 13-015 Bressler Research Building, 655 West Baltimore Street, Baltimore, MD 21201, USA
e-mail: tblanchard@som.umaryland.edu

S. J. Czinn
e-mail: sczinn@som.umaryland.edu

S. Backert (ed.), *Molecular Mechanisms of Inflammation: Induction, Resolution and Escape by Helicobacter pylori*, Current Topics in Microbiology and Immunology 421, https://doi.org/10.1007/978-3-030-15138-6_1

against naturally acquired infection compared to control volunteers one year post-immunization. Efficacy was reduced, but still above 50% at three years. This study provided new insight into the strategies for developing an improved vaccine for widespread use in countries with high infection rates and where gastric cancer (GC) remains one of the most common causes of death due to cancer.

1 Introduction

Helicobacter pylori is a microaerophilic Gram-negative, spiral-shaped bacterial pathogen that resides in the human stomach, aided in large part by the production of urease against the highly acidic environment. *H. pylori* is one of the most prevalent global pathogens and colonizes an estimated 50% of the world's population (Parkin 2004). Prevalence varies widely throughout the world from 10% to greater than 70% (Graham et al. 1991; Hunt et al. 2011). Infection is primarily, but not exclusively, acquired in early childhood and spreads through oral–oral and fecal–oral transmission pathways. It lasts for the entire life of the host, and while most infected individuals remain asymptomatic, 10-20% of infected people develop peptic ulcer disease (PUD), ~1% develop gastric adenocarcinoma, and less than 1% develop mucosa-associated lymphoid tissue (MALT) lymphoma (Posselt et al. 2013). However, gastric cancer (GC) remains the fourth leading cause of death due to cancer worldwide and for large geographic regions including South America, Eastern Europe, and the Far East levels range from 20 to 40 per 100,000 people (Yamaoka et al. 2008). *H. pylori* may be responsible for 75% of all GC cases throughout the world (de Martel et al. 2012). The World Health Organization classified *H. pylori* as a definite class I carcinogen in 1994 (Bouvard et al. 2009), and this classification was confirmed in 2012 (IARC 2012). In this chapter, we review our current knowledge about the host immunity in response to *H. pylori* infection and the clinical vaccine trials that have elucidated potential strategies for developing an efficacious vaccine against *H. pylori*.

2 Host Immune Response to *H. pylori* Infection

H. pylori colonizes the surface of the gastric mucosa, where bacterial antigens are recognized and presented by antigen-presenting cells (APCs), including dendritic cells (DCs) and macrophages (Lee and Buck 1996). The activated APCs stimulate naïve $CD4^+$ T cells and induce antigen-specific responses in Th1 cells (D'Elios et al. 1997; Bimczok et al. 2010; Eaton et al. 2001) and Th17 cells (Rolig et al. 2011; Zhuang et al. 2011) by secreting cytokines such as TGF-β, IL-6, IL-12, IL-21, and IL-23. Th1 cells secrete IFN-γ and regulate cellular immunity, and Th17 cells secrete IL-17 and maintain mucosal barriers and contribute to pathogen clearance at mucosal surfaces. Several studies revealed significant expression of IL-12 and

IL-23 in *H. pylori*-stimulated macrophages and IFN-γ and IL-17 in gastric tissue of mice and humans with *H. pylori* infection (Shi et al. 2010; Shimizu et al. 2004; Bamford et al. 1998; Lindholm et al. 1998). Th1 and Th17 cell subset differentiation along with their signature cytokines are critical for proper control of *H. pylori* infection, whereas B cells and antibodies have been shown not to be required for control of *H. pylori* (Ermak et al. 1998). Additionally, Th1 and Th17 cell subtypes have been implicated in initiating and promoting the immunopathology in the chronically infected gastric mucosa due to inflammation that manifests histologically as atrophic gastritis, compensatory epithelial hyperplasia, and intestinal metaplasia in experimentally infected animals and symptomatic human carriers (Shi et al. 2010; Stoicov et al. 2009). The Th17/IL-17 pathway modulates cellular responses and promotes gastric epithelial cells (GECs), stromal cells, endothelial cells, and lamina propria mononuclear cells that express the IL-17 receptor to release cytokines such as IL-1, IL-6, IL-8, and TNF-α, which then attract neutrophils to attack *H. pylori* (Mizuno et al. 2005; DeLyria et al. 2009) and are also involved in the pathogenesis of *H. pylori*-induced gastric adenocarcinoma (Shi et al. 2010; Stoicov et al. 2009).

Whereas *H. pylori* infection promotes Th1 and Th17 cell responses and induces gastric mucosal inflammation leading to gastric adenocarcinoma, a T regulatory (Treg) cell response is also observed, which drives immune tolerance and suppresses Th1- and Th17-mediated immunity against *H. pylori* infection (Robinson et al. 2008; Harris et al. 2008; Arnold et al. 2011b; Oertli et al. 2013). The induction of Treg cells by *H. pylori* is in large part due to its lack of pathogen-associated molecular patterns (PAMPs) as illustrated in Fig. 1. While the T effector cell response is generally driven in symptomatic patients with *H. pylori*-associated diseases, a Treg-predominant response with high levels of IL-10 and TGF-β production tends to be involved in asymptomatic *H. pylori* carriers (Robinson et al. 2008; Matsumoto et al. 2005). Experimental murine infection suggests that the age at the time of first exposure against *H. pylori* influences the immune response with respect to pro-inflammatory T cells or Treg cells (Arnold et al. 2011b). Mice infected during adulthood develop chronic gastritis and preneoplastic lesions over time, but mice infected during the neonatal period mimic the asymptomatic human carrier and are protected against gastric pathology (Arnold et al. 2011b). Moreover, the neonatal infection model generates immune tolerance toward *H. pylori* that is driven by Treg cells, which suppress Th1 and Th17 cell activity (Arnold et al. 2011b).

Two *H. pylori* virulence determinants, the vacuolating cytotoxin A (VacA) and the γ-glutamyl-transpeptidase (GGT), are suggested to promote the differentiation of Treg cells over Th1 and Th17 cells (Oertli et al. 2013). VacA and GGT promote tolerant DCs to *H. pylori*, which retain a semi-mature phenotype and prime Treg differentiation (Oertli et al. 2012, 2013; Oertli and Muller 2012). Additionally, in the presence of VacA, DCs induced FoxP3 and CD25 expression in T cells, thereby promoting a bias toward regulatory T cell responses (Oertli et al. 2013; Maldonado and von Andrian 2010). Conversely, VacA deficiency leads to higher numbers of Th1 and Th17 cells and higher expression of IFN-γ and IL-17, which cause severe inflammation, atrophic gastritis, intestinal metaplasia, and epithelial hyperplasia (Oertli et al. 2013).

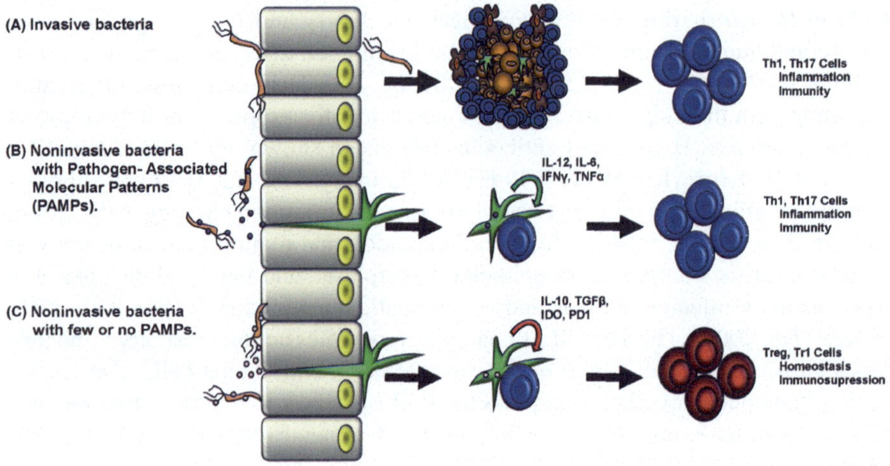

Fig. 1 Host immunity to mucosal bacteria varies with bacterial phenotype. The composition of mucosal bacteria and the expression of specific macromolecules influence the nature of the host response. An aberrant immune response to commensal bacteria in the alimentary tract can lead to inflammatory bowel diseases. **a** Any bacteria that are invasive will induce an immune response as these bacteria or bacterial components are ultimately filtered through the lymph tissue. The presence of these new, complex, and non-host elements will generate a T helper cell response capable of promoting a multi-faceted immune response. In the case of bacteria, this response would most likely be dominated by Th1 and/or Th17 cells. Repeated exposure to the same bacteria would evoke a strong secondary immune response. **b** Bacteria that remain in the lumen of the mucosa reside "outside" of the tissue. These bacteria are sampled by the immune system through dendritic cell (DC) surveillance. Macromolecules from resident bacteria are taken up by DCs which can then present bacterial peptides to T cells. The nature of the DC response is highly dependent upon the presence of pathogen-associated molecular patterns (PAMPs) such as flagellin protein, LPS, peptidoglycan, and others. If these PAMPs contain the appropriate consensus sequence or structure, they elicit maturation of the DC. The DC then activates T helper cells in the context of cytokines that promote the development of pro-inflammatory Th1 or Th17 cells. These T cells are then poised to promote a specific immune response to eradicate the potential pathogen. **c** Many bacteria that reside at the mucosa lack PAMPs with the ability to induce maturation in DC. In these cases, the DC presents bacterial peptides to T cells but in the context of cytokines and receptors that promote the activation of Treg cells. These Tregs have the ability to limit the host response to what the host treats as a commensal organism. *H. pylori* lacks PAMPs that induce maturation in DCs and are known to induce Treg activity, contributing to its ability to remain at the gastric mucosa

Studies support a crucial role for Treg cells in maintaining a benign interaction between *H. pylori* and its host. Additionally, immunomodulation and tolerance induced by Treg cells in response to *H. pylori* infection not only ensures persistent *H. pylori* infection but also contributes to protect mice from experimental allergic asthma (Oertli et al. 2013; Engler et al. 2014; Arnold et al. 2011a; Koch et al. 2015). *H. pylori*-induced Treg cell activity suppressed allergen-specific immune responses in the lung, airway hyper-responsiveness, tissue inflammation and goblet cell metaplasia, and prevention of allergen-induced pulmonary and bronchoalveolar infiltration with eosinophils, Th2 cells, and Th17 cells (Arnold et al. 2011a). Conversely, the depletion of Treg cells abrogates tolerance and asthma protection (Arnold et al. 2011a).

The decline of the prevalence of *H. pylori* infection in Western countries correlates with increases in childhood asthma and allergic diseases, and several epidemiological studies have reported an inverse correlation between asthma incidence and *H. pylori* infection (Blaser et al. 2008; Chen and Blaser 2007, 2008; Reibman et al. 2008). The relationship between asthma and *H. pylori* infection, however, remains unclear as outcomes are different depending on the population, age, and other factors (den Hollander et al. 2016; Wang et al. 2013, 2017; Zhou et al. 2013). A detailed understanding of the Treg response and tolerance promoted by *H. pylori* infection may lead to alternative strategies to antimicrobial eradication therapy. Immunomodulation of the response against *H. pylori* infection could be used to either reduce immunopathology, or conversely, promote a more aggressive inflammatory response with the potential to eradicate *H. pylori*.

3 Obstacles in Developing *H. pylori* Vaccine

H. pylori infection can be treated by established antimicrobial therapies. These therapies can vary, but in general consist of combination therapies including two antibiotics and a proton pump inhibitor (NIH 1994). Eradication therapy against *H. pylori* infection can lead to significant improvement of gastric atrophy and may prevent the development of GC (Ito et al. 2009; Kato et al. 2007; Lee et al. 2016; Ma et al. 2012; Mabe et al. 2009; Take et al. 2015). However, successful rates of eradication therapy are declining in recent years, and antibiotic resistance is thought to be the main cause of eradication failure (Kobayashi et al. 2007; Murakami et al. 2016; Nishizawa et al. 2015). In addition, these therapies are taken multiple times per day for a minimum of 7–14 days and are often accompanied by side effects including diarrhea and nausea. Patient compliance is often poor, which induces eradication failure as well. Moreover, antimicrobial eradication is not expected to protect against reinfection, although this would likely occur at a relatively low rate in adults. It is not fiscally or practically possible to treat one half of the world's population with antimicrobial agents in an effort to cure PUD or prevent GC, and it would also lead to the development of antibiotic resistance in both *H. pylori* and other human pathogens. Given the limitations of antimicrobial therapy and analyses that indicate the use of a prophylactic *H. pylori* vaccine could be cost-effective (Rupnow et al. 2009), a concerted effort has been made toward vaccine development.

Development of a vaccine against *H. pylori* is technically challenging given its niche in the gastric mucosa. The bacteria reside in the gastric mucous and in close association with the surface of GECs. Unlike some other enteric pathogens such as Salmonella which cross the gut epithelium, *H. pylori* is generally not invasive per se. It, therefore, remains sequestered from many immune effector mechanisms. Neutrophils are recruited and can cross the epithelium to form crypt abscesses, and similar to the intestinal mucosa the gastric mucosa contains secreted antibodies and antimicrobial peptides (see also Chapter "Carbohydrate-Dependent and Antimicrobial Peptide Defense Mechanisms Against *Helicobacter pylori* Infections" of this book). But these mechanisms appear insufficient to eradicate *H. pylori* infection. In addition, as described above, *H. pylori* induces Treg cells that are dominant which results in reduced inflammation, suppression of Th1- and Th17-mediated immunity, and persistent *H. pylori* infection. In the early days of mucosal immunology research, there was a consensus that traditional systemic immunizations were ineffective at inducing protection against mucosal pathogens and that such protection would have to be generated locally through mucosal immunization. This presented researchers with a technical challenge as protein antigens were known to stimulate weak immune response or even tolerance at mucosal surfaces. Although such difficulties could in theory be overcome through the use of adjuvants, no safe and efficacious adjuvants for use in mucosal immunization existed. This limitation contributes to the lack of vaccines against many enteric pathogens and venereal diseases.

4 Establishing the Principles for *H. pylori* Vaccination

One of the earliest and still most successful experimental mucosal adjuvants is exotoxins from bacteria such as cholera toxin (CT) from *Vibrio cholerae* and the *Escherichia coli* heat labile toxin (LT). These related toxins are composed of a pentameric ring of B subunits that bind to GM1 gangliosides present on epithelial cells and an A subunit that exhibits enzymatic activity (Spangler 1992). The B subunits, therefore, target the toxin to the cell surface, where the A subunit activates a G protein, which ultimately leads to the continuous production of cAMP, resulting in the efflux of ions and water. These exotoxins, however, can be given orally to animals in small doses where they are not only immunogenic, but when mixed in solution with an unrelated protein antigen confer potent immunogenicity to that protein when applied by mucosal immunization (Elson and Ealding 1984; Lycke and Holmgren 1986; Nedrud et al. 1987). In its early days of use, CT was applied in mice with great effectiveness to study host immunity to Sendai virus as a model for human influenza (Nedrud et al. 1987). The protocol developed to induce immunity to Sendai virus in the lungs was extended to models of *H. pylori* to determine whether similar protection could be induced against the gastric pathogen (Czinn and Nedrud 1991).

The basic protocol proved effective and has been used either as developed or with slight modification by numerous laboratories for over 25 years (Blanchard and Czinn 2017; Blanchard and Nedrud 2010). The protocol is used for both CT and LT. In the original protocol, mice received 1 mg *H. pylori* whole-cell lysate antigen mixed with 10 μg CT adjuvant by oral gavage. The immunization was given in four weekly doses. When serum and intestinal washes were examined, a fivefold increase in anti-*H. pylori* serum IgA titer and a 16-fold increase in intestinal IgA compared with mice immunized without CT was achieved. Anti-*H. pylori* IgG levels were also significantly increased. The immunogenicity study was also extended to ferrets. Most ferret's stomachs are naturally colonized by the closely related species *H. mustelae*

(Fox et al. 1990). Oral immunization of the ferrets with 7 mg *H. pylori* lysate and 60 μg CT doses also induced a significant increase in anti-Helicobacter antibody levels. This study demonstrated that oral immunization combined with experimental exotoxin adjuvants could be used to manipulate immunity at the gastric mucosa against *H. pylori* in animals.

Although this initial study demonstrated the utility of experimental oral immunization to induce mucosal immune responses against *H. pylori*, the development of a mouse model for gastric Helicobacter infection was an important step in vaccine development. A species isolated from domestic cats, *H. felis,* was shown to infect mice and to induce histologic gastritis similar to that observed in *H. pylori*-infected humans (Lee et al. 1990). The *H. felis* mouse model served as the original challenge model for Helicobacter immunization. Mice were given the immunization protocol described above, challenged 7–10 d after the final immunization, and then examined for 7 days after challenge (Czinn et al. 1993). Significant increases in serum and mucosal anti-*H. felis* antibody titers including a fourfold increase in gastric IgA, and an eightfold increase in intestinal IgA were achieved in mice, which received antigen plus CT adjuvant. When gastric tissue was tested for infection, 76% of immunized and challenged mice were determined to be protected from infection, whereas only 22% of control mice were bacteria free. A similar experiment was performed by a second group using the *H. felis* mouse model and a similar immunization protocol to achieve comparable results (Chen et al. 1992). Subsequent to these initial studies, mouse-adapted *H. pylori* strains have been developed and used along with the *H. felis* model to extensively study and characterize *H. pylori* pathogenesis, identify the immune mechanisms that contribute to protective immunity, and to test multiple variations and new strategies for vaccinating against *H. pylori* (Blanchard and Nedrud 2010). The mouse models have been used to demonstrate that immunity against *H. pylori* can be achieved with almost any candidate *H. pylori* protein antigens, by multiple routes of immunization and with many alternative adjuvants and delivery mechanisms. The model has also been useful to demonstrate that therapeutic immunization is as efficacious as prophylactic immunization and that immunity is largely based on Th1- and Th17-associated cellular immune responses. The widespread success on immunization in these animals supports a model in which infection induces a Treg dominant non-protective inflammatory response, while vaccination induces protective T helper cells in organized lymphoid tissue which are recruited to the gastric mucosa upon *H. pylori* challenge to promote more severe inflammation and often protective immunity (Fig. 2). Despite such encouraging results in small animal models, clinical trials in humans have shown less favorable results.

5 Varied Vaccination Strategies Employed in Clinical Trials

The first published clinical trial was reported in 1999 (Michetti et al. 1999). The investigators used purified recombinant *H. pylori* urease as antigen utilizing an oral immunization strategy based on the original description by Czinn and Nedrud (1991).

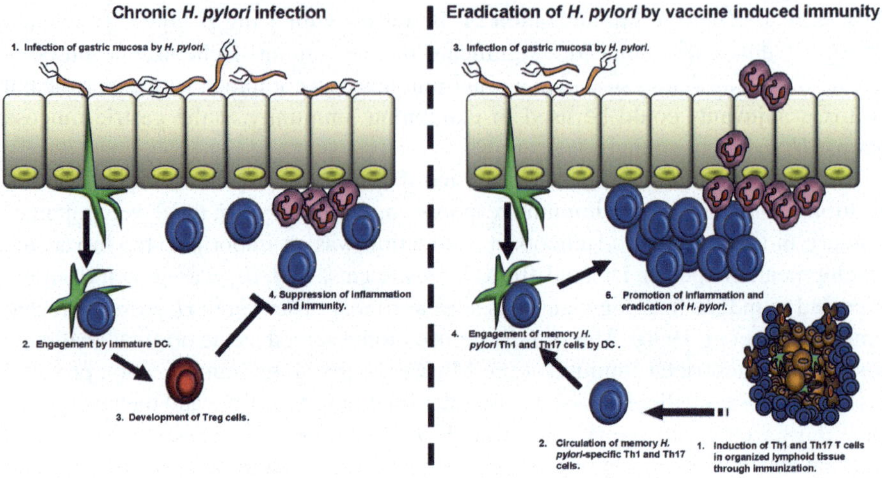

Fig. 2 Immunization against *H. pylori* induces a host immune response distinct from the response induced by chronic infection. (Left panel) During chronic *H. pylori* infection, bacterial antigens are sampled by dendritic cells (DCs) (1) for presentation to T cells (2). *H. pylori* lacks PAMPs with the ability to induce maturation of DCs, and, therefore, the interaction of the DCs with T cells promotes Treg activity (3). The Treg cell response is specific for *H. pylori* and prevents or limits the host immune response against *H. pylori* (4). Although there are some T helper cell activity and effector responses such as the recruitment of neutrophils, this response fails to limit *H. pylori* infection. (Right panel) Vaccination of the host results in *H. pylori* antigens activating T helper cells in organized lymph tissue (1). The incorporation of an adjuvant with the bacterial antigen, and the environment within the lymph nodes, promotes Th1 and Th17 cell activation. The expanded *H. pylori*-specific Th1 and Th17 cells then travel throughout the body via the circulatory system (2). When *H. pylori* colonizes the gastric mucosa (3) bacterial antigens are sampled by DCs for presentation to T cells. In the immunized host, there is now a population of memory Th1 and Th17 cells that can be activated (4) resulting in a potent pro-inflammatory response in which both lymphocytes and polymorphonuclear cells are amplified in number and activity. These cells can then orchestrate the limitation and in some cases the eradication of *H. pylori* (5)

Since many parts of the world have a high prevalence of *H. pylori* infection, there was a strong interest in a therapeutic vaccine. Previous studies in mice and ferrets demonstrated success when applying a *Helicobacter* vaccine therapeutically (Corthesy-Theulaz et al. 1995; Cuenca et al. 1996; Doidge et al. 1994). Therefore, the investigators screened for *H. pylori*-infected adults to receive a therapeutic vaccine (Michetti et al. 1999). Groups of subjects were confirmed as *H. pylori*-infected using the ^{13}C urea breath test and by serology. Subjects were immunized orally with giving either 180, 60, or 20 mg recombinant *H. pylori* urease plus 5 μg LT. The urease protein has already been established as an excellent vaccine candidate in mice in many immunization experiments which began with the first demonstration of induced protective immunity using a subunit vaccine instead of bacterial lysate antigen (Michetti et al. 1994). Subjects were given four weekly doses, and then gastric biopsies were collected by endoscopy 1 and 6 months after the final immunization to evaluate the host immune response and bacterial load. The results were compared with subjects receiving either LT without antigen, or neither LT or antigen.

Despite some elements of success, the study was widely viewed as disappointing since *H. pylori* were not eradicated by any of the tested urease doses. More discouraging were the noted side effects to the LT adjuvant reported by large numbers of subjects. Diarrhea was reported by 62.5% of subjects within 12 h of receiving the first dose and lasted as long as 24 h in some subjects. Episodes of diarrhea did decline with successive doses of vaccine. Since its occurrence was independent of urease dose and since bacterial exotoxins are known to induce diarrhea, this side effect was attributable to the LT. A subgroup of 12 subjects was initially given 10 μg doses of LT adjuvant and half of them experienced diarrhea that was significant enough to interfere with daily activities or lasted more than 1 d, prompting the investigators to switch these subjects to 5 μg LT in subsequent doses. These results highlight the inherent problem in using exotoxin adjuvants for clinical applications. Humans appear to be much more sensitive to CT and LT than rodents. The investigators attempted to remedy this problem by performing subsequent studies to test a range of lower LT doses and alternative mucosal routes of immunization to improve safety but were unable to derive a satisfactory protocol to attempt another clinical vaccine trial (Banerjee et al. 2002; Sougioultzis et al. 2002). The lack of protective immunity achieved in this study and the side effects attributable to LT overshadowed the more positive aspects of the study (Michetti et al. 1999). The measure of immune responsiveness was determined by measuring the levels of anti-urease antibody titers and by enumerating the number of circulating urease-specific B cells. By these metrics, the oral vaccine was demonstrated to have immunogenicity. Significant increases in urease-specific serum IgA was noted in subjects receiving either 180 or 60 mg urease antigen compared with control subjects. Additionally, all doses of urease induced significantly greater numbers of urease-specific IgA positive antibody-secreting cells. But there were also positive elements with respect to efficacy. A significant reduction in bacterial load was achieved in subjects receiving the 20 mg dose of urease and when all subjects receiving urease were combined. Therefore, this study demonstrated that oral immunization could be used to positively impact mucosal immunity at the gastric mucosa as demonstrated by the reduced bacterial load. Given the predicted difficulties in immunizing against a mucosal pathogen, such a result could be categorized as encouraging despite the lack of efficacy compared with studies in the mouse model.

Four additional *H. pylori* vaccine clinical studies have been reported. Three of these studies resulted in largely negative results: a killed, whole-cell therapeutic vaccine administered orally in combination with a nontoxic LT variant containing an amino acid substitution (Kotloff et al. 2001), an oral prophylactic attenuated Salmonella vaccine expressing *H. pylori* proteins (Aebischer et al. 2008), and a prophylactic trivalent protein vaccine mixed with alum given intramuscularly (Malfertheiner et al. 2018). Although in some instances clearance of bacteria was observed in some subjects, the numbers did not exceed the number of control subjects that also cleared the infection. These three studies employed strategies common in the study of mucosal immunity in mice. The failure of these models to achieve the limited success reported by Michetti et al. (1999) suggests the use of the LT exotoxin seems to provide an advantage over other immunization strategies. The biggest challenge, therefore, remains developing a vaccine with the adjuvant properties of CT or LT, while eliminating its toxicity.

6 Unique Aspects of a Successful Clinical Trial

The results of another study performed over 10 years ago, but reported in 2015, provided additional evidence for the utility of the LT adjuvant and revealed the best evidence yet for the potential for an efficacious vaccine against *H. pylori* (Zeng et al. 2015; Sutton and Boag 2018). The study was the first large-scale phase III clinical study on a *H. pylori* subunit vaccine. It was conducted in China, and the vaccine was demonstrated to have a vaccine efficacy of 71.8% (95% confidence interval 48.2–85.6) one year after vaccination (Zeng et al. 2015). Efficacy was shown to drop to 55% by 24 months and remained consistent at 36 months (55.8%). These results are highly significant for the long-term goal of immunizing against childhood infection of *H. pylori*. The study strategy utilized a number of changes that distinguish it from previous *H. pylori* clinical vaccine trials. It is difficult to attribute its success to any particular factor, but each had the potential to positively impact the results and suggest elements that might be incorporated in future trials.

The study by Zeng et al. (2015) was performed in children ranging from 6 to 15 years of age. The four other published *H. pylori* vaccine trials were all performed in adults (Aebischer et al. 2008; Kotloff et al. 2001; Malfertheiner et al. 2018; Michetti et al. 1999). Given its success, the question arises as to how much better it may possibly have performed if the vaccine had been administered even earlier in life. Published data indicate that *H. pylori* infection occurs within the first 5 years of life in many societies (Feldman et al. 1998). It is possible that vaccine delivery during the first several years of life when many other childhood vaccines are administered may improve long-term outcomes. Infection can occur, however, throughout life. Therefore, if a vaccine could be given when applied to teenagers and adults, it would find greater utility. Although not tested in the study by Zeng et al. (2015), if the tested vaccine were capable of inducing bacterial clearance when administered therapeutically, it would increase effectiveness in that the infected pool of hosts within a given population could be reduced, thereby limiting opportunity for spread. Additional studies with this construct to determine whether the age of application is important for vaccine efficacy would be of great value and perhaps increase its utility.

Another unique aspect of the study was the nature of the *H. pylori* challenge. It was not the first trial to be administered prophylactically. The two other trials that did use a prophylactic delivery, however, employed an experimental challenge using well-characterized strains of *H. pylori* (Aebischer et al. 2008; Graham et al. 2004; Malfertheiner et al. 2018). The doses used for this experimental challenge were undoubtedly orders of magnitude higher than what occurs naturally. Zeng et al.

(2015) tested the efficacy of their vaccine against a naturally acquired *H. pylori* infection. The possibility that infectious challenge dose plays an important role in vaccine efficacy bears consideration. The scale of the study also distinguishes it from the other four clinical vaccine trials. The study involved a number of participants over two orders of magnitude higher than the average of the other four studies. Although almost 6000 subjects were screened for eligibility, 2232 eligible subjects were each eventually assigned to vaccine and placebo groups. Ultimately, 2199 subjects successfully completed the three-dose vaccination schedule with another 2204 receiving a placebo.

Zeng et al. (2015) also used a different approach with regard to the adjuvant. They used the urease B protein antigen genetically fused to the exotoxin B subunit of *E. coli* LT. With the use of this recombinant protein, the investigators, therefore, chose an antigen with an extensive history of use in animal and clinical trials fused to a mucosal adjuvant subunit. The number of protein antigens tested in animal models as candidate vaccines for *H. pylori* is extensive and includes many that play a role in pathogenesis including urease, CagA, VacA, catalase, and others (Blanchard and Nedrud 2010). Remarkably, almost every antigen that has been studied shows promise, at least in rodent models, and no single protein seems to be superior to the others.

There may be some benefits to using a multivalent vaccine. Ferrero et al. demonstrated improved efficacy when combining two protein antigens compared to a single antigen in mice (Ferrero et al. 1995). More recently, Guo et al. (2017) tested a multivalent epitope-based therapeutic vaccine with seven antigenic fragments from four *H. pylori* adhesins (urease, Lpp20, HpaA, and CagL) referred to as CFAdE using the Mongolian gerbil model and using *Lycium barbarum* polysaccharide adjuvant (PA). This vaccine was compared with whole *H. pylori* lysate with PA, recombinant urease with PA, or PA alone. *H. pylori*-infected gerbils were immunized intragastrically four times at one-week intervals. Two weeks after the final vaccination, the rodents were evaluated. The oral therapeutic immunization with CFAdE plus PA, *H. pylori* lysate with PA, or urease with PA all dramatically decreased the *H. pylori* loads in the stomachs and reduced the infiltration of leukocytes and neutrophils compared with oral therapeutic immunization with PA alone. Of note, the CFAdE vaccine-induced eradication levels were comparable to the *H. pylori* lysate plus PA-immunized group, and better than the monovalent urease plus PA vaccine.

It is possible, therefore, that the inclusion of additional proteins may improve the efficacy of the vaccine tested in China. However, the use of urease is well supported by numerous studies including the initial clinical trial by Michetti et al. (1999). Although their results were a dramatic improvement compared to the other clinical trials, Zeng and co-workers utilized a strategy that at its core is remarkably similar to the protocol originally described by Czinn and Nedrud (1991) and Czinn et al. (1993). They administered an oral vaccine with multiple doses spread out over a 28-day period, and their construct included a protein antigen combined with a bacterial exotoxin mucosal adjuvant. The manner in which the mucosal adjuvant was incorporated, however, made their results less predictable.

Zeng et al. (2015) limited their adjuvant to the B subunit of *E. coli* LT as part of their fusion protein. It, therefore, lacked the A subunit which is responsible for the toxins enzymatic activity and its mucosal adjuvanticity. Since the A subunit is also responsible for the protein's toxicity it is common for investigators to attempt to minimize its function. Unfortunately, purified B subunit in the absence of contaminating A subunit is indeed safe, but lacks adjuvanticity (Blanchard et al. 1998; Lycke et al. 1991, 1992; Vajdy and Lycke 1992). Investigators have attempted to circumvent this problem by developing holotoxin constructs that have introduced amino acid point mutations in the A subunit. The goal is to retain adjuvanticity while minimizing the toxicity. Since the role of the B subunit is to bind to the epithelium, the construct employed by Zeng et al. would potentially bind to the gastrointestinal epithelium but would lack adjuvancity. This is an attractive strategy and has been used in many experimental models to induce immunity. It has also been used, however, to induce immunologic tolerance to a fusion protein to treat autoimmune diseases in animal models (Arakawa et al. 1998; Bergerot et al. 1997; Petersen et al. 2003; Ploix et al. 1999; Sun et al. 1994, 1996).

The importance of the *E. coli* LT B subunit in the study by Zeng et al. (2015) cannot be determined. They observed the induction of significant levels of urease B-specific serum IgG and salivary IgA titers within one month of vaccination. At three years, the serum IgG titers had decreased, but remained fourfold higher for immunized subjects compared with those receiving placebo. Animal models have been used to demonstrate that antibodies are not required for vaccine-induced protective immunity against *H. pylori* (Blanchard et al. 1999; Ermak et al. 1998), and they do not correlate with immunity to *H. pylori*. Increases in antigen-specific antibodies can be used, however, as a measure of vaccine immunogenicity. The groups of subjects vaccinated with the urease B-LT B subunit fusion protein were not compared to a group receiving the urease B protein in the absence of adjuvant. Therefore, without a direct comparison to such a control, it is impossible to determine whether the antibody titers and protection achieved was due to the presence of the fused enterotoxin B subunit or would have been achieved by 15 mg doses of urease B protein in the absence of the LT B subunit.

Finally, the safety of the vaccine represents an important factor in *H. pylori* vaccine development. Seven percent of subjects receiving the experimental vaccine did report dealing with adverse reactions, primarily within the first few days of immunization (Zeng et al. 2015). The numbers were similar in control subjects receiving the placebo. Subjects' most frequent complaints included vomiting and to a lesser extent fever and headache, but all side effects resolved within 24 h. The application of the vaccine involved fasting for a minimum of 2 h and ingesting a buffer solution consisting sodium bicarbonate and sodium citrate prior. The vaccine or placebo was then given as an 80-ml dose. More subjects receiving the vaccine experienced bloating than the placebo control group ($P = 0.0427$), so it is difficult to attribute this side effect to the buffers taken prior to vaccine, but this condition was mild and quickly resolved.

7 Concluding Remarks

Despite the promise in *H. pylori* vaccination demonstrated by Zeng et al. (2015), there are currently no advanced phase clinical trials underway and large biotechnology companies are not engaged in *H. pylori* vaccine programs, although several smaller companies have vaccine candidates in early development (Sutton and Boag 2018). The decision of whether or not to promote *H. pylori* vaccination is complex. In fact, infection with *H. pylori* is associated with several significant health benefits, at least in the Western hemisphere. Published analyses indicate that there is an inverse correlation between *H. pylori* infection and esophageal pathologies, Barrett's esophagus (Fischbach et al. 2012; Rokkas et al. 2007; Rubenstein et al. 2014), Barrett's metaplasia (Sonnenberg et al. 2010), esophageal adenocarcinoma (Rokkas et al. 2007), and esophageal eosinophilia (Dellon et al. 2011) as well as allergic asthma. In the lower gastrointestinal tract, a meta-analysis on published studies indicates that *H. pylori* may provide protection from inflammatory bowel disease (Luther et al. 2010). Therefore, before widespread vaccination against *H. pylori* is adopted or promoted, it is important to weigh the benefits against the detriments associated with increasing other chronic diseases with high morbidity.

One compromise may be to aim at immune regulation and not at bacterial eradication. Several recent studies demonstrate immunomodulatory effects due to *H. pylori* vaccination in rodents. Vaccination was shown to reduce the infiltration of lymphocytes and neutrophils into the *H. pylori*-infected gastric mucosa in immunized mice (Sun et al. 2018; Mahboubi et al. 2017; Liu et al. 2016; Li et al. 2015) and Mongolian gerbils (Guo et al. 2017) accompanied by significant reduction of *H. pylori* burden. Since *H. pylori*-associated diseases are thought to be an outgrowth of chronic inflammation due to *H. pylori* infection, an immunomodulatory vaccine, which reduces the bacterial load of *H. pylori* colonization and suppresses the chronic inflammation, might limit the development *H. pylori*-associated diseases while maintaining the benefits associated with infection. However, in populations where GC remains prevalent and associated with high morbidity and mortality vaccination would provide an overall advantage. Additionally, PUD is still associated with high morbidity and mortality in many low- and middle-income countries. Today, a diagnosis of PUD accompanied by *H. pylori* infection can be permanently cured in most cases with antimicrobial therapy. However, where health care is scarce or therapy is prohibitively expensive, a successful vaccine could prevent a lifetime of pain and suffering and even death.

The vaccine trials performed to date indicate that if appropriately designed, a vaccine against *H. pylori* is possible. While many barriers remain, the vaccine developed by Zeng et al. (2015) provides encouragement that such a vaccine is within reach. In certain countries, such a vaccine would have a profound impact in reducing the incidence, and the morbidity and mortality associated with GC. Further research is warranted in the interest of developing a vaccine with improved efficacy or with the ability to reduce the immunopathology associated with infection. The development of such a vaccine will require a significant commitment in resources from the biotechnology industry or state governments.

References

Aebischer T, Bumann D, Epple HJ, Metzger W, Schneider T, Cherepnev G et al (2008) Correlation of T cell response and bacterial clearance in human volunteers challenged with *Helicobacter pylori* revealed by randomised controlled vaccination with Ty21a-based Salmonella vaccines. Gut 57(8):1065–1072. https://doi.org/10.1136/gut.2007.145839

Arakawa T, Yu J, Chong DK, Hough J, Engen PC, Langridge WH (1998) A plant-based cholera toxin B subunit-insulin fusion protein protects against the development of autoimmune diabetes. Nat Biotechnol 16(10):934–938. https://doi.org/10.1038/nbt1098-934

Arnold IC, Dehzad N, Reuter S, Martin H, Becher B, Taube C et al (2011a) *Helicobacter pylori* infection prevents allergic asthma in mouse models through the induction of regulatory T cells. J Clin Invest 121(8):3088–3093. https://doi.org/10.1172/jci45041

Arnold IC, Lee JY, Amieva MR, Roers A, Flavell RA, Sparwasser T et al (2011b) Tolerance rather than immunity protects from *Helicobacter pylori*-induced gastric preneoplasia. Gastroenterology 140(1):199–209. https://doi.org/10.1053/j.gastro.2010.06.047

Bamford KB, Fan X, Crowe SE, Leary JF, Gourley WK, Luthra GK et al (1998) Lymphocytes in the human gastric mucosa during *Helicobacter pylori* have a T helper cell 1 phenotype. Gastroenterology 114(3):482–492

Banerjee S, Medina-Fatimi A, Nichols R, Tendler D, Michetti M, Simon J et al (2002) Safety and efficacy of low dose *Escherichia coli* enterotoxin adjuvant for urease based oral immunisation against *Helicobacter pylori* in healthy volunteers. Gut 51(5):634–640

Bergerot I, Ploix C, Petersen J, Moulin V, Rask C, Fabien N et al (1997) A cholera toxoid-insulin conjugate as an oral vaccine against spontaneous autoimmune diabetes. Proc Natl Acad Sci U S A 94(9):4610–4614

Bimczok D, Clements RH, Waites KB, Novak L, Eckhoff DE, Mannon PJ et al (2010) Human primary gastric dendritic cells induce a Th1 response to *H. pylori*. Mucosal Immunol 3(3):260–269. https://doi.org/10.1038/mi.2010.10

Blanchard TG, Czinn SJ (2017) Identification of *Helicobacter pylori* and the evolution of an efficacious childhood vaccine to protect against gastritis and peptic ulcer disease. Pediatr Res 81(1–2):170–176. https://doi.org/10.1038/pr.2016.199

Blanchard TG, Czinn SJ, Redline RW, Sigmund N, Harriman G, Nedrud JG (1999) Antibody-independent protective mucosal immunity to gastric *Helicobacter infection* in mice. Cell Immunol 191(1):74–80. https://doi.org/10.1006/cimm.1998.1421

Blanchard TG, Lycke N, Czinn SJ, Nedrud JG (1998) Recombinant cholera toxin B subunit is not an effective mucosal adjuvant for oral immunization of mice against *Helicobacter felis*. Immunology 94(1):22–27

Blanchard TG, Nedrud JG (2010) *Helicobacter pylori* vaccines. In: Sutton P and Mitchell H (eds) Helicobacter pylori in the 21st century, vol 17. CABI, Wallingford, UK, pp 167–189

Blaser MJ, Chen Y, Reibman J (2008) Does *Helicobacter pylori* protect against asthma and allergy? Gut 57(5):561–567. https://doi.org/10.1136/gut.2007.133462

Bouvard V, Baan R, Straif K, Grosse Y, Secretan B, El Ghissassi F et al (2009) A review of human carcinogens–Part B: biological agents. Lancet Oncol 10(4):321–322

Chen M, Lee A, Hazell S (1992) Immunisation against gastric *Helicobacter* infection in a mouse/*Helicobacter felis* model. Lancet 339(8801):1120–1121

Chen Y, Blaser MJ (2007) Inverse associations of *Helicobacter pylori* with asthma and allergy. Arch Intern Med 167(8):821–827. https://doi.org/10.1001/archinte.167.8.821

Chen Y, Blaser MJ (2008) *Helicobacter pylori* colonization is inversely associated with childhood asthma. J Infect Dis 198(4):553–560. https://doi.org/10.1086/590158

Corthesy-Theulaz I, Porta N, Glauser M, Saraga E, Vaney AC, Haas R et al (1995) Oral immunization with *Helicobacter pylori* urease B subunit as a treatment against *Helicobacter* infection in mice. Gastroenterology 109(1):115–121

Cuenca R, Blanchard TG, Czinn SJ, Nedrud JG, Monath TP, Lee CK et al (1996) Therapeutic immunization against *Helicobacter mustelae* in naturally infected ferrets. Gastroenterology 110(6):1770–1775

Czinn SJ, Cai A, Nedrud JG (1993) Protection of germ-free mice from infection by *Helicobacter felis* after active oral or passive IgA immunization. Vaccine 11(6):637–642

Czinn SJ, Nedrud JG (1991) Oral immunization against *Helicobacter pylori*. Infect Immun 59(7):2359–2363

de Martel C, Ferlay J, Franceschi S, Vignat J, Bray F, Forman D et al (2012) Global burden of cancers attributable to infections in 2008: a review and synthetic analysis. Lancet Oncol 13(6):607–615. https://doi.org/10.1016/s1470-2045(12)70137-7

D'Elios MM, Manghetti M, De Carli M, Costa F, Baldari CT, Burroni D et al (1997) T helper 1 effector cells specific for *Helicobacter pylori* in the gastric antrum of patients with peptic ulcer disease. J Immunol 158(2):962–967

Dellon ES, Peery AF, Shaheen NJ, Morgan DR, Hurrell JM, Lash RH et al (2011) Inverse association of esophageal eosinophilia with *Helicobacter pylori* based on analysis of a US pathology database. Gastroenterology 141(5):1586–1592. https://doi.org/10.1053/j.gastro.2011.06.081

DeLyria ES, Redline RW, Blanchard TG (2009) Vaccination of mice against *H. pylori* induces a strong Th-17 response and immunity that is neutrophil dependent. Gastroenterology 136(1):247–256. https://doi.org/10.1053/j.gastro.2008.09.017

den Hollander WJ, Sonnenschein-van der Voort AM, Holster IL, de Jongste JC, Jaddoe VW, Hofman A et al (2016) *Helicobacter pylori* in children with asthmatic conditions at school age, and their mothers. Aliment Pharmacol Ther 43(8):933–943. https://doi.org/10.1111/apt.13572

Doidge C, Crust I, Lee A, Buck F, Hazell S, Manne U (1994) Therapeutic immunisation against *Helicobacter* infection. Lancet 343(8902):914–915

Eaton KA, Mefford M, Thevenot T (2001) The role of T cell subsets and cytokines in the pathogenesis of *Helicobacter pylori* gastritis in mice. J Immunol 166(12):7456–7461

Elson CO, Ealding W (1984) Generalized systemic and mucosal immunity in mice after mucosal stimulation with cholera toxin. J Immunol 132(6):2736–2741

Engler DB, Reuter S, van Wijck Y, Urban S, Kyburz A, Maxeiner J et al (2014) Effective treatment of allergic airway inflammation with *Helicobacter pylori* immunomodulators requires BATF3-dependent dendritic cells and IL-10. Proc Natl Acad Sci U S A 111(32):11810–11815. https://doi.org/10.1073/pnas.1410579111

Ermak TH, Giannasca PJ, Nichols R, Myers GA, Nedrud J, Weltzin R et al (1998) Immunization of mice with urease vaccine affords protection against *Helicobacter pylori* infection in the absence of antibodies and is mediated by MHC class II-restricted responses. J Exp Med 188(12):2277–2288

Feldman RA, Eccersley AJ, Hardie JM (1998) Epidemiology of *Helicobacter pylori:* acquisition, transmission, population prevalence and disease-to-infection ratio. Br Med Bull 54(1):39–53

Ferrero RL, Thiberge JM, Kansau I, Wuscher N, Huerre M, Labigne A (1995) The GroES homolog of *Helicobacter pylori* confers protective immunity against mucosal infection in mice. Proc Natl Acad Sci U S A 92(14):6499–6503

Fischbach LA, Nordenstedt H, Kramer JR, Gandhi S, Dick-Onuoha S, Lewis A et al (2012) The association between Barrett's esophagus and *Helicobacter pylori* infection: a meta-analysis. Helicobacter 17(3):163–175. https://doi.org/10.1111/j.1523-5378.2011.00931.x

Fox JG, Correa P, Taylor NS, Lee A, Otto G, Murphy JC et al (1990) *Helicobacter mustelae*-associated gastritis in ferrets. An animal model of *Helicobacter pylori* gastritis in humans. Gastroenterology 99(2):352–361

Graham DY, Malaty HM, Evans DG, Evans DJ Jr, Klein PD, Adam E (1991) Epidemiology of *Helicobacter pylori* in an asymptomatic population in the United States. Effect of age, race, and socioeconomic status. Gastroenterology 100(6):1495–1501

Graham DY, Opekun AR, Osato MS, El-Zimaity HM, Lee CK, Yamaoka Y et al (2004) Challenge model for *Helicobacter pylori* infection in human volunteers. Gut 53(9):1235–1243. https://doi.org/10.1136/gut.2003.037499

Guo L, Yin R, Xu G, Gong X, Chang Z, Hong D et al (2017) Immunologic properties and therapeutic efficacy of a multivalent epitope-based vaccine against four *Helicobacter pylori* adhesins (urease, Lpp20, HpaA, and CagL) in Mongolian gerbils. Helicobacter 22(6). https://doi.org/10.1111/hel.12428

Harris PR, Wright SW, Serrano C, Riera F, Duarte I, Torres J et al (2008) *Helicobacter pylori* gastritis in children is associated with a regulatory T-cell response. Gastroenterology 134(2):491–499. https://doi.org/10.1053/j.gastro.2007.11.006

Hunt RH, Xiao SD, Megraud F, Leon-Barua R, Bazzoli F, van der Merwe S et al (2011) *Helicobacter pylori* in developing countries. World Gastroenterology Organisation Global Guideline. J Gastrointestin Liver Dis 20(3):299–304

IARC (2012) Biological agents. Volume 100 B. A review of human carcinogens. IARC Monogr Eval Carcinog Risks Hum 100(Pt B):1–441

Ito M, Takata S, Tatsugami M, Wada Y, Imagawa S, Matsumoto Y et al (2009) Clinical prevention of gastric cancer by *Helicobacter pylori* eradication therapy: a systematic review. J Gastroenterol 44(5):365–371. https://doi.org/10.1007/s00535-009-0036-8

Kato M, Asaka M, Ono S, Nakagawa M, Nakagawa S, Shimizu Y et al (2007) Eradication of *Helicobacter pylori* for primary gastric cancer and secondary gastric cancer after endoscopic mucosal resection. J Gastroenterol 42(Suppl 17):16–20. https://doi.org/10.1007/s00535-006-1928-5

Koch KN, Hartung ML, Urban S, Kyburz A, Bahlmann AS, Lind J, Backert S, Taube C, Müller A (2015) *Helicobacter* urease-induced activation of the TLR2/NLRP3/IL-18 axis protects against asthma. J Clin Invest 125:3297–3302. https://doi.org/10.1172/JCI79337

Kobayashi I, Murakami K, Kato M, Kato S, Azuma T, Takahashi S et al (2007) Changing antimicrobial susceptibility epidemiology of *Helicobacter pylori* strains in Japan between 2002 and 2005. J Clin Microbiol 45(12):4006–4010. https://doi.org/10.1128/jcm.00740-07

Kotloff KL, Sztein MB, Wasserman SS, Losonsky GA, DiLorenzo SC, Walker RI (2001) Safety and immunogenicity of oral inactivated whole-cell *Helicobacter pylori* vaccine with adjuvant among volunteers with or without subclinical infection. Infect Immun 69(6):3581–3590. https://doi.org/10.1128/iai.69.6.3581-3590.2001

Lee A, Buck F (1996) Vaccination and mucosal responses to *Helicobacter pylori* infection. Aliment Pharmacol Ther 10(Suppl 1):129–138

Lee A, Fox JG, Otto G, Murphy J (1990) A small animal model of human *Helicobacter pylori* active chronic gastritis. Gastroenterology 99(5):1315–1323

Lee YC, Chiang TH, Chou CK, Tu YK, Liao WC, Wu MS et al (2016) Association between *Helicobacter pylori* eradication and gastric cancer incidence: A systematic review and meta-analysis. Gastroenterology 150(5):1113–1124.e1115. https://doi.org/10.1053/j.gastro.2016.01.028

Li B, Chen L, Sun H, Yang W, Hu J, He Y et al (2015) Immunodominant epitope-specific Th1 but not Th17 responses mediate protection against *Helicobacter pylori* infection following UreB vaccination of BALB/c mice. Sci Rep 5:14793. https://doi.org/10.1038/srep14793

Lindholm C, Quiding-Jarbrink M, Lonroth H, Hamlet A, Svennerholm AM (1998) Local cytokine response in *Helicobacter pylori*-infected subjects. Infect Immun 66(12):5964–5971

Liu W, Tan Z, Xue J, Luo W, Song H, Lv X et al (2016) Therapeutic efficacy of oral immunization with a non-genetically modified *Lactococcus lactis*-based vaccine CUE-GEM induces local immunity against *Helicobacter pylori* infection. Appl Microbiol Biotechnol 100(14):6219–6229. https://doi.org/10.1007/s00253-016-7333-y

Luther J, Dave M, Higgins PD, Kao JY (2010) Association between *Helicobacter pylori* infection and inflammatory bowel disease: a meta-analysis and systematic review of the literature. Inflamm Bowel Dis 16(6):1077–1084. https://doi.org/10.1002/ibd.21116

Lycke N, Holmgren J (1986) Strong adjuvant properties of cholera toxin on gut mucosal immune responses to orally presented antigens. Immunology 59(2):301–308

Lycke N, Karlsson U, Sjolander A, Magnusson KE (1991) The adjuvant action of cholera toxin is associated with an increased intestinal permeability for luminal antigens. Scand J Immunol 33(6):691–698

Lycke N, Tsuji T, Holmgren J (1992) The adjuvant effect of *Vibrio cholerae* and *Escherichia coli* heat-labile enterotoxins is linked to their ADP-ribosyltransferase activity. Eur J Immunol 22(9):2277–2281. https://doi.org/10.1002/eji.1830220915

Ma JL, Zhang L, Brown LM, Li JY, Shen L, Pan KF et al (2012) Fifteen-year effects of *Helicobacter pylori*, garlic, and vitamin treatments on gastric cancer incidence and mortality. J Natl Cancer Inst 104(6):488–492. https://doi.org/10.1093/jnci/djs003

Mabe K, Takahashi M, Oizumi H, Tsukuma H, Shibata A, Fukase K et al (2009) Does *Helicobacter pylori* eradication therapy for peptic ulcer prevent gastric cancer? World J Gastroenterol 15(34):4290–4297

Mahboubi M, Falsafi T, Sadeghizadeh M, Mahjoub F (2017) The role of outer inflammatory protein A (OipA) in vaccination of theC57BL/6 mouse model infected by *Helicobacter pylori*. Turk J Med Sci 47(1):326–333. https://doi.org/10.3906/sag-1505-108

Maldonado RA, von Andrian UH (2010) How tolerogenic dendritic cells induce regulatory T cells. Adv Immunol 108:111–165. https://doi.org/10.1016/b978-0-12-380995-7.00004-5

Malfertheiner P, Selgrad M, Wex T, Romi B, Borgogni E, Spensieri F, et al (2018) Efficacy, immunogenicity, and safety of a parenteral vaccine against *Helicobacter pylori* in healthy volunteers challenged with a Cag-positive strain: a randomised, placebo-controlled phase 1/2 study. Lancet Gastroenterol Hepatol https://doi.org/10.1016/s2468-1253(18)30125-0

Matsumoto Y, Blanchard TG, Drakes ML, Basu M, Redline RW, Levine AD et al (2005) Eradication of *Helicobacter pylori* and resolution of gastritis in the gastric mucosa of IL-10-deficient mice. Helicobacter 10(5):407–415. https://doi.org/10.1111/j.1523-5378.2005.00349.x

Michetti P, Corthesy-Theulaz I, Davin C, Haas R, Vaney AC, Heitz M et al (1994) Immunization of BALB/c mice against *Helicobacter felis* infection with *Helicobacter pylori* urease. Gastroenterology 107(4):1002–1011

Michetti P, Kreiss C, Kotloff KL, Porta N, Blanco JL, Bachmann D et al (1999) Oral immunization with urease and *Escherichia coli* heat-labile enterotoxin is safe and immunogenic in *Helicobacter pylori*-infected adults. Gastroenterology 116(4):804–812

Mizuno T, Ando T, Nobata K, Tsuzuki T, Maeda O, Watanabe O et al (2005) Interleukin-17 levels in *Helicobacter pylori*-infected gastric mucosa and pathologic sequelae of colonization. World J Gastroenterol 11(40):6305–6311

Murakami K, Sakurai Y, Shiino M, Funao N, Nishimura A, Asaka M (2016) Vonoprazan, a novel potassium-competitive acid blocker, as a component of first-line and second-line triple therapy for *Helicobacter pylori* eradication: a phase III, randomised, double-blind study. Gut 65(9):1439–1446. https://doi.org/10.1136/gutjnl-2015-311304

Nedrud JG, Liang XP, Hague N, Lamm ME (1987) Combined oral/nasal immunization protects mice from Sendai virus infection. J Immunol 139(10):3484–3492

NIH Consensus Conference (1994) *Helicobacter pylori* in peptic ulcer disease. J Am Med Assoc 272:65–69

Nishizawa T, Maekawa T, Watanabe N, Harada N, Hosoda Y, Yoshinaga M et al (2015) Clarithromycin versus metronidazole as first-line *Helicobacter pylori* eradication: a multicenter, prospective, randomized controlled study in Japan. J Clin Gastroenterol 49(6):468–471. https://doi.org/10.1097/mcg.0000000000000165

Oertli M, Muller A (2012) *Helicobacter pylori* targets dendritic cells to induce immune tolerance, promote persistence and confer protection against allergic asthma. Gut Microbes 3(6):566–571. https://doi.org/10.4161/gmic.21750

Oertli M, Noben M, Engler DB, Semper RP, Reuter S, Maxeiner J et al (2013) *Helicobacter pylori* gamma-glutamyl transpeptidase and vacuolating cytotoxin promote gastric persistence and immune tolerance. Proc Natl Acad Sci U S A 110(8):3047–3052. https://doi.org/10.1073/pnas.1211248110

Oertli M, Sundquist M, Hitzler I, Engler DB, Arnold IC, Reuter S et al (2012) DC-derived IL-18 drives Treg differentiation, murine *Helicobacter pylori*-specific immune tolerance, and asthma protection. J Clin Invest 122(3):1082–1096. https://doi.org/10.1172/jci61029

Parkin DM (2004) International variation. Oncogene 23(38):6329–6340. https://doi.org/10.1038/sj.onc.1207726

Petersen JS, Bregenholt S, Apostolopolous V, Homann D, Wolfe T, Hughes A et al (2003) Coupling of oral human or porcine insulin to the B subunit of cholera toxin (CTB) overcomes critical antigenic differences for prevention of type I diabetes. Clin Exp Immunol 134(1):38–45

Ploix C, Bergerot I, Durand A, Czerkinsky C, Holmgren J, Thivolet C (1999) Oral administration of cholera toxin B-insulin conjugates protects NOD mice from autoimmune diabetes by inducing CD4+ regulatory T-cells. Diabetes 48(11):2150–2156

Posselt G, Backert S, Wessler S (2013) The functional interplay of *Helicobacter pylori* factors with gastric epithelial cells induces a multi-step process in pathogenesis. Cell Commun Signal 11:77. https://doi.org/10.1186/1478-811X-11-77

Reibman J, Marmor M, Filner J, Fernandez-Beros ME, Rogers L, Perez-Perez GI et al (2008) Asthma is inversely associated with *Helicobacter pylori* status in an urban population. PLoS ONE 3(12):e4060. https://doi.org/10.1371/journal.pone.0004060

Robinson K, Kenefeck R, Pidgeon EL, Shakib S, Patel S, Polson RJ et al (2008) *Helicobacter pylori*-induced peptic ulcer disease is associated with inadequate regulatory T cell responses. Gut 57(10):1375–1385. https://doi.org/10.1136/gut.2007.137539

Rokkas T, Pistiolas D, Sechopoulos P, Robotis I, Margantinis G (2007) Relationship between *Helicobacter pylori* infection and esophageal neoplasia: a meta-analysis. Clin Gastroenterol Hepatol 5(12):1413–1417. https://doi.org/10.1016/j.cgh.2007.08.010

Rolig AS, Carter JE, Ottemann KM (2011) Bacterial chemotaxis modulates host cell apoptosis to establish a T-helper cell, type 17 (Th17)-dominant immune response in *Helicobacter pylori* infection. Proc Natl Acad Sci U S A 108(49):19749–19754. https://doi.org/10.1073/pnas.1104598108

Rubenstein JH, Inadomi JM, Scheiman J, Schoenfeld P, Appelman H, Zhang M et al (2014) Association between *Helicobacter pylori* and Barrett's esophagus, erosive esophagitis, and gastroesophageal reflux symptoms. Clin Gastroenterol Hepatol 12(2):239–245. https://doi.org/10.1016/j.cgh.2013.08.029

Rupnow MFT, Chang AH, Shachter RD, Owens DK, Parsonnet J (2009) Cost-effectiveness of a potential prophylactic *Helicobacter pylori* vaccine in the United States. J Infect Dis 200(8):1311–1317. https://doi.org/10.1086/605845

Shi Y, Liu XF, Zhuang Y, Zhang JY, Liu T, Yin Z et al (2010) *Helicobacter pylori*-induced Th17 responses modulate Th1 cell responses, benefit bacterial growth, and contribute to pathology in mice. J Immunol 184(9):5121–5129. https://doi.org/10.4049/jimmunol.0901115

Shimizu T, Haruna H, Ohtsuka Y, Kaneko K, Gupta R, Yamashiro Y (2004) Cytokines in the gastric mucosa of children with *Helicobacter pylori* infection. Acta Paediatr 93(3):322–326

Sonnenberg A, Lash RH, Genta RM (2010) A national study of *Helicobactor pylori* infection in gastric biopsy specimens. Gastroenterology 139(6):1894–1901. https://doi.org/10.1053/j.gastro.2010.08.018

Sougioultzis S, Lee CK, Alsahli M, Banerjee S, Cadoz M, Schrader R et al (2002) Safety and efficacy of *E. coli* enterotoxin adjuvant for urease-based rectal immunization against *Helicobacter pylori*. Vaccine 21(3–4):194–201

Spangler BD (1992) Structure and function of cholera toxin and the related *Escherichia coli* heat-labile enterotoxin. Microbiol Rev 56(4):622–647

Stoicov C, Fan X, Liu JH, Bowen G, Whary M, Kurt-Jones E et al (2009) T-bet knockout prevents *Helicobacter felis*-induced gastric cancer. J Immunol 183(1):642–649. https://doi.org/10.4049/jimmunol.0900511

Sun H, Yuan H, Tan R, Li B, Guo G, Zhang J et al (2018) Immunodominant antigens that induce Th1 and Th17 responses protect mice against *Helicobacter pylori* infection. Oncotarget 9(15):12050–12063. https://doi.org/10.18632/oncotarget.23927

Sun JB, Holmgren J, Czerkinsky C (1994) Cholera toxin B subunit: an efficient transmucosal carrier-delivery system for induction of peripheral immunological tolerance. Proc Natl Acad Sci U S A 91(23):10795–10799

Sun JB, Rask C, Olsson T, Holmgren J, Czerkinsky C (1996) Treatment of experimental autoimmune encephalomyelitis by feeding myelin basic protein conjugated to cholera toxin B subunit. Proc Natl Acad Sci U S A 93(14):7196–7201

Sutton P, Boag JM (2018) Status of vaccine research and development for *Helicobacter pylori*. Vaccine. https://doi.org/10.1016/j.vaccine.2018.01.001

Take S, Mizuno M, Ishiki K, Hamada F, Yoshida T, Yokota K et al (2015) Seventeen-year effects of eradicating *Helicobacter pylori* on the prevention of gastric cancer in patients with peptic ulcer; a prospective cohort study. J Gastroenterol 50(6):638–644. https://doi.org/10.1007/s00535-014-1004-5

Vajdy M, Lycke NY (1992) Cholera toxin adjuvant promotes long-term immunological memory in the gut mucosa to unrelated immunogens after oral immunization. Immunology 75(3):488–492

Wang Q, Yu C, Sun Y (2013) The association between asthma and *Helicobacter pylori*: a meta-analysis. Helicobacter 18(1):41–53. https://doi.org/10.1111/hel.12012

Wang YC, Lin TY, Shang ST, Chen HJ, Kao CH, Wu CC et al (2017) *Helicobacter* pylori infection increases the risk of adult-onset asthma: a nationwide cohort study. Eur J Clin Microbiol Infect Dis 36(9):1587–1594. https://doi.org/10.1007/s10096-017-2972-1

Yamaoka Y, Kato M, Asaka M (2008) Geographic differences in gastric cancer incidence can be explained by differences between *Helicobacter pylori* strains. Intern Med 47(12):1077–1083

Zeng M, Mao XH, Li JX, Tong WD, Wang B, Zhang YJ et al (2015) Efficacy, safety, and immunogenicity of an oral recombinant *Helicobacter pylori* vaccine in children in China: a randomised, double-blind, placebo-controlled, phase 3 trial. Lancet 386(10002):1457–1464. https://doi.org/10.1016/s0140-6736(15)60310-5

Zhou X, Wu J, Zhang G (2013) Association between *Helicobacter pylori* and asthma: a meta-analysis. Eur J Gastroenterol Hepatol 25(4):460–468. https://doi.org/10.1097/MEG.0b013e32835c280a

Zhuang Y, Shi Y, Liu XF, Zhang JY, Liu T, Fan X et al (2011) *Helicobacter pylor*i-infected macrophages induce Th17 cell differentiation. Immunobiology 216(1–2):200–207. https://doi.org/10.1016/j.imbio.2010.05.005

Impact of *Helicobacter pylori* Virulence Factors on the Host Immune Response and Gastric Pathology

Sundus Javed, Emma C. Skoog and Jay V. Solnick

Contents

Abstract *Helicobacter pylori* chronically infects nearly half the world's population, yet most of those infected remain asymptomatic throughout their lifetime. The outcome of infection—peptic ulcer disease or gastric cancer versus asymptomatic colonization—is a product of host genetics, environmental influences, and differences

Sundus Javed and Emma C. Skoog—These authors contributed equally.

S. Javed · E. C. Skoog · J. V. Solnick
Department of Medicine, Department of Microbiology & Immunology, Center for Comparative Medicine, University of California, Davis School of Medicine, Davis, CA 95616, USA

J. V. Solnick (✉)
Center for Comparative Medicine, University of California, Davis, Davis, CA 95616, USA
e-mail: jvsolnick@ucdavis.edu

Present Address
S. Javed
Department of Biosciences, COMSATS Institute of Information Technology, Islamabad, Pakistan

S. Backert (ed.), *Molecular Mechanisms of Inflammation: Induction, Resolution and Escape by Helicobacter pylori*, Current Topics in Microbiology and Immunology 421, https://doi.org/10.1007/978-3-030-15138-6_2

in bacterial virulence factors. Here, we review the current understanding of the *cag* pathogenicity island (*cag*PAI), the vacuolating cytotoxin (VacA), and a large family of outer membrane proteins (OMPs), which are among the best understood *H. pylori* virulence determinants that contribute to disease. Each of these virulence factors is characterized by allelic and phenotypic diversity that is apparent within and across individuals, as well as over time, and modulates inflammation. From the bacterial perspective, inflammation is probably a necessary evil because it promotes nutrient acquisition, but at the cost of reduction in bacterial load and therefore decreases the chance of transmission to a new host. The general picture that emerges is one of a chronic bacterial infection that is dependent on both inducing and carefully regulating the host inflammatory response. A better understanding of these regulatory mechanisms may have implications for the control of chronic inflammatory diseases that are increasingly common causes of human morbidity and mortality.

1 Introduction

Inflammation, often called chronic active gastritis, is the hallmark of infection with *Helicobacter pylori*, occurring uniformly in all those that are infected. In most cases, this is asymptomatic and remains so throughout the lifetime of the host. However, about 10% of those infected will develop peptic ulcer and 1–3% will progress to gastric cancer (GC), which is the second most common cause of cancer death worldwide (Bray et al. 2018). The outcome of infection depends to a large extent upon the relative balance of Th1, Th17, and regulatory T cells (Treg), which together represent the predominant adaptive immune response to *H. pylori*. Th1 and Th17 responses control the infection—though do not eliminate it—but it comes at the expense of immunopathology and disease (Arnold et al. 2011). On the other hand, a Treg response induces tolerance and is likely also responsible for the protective effects of *H. pylori* against asthma and some other inflammatory diseases. For example, those with peptic ulcer disease (PUD) have a threefold greater Th1 and twofold lower Treg response than asymptomatic controls (Robinson et al. 2008), and there is an inverse relationship between the number of gastric Treg cells and the degree of gastritis (Harris et al. 2008). Mechanistic studies have also demonstrated that Treg-mediated immune tolerance protects neonatal mice from pre-neoplastic pathology (Arnold et al. 2011), an observation that is consistent with the Treg response and lack of disease from *H. pylori* in children (Harris et al. 2008).

What then determines the relative balance of Th1/Th17 and Treg immune responses to *H. pylori* infection? If we could understand this sufficiently, we might better target our screening and treatment interventions to those most likely to benefit, which is important because nearly half the world's population is infected and universal treatment is probably impractical and may also be harmful. The answer likely rests on the same three-legged stool that is the case for most infectious diseases: host genetics or epigenetics, microbial genetics, and environmental factors. Host genetic polymorphisms associated with increased risk of *H. pylori*-associated GC include

IL-1β and TNF-α, among others. Risk of acquiring *H. pylori*, though not disease, has been associated with polymorphisms in TLR1 (Mayerle et al. 2013). Recent evidence suggests that environmental factors may also affect the outcome of *H. pylori* infection. For example, both high salt and low iron in the diet enhance *H. pylori* virulence (Gaddy et al. 2013; Noto et al. 2013). Observational studies in humans (Yang et al. 2016) and experimental studies in gnotobiotic mice (Lertpiriyapong et al. 2014) have also demonstrated that the commensal microbiota may affect the development of *H. pylori*-associated GC, including chronic infection with Epstein–Barr virus (Borozan et al. 2018).

But the most well-characterized factors that contribute to PUD or to GC, rather than asymptomatic infection, are bacterial virulence genes. There is no consensus regarding the definition of exactly what constitutes an *H. pylori* virulence gene. Some would include factors such as urease or flagella, which are necessary for colonization, but are not implicated in disease per se. Here we will focus on genetic loci that are present or demonstrate allelic variants in strains that cause disease rather than colonization. The best studied of these are the vacuolating cytotoxin (VacA), the cytotoxin-associated gene A (CagA) and the related *cagA* pathogenicity island (*cag*PAI), and a large family of outer membrane proteins (OMPs). It was noticed very early that *H. pylori* strains recovered from patients with PUD induced vacuolization in susceptible host cells, and when the gene was identified, it was called *vacA* (Cover et al. 1992). Although *vacA* is present in all *H. pylori* strains, there is extensive allelic variation that in part determines the host response and disease outcome (Posselt et al. 2013). Cytotoxic strains also express a high molecular weight immunodominant antigen, which was called CagA because it was found only in cytotoxin-positive strains. DNA sequencing of genes adjacent to *cagA* revealed the presence of the *cag*PAI (Covacci et al. 1997), which encodes a type IV secretion system (T4SS) that injects CagA and other effectors into host cells (Backert et al. 2010). *H. pylori* also encodes an unusually large family of OMPs, some of which have been characterized as adhesins that contribute to the disease outcome (Alm et al. 2000). While these are not the only *H. pylori* virulence factors that have been identified, they are the best understood, and they are the focus of this review.

2 Vacuolating Cytotoxin

2.1 Heterogeneity and the Epidemiology of Gastric Disease

VacA was first identified and named after its ability to induce vacuolization in cultured epithelial cells. All *H. pylori* strains have the *vacA* gene, which is expressed as a 140-kDa protein that is cleaved into p33 and p55 kDa segments that are both required for toxicity (McClain et al. 2017). However, only some strains produce a VacA with cytotoxic activity, which is largely attributed to differences in amino acid sequence in three specific regions called “s” (signal), “i” (intermediate), and “m” (middle) that can each be classified into two main families. Since *H. pylori* is naturally competent, it is not surprising that strains can be found with *vacA* alleles that contain nearly all s, i, and m combinations. However, strains with s1, i1, and m1 alleles demonstrate greater vacuolating activity in cell culture, and they are more often associated with peptic ulcer and GC or its precursor lesions than strains carrying s2, i2, and m2 alleles (McClain et al. 2017). A recent meta-analysis that included 25 studies suggested that GC risk is most strongly associated with s1m1 *vacA* alleles (Pormohammad et al. 2018), which is consistent with the observation that these strains are also very common in East Asia where GC is highly prevalent.

The mechanistic explanation for the allelic variation in vacuolation and disease risk is not well understood. The best characterized of the *vacA* alleles is the difference between s1 and s2, which are located in the signal peptide at the 5' end of *vacA* that encodes the p33 domain. The s2 allele affects the hydrophobicity of the secreted protein in a way that prevents formation of anion channels and vacuolization (McClain et al. 2001). The s2 allele is also less transcribed leading to lower amounts of VacA. The i region is located in the p33 domain, though little is known about how it affects the function of VacA. In s1m2 strains, the i region determines the vacuolating activity: The i1 genotype can induce vacuolation, but the i2 genotype and the rarely found i3 genotype cannot (Chung et al. 2010; Rhead et al. 2007). The m region is involved in channel-forming properties and VacA binding to host cells. The m1 allele binds to low-density lipoprotein receptor-related protein 1 (LRP1), which results in an accumulation of reactive oxygen species leading to autophagy and apoptosis (Yahiro et al. 2012). The lack of binding of the m2 allele can also lead to less vacuolating activity. This depends on the host cell, since VacA has been suggested to bind to several other host receptors, which include sphingomyelin in lipid rafts (Gupta et al. 2008), fibronectin (Hennig et al. 2005), receptor-like protein tyrosine phosphatase alpha and beta (Yahiro et al. 2003), EGFR on epithelial cells (Seto et al. 1998), as well as β2-integrin on T cells (Sewald et al. 2008).

2.2 Animal Models and Cell Biology

Determining the contribution of VacA to the development of gastric disease in humans is complicated because strains with *vacA* alleles that are associated with disease also frequently carry other virulence factors. For example, the s1 allele is associated with the *cag*PAI and with particular OMPs such as BabA and OipA, which are also linked to disease. Animal models and in vitro studies provide an opportunity to dissect the particular role of VacA independently of other virulence factors. The VacA toxin is secreted from *H. pylori* through an autotransporter pathway and binds to the surface of host cells within lipid rafts (McClain et al. 2017). From there, it can be internalized into endosomes or integrated into the plasma membrane to form anion-selective channels. It can also interact with the membranes of mitochondria, the Golgi apparatus, and the endoplasmic reticulum. Intragastric inoculation of mice with sonicates from toxin-positive strains, or of purified cytotoxin, results in epithelial cell damage (Ghiara et al. 1995; Telford et al. 1994), which suggests a key role in virulence. In vitro, VacA also targets mitochondria to induce apoptotic cell death (Jain et al. 2011). A more physiologic approach is to challenge animals with wild-type or VacA mutants and examine gastric pathology. Studies in gnotobiotic piglets, gerbils, and mice found no differences in gastritis compared to wild type (Eaton et al. 1997; Salama et al. 2001; Wirth et al. 1998), though prolonged colonization in gerbils led to an increase in gastric ulceration (Ogura et al. 2000). More recently, it has been demonstrated that isogenic strains expressing the s1m1 allele induce more gastric inflammation and spasmolytic polypeptide expressing metaplasia, which is thought to be a precursor to GC (Winter et al. 2014).

However, our understanding of VacA biology took an unexpected turn, when it was demonstrated in competition experiments that deletion of *vacA* actually impairs the capacity of *H. pylori* to colonize mice (Salama et al. 2001), which suggested that in vivo VacA must function differently, or have pleotropic effects. This has now been clarified in a series of studies from several laboratories, demonstrating that VacA regulates the immune response to *H. pylori* by inhibition of phagosome maturation (Zheng and Jones 2003), blocking T-cell activation (Gebert et al. 2003), and promoting regulatory over effector T cells (Oertli et al. 2013). VacA has also been shown to have antagonistic effects to CagA-mediated inhibition of apoptosis and stimulation of the immune system (Yokoyama et al. 2005). The immunoregulatory effects of VacA are particularly apparent with strains encoding the s2i2 allele, which actually colonizes more effectively than both the null mutant and the more inflammatory s1i1 allele (Winter et al. 2014). In fact, so profound are the immunosuppression effects of VacA that it is perhaps best considered an immunoregulatory protein, rather than a cytotoxin, and if this effect was discovered first, it might well be named differently.

3 Cag Pathogenicity Island

3.1 Epidemiology

The *H. pylori cag*PAI is a ~40-kb chromosomal DNA insertion element found in so-called type I strains that are associated with PUD and GC, rather than asymptomatic gastritis. These strains are especially common in East Asia and Latin America, where the incidence of GC is highest. However, the relationship of the *cag*PAI to disease is complex because type I strains also typically express *babA* and the virulent s1 form of *vacA*, while *cagA*-negative strains from the United States and Western Europe may lack *babA* expression and contain s2 *vacA* alleles. Moreover, geographic variation seems to play an important role in the virulence profile of *H. pylori* strains. Multilocus sequence typing (MLST) has identified seven populations and subpopulations of *H. pylori* with a distinct geographic distribution traceable to the ancestral population arising from Africa, Central Asia, and East Asia (Falush et al. 2003). *H. pylori* virulence may be best understood when the ancestral bacterial population is analyzed in the context of host ancestry. Individuals harboring strains with a different ancestry than the host have a greater risk of developing gastric lesions compared to those infected with similar ancestral percentage as the host (Kodaman et al. 2014). This suggests that the relationship between microbial and host ancestry, not geographic location per se, is driving disease severity. Nevertheless, the *cag*PAI is the *H. pylori* virulence locus that is most clearly associated with disease.

3.2 Genetic Organization

The *cag*PAI typically encodes about 31 contiguous genes, though in some strains it may be disrupted, and in others, *cagA* may be off the island. Gene content and gene order on the *cag*PAI are highly conserved and organized into several operons that are also consistent across strains (Sharma et al. 2010; Ta et al. 2012). Sequence analysis demonstrates that the *cag*PAI phylogeny parallels that of housekeeping genes, suggesting that it was likely acquired only once since humans migrated out of Africa. However, there is considerable sequence variability in several component genes, particularly those encoding cell surface proteins that frequently show signs of diversifying selection (Olbermann et al. 2010), most dramatically *cagA* and *cagY*. An early systemic mutagenesis approach to determine the genes essential for *cag*PAI function identified 17 genes required for translation of the CagA oncoprotein (Fischer et al. 2001), 14 of which were also necessary for induction of IL-8 in cell culture. Subsequent studies have modified these results somewhat by identifying polar effects and strain-dependent phenotypes (Backert et al. 2015). Some of the essential genes are thought to be required for assembly of the T4SS, while others are necessary for function but not formation of the pilus (Table 1).

3.3 T4SS

T4SSs are multiprotein structures on the cell surface that are exploited by many Gram-positive and Gram-negative bacteria to inject effector molecules into host cells, where they disrupt or co-opt host cell signaling pathways (Grohmann et al. 2018). The *H. pylori cag*PAI encodes orthologs of all 11 VirB and VirD4 proteins that are found in the canonical *Agrobacterium tumefaciens* T4SS, as well as several additional components that are essential for its function (Table 1). Though it was discovered more than 30 years ago by DNA sequence analysis, only recently has the T4SS structure been convincingly visualized. Even with current methods, the pili can be found on only about 5% of bacterial cells and only after contact with gastric epithelial cells (GECs) (Chang et al. 2018; Johnson et al. 2014; Shaffer et al. 2011; Skoog et al. 2018). Although the critical host factors are unknown, T4SS pili are increased under iron-limiting conditions (Noto et al. 2013) and decreased upon depletion of zinc (Gaddy et al. 2014). The diameter of the *H. pylori* T4SS pilus has been variously

Table 1 Summary of *Cag*PAI gene characterization

CagPAI		Gene designation			Subcellular localization	Function				
cag/vir Gene designation[a]		Gene ID	26695	J99		Putative function	Structural component	IL-8 induction	CagA translocation	Pilus formation
cagζ	*cag1*	HPY0520	HP0520	JHP0469	IM	Accessory factor	–	+	+	–
cagε	*cag2*	HPY0521	HP0521	JHP0470	C	Accessory factor	–	+	+	–
cagδ	*cag3*	HPY0522	HP0522	JHP0471	OM	OM complex	+	–	–	+
cagγ	*cag4/virB1*	HPY0523	HP0523	JHP0472	PP	PGN hydrolase	–	–	–	–
cagβ	*cag5/virD4*	HPY0524	HP0524	JHP0473	IM	Coupling factor	–	+	–	+
cagα	*virB11*	HPY0525	HP0525	JHP0474	C, IM	NTPase	–	–	–	+
cagZ	*cag6*	HPY0526	HP0526	JHP0475	C, IM	cagβ stabilization		+	+	–
cagY	*cag7/virB10*	HPY1753	HP0527	JHP0476	IM, OM, S	Integrin binding	+	+	+	+
cagX	*cag8/virB9*	HPY0528	HP0528	JHP0477	IM, OM, S	Integrin binding	+	+	–	+
cagW	*cag9/virB6*	HPY0529	HP0529	JHP0478	IM	IM channel	+	–	–	–
cagV	*cag10/virB8*	HPY0530	HP0530	JHP0479	IM	Core complex	+	+	–	+
cagU	*cag11*	HPY0531	HP0531	JHP0480	IM	Accessory factor	–	–	–	–
cagT	*cag12/virB7*	HPY0532	HP0532	JHP0481	OM, S	Core complex, lipoprotein	+	+	–	+

(continued)

Table 1 (continued)

CagPAI		Gene designation			Subcellular localization	Function				
cag/vir Gene designation[a]		Gene ID	26695	J99		Putative function	Structural component	IL-8 induction	CagA translocation	Pilus formation
cagS	*cag13*	HPY0534	HP0534	JHP0482	C	Accessory factor	–	+	+	–
cagQ	*cag14*	HPY0535	HP0535	JHP0483	IM	Accessory factor	–	+	+	–
cagP	*cag15*	HPY0536	HP0536	JHP0484	IM	Accessory factor	–	+	+	–
cagM	*cag16*	HPY0537	HP0537	JHP0485	OM	OM complex	+	–	–	+
cagN	*cag17*	HPY0538	HP0538	JHP0486	PP, IM	Accessory factor	–	+	+	–
cagL	*cag18/virB5*	HPY0539	HP0539	JHP0487	PP, S	Integrin binding	+	–	–	+
cagI	*cag19*	HPY0540	HP0540	JHP0488	PP, S	Integrin binding	+	–, +	–	+
cagH	*cag20*	HPY1714	HP0541	JHP0489	IM	Pilus biogenesis	+	–	–	+
cagG	*cag21*	HPY0542	HP0542	JHP0490	PP	Accessory factor	–	+	–, +	–
cagF	*cag22/virB3/B4*	HPY0543	HP0543	JHP0491	C, IM	*cagA* chaperone	–	+	–	–
cagE	*cag23*	HPY0544	HP0544	JHP0492	IM	NTPase	–	–	–	–
cagD	*cag24*	HPY0545	HP0545	JHP0493	IM, PP, S	Accessory factor	–	+	+	–
cagC	*cag25/virB2*	HPY0546	HP0546	JHP0494	IM, OM, S	Pilus subunit	–	–	–	–, +
cagA	*cag26*	HPY1758	HP0547	JHP0495	C, S	Translocated effector	–	+	–	–

Modified from Backert et al. (2017)

IM inner membrane; *OM* outer membrane; *C* cytoplasm; *PP* periplasmic; *S* surface exposed

[a] Alphabetical and numerical *cag* gene designations are used interchangeably; both are listed for cross-reference

reported to range from 14–70 nm, perhaps due to different bacterial culture conditions and EM staining procedures. More recently, cryotomographic imaging found that each bacterial cell produced 1 to 10 "tubes" that measured 37 nm in diameter and corresponded to what others have called T4SS "pili" (Chang et al. 2018). These studies also identified a periplasmic cone-like structure with two concentric rings, which was proposed to represent the nascent T4SS based on structural similarity to the core complex of *H. pylori* T4SS (Frick-Cheng et al. 2016) and the *dot/icm* T4SS in *Legionella pneumophila* (Ghosal et al. 2017). Because both the T4SS core complex and the pili or tubes have been associated with CagX, CagY, and CagT, it was suggested that they may represent different states of the same secretion apparatus, with the T4SS structure representing a "pre-extension" state that assembles into a tube in response to host cell contact (Chang et al. 2018). This interesting hypothesis remains to be tested.

Recent studies have also begun to reveal additional ultrastructural details of the T4SS and its protein components. The major pilin subunit is predicted to be CagC based on similarity to VirB2 and its presence on the bacterial surface. However, this is inconsistent with the observation that *cagC* mutants have similar numbers of pili as wild-type bacteria (Johnson et al. 2014); hence, identification of the pilin subunit remains unresolved. The length and thickness of the pilus appears to be regulated by CagH, since mutants are hyperpiliated and produce thicker and longer pili (Shaffer et al. 2011). The T4SS core complex spans the periplasmic space and is composed of CagM, CagT, Cag3, CagX, and CagY (Frick-Cheng et al. 2016), each of which is required for *cag* T4SS activity. In 2007, it was first reported that T4SS function also requires host cell β1-integrin (Kwok et al. 2007), which is expressed basolaterally but becomes accessible to *H. pylori* via disruption of cell junctions by the serine protease, HtrA (Tegtmeyer et al. 2017b). In vitro, β1-integrin binds CagA, CagI, CagL, and the C-terminus of CagY (Jimenez-Soto et al. 2009; Koelblen et al. 2017; Kwok et al. 2007; Barden et al. 2013; Conradi et al. 2012). Yet, it is the highly variable middle repeat region of CagY that decorates the bacterial surface (Jimenez-Soto et al. 2009; Rohde et al. 2003; Tanaka et al. 2003), even in the absence of all other *cag*PAI genes, and is necessary for whole-cell binding to β1-integrin (Skoog et al. 2018). This suggests the hypothesis that a particular motif structure in the hypervariable CagY middle repeat region promotes surface expression of the CagY C-terminus, which binds β1-integrin and serves as a nucleation signal to promote expression of a functional T4SS "tube." Cells with functional or non-functional *cagY* variants may then be acted upon and selected by environmental factors at the gastric mucosal surface or deep in the glands (Fig. 1). However, very recently, it was reported that CagA translocation was unaffected by deletion of αβ-integrin heterodimers by CRISPR/Cas-9 technology in AGS and Kato-III cells (Zhao et al. 2018). This surprising result will no doubt prompt confirmatory studies by others.

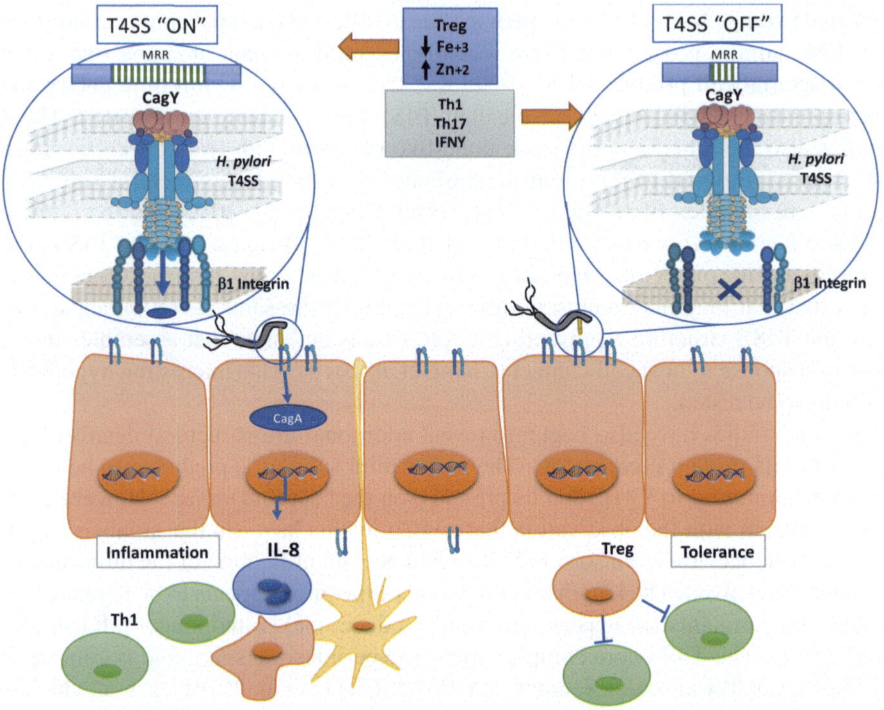

Fig. 1 CagY-dependent regulation of the T4SS. CagY is an essential component of the *H. pylori* T4SS. At the 3′ terminus, *cagY* has homology to other VirB10 orthologs found in all bacterial T4SS, but there is a large middle repeat region (MRR) containing numerous direct DNA repeats that is unique to the *H. pylori virB10*. Under Th1 immune pressure, MRR variants are selected that arise from in-frame recombination, typically reducing the size of CagY. Although these variants continue to make T4SS pili, they do not bind integrin and they are functionally "off," with reduced induction of IL-8 and translocation of T4SS effectors (CagA, peptidoglycan, HBP, DNA). This process is reversible and may be promoted by factors that favor enhanced T4SS activity, such as low Fe^{+3}, high Zn^{+2}, and a Treg immune response. Although a very recent work suggests that CEACAMs but not integrins are essential for CagA translocation (Zhao et al. 2018), this surprising observation remains to be verified

3.4 T4SS Effectors

3.4.1 CagA

CagA is a highly immunogenic protein associated with type I strains that is associated epidemiologically with GC and is the only known bacterial oncoprotein. When expressed transgenically in mice, it induces neoplasia (Ohnishi et al. 2008). Upon contact, CagA enters the host cell and is phosphorylated by host kinases on tyrosine residues (Selbach et al. 2002a; Tammer et al. 2007). CagA entry requires an intact T4SS, but it remains unclear whether it is injected syringe-like, or presented to the host cell surface and then actively internalized by the host plasma membrane (Murata-Kamiya et al. 2010). The CagA C-terminus contains the tyrosine phosphorylation motifs, Glu-Pro-Ile-Tyr-Ala (EPIYA), which differ in number and sequence (Backert and Blaser 2016). Four distinct motifs (denoted EPIYA-A-D) have been identified, based on the characteristics of the amino acids that flank tyrosine (Higashi et al. 2002; Xia et al. 2009). East Asian *H. pylori* strains typically contain EPIYA-A and EPIYA-B motifs, followed by EPIYA-D, whereas Western *H. pylori* strains usually contain EPIYA-A and EPIYA-B followed by one to three repeats of EPIYA-C (Lind et al. 2014, 2016). Qualitative and quantitative differences in the EPIYA motifs are linked epidemiologically to the relative risk of *H. pylori*-associated GC and PUD (Li et al. 2017). CagA is highly polymorphic, due not only to different combinations of EPIYA, but also to duplication of regions within the gene (Covacci et al. 1993; Zhang et al. 2015) and even copy number variation (Draper et al. 2017; Jang et al. 2017). Expression of CagA is increased by salt (Gaddy et al. 2013; Loh et al. 2007), which may in part explain the epidemiologic link between dietary salt and GC.

3.4.2 Peptidoglycan

H. pylori is for the most part a mucosal pathogen, yet its LPS and flagellin are not recognized by surface expression of TLR4 and TLR5, respectively. How then do GECs generate an innate immune response to *H. pylori*? This dilemma was resolved, at least in part, with the observation that *H. pylori* induction of NF-κB was dependent upon the intracellular recognition protein Nod1 (Viala et al. 2004), which recognizes peptidoglycan. Nod1-deficient mice are more susceptible to infection with wild-type *H. pylori,* but not strains lacking the *cag*PAI, suggesting that internalization of peptidoglycan is dependent on an intact T4SS. However, small amounts of peptidoglycan can be internalized even in the absence of a T4SS, which may result from internalization of outer membrane vesicles (Olofsson et al. 2010). Recognition of peptidoglycan by Nod1 has also been shown to mediate innate defense against *H. pylori* by induction of β-defensins (Grubman et al. 2010).

3.4.3 Chromosomal DNA

H. pylori also appears to stimulate innate immune responses via T4SS-dependent translocation of chromosomal DNA into host cells, where it stimulates TLR9 (Varga et al. 2016). This observation mirrors the classical translocation of DNA by the T4SS of *A. tumefaciens*, though it differs in that the *H. pylori* DNA is chromosomal, while in *A. tumefaciens* it is episomal.

3.4.4 ADP-heptose (ADP-hep)

Three groups independently reported that another effector of the *cag*T4SS is heptose-1,7-bisphosphate (HBP), an important intermediate metabolite of LPS inner heptose core (Stein et al. 2017; Zimmermann et al. 2017; Gall et al. 2017). The ALPK1-TIFA axis was proposed to play a central role in induction of IL-8 transcription in response to *H. pylori* infection through HBP-mediated NF-κB activation. However, more recent results suggest that the bacterial pathogen-associated molecular pattern (PAMP) that is the actual ligand for ALPK1 is ADP-hep, not HBP (Pfannkuch et al. 2018). Analysis of *H. pylori* extracts by mass spectrometry revealed only small amounts of HBP that were insufficient to induce NF-κB activation. In contrast, ADP-hep is a prominent compound in *H. pylori* lysates and can activate NF-κB activation via the ALPK1-TIFA axis when applied extracellularly without transfection or permeabilization of the cell membrane, which is required for HBP. It appears that ADP-hep can be converted to HBP intracellularly by host cell adenylyltranferases (Zhou et al. 2018). The exact mechanism of uptake for ADP-hep remains unclear, but recognition of ADP-hep by ALPK1 appears to be a form of innate sensing in diverse bacteria, including some like *H. pylori*, which have injection systems, and others that are extracellular and do not (Zhou et al. 2018).

3.5 *Hijacking Host Signaling*

H. pylori infection of GECs activates the NF-κB inflammatory pathway and induces IL-8, a chemokine that recruits neutrophils to the site of infection and promotes chronic inflammation. This typically requires a functional T4SS, but can occur by CagA-dependent and CagA-independent mechanisms (Backert et al. 2017; Backert and Tegtmeyer 2017). Upon intracellular delivery of CagA by the T4SS, or perhaps by endocytosis of outer membrane vesicles (Olofsson et al. 2010), CagA is phosphorylated on EPIYA motifs by c-Src and c-Abl tyrosine kinases (Backert et al. 2008; Mueller et al. 2012). Phosphorylated CagA interacts with numerous host cell proteins that contain Src homology 2 (SH2) domains, which are phosphotyrosine-specific binding sites that mediate crucial pathways for cell signaling (Selbach et al. 2009). This suggests that phospho-CagA mimics tyrosine-phosphorylated host proteins and acts as a sort of "master key" that alters host cell signaling (Tegtmeyer et al. 2017a). *H. pylori* can also disrupt signal transduction in GECs by unphosphorylated CagA and by T4SS-dependent but CagA-independent effects, which result from injection of HBP, DNA, peptidoglycan, and perhaps other unknown T4SS effectors.

4 Outer Membrane Proteins (OMPs)

4.1 Paralogous OMP Gene Families

Compared to other bacteria, *H. pylori* has a remarkable number of outer membrane proteins. The genome of *H. pylori* contains over 50 OMPs, representing up to 4% of its coding potential. Some of them have been identified as porins, iron transporters, or adhesins (Alm et al. 2000). The best understood group of OMPs belongs to the Helicobacter outer membrane protein (Hop) family, but there are also the hop-related (Hor) proteins, the Helicobacter OMP family (Hof) proteins, and the Helicobacter outer membrane (Hom) family proteins, which are grouped based on sequence similarities (Alm et al. 2000). Recent structural analysis suggests that the Hop family can be further divided into three monophyletic groups that form a novel class of secretion proteins (Coppens et al. 2018). The OMPs typically contain a β-barrel structure that spans the outer membrane and conserved N- and C-terminal regions, while the surface-exposed regions carry more diversity. The conserved regions allow in some cases for gene duplication or gene conversion, where one gene is replaced by its paralog. Duplicated paralogs within a strain are often more closely related to one another than either is to its ortholog in other strains, suggesting co-evolution and diversification during the often lifelong infection within a given host. Many *H. pylori* OMPs can also undergo phase variation by slipped strand mispairing, altering the number of dinucleotide CT repeats in their 5' coding region to turn on or off gene expression (Alm et al. 2000).

4.2 H. pylori *OMPs and Clinical Disease*

Bacterial attachment to the gastric mucosal surface is the first step in colonization. Several OMPs serve as adhesins that mediate attachment to structures present on the secreted mucus and the epithelial cell surface. The best characterized of these are the Hop family members, designated most commonly as BabA (HopS), SabA (HopP), AlpA (HopC), AlpB (HopB), OipA (HopH), LabA (HopD), and HopQ. Most known adhesins on the bacterial surface are variably present in 40–80% of strains, or show allelic variation, with the exception of AlpA and AlpB. This raises the possibility that a particular repertoire of OMPs may be associated with disease, rather than asymptomatic colonization. In some studies, gastric pathology has been associated with expression of the blood group antigen-binding adhesin, BabA (Chen et al. 2013; Gerhard et al. 1999; Prinz et al. 2001). This might be explained in part by the association of attachment with potentiation of the *H. pylori* T4SS activity and expression of the neutrophil chemoattractant IL-8 (Ishijima et al. 2011), though this has not been generally observed (Odenbreit et al. 2002). BabA may also be associated with disease because attachment affects the expression of other virulence genes, or because its presence is correlated with other virulence loci, most notably the *cag*PAI. However, other studies have found no association of gastric disease with BabA expression (Mattar et al. 2005; Mizushima et al. 2001), or have identified associations with other OMPs or some combination of them, including SabA (Su et al. 2016; Yamaoka et al. 2006), HopQ (Yakoob et al. 2016; Yamaoka et al. 2006), and HomB (Jung et al. 2009; Talebi Bezmin Abadi et al. 2011). Discrepant findings likely result in part from differences in patient populations and from differences in the methods used to detect OMP expression, which variously include PCR, identification of an open reading frame, immunoblotting, and mass spectrometry.

4.3 OMPs and Pathogenesis

4.3.1 BabA

The best characterized *H. pylori* adhesin is BabA (HopS), which binds fucosylated blood group antigens that are expressed on GECs and the overlying mucus (Borén et al. 1993). Binding was initially thought to be specific for the Lewis B (Leb) antigen (Borén et al. 1993) found in blood group O individuals. Later, it was realized that while this is true for strain P466 used in the early work, as well as in some others designated "specialists," most strains are actually "generalists" that bind LeB, but also ALeb and BLeb (Aspholm-Hurtig et al. 2004). X-ray structural analysis of BabA isoforms recently demonstrated that the carbohydrate-binding domain consists of two diversity loops, designated DL1 and DL2, which are under strong selective pressure and provide rapid adaptation to binding affinity for ABO (generalist) versus O (specialist) blood group binding, even with single amino acid substitutions (Moonens et al. 2016). Another early observation that was later corrected is that BabA is encoded by what was called *babA*2, because in strain 17,875 there is a duplication called *babA*1 that lacked 10 nucleotides needed for the translation initiation codon (Ilver et al. 1998). This has led to some confusion in studies that use primers for detection of *babA*2 as a measure of BabA expression. In fact, most strains have one copy of *babA* and one or sometimes two *babA* paralogs with unknown function (Colbeck et al. 2006), called *babB* (*hopT*) and *babC* (*hopU*). Occasional strains lack *babA* and have duplicate copies of *babB*, or have *babB* and babC, while a few, like 17875, have two copies of *babA*. Some strains, like the common laboratory strain 26695, express a BabA that cannot bind Leb.

Gene conversion between these *babA* paralogs has been observed during experimental infection of gerbils and rhesus macaques and typically occurs by replacement of *babA* with a duplicate copy of *babB* that eliminates binding to blood group antigens (Solnick et al. 2004; Styer et al. 2010). BabA expression is also lost during experimental infection of mice, but it occurs by phase variation rather than gene conversion (Kable et al. 2017; Styer et al. 2010). Characterization of sequential clinical isolates suggests that changes in BabA and loss of Leb binding also occur in human beings (Nell et al. 2014). Presumably, this serves to promote adaptation to changes in the gastric environment, but the selective pressures are unknown. Unlike regulation of the T4SS, BabA expression in mice is not affected by adaptive immunity or by Toll-like receptor signaling (Kable et al. 2017). Surprisingly, loss of BabA expression is also not dependent on Leb expression, nor even on the capacity of BabA to bind Leb, suggesting the possibility that BabA may have other unrecognized functions beyond attachment to blood group antigens (Hansen et al. 2017; Kable et al. 2017). In monkeys, there is independent selective pressure to delete BabA and to overexpress BabB, which confers a fitness advantage, but again the selective pressures are unknown (Hansen et al. 2017).

Genomic change resulting from gene conversion, phase variation, or mutation may enable *H. pylori* to modify blood group antigen binding, but it is insufficient to explain how *H. pylori* can adapt to epithelial cell turnover that occurs every few days. If binding were irreversible, then *H. pylori* would be rapidly released from the epithelial surface into the lumen of the stomach and shed into the lower gastrointestinal tract. The solution to this dilemma appears to be that BabA binding to Leb is reversibly acid sensitive, so that BabA can detach from its ligand when the pH decreases, but then re-attach when pH is neutralized (Bugaytsova et al. 2017). This acid sensitivity may also play a role in colonization preferences for gastric antrum or corpus, which can change with the development of gastric atrophy after years of infection. *H. pylori* with an acid-sensitive BabA localizes predominantly to the antrum where there are no acid-secreting parietal cells and the mucus layer is thicker and more protective against acid and digestive enzymes. The acid sensitivity of BabA may also be associated with global disease patterns, as populations more prone to peptic ulcer and low gastric pH carry strains with a highly acid-sensitive BabA, while those with a higher risk of atrophic gastritis and GC more often carry strains with acid-resistant BabA that allows for a pan-gastric colonization (Bugaytsova et al. 2017).

4.3.2 SabA

It has been known for more than 30 years that *H. pylori* has the capacity to agglutinate erythrocytes, which is thought to play a role in disease pathogenesis. Although originally identified as HpaA, we now know that the *H. pylori* hemagglutinin is the sialic acid-binding adhesin, SabA (Aspholm et al. 2006), which binds sialylated Lewis antigens, particularly sLex (Aspholm et al. 2006; Mahdavi et al. 2002). Sialylated glycans are usually not prominent in the gastric mucosa, but their expression increases upon *H. pylori* infection in cell lines (Marcos et al. 2008) and in vivo in the rhesus monkey model (Linden et al. 2008; Marcos et al. 2008). Expression of SabA is not universal in all *H. pylori* strains. Like BabA, expression of SabA is modulated by gene conversion and phase variation (Talarico et al. 2012), but also by variability in the length of a polynucleotide tract in the promoter region (Aberg et al. 2014). Some evidence suggests that SabA expression is associated with GC (Yamaoka et al. 2006), perhaps in association with other OMPs (Su et al. 2016). Crystal structures of the extracellular domain of SabA from *H. pylori* 26695 (Pang et al. 2014) and J99 (Coppens et al. 2018) are in good agreement and reveal an L-shape that resembles a golf putter, with the head region featuring loops that vary in sequence among *H. pylori* strains.

4.3.3 OipA

HopH is designated outer inflammatory protein A (OipA) because a deletion mutant shows ~50% reduction in IL-8 after co-culture with GECs (Yamaoka et al. 2000) and may also reduce translocation of CagA (Horridge et al. 2017). While this is not true of all strains, the exceptions are typically phase variants of *oipA* in which the protein is not expressed (Yamaoka et al. 2000). Like BabA, OipA is typically present and expressed only in strains that have the *cag*PAI (Dossumbekova et al. 2006), though *H. pylori* can induce low to intermediate levels of IL-8 even when the *cag*PAI is deleted, so long as OipA is still functional (Tabassam et al. 2008; Yamaoka et al. 2000). A role for OipA in inflammation and carcinogenesis is also supported by the capacity of OipA to phosphorylate focal adhesion kinase (FAK) and its downstream pathways, which mediate OipA-dependent IL-8 induction (Tabassam et al. 2008, 2011). OipA also mediates attachment to host cells, but the ligand is unknown (Dossumbekova et al. 2006; Horridge et al. 2017).

4.3.4 HopQ

All *H. pylori* strains studied to date express HopQ, but there are two allelic variants, which have ~70% amino acid sequence identity (Cao and Cover 2002). Type I is found most commonly in the more virulent East Asian strains carrying the *cag*PAI and type S1 *vacA*, while Western strains are more commonly type II (Cao et al. 2005). HopQ is required for *H. pylori*-induced activation of NF-κB, translocation of CagA, and induction of IL-8 (Belogolova et al. 2013). Recently, HopQ has been identified as an adhesin for carcinoembryonic antigen-related cell adhesion molecules (CEACAMs), a large immunoglobulin superfamily with wide tissue distribution (Javaheri et al. 2016; Koniger et al. 2016). The binding is species-specific for human and rat, but not to murine, bovine, or canine CEACAM1 (Javaheri et al. 2016). Strains that carry HopQ type II can also bind CEACAM5, but show only weak binding to CEACAM1 (Koniger et al. 2016; Moonens et al. 2018). Although HopQ has a similar genetic organization to BabA and SabA, with an insertion domain that interacts with their glycan ligands, HopQ binding to CEACAM is not mediated by the *hopQ* insertion domain, and glycosylation of CEACAM plays a minimal role (Bonsor et al. 2018). Unlike the binding of BabA to Leb, HopQ binding to CEACAM is not reversible after being exposed to low pH that disrupts the binding (Bonsor et al. 2018). Recent structural studies have shown that HopQ interacts with CEACAM monomers at their dimer interface and induces monomerization (Bonsor et al. 2018), which abrogates CEACAM-dependent cell adhesion and signaling. No structural or signaling feature has been found that links HopQ–CEACAM interactions to CagA translocation, except for bringing the bacteria and the host cell in close contact. This suggests that

HopQ does not induce NF-κB activation, but affects its activation level by promoting the functionality of the T4SS. This effect seems to depend on the postulated HopQ interaction partners, CEACAM1, 5, and 6. These results demonstrate unambiguously that direct binding of components of the T4SS to β1-integrin receptors is not sufficient for CagA translocation, and are consistent with the recent observation that integrins, but not CEACAMs, are dispensable for CagA translocation (Zhao et al. 2018), while in other cells, both integrin and CEACAMs are required (Tegtmeyer et al. 2019).

4.3.5 AlpAB

Adherence-associated lipoprotein A and B (HopC and HopB) were first identified as adhesins involved in binding to human gastric tissue sections (Odenbreit et al. 1999). Subsequent work identified the cell receptor for AlpAB as laminin (Senkovich et al. 2011), a glycoprotein expressed on the basolateral membrane. AlpA and AlpB are paralogous genes with 46% amino acid identity that are organized in an operon (Odenbreit et al. 1999). Both alleles are present in all clinical isolates studied to date. Not surprisingly, deletion of AlpA leads to a polar loss of AlpB expression, so the specific effect of AlpA has been difficult to determine. Most studies suggest that AlpAB is important for successful infection, as mutants show reduced or absent colonization in mice, gerbils, and guinea pigs (de Jonge et al. 2004; Lu et al. 2007; Sugimoto et al. 2011). A deletion mutant of AlpB alone was also cleared from stomachs of guinea pigs within 3 weeks and resulted in a marked decrease in antibody response (de Jonge et al. 2004). However, an effect of AlpAB on colonization has not been seen uniformly (Senkovich et al. 2011). The effect of AlpAB on the host inflammatory response has been even more controversial. Deletion of AlpAB led to a reduction in the capacity to induce IL-6 (Lu et al. 2007), as well as IL-8 in some studies (Selbach et al. 2002b), but not in others (de Jonge et al. 2004; Odenbreit et al. 2002). As is so often the case with *H. pylori*, the discrepant results likely depend on strain differences. When a large population of strains was examined, deletion of AlpAB reduced the capacity to induce IL-8 in East Asian but not in Western strains, though it uniformly reduced the capacity of *H. pylori* to induce IL-6 (Lu et al. 2007). In vivo in the mouse model, deletion of AlpAB reduced IL-6 and CXCL-1 (previously called keratinocyte chemokine, KC), but gerbils infected with an AlpAB deletion mutant actually showed increased neutrophilic and lymphocytic infiltration using strain SS1. These effects were independent from the *cag*PAI as the SS1 strain has a non-functional T4SS. The reason for different results between studies and strains is unknown, but may reflect polymorphism in AlpAB proteins or in their interaction with different hosts.

4.3.6 HopZ

Hop Z is found in all known *H. pylori* strains, though it is highly polymorphic and exists as two distinct alleles that are <80% identical at the amino acid level from one strain to another. Like several other OMPs, HopZ expression is regulated by slipped strand mispairing in a CT repeat region located in the signal peptide coding region (Peck et al. 1999). HopZ may itself also regulate expression of some pH-dependent genes, including SabA (Giannakis et al. 2009). HopZ is found on the bacterial surface and mediates attachment to GECs (Peck et al. 1999), but the receptor is unknown. Deletion of *hopZ* reduces *H. pylori* fitness in hypochlorhydric gnotobiotic mice without expression of the H+/K+ ATPase, but not in wild-type mice (Giannakis et al. 2009). Re-isolation of *H. pylori* within individuals over time after acute infection, either from experimental challenge or from transmission within families, suggests that there is selective pressure for an ON status of HopZ expression. In contrast, ON versus OFF status appears stable during chronic infection, suggesting a role for HopZ early during acute infection. Since acute infection has been associated with hypochlorhydria, this finding is consistent with reduced fitness of a *hopZ* deletion mutant in hypochlorhydric mice (Giannakis et al. 2009).

4.3.7 LabA

MUC5AC gastric mucin expressing LacdiNAc was found to be correlated with *H. pylori* localization in humans, and all strains tested adhered significantly to this glycan motif (Rossez et al. 2014). The adhesin was subsequently identified and called LacdiNAc-binding adhesin A (LabA, HopD). LacdiNAc decoration on gastric MUC5AC may be in competition with complex glycosylation such as the Le(b) and H type 1, as well as complex sialylated structures that serve as receptors for BabA and SabA, respectively (Kenny et al. 2012).

5 Modulation of the Host Response: Bacterial Virulence from the Bacterial Perspective

The inflammatory response to infection is traditionally thought of as a host defense to control microbial replication, and bacteria in turn frequently elaborate virulence factors that inhibit host immunity, promote colonization, and increase the likelihood of transmission. Often, these virulence factors are expressed on mobile genetic elements, such as plasmids, phage, or pathogenicity islands. For example, T3SSs or T4SSs found in many bacteria secrete effectors that induce apoptosis in macrophages, inhibit phagocytosis, and decrease inflammatory cytokines and chemokines (Byndloss et al. 2017). But it is increasingly appreciated that from the bacterial perspective, inflammation may also be beneficial. T3SS-dependent inflammation in the gut

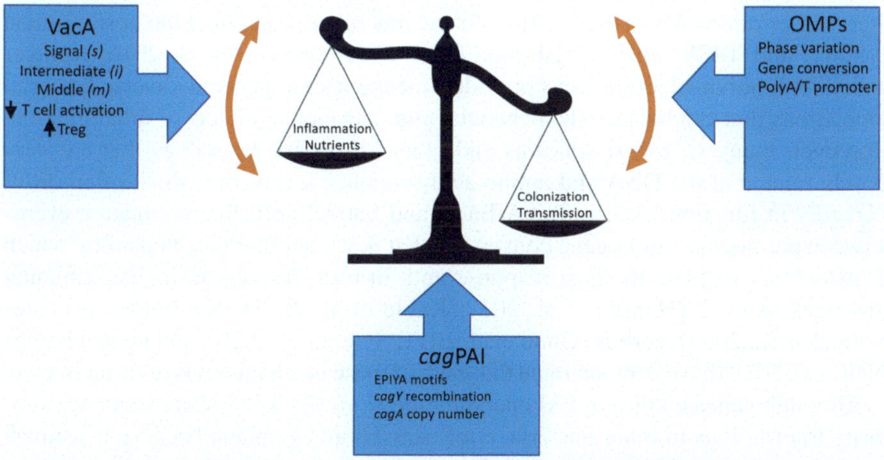

Fig. 2 ***H. pylori*** **virulence factors balance inflammation and colonization**. The hallmark of chronic *H. pylori* infection is inflammation, which probably enhances nutrient availability on the one hand, but comes at the cost of decreased colonization and therefore decreased possibility of transmission to a new host. Although *H. pylori* virulence factors are most often viewed in the context of their contribution to gastric pathology, from the bacterial perspective, they may be mechanisms to tune the host immune response so as to regulate these competing pressures

is actually *required* for Salmonella infection, because it enables *Salmonella* to compete with the intestinal microbiota by exploiting novel systems for respiration and for acquiring iron and other metals (Baumler and Sperandio 2016). In *Mycobacteria* infection, granuloma formation that has long been thought of as a host response to control bacterial spread actually limits immune cell access and promotes chronic infection (Cronan et al. 2016). These fundamental observations in animal models are supported by the growing recognition that immune suppression with corticosteroids has a role in the treatment of many infectious diseases. From the bacterial perspective, the optimal strategy may be to balance the benefits and the risks of the host immune response in a manner that promotes colonization, fitness, and transmission to a new host (Fig. 2).

Viewed in this context, it is not surprising that *H. pylori* virulence factors may not only promote, but also regulate inflammation, achieving a sort of "immunological sweet spot" that is probably critical to maintain chronic infection. This is perhaps best understood for VacA, where particular allelic variants are immunoregulatory. Like other bacteria with small genomes and a restricted niche (Alm et al. 2006), *H. pylori* has relatively few two-component regulatory systems and instead often uses DNA-based mechanisms to respond to the environment. This is particularly true for many of the *H. pylori* adhesins, which are commonly phase variable via a poly-CT tract in the 5′ coding region, and so can toggle with high frequency between ON and OFF states. Many also have long homopolymeric adenine or thymine tracts in the promoter region, which are prone to replication errors that change the length of the tract and alter transcription (Aberg et al. 2014). Since attachment can affect the host immune response and T4SS function (Ishijima et al. 2011; Javaheri et al. 2016; Koniger et al. 2016), polynucleotide tracts provide a mechanism to generate diverse bacterial populations that can balance the beneficial and detrimental effects of inflammation. Moreover, many *H. pylori* adhesins and other OMPs have paralogs that are very closely related at the DNA and amino acid sequence levels, but which often differ markedly in function, such as BabA/BabB and SabA/SabB. Recombination events between paralogs result in gene conversion that is yet another mechanism by which *H. pylori* can regulate the host response and, in turn, respond to it. Experimental infections in mice (Hansen et al. 2017; Kable et al. 2017), non-human primates (Solnick et al. 2004), gerbils (Ohno et al. 2011; Styer et al. 2010), and human beings (Nell et al. 2018) have demonstrated that each of these mechanisms is relevant in vivo.

Although genes on the *cag*PAI encoding the *H. pylori* T4SS share sequence similarity to orthologs in other bacteria, *cagY* stands out as unique because it is much larger than other *virB10* orthologs, and it contains an extraordinary number of direct DNA repeats that are predicted to encode in-frame insertions or deletions (Liu et al. 1999). When this was first recognized, it was proposed that repeat-mediated recombination in *cagY* might be a mechanism of immune evasion to avoid host recognition of a surface component of the T4SS pilus (Aras et al. 2003), although surprisingly CagY did not elicit a humoral immune response in naturally infected human beings. More recently, we discovered that CagY recombination serves not simply to avoid the host immune response, but rather to modulate it (Barrozo et al. 2013). During *H. pylori* infection of C57Bl/6 mice, *cagY* frequently undergoes CD4+T-cell- and interferon gamma (IFNγ)-dependent recombination that eliminates T4SS function (Barrozo et al. 2016). Similar results sometimes occur during infection of non-human primates (Barrozo et al. 2013) and in human beings (Barrozo et al. 2016). However, in non-human primates, *cagY* recombination can also cause gain of function in the T4SS, which implies that CagY serves as a sort of molecular rheostat that, from the bacterial perspective, "tunes" the host response to achieve the optimal level of inflammation that maximizes benefits and minimizes risks. CagY recombination occurs at a very high frequency—variants can be found easily by screening only a few dozen colonies—and presumably the immune system selects from among this random diversity those variants that are most fit. Since recombination in the *cagY* MRR that reduces T4SS function also produces commensurate reduction in whole-cell binding to $\alpha_5\beta_1$-integrin (Skoog et al. 2018), it is tempting to speculate that alteration in integrin binding is the mechanism by which CagY recombination modulates T4SS function. However, surface plasmon resonance studies demonstrated that the C-terminus but not the MRR binds integrin (Koelblen et al. 2017), so the details remain unclear.

Despite these many DNA-based mechanisms by which the *H. pylori* T4SS function can be modulated or reversibly inactivated, it is typically active in isolates from naturally infected human beings (Olbermann et al. 2010; Reyes-Leon et al. 2007) and non-human primates (Skoog et al. 2016), though to varying degrees. This begs the question of why T4SS function is commonly lost in experimental but not natural infection. The most likely explanation appears to be immune tolerance. Natural infection in both human beings and socially housed non-human primates occurs very early in life, often during the first year and sometimes even the first few weeks, when the immune system is not fully mature. Studies in neonatal mice demonstrate that under these circumstances, T4SS function is maintained because of Treg-mediated immunological tolerance (Arnold et al. 2011), which controls CD4+ T cells that would otherwise inhibit colonization and drive off T4SS (Barrozo et al. 2016). This conclusion is also consistent the observation that in countries where *H. pylori* prevalence is low and acquired later in life, the *cag*PAI is rapidly being lost (Perez-Perez et al. 2002).

Since T4SS function is typically retained in natural *H. pylori* isolates, yet *cagY* and the MRR are ubiquitous, there must be situations where loss of T4SS function is advantageous to the bacterium. But when might *cagY* recombination and loss of T4SS function occur during natural human infection? It may be soon after acquisition, as in mice and monkeys, or during some environmental events, such as a change in pH, which from the bacterial perspective tips the balance against inflammation. Another possibility is a transient, systemic infection that enhances non-specific innate immunity that would promote bacterial clearance, but could be at least partially overcome by downregulation of T4SS-mediated inflammation. Conceptual support for this idea comes from experiments in which mice latently infected with herpesviruses are resistant to infection with *Listeria monocytogenes* and *Yersinia pestis* (Barton et al. 2007). Protection is not antigen specific but is mediated by viral induction of IFNγ and systemic activation of macrophages, upregulating innate immunity against subsequent bacterial infection. If so, the natural, homeostatic state of the T4SS may be on, and it might be challenging to "catch it in the act" of turning off in natural infection, because acute *H. pylori* infection is rarely detected in human beings, and the hypothetical environmental events are unknown.

6 Concluding Remarks

The sine qua non of *H. pylori* infection is chronic gastritis, which partially controls the bacterial burden, but in some individuals comes at the cost of clinical disease. Using the paradigm of tolerance versus resistance as the host response to infection (Medzhitov et al. 2012), disease results from resistance rather than from bacterial infection per se. For the bacterium, resistance is probably a necessary evil—necessary because it promotes nutrient acquisition but evil because it leads to reduction in bacterial load and therefore decreased chance of transmission to a new host. The solution to this dilemma is to have the *cag*PAI, which promotes inflammation, but also to have numerous, redundant mechanisms to control it. These include many discussed here with respect to VacA, OMPs, and CagY-dependent regulation of the *cag*PAI, but also others, such as γ-glutamyl transpeptidase (Oertli et al. 2013) and the recently described role of eosinophils in homeostasis and immune regulation (Arnold et al. 2018). While the study of *H. pylori* began with efforts to characterize its virulence factors so as to better treat or prevent infection, or perhaps develop biomarkers to identify pathogenic strains, better understanding of its regulatory mechanisms may in the end have a greater payoff for the control of asthma and other chronic inflammatory diseases that are increasingly common causes of human morbidity and mortality.

Acknowledgements Work in the Solnick laboratory is supported by Public Health Service Grant R01 AI08713 from the NIH to JVS.

References

Aberg A, Gideonsson P, Vallstrom A, Olofsson A, Ohman C, Rakhimova L, Boren T, Engstrand L, Brannstrom K, Arnqvist A (2014) A repetitive DNA element regulates expression of the *Helicobacter pylori* sialic acid binding adhesin by a rheostat-like mechanism. PLoS Pathog 10:e1004234. https://doi.org/10.1371/journal.ppat.1004234

Alm E, Huang K, Arkin A (2006) The evolution of two-component systems in bacteria reveals different strategies for niche adaptation. PLoS Comput Biol 2:e143. https://doi.org/10.1371/journal.pcbi.0020143

Alm RA, Bina J, Andrews BM, Doig P, Hancock RE, Trust TJ (2000) Comparative genomics of *Helicobacter pylori*: analysis of the outer membrane protein families. Infect Immun 68:4155–4168. https://doi.org/10.1128/iai.68.7.4155-4168.2000

Aras RA, Fischer W, Perez-Perez GI, Crosatti M, Ando T, Haas R, Blaser MJ (2003) Plasticity of repetitive DNA sequences within a bacterial (Type IV) secretion system component. J Exp Med 198:1349–1360. https://doi.org/10.1084/jem.20030381

Arnold IC, Artola-Boran M, Tallon de Lara P, Kyburz A, Taube C, Ottemann K, van den Broek M, Yousefi S, Simon HU, Muller A (2018) Eosinophils suppress Th1 responses and restrict bacterially induced gastrointestinal inflammation. J Exp Med 215:2055–2072. https://doi.org/10.1084/jem.20172049

Arnold IC, Lee JY, Amieva MR, Roers A, Flavell RA, Sparwasser T, Muller A (2011) Tolerance rather than immunity protects from *Helicobacter pylori*-induced gastric preneoplasia. Gastroenterology 140:199–209. https://doi.org/10.1053/j.gastro.2010.06.047

Aspholm-Hurtig M, Dailide G, Lahmann M, Kalia A, Ilver D, Roche N, Vikstrom S, Sjostrom R, Linden S, Backstrom A, Lundberg C, Arnqvist A, Mahdavi J, Nilsson UJ, Velapatino B, Gilman RH, Gerhard M, Alarcon T, Lopez-Brea M, Nakazawa T, Fox JG, Correa P, Dominguez-Bello MG, Perez-Perez GI, Blaser MJ, Normark S, Carlstedt I, Oscarson S, Teneberg S, Berg DE, Boren T (2004) Functional adaptation of BabA, the *H. pylori* ABO blood group antigen binding adhesin. Science 305:519–522. https://doi.org/10.1126/science.1098801

Aspholm M, Olfat FO, Norden J, Sonden B, Lundberg C, Sjostrom R, Altraja S, Odenbreit S, Haas R, Wadstrom T, Engstrand L, Semino-Mora C, Liu H, Dubois A, Teneberg S, Arnqvist A, Boren T (2006) SabA is the *H. pylori* hemagglutinin and is polymorphic in binding to sialylated glycans. PLoS Pathog 2: e110. https://doi.org/10.1371/journal.ppat.0020110

Backert S, Feller SM, Wessler S (2008) Emerging roles of Abl family tyrosine kinases in microbial pathogenesis. Trends Biochem Sci 33(2):80–90. https://doi.org/10.1016/j.tibs.2007.10.006

Backert S, Blaser MJ (2016) The role of CagA in the gastric biology of *Helicobacter pylori*. Cancer Res 76(14):4028–4031. https://doi.org/10.1158/0008-5472.CAN-16-1680

Backert S, Haas R, Gerhard M, Naumann M (2017) The *Helicobacter pylori* type IV secretion system encoded by the *cag* pathogenicity Island: architecture, function, and signaling. Curr Top Microbiol Immunol 413:187–220. https://doi.org/10.1007/978-3-319-75241-9_8

Backert S, Tegtmeyer N (2017) Type IV secretion and signal transduction of *Helicobacter pylori* CagA through interactions with host cell receptors. Toxins (Basel) 9:115. https://doi.org/10.3390/toxins9040115

Backert S, Tegtmeyer N, Fischer W (2015) Composition, structure and function of the *Helicobacter pylori* cag pathogenicity island encoded type IV secretion system. Future Microbiol 10:955–965. https://doi.org/10.2217/fmb.15.32

Backert S, Tegtmeyer N, Selbach M (2010) The versatility of *Helicobacter pylori* CagA effector protein functions: the master key hypothesis. Helicobacter 15(3):163–176. https://doi.org/10.1111/j.1523-5378.2010.00759.x

Barden S, Lange S, Tegtmeyer N, Conradi J, Sewald N, Backert S, Niemann HH (2013) A helical RGD motif promoting cell adhesion: crystal structures of the *Helicobacter pylori* type IV secretion system pilus protein CagL. Structure 21(11):1931–1941. https://doi.org/10.1016/j.str.2013.08.018

Barrozo RM, Cooke CL, Hansen LM, Lam AM, Gaddy JA, Johnson EM, Cariaga TA, Suarez G, Peek RM Jr, Cover TL, Solnick JV (2013) Functional plasticity in the type IV secretion system of *Helicobacter pylori*. PLoS Pathog 9:e1003189. https://doi.org/10.1371/journal.ppat.1003189

Barrozo RM, Hansen LM, Lam AM, Skoog EC, Martin ME, Cai LP, Lin Y, Latoscha A, Suerbaum S, Canfield DR, Solnick JV (2016) CagY is an immune-sensitive regulator of the *Helicobacter pylori* type IV secretion system. Gastroenterology 151:1164–1175. https://doi.org/10.1053/j.gastro.2016.08.014

Barton ES, White DW, Cathelyn JS, Brett-McClellan KA, Engle M, Diamond MS, Miller VL, Virgin HW (2007) Herpesvirus latency confers symbiotic protection from bacterial infection. Nature 447:326–329. https://doi.org/10.1038/nature05762

Baumler AJ, Sperandio V (2016) Interactions between the microbiota and pathogenic bacteria in the gut. Nature 535:85–93. https://doi.org/10.1038/nature18849

Belogolova E, Bauer B, Pompaiah M, Asakura H, Brinkman V, Ertl C, Bartfeld S, Nechitaylo TY, Haas R, Machuy N, Salama N, Churin Y, Meyer TF (2013) *Helicobacter pylori* outer membrane protein HopQ identified as a novel T4SS-associated virulence factor. Cell Microbiol 15:1896–1912. https://doi.org/10.1111/cmi.12158

Bonsor DA, Zhao Q, Schmidinger B, Weiss E, Wang J, Deredge D, Beadenkopf R, Dow B, Fischer W, Beckett D, Wintrode PL, Haas R, Sundberg EJ (2018) The *Helicobacter pylori* adhesin protein HopQ exploits the dimer interface of human CEACAMs to facilitate translocation of the oncoprotein CagA. EMBO J 37:e98664. https://doi.org/10.15252/embj.201798664

Borén T, Falk P, Roth KA, Larson G, Normark S (1993) Attachment of *Helicobacter pylori* to human gastric epithelium mediated by blood group antigens. Science 262:1892–1895. https://doi.org/10.1126/science.8018146

Borozan I, Zapatka M, Frappier L, Ferretti V (2018) Analysis of Epstein-Barr virus genomes and expression profiles in gastric adenocarcinoma. J Virol 92:e01239–01217. https://doi.org/10.1128/jvi.01239-17

Bray F, Ferlay J, Soerjomataram I, Siegel RL, Torre LA, Jemal A (2018) Global cancer statistics 2018: GLOBOCAN estimates of incidence and mortality worldwide for 36 cancer in 185 countries. CA Cancer J Clin 68:394–424. https://doi.org/10.3322/caac.21492

Bugaytsova JA, Björnham O, Chernov YA, Gideonsson P, Henriksson S, Mendez M, Sjöström R, Mahdavi J, Shevtsova A, Ilver D, Moonens K, Quintana-Hayashi MP, Moskalenko R, Aisenbrey C, Bylund G, Schmidt A, Åberg A, Brännström K, Königer V, Vikström S, Rakhimova L, Hofer A, Ögren J, Liu H, Goldman MD, Whitmire JM, Ådén J, Younson J, Kelly CG, Gilman RH, Chowdhury A, Mukhopadhyay AK, Nair GB, Papadakos KS, Martinez-Gonzalez B, Sgouras DN, Engstrand L, Unemo M, Danielsson D, Suerbaum S, Oscarson S, Morozova-Roche LA, Olofsson A, Gröbner G, Holgersson J, Esberg A, Strömberg N, Landström M, Eldridge AM, Chromy BA, Hansen LM, Solnick JV, Lindén SK, Haas R, Dubois A, Merrell DS, Schedin S, Remaut H, Arnqvist A, Berg DE, Borén T (2017) *Helicobacter pylori* adapts to chronic infection and gastric disease via pH-responsive BabA-mediated adherence. Cell Host Microbe 21:376–389. https://doi.org/10.1016/j.chom.2017.02.013

Byndloss MX, Rivera-Chavez F, Tsolis R, Baumler AJ (2017) How bacterial pathogens use type III and type IV secretion systems to facilitate their transmission. Curr Opin Microbiol 35:1–7

Cao P, Cover TL (2002) Two different families of hopQ alleles in *Helicobacter pylori*. J Clin Microbiol 40:4504–4511. https://doi.org/10.1128/jcm.40.12.4504-4511.2002

Cao P, Lee KJ, Blaser MJ, Cover TL (2005) Analysis of *hopQ* alleles in East Asian and Western strains of *Helicobacter pylori*. FEMS Microbiol Lett 251:37–43. https://doi.org/10.1016/j.femsle.2005.07.023

Chang YW, Shaffer CL, Rettberg LA, Ghosal D, Jensen GJ (2018) In Vivo structures of the *Helicobacter pylori* cag Type IV secretion system. Cell Rep 23:673–681. https://doi.org/10.1016/j.celrep.2018.03.085

Chen MY, He CY, Meng X, Yuan Y (2013) Association of *Helicobacter pylori* babA2 with peptic ulcer disease and gastric cancer. World J Gastroenterol 19:4242–4251. https://doi.org/10.3748/wjg.v19.i26.4242

Chung C, Olivares A, Torres E, Yilmaz O, Cohen H, Perez-Perez G (2010) Diversity of VacA intermediate region among *Helicobacter pylori* strains from several regions of the world. J Clin Microbiol 48:690–696. https://doi.org/10.1128/jcm.01815-09

Colbeck JC, Hansen LM, Fong JM, Solnick JV (2006) Genotypic profile of the outer membrane proteins BabA and BabB in clinical isolates of *Helicobacter pylori*. Infect Immun 74:4375–4378. https://doi.org/10.1128/iai.00485-06

Conradi J, Tegtmeyer N, Woźna M, Wissbrock M, Michalek C, Gagell C, Cover TL, Frank R, Sewald N, Backert S (2012) An RGD helper sequence in CagL of *Helicobacter pylori* assists in interactions with integrins and injection of CagA. Front Cell Infect Microbiol. 2:70. https://doi.org/10.3389/fcimb.2012.00070

Coppens F, Castaldo G, Debraekeleer A, Subedi S, Moonens K, Lo A, Remaut H (2018) Hop-family Helicobacter outer membrane adhesins form a novel class of Type 5-like secretion proteins with an interrupted beta-barrel domain. Mol Microbiol https://doi.org/10.1111/mmi.14075

Covacci A, Censini S, Bugnoli M, Petracca R, Burroni D, Macchia G, Massone A, Papini E, Xiang Z, Figura N (1993) Molecular characterization of the 128-kDa immunodominant antigen of *Helicobacter pylori* associated with cytotoxicity and duodenal ulcer. Proc Natl Acad Sci U S A 90:5791–5795. https://doi.org/10.1073/pnas.90.12.5791

Cover TL, Blaser MJ (1992) Purification and characterization of the vacuolating toxin from *Helicobacter pylori*. J Biol Chem 267:10570–10575

Covacci A, Falkow S, Berg DE, Rappuoli R (1997) Did the inheritance of a pathogenicity Island modify the virulence of *Helicobacter pylori*? Trends Microbiol 5:205–208. https://doi.org/10.1016/S0966-842X(97)01035-4

Cronan MR, Beerman RW, Rosenberg AF, Saelens JW, Johnson MG, Oehlers SH, Sisk DM, Jurcic Smith KL, Medvitz NA, Miller SE, Trinh LA, Fraser SE, Madden JF, Turner J, Stout JE, Lee S, Tobin DM (2016) *Macrophage epithelial* reprogramming underlies mycobacterial granuloma formation and promotes infection. Immunity 45:861–876. https://doi.org/10.1016/j.immuni.2016.09.014

de Jonge R, Durrani Z, Rijpkema SG, Kuipers EJ, van Vliet AH, Kusters JG (2004) Role of the *Helicobacter pylori* outer-membrane proteins AlpA and AlpB in colonization of the guinea pig stomach. J Med Microbiol 53:375–379. https://doi.org/10.1099/jmm.0.45551-0

Dossumbekova A, Prinz C, Mages J, Lang R, Kusters JG, Van Vliet AH, Reindl W, Backert S, Saur D, Schmid RM, Rad R (2006) *Helicobacter pylori* HopH (OipA) and bacterial pathogenicity: genetic and functional genomic analysis of hopH gene polymorphisms. J Infect Dis 194:1346–1355. https://doi.org/10.1086/508426

Draper JL, Hansen LM, Bernick DL, Abedrabbo S, Underwood JG, Kong N, Huang BC, Weis AM, Weimer BC, van Vliet AH, Pourmand N, Solnick JV, Karplus K, Ottemann KM (2017) Fallacy of the unique genome: sequence diversity within single *Helicobacter pylori* strains. MBio 8:e02321–16. https://doi.org/10.1128/mbio.02321-16

Eaton KA, Cover TL, Tummuru MK, Blaser MJ, Krakowka S (1997) Role of vacuolating cytotoxin in gastritis due to *Helicobacter pylori* in gnotobiotic piglets. Infect Immun 65:3462–3464

Falush D, Wirth T, Linz B, Pritchard JK, Stephens M, Kidd M, Blaser MJ, Graham DY, Vacher S, Perez-Perez GI, Yamaoka Y, Mégraud F, Otto K, Reichard U, Katzowitsch E, Wang X, Achtman

M, Suerbaum S (2003) Traces of human migrations in *Helicobacter pylori* populations. Science 299:1582–1585. https://doi.org/10.1126/science.1080857

Fischer W, Püls J, Buhrdorf R, Gebert B, Odenbreit S, Haas R (2001) Systematic mutagenesis of the *Helicobacter pylori* cag pathogenicity island: essential genes for CagA translocation in host cells and induction of interleukin-8. Mol Microbiol 42:1337–1348. https://doi.org/10.1046/j.1365-2958.2001.02714.x

Frick-Cheng AE, Pyburn TM, Voss BJ, McDonald WH, Ohi MD, Cover TL (2016) Molecular and structural analysis of the *Helicobacter pylori* cag type IV secretion system core complex. MBio 7:e02001–02015. https://doi.org/10.1128/mbio.02001-15

Gaddy JA, Radin JN, Loh JT, Piazuelo MB, Kehl-Fie TE, Delgado AG, Ilca FT, Peek RM, Cover TL, Chazin WJ, Skaar EP, Scott Algood HM (2014) The host protein calprotectin modulates the *Helicobacter pylori* cag type IV secretion system via zinc sequestration. PLoS Pathog 10:e1004450. https://doi.org/10.1371/journal.ppat.1004450

Gaddy JA, Radin JN, Loh JT, Zhang F, Washington MK, Peek RM, Algood HM, Cover TL (2013) High dietary salt intake exacerbates *Helicobacter pylori*-induced gastric carcinogenesis. Infect Immun 81:2258–2267. https://doi.org/10.1128/iai.01271-12

Gall A, Gaudet RG, Gray-Owen SD, Salama NR (2017) TIFA signaling in gastric epithelial cells initiates the cag type 4 secretion system-dependent innate immune response to *Helicobacter pylori* Infection. MBio 8:e01168–17. https://doi.org/10.1128/mbio.01168-17

Gebert B, Fischer W, Weiss E, Hoffmann R, Haas R (2003) *Helicobacter pylori* vacuolating cytotoxin inhibits T lymphocyte activation. Science 301:1099–1102. https://doi.org/10.1126/science.1086871

Gerhard M, Lehn N, Neumayer N, Boren T, Rad R, Schepp W, Miehlke S, Classen M, Prinz C (1999) Clinical relevance of the *Helicobacter pylori* gene for blood-group antigen-binding adhesin. Proc Natl Acad Sci U S A 96:12778–12783. https://doi.org/10.1073/pnas.96.22.12778

Ghiara P, Marchetti M, Blaser MJ, Tummuru MK, Cover TL, Segal ED, Tompkins LS, Rappuoli R (1995) Role of the *Helicobacter pylori* virulence factors vacuolating cytotoxin, CagA, and urease in a mouse model of disease. Infect Immun 63:4154–4160

Ghosal D, Chang YW, Jeong KC, Vogel JP, Jensen GJ (2017) In situ structure of the Legionella Dot/Icm type IV secretion system by electron cryotomography. EMBO Rep 18:726–732. https://doi.org/10.15252/embr.201643598

Giannakis M, Backhed HK, Chen SL, Faith JJ, Wu M, Guruge JL, Engstrand L, Gordon JI (2009) Response of gastric epithelial progenitors to *Helicobacter pylori* Isolates obtained from Swedish patients with chronic atrophic gastritis. J Biol Chem 284:30383–30394. https://doi.org/10.1074/jbc.m109.052738

Grohmann E, Christie PJ, Waksman G, Backert S (2018) Type IV secretion in Gram-negative and Gram-positive bacteria. Mol Microbiol 107:455–471. https://doi.org/10.1111/mmi.13896

Grubman A, Kaparakis M, Viala J, Allison C, Badea L, Karrar A, Boneca IG, Le Bourhis L, Reeve S, Smith IA, Hartland EL, Philpott DJ, Ferrero RL (2010) The innate immune molecule, NOD1, regulates direct killing of *Helicobacter pylori* by antimicrobial peptides. Cell Microbiol 12:626–639. https://doi.org/10.1111/j.1462-5822.2009.01421.x

Gupta VR, Patel HK, Kostolansky SS, Ballivian RA, Eichberg J, Blanke SR (2008) Sphingomyelin functions as a novel receptor for *Helicobacter pylori* VacA. PLoS Pathog 4:e1000073. https://doi.org/10.1371/journal.ppat.1000073

Hansen LM, Gideonsson P, Canfield DR, Boren T, Solnick JV (2017) Dynamic expression of the BabA adhesin and Its BabB paralog during *Helicobacter pylori* infection in rhesus macaques. Infect Immun 85:e00094–00017. https://doi.org/10.1128/iai.00094-17

Harris PR, Wright SW, Serrano C, Riera F, Duarte I, Torres J, Pena A, Rollan A, Viviani P, Guiraldes E, Schmitz JM, Lorenz RG, Novak L, Smythies LE, Smith PD (2008) *Helicobacter pylori* gastritis in children is associated with a regulatory T-cell response. Gastroenterology 134:491–499. https://doi.org/10.1053/j.gastro.2007.11.006

Hennig EE, Godlewski MM, Butruk E, Ostrowski J (2005) *Helicobacter pylori* VacA cytotoxin interacts with fibronectin and alters HeLa cell adhesion and cytoskeletal organization in vitro. FEMS Immunol Med Microbiol 44:143–150. https://doi.org/10.1016/j.femsim.2004.10.020

Higashi H, Tsutsumi R, Fujita A, Yamazaki S, Asaka M, Azuma T, Hatakeyama M (2002) Biological activity of the *Helicobacter pylori* virulence factor CagA is determined by variation in the tyrosine phosphorylation sites. Proc Natl Acad Sci U S A 99:14428–14433. https://doi.org/10.1073/pnas.222375399

Horridge DN, Begley AA, Kim J, Aravindan N, Fan K, Forsyth MH (2017) Outer inflammatory protein a (OipA) of *Helicobacter pylori* is regulated by host cell contact and mediates CagA translocation and interleukin-8 response only in the presence of a functional cag pathogenicity island type IV secretion system. Pathog Dis 75:113. https://doi.org/10.1093/femspd/ftx113

Ilver D, Arnqvist A, Ogren J, Frick IM, Kersulyte D, Incecik ET, Berg DE, Covacci A, Engstrand L, Borén T (1998) *Helicobacter pylori* adhesin binding fucosylated histo-blood group antigens revealed by retagging. Science 279:373–377. https://doi.org/10.1126/science.279.5349.373

Ishijima N, Suzuki M, Ashida H, Ichikawa Y, Kanegae Y, Saito I, Boren T, Haas R, Sasakawa C, Mimuro H (2011) BabA-mediated adherence is a potentiator of the *Helicobacter pylori* Type IV secretion system activity. J Biol Chem 286:25256–25264. https://doi.org/10.1074/jbc.m111.233601

Jain P, Luo ZQ, Blanke SR (2011) *Helicobacter pylori* vacuolating cytotoxin A (VacA) engages the mitochondrial fission machinery to induce host cell death. Proc Natl Acad Sci U S A 108:16032–16037. https://doi.org/10.1073/pnas.1105175108

Jang S, Su H, Blum FC, Bae S, Choi YH, Kim A, Hong YA, Kim J, Kim JH, Gunawardhana N, Jeon YE, Yoo YJ, Merrell DS, Ge L, Cha JH (2017) Dynamic expansion and contraction of cagA copy number in *Helicobacter pylori* Impact development of gastric disease. MBio 8:e01779–16. https://doi.org/10.1128/mbio.01779-16

Javaheri A, Kruse T, Moonens K, Mejias-Luque R, Debraekeleer A, Asche CI, Tegtmeyer N, Kalali B, Bach NC, Sieber SA, Hill DJ, Koniger V, Hauck CR, Moskalenko R, Haas R, Busch DH, Klaile E, Slevogt H, Schmidt A, Backert S, Remaut H, Singer BB, Gerhard M (2016) *Helicobacter pylori* adhesin HopQ engages in a virulence-enhancing interaction with human CEACAMs. Nat Microbiol 2:16189. https://doi.org/10.1038/nmicrobiol.2016.189

Jimenez-Soto LF, Kutter S, Sewald X, Ertl C, Weiss E, Kapp U, Rohde M, Pirch T, Jung K, Retta SF, Terradot L, Fischer W, Haas R (2009) *Helicobacter pylori* type IV secretion apparatus exploits β1 integrin in a novel RGD-independent manner. PLoS Pathog 5:e1000684. https://doi.org/10.1371/journal.ppat.1000684

Johnson EM, Gaddy JA, Voss BJ, Hennig EE, Cover TL (2014) Genes required for assembly of pili associated with the *Helicobacter pylori* cag type IV secretion system. Infect Immun 82:3457–3470. https://doi.org/10.1128/iai.01640-14

Jung SW, Sugimoto M, Graham DY, Yamaoka Y (2009) The homB status of *Helicobacter pylori* as a novel marker to distinguish gastric cancer from duodenal ulcer. J Clin Microbiol 37:3241–3245. https://doi.org/10.1128/jcm.00293-09

Kable ME, Hansen LM, Styer CM, Deck SL, Rakhimova O, Shevtsova A, Eaton KA, Martin ME, Gideonsson P, Boren T, Solnick JV (2017) Host determinants of expression of the *Helicobacter pylori* BabA adhesin. Sci Rep 7:46499. https://doi.org/10.1038/srep46499

Kenny DT, Skoog EC, Linden SK, Struwe WB, Rudd PM, Karlsson NG (2012) Presence of terminal N-acetylgalactosaminebeta1-4 N-acetylglucosamine Residues on O-linked oligosaccharides from gastric MUC5AC: involvement in *Helicobacter pylori* colonization? Glycobiology 22:1077–1085. https://doi.org/10.1093/glycob/cws076

Kodaman N, Pazos A, Schneider BG, Piazuelo MB, Mera R, Sobota RS, Sicinschi LA, Shaffer CL, Romero-Gallo J, de Sablet T, Harder RH, Bravo LE, Peek RM, Wilson KT, Cover TL, Williams SM, Correa P (2014) Human and *Helicobacter pylori* coevolution shapes the risk of gastric disease. Proc Natl Acad Sci U S A 111:1455–1460. https://doi.org/10.1073/pnas.1318093111

Koelblen T, Bergé C, Cherrier MV, Brillet K, Jimenez-Soto L, Ballut L, Takagi J, Montserret R, Rousselle P, Fischer W, Haas R, Fronzes R, Terradot L (2017) Molecular dissection of protein-

protein interactions between integrin α5β1 and the *Helicobacter pylori* Cag type IV secretion system. FEBS J 284:4143–4157. https://doi.org/10.1111/febs.14299

Koniger V, Holsten L, Harrison U, Busch B, Loell E, Zhao Q, Bonsor DA, Roth A, Kengmo-Tchoupa A, Smith SI, Mueller S, Sundberg EJ, Zimmermann W, Fischer W, Hauck CR, Haas R (2016) *Helicobacter pylori* exploits human CEACAMs via HopQ for adherence and translocation of CagA. Nat Microbiol 2:16188. https://doi.org/10.1038/nmicrobiol.2016.188

Kwok T, Zabler D, Urman S, Rohde M, Hartig R, Wessler S, Misselwitz R, Berger J, Sewald N, Konig W, Backert S (2007) Helicobacter exploits integrin for type IV secretion and kinase activation. Nature 449:862–866. https://doi.org/10.1038/nature06187

Lertpiriyapong K, Whary MT, Muthupalani S, Lofgren JL, Gamazon ER, Feng Y, Ge Z, Wang TC, Fox JG (2014) Gastric colonisation with a restricted commensal microbiota replicates the promotion of neoplastic lesions by diverse intestinal microbiota in the *Helicobacter pylori* INS-GAS mouse model of gastric carcinogenesis. Gut 63:54–63. https://doi.org/10.1136/gutjnl-2013-305178

Li Q, Liu J, Gong Y, Yuan Y (2017) Association of CagA EPIYA-D or EPIYA-C phosphorylation sites with peptic ulcer and gastric cancer risks: a meta-analysis. Medicine (Baltimore) 96:e6620. https://doi.org/10.1097/md.0000000000006620

Lind J, Backert S, Pfleiderer K, Berg DE, Yamaoka Y, Sticht H, Tegtmeyer N (2014) Systematic analysis of phosphotyrosine antibodies recognizing single phosphorylated EPIYA-motifs in CagA of Western-type *Helicobacter pylori* strains. PLoS One 9(5):e96488. https://doi.org/10.1371/journal.pone.0096488

Lind J, Backert S, Hoffmann R, Eichler J, Yamaoka Y, Perez-Perez GI, Torres J, Sticht H, Tegtmeyer N (2016) Systematic analysis of phosphotyrosine antibodies recognizing single phosphorylated EPIYA-motifs in CagA of East Asian-type *Helicobacter pylori* strains. BMC Microbiol 16(1):201. https://doi.org/10.1186/s12866-016-0820-6

Linden S, Mahdavi J, Semino-Mora C, Olsen C, Carlstedt I, Boren T, Dubois A (2008) Role of ABO secretor status in mucosal innate immunity and *H. pylori* infection. PLoS Pathog 4:e2. https://doi.org/10.1371/journal.ppat.0040002

Liu G, McDaniel TK, Falkow S, Karlin S (1999) Sequence anomalies in the Cag7 gene of the *Helicobacter pylori* pathogenicity island. Proc Natl Acad Sci U S A 96:7011–7016. https://doi.org/10.1073/pnas.96.12.7011

Loh JT, Torres VJ, Cover TL (2007) Regulation of *Helicobacter pylori* cagA expression in response to salt. Cancer Res 67:4709–4715. https://doi.org/10.1158/0008-5472.can-06-4746

Lu H, Wu JY, Beswick EJ, Ohno T, Odenbreit S, Haas R, Reyes VE, Kita M, Graham DY, Yamaoka Y (2007) Functional and intracellular signaling differences associated with the *Helicobacter pylori* AlpAB adhesin from Western and East Asian strains. J Biol Chem 282:6242–6254. https://doi.org/10.1074/jbc.m611178200

Mahdavi J, Sonden B, Hurtig M, Olfat FO, Forsberg L, Roche N, Angstrom J, Larsson T, Teneberg S, Karlsson KA, Altraja S, Wadstrom T, Kersulyte D, Berg DE, Dubois A, Petersson C, Magnusson KE, Norberg T, Lindh F, Lundskog BB, Arnqvist A, Hammarstrom L, Boren T (2002) *Helicobacter pylori* SabA adhesin in persistent infection and chronic inflammation. Science 297:573–578. https://doi.org/10.1126/science.1069076

Marcos NT, Magalhaes A, Ferreira B, Oliveira MJ, Carvalho AS, Mendes N, Gilmartin T, Head SR, Figueiredo C, David L, Santos-Silva F, Reis CA (2008) *Helicobacter pylori* induces beta3GnT5 in human gastric cell lines, modulating expression of the SabA ligand sialyl-Lewis x. J Clin Invest 118:2325–2336. https://doi.org/10.1172/jci34324

Mattar R, dos Santos AF, Eisig JN, Rodrigues TN, Silva FM, Lupinacci RM, Iriya K, Carrilho FJ (2005) No correlation of babA2 with vacA and cagA genotypes of *Helicobacter pylori* and grading of gastritis from peptic ulcer disease patients in Brazil. Helicobacter 10:601–608. https://doi.org/10.1111/j.1523-5378.2005.00360.x

Mayerle J, den Hoed CM, Schurmann C, Stolk L, Homuth G, Peters MJ, Capelle LG, Zimmermann K, Rivadeneira F, Gruska S, Völzke H, de Vries AC, Völker U, Teumer A, van Meurs JB, Steinmetz I, Nauck M, Ernst F, Weiss FU, Hofman A, Zenker M, Kroemer HK, Prokisch H, Uitterlinden AG,

Lerch MM, Kuipers EJ (2013) Identification of genetic loci associated with Helicobacter pylori serologic status. J Am Med Assoc 309:1012–1920. https://doi.org/10.1001/jama.2013.4350
McClain MS, Beckett AC, Cover TL (2017) *Helicobacter pylori* vacuolating toxin and gastric cancer. Toxins (Basel) 9:E316. https://doi.org/10.3390/toxins9100316
McClain MS, Cao P, Iwamoto H, Vinion-Dubiel AD, Szabo G, Shao Z, Cover TL (2001) A 12-amino-acid segment, present in type s2 but not type s1 *Helicobacter pylori* VacA proteins, abolishes cytotoxin activity and alters membrane channel formation. J Bacteriol 183:6499–6508. https://doi.org/10.1128/jb.183.22.6499-6508.2001
Medzhitov R, Schneider DS, Soares MP (2012) Disease tolerance as a defense strategy. Science 335:936–941. https://doi.org/10.1126/science.1214935
Mizushima T, Sugiyama T, Komatsu Y, Ishizuka J, Kato M, Asaka M (2001) Clinical relevance of the babA2 genotype of *Helicobacter pylori* in Japanese clinical isolates. J Clin Microbiol 39:2463–2465. https://doi.org/10.1128/jcm.39.7.2463-2465.2001
Moonens K, Gideonsson P, Subedi S, Bugaytsova J, Romao E, Mendez M, Norden J, Fallah M, Rakhimova L, Shevtsova A, Lahmann M, Castaldo G, Brannstrom K, Coppens F, Lo AW, Ny T, Solnick JV, Vandenbussche G, Oscarson S, Hammarstrom L, Arnqvist A, Berg DE, Muyldermans S, Boren T, Remaut H (2016) Structural Insights into polymorphic ABO glycan binding by *Helicobacter pylori*. Cell Host Microbe 19:55–66. https://doi.org/10.1016/j.chom.2015.12.004
Moonens K, Hamway Y, Neddermann M, Reschke M, Tegtmeyer N, Kruse T, Kammerer R, Mejías-Luque R, Singer BB, Backert S, Gerhard M, Remaut H (2018) *Helicobacter pylori* adhesin HopQ disrupts *trans* dimerization in human CEACAMs. EMBO J 37(13):e98665. https://doi.org/10.15252/embj.201798665
Mueller D, Tegtmeyer N, Brandt S, Yamaoka Y, De Poire E, Sgouras D, Wessler S, Torres J, Smolka A, Backert S (2012) c-Src and c-Abl kinases control hierarchic phosphorylation and function of the CagA effector protein in Western and East Asian *Helicobacter pylori* strains. J Clin Invest 122:1553–1566. https://doi.org/10.1172/jci61143
Murata-Kamiya N, Kikuchi K, Hayashi T, Higashi H, Hatakeyama M (2010) *Helicobacter pylori* exploits host membrane phosphatidylserine for delivery, localization, and pathophysiological action of the CagA oncoprotein. Cell Host Microbe 7:399–411. https://doi.org/10.1016/j.chom.2010.04.005
Nell S, Estibariz I, Krebes J, Bunk B, Graham DY, Overmann J, Song Y, Sproer C, Yang I, Wex T, Korlach J, Malfertheiner P, Suerbaum S (2018) Genome and methylome variation in *Helicobacter pylori* with a cag pathogenicity Island during early stages of human infection. Gastroenterology 154:612–623.e617. https://doi.org/10.1053/j.gastro.2017.10.014
Nell S, Kennemann L, Schwarz S, Josenhans C, Suerbaum S (2014) Dynamics of Lewis b binding and sequence variation of the babA adhesin gene during chronic *Helicobacter pylori* infection in humans. MBio 5:e02281–02214. https://doi.org/10.1128/mbio.01233-15
Noto JM, Gaddy JA, Lee JY, Piazuelo MB, Friedman DB, Colvin DC, Romero-Gallo J, Suarez G, Loh J, Slaughter JC, Tan S, Morgan DR, Wilson KT, Bravo LE, Correa P, Cover TL, Amieva MR, Peek RM Jr (2013) Iron deficiency accelerates *Helicobacter pylori*-induced carcinogenesis in rodents and humans. J Clin Invest 123:479–492. https://doi.org/10.1172/jci64373
Odenbreit S, Kavermann H, Puls J, Haas R (2002) CagA tyrosine phosphorylation and interleukin-8 induction by *Helicobacter pylori* are independent from AlpAB, HopZ and Bab group outer membrane proteins. Int J Med Microbiol 292:257–266. https://doi.org/10.1078/1438-4221-00205
Odenbreit S, Till M, Hofreuter D, Faller G, Haas R (1999) Genetic and functional characterization of the AlpAB gene locus essential for the adhesion of *Helicobacter pylori* to human gastric tissue. Mol Microbiol 31:1537–1548. https://doi.org/10.1046/j.1365-2958.1999.01300.x
Oertli M, Noben M, Engler DB, Semper RP, Reuter S, Maxeiner J, Gerhard M, Taube C, Muller A (2013) *Helicobacter pylori* gamma-glutamyl transpeptidase and vacuolating cytotoxin promote gastric persistence and immune tolerance. Proc Natl Acad Sci U S A 110:3047–3052. https://doi.org/10.1073/pnas.1211248110

Ogura K, Maeda S, Nakao M, Watanabe T, Tada M, Kyutoku T, Yoshida H, Shiratori Y, Omata M (2000) Virulence factors of *Helicobacter pylori* responsible for gastric diseases in *Mongolian gerbil*. J Exp Med 192:1601–1610. https://doi.org/10.1084/jem.192.11.1601

Ohnishi N, Yuasa H, Tanaka S, Sawa H, Miura M, Matsui A, Higashi H, Musashi M, Iwabuchi K, Suzuki M, Yamada G, Azuma T, Hatakeyama M (2008) Transgenic expression of *Helicobacter pylori* CagA induces gastrointestinal and hematopoietic neoplasms in mouse. Proc Natl Acad Sci U S A 105:1003–1008. https://doi.org/10.1073/pnas.0711183105

Ohno T, Vallstrom A, Rugge M, Ota H, Graham DY, Arnqvist A, Yamaoka Y (2011) Effects of blood group antigen-binding adhesin expression during *Helicobacter pylori* infection of *Mongolian gerbils*. J Infect Dis 203:726–735. https://doi.org/10.1093/infdis/jiq090

Olbermann P, Josenhans C, Moodley Y, Uhr M, Stamer C, Vauterin M, Suerbaum S, Achtman M, Linz B (2010) A global overview of the genetic and functional diversity in the *Helicobacter pylori* cag pathogenicity Island. PLoS Genet 6:e1001069. https://doi.org/10.1371/journal.pgen.1001069

Olofsson A, Vallstrom A, Petzold K, Tegtmeyer N, Schleucher J, Carlsson S, Haas R, Backert S, Wai SN, Grobner G, Arnqvist A (2010) Biochemical and functional characterization of *Helicobacter pylori* vesicles. Mol Microbiol 77:1539–1555. https://doi.org/10.1111/j.1365-2958.2010.07307.x

Pang SS, Nguyen ST, Perry AJ, Day CJ, Panjikar S, Tiralongo J, Whisstock JC, Kwok T (2014) The three-dimensional structure of the extracellular adhesion domain of the sialic acid-binding adhesin SabA from *Helicobacter pylori*. J Biol Chem 289:6332–6340. https://doi.org/10.1074/jbc.m113.513135

Peck B, Ortkamp M, Diehl KD, Hundt E, Knapp B (1999) Conservation, localization and expression of HopZ, a protein involved in adhesion of *Helicobacter pylori*. Nucleic Acids Res 27:3325–3333. https://doi.org/10.1093/nar/27.16.3325

Perez-Perez GI, Salomaa A, Kosunen TU, Daverman B, Rautelin H, Aromaa A, Knekt P, Blaser MJ (2002) Evidence that cagA(+) *Helicobacter pylori* strains are disappearing more rapidly than cagA(−) strains. Gut 50:295–298. https://doi.org/10.1136/gut.50.3.295

Pfannkuch L, Hurwitz R, Traulsen J, Kosma P, Schmid M, Meyer TF (2018) ADP heptose, a novel pathogen-associated moledular pattern associated with *Helicobacter pylori* type 4 secretion. bioRxiv https://doi.org/10.1101/405951

Pormohammad A, Ghotaslou R, Leylabadlo HE, Nasiri MJ, Dabiri H, Hashemi A (2018) Risk of gastric cancer in association with *Helicobacter pylori* different virulence factors: a systematic review and meta-analysis. Microb Pathog 118:214–219. https://doi.org/10.1016/j.micpath.2018.03.004

Posselt G, Backert S, Wessler S (2013) The functional interplay of *Helicobacter pylori* factors with gastric epithelial cells induces a multi-step process in pathogenesis. Cell Commun Signal 11:77. https://doi.org/10.1186/1478-811X-11-77

Prinz C, Schoniger M, Rad R, Becker I, Keiditsch E, Wagenpfeil S, Classen M, Rosch T, Schepp W, Gerhard M (2001) Key importance of the *Helicobacter pylori* adherence factor blood group antigen binding adhesin during chronic gastric inflammation. Cancer Res 61:1903–1909

Reyes-Leon A, Atherton JC, Argent RH, Puente JL, Torres J (2007) Heterogeneity in the activity of Mexican *Helicobacter pylori* strains in gastric epithelial cells and its association with diversity in the cagA gene. Infect Immun 75:3445–3454. https://doi.org/10.1128/iai.01951-06

Rhead JL, Letley DP, Mohammadi M, Hussein N, Mohagheghi MA, Eshagh Hosseini M, Atherton JC (2007) A new *Helicobacter pylori* vacuolating cytotoxin determinant, the intermediate region, is associated with gastric cancer. Gastroenterology 133:926–936. https://doi.org/10.1053/j.gastro.2007.06.056

Robinson K, Kenefeck R, Pidgeon EL, Shakib S, Patel S, Polson RJ, Zaitoun AM, Atherton JC (2008) *Helicobacter pylori*-induced peptic ulcer disease is associated with inadequate regulatory T cell responses. Gut 57:1375–1385. https://doi.org/10.1136/gut.2007.137539

Rohde M, Puls J, Buhrdorf R, Fischer W, Haas R (2003) A novel sheathed surface organelle of the *Helicobacter pylori* cag type IV secretion system. Mol Microbiol 49:219–234. https://doi.org/10.1046/j.1365-2958.2003.03549.x

Rossez Y, Gosset P, Boneca IG, Magalhaes A, Ecobichon C, Reis CA, Cieniewski-Bernard C, Chevalier Curt MJ, Leonard R, Maes E, Sperandio B, Slomianny C, Sansonetti PJ, Michalski

JC, Robbe-Masselot C (2014) The LacdiNAc specific adhesin LabA mediates adhesion of *Helicobacter pylori* to human gastric mucosa. J Infect Dis 210:1286–1295. https://doi.org/10.1093/infdis/jiu239
Salama NR, Otto G, Tompkins L, Falkow S (2001) Vacuolating cytotoxin of *Helicobacter pylori* plays a role during colonization in a mouse model of infection. Infect Immun 69:730–736. https://doi.org/10.1128/iai.69.2.730-736.2001
Selbach M, Moese S, Hauck CR, Meyer TF, Backert S (2002a) Src is the kinase of the *Helicobacter pylori* CagA protein in vitro and in vivo. J Biol Chem 277(9):6775–8
Selbach M, Moese S, Meyer TF, Backert S (2002b) Functional analysis of the *Helicobacter pylori* cag pathogenicity island reveals both VirD4-CagA-dependent and VirD4-CagA-independent mechanisms. Infect Immun 70:665–671. https://doi.org/10.1128/iai.70.2.665-671.2002
Selbach M, Paul FE, Brandt S, Guye P, Daumke O, Backert S, Dehio C, Mann M (2009) Host cell interactome of tyrosine-phosphorylated bacterial proteins. Cell Host Microbe 5:397–403. https://doi.org/10.1016/j.chom.2009.03.004
Senkovich OA, Yin J, Ekshyyan V, Conant C, Traylor J, Adegboyega P, McGee DJ, Rhoads RE, Slepenkov S, Testerman TL (2011) *Helicobacter pylori* AlpA and AlpB bind host laminin and influence gastric inflammation in gerbils. Infect Immun 79:3106–3116. https://doi.org/10.1128/iai.01275-10
Seto K, Hayashi-Kuwabara Y, Yoneta T, Suda H, Tamaki H (1998) Vacuolation induced by cytotoxin from *Helicobacter pylori* is mediated by the EGF receptor in HeLa cells. FEBS Lett 431:347–350. https://doi.org/10.1016/s0014-5793(98)00788-1
Sewald X, Gebert-Vogl B, Prassl S, Barwig I, Weiss E, Fabbri M, Osicka R, Schiemann M, Busch DH, Semmrich M, Holzmann B, Sebo P, Haas R (2008) Integrin subunit CD18 Is the T-lymphocyte receptor for the *Helicobacter pylori* vacuolating cytotoxin. Cell Host Microbe 3:20–29. https://doi.org/10.1016/j.chom.2007.11.003
Shaffer CL, Gaddy JA, Loh JT, Johnson EM, Hill S, Hennig EE, McClain MS, McDonald WH, Cover TL (2011) *Helicobacter pylori* exploits a unique repertoire of type IV secretion system components for pilus assembly at the bacteria-host cell interface. PLoS Pathog 7:e1002237. https://doi.org/10.1371/journal.ppat.1002237
Sharma CM, Hoffmann S, Darfeuille F, Reignier J, Findeiss S, Sittka A, Chabas S, Reiche K, Hackermüller J, Reinhardt R, Stadler PF, Vogel J (2010) The primary transcriptome of the major human pathogen *Helicobacter pylori*. Nature 464:250–255. https://doi.org/10.1038/nature08756
Skoog EC, Deck SL, Entwistle HD, Hansen LM, Solnick JV (2016) Characterization of the Cag pathogenicity island in *Helicobacter pylori* from naturally infected rhesus macaques. FEMS Microbiol Lett 363:275. https://doi.org/10.1093/femsle/fnw275
Skoog EC, Morikis VA, Martin ME, Foster GA, Cai LP, Hansen LM, Li B, Gaddy JA, Simon SI, Solnick JV (2018) CagY-dependent regulation of type IV secretion in *Helicobacter pylori* is associated with alterations in integrin binding. MBio 9:e00717–00718. https://doi.org/10.1101/294769
Solnick JV, Hansen LM, Salama NR, Boonjakuakul JK, Syvanen M (2004) Modification of *Helicobacter pylori* outer membrane protein expression during experimental infection of rhesus macaques. Proc Natl Acad Sci U S A 101:2106–2111. https://doi.org/10.1086/423142
Stein SC, Faber E, Bats SH, Murillo T, Speidel Y, Coombs N, Josenhans C (2017) *Helicobacter pylori* modulates host cell responses by CagT4SS-dependent translocation of an intermediate metabolite of LPS inner core heptose biosynthesis. PLoS Pathog 13:e1006514. https://doi.org/10.1371/journal.ppat.1006514
Styer CM, Hansen LM, Cooke CL, Gundersen AM, Choi SS, Berg DE, Benghezal M, Marshall BJ, Peek RM Jr, Boren T, Solnick JV (2010) Expression of the BabA adhesin during experimental infection with *Helicobacter pylori*. Infect Immun 78:1593–1600. https://doi.org/10.1128/iai.01297-09
Su YL, Huang HL, Huang BS, Chen PC, Chen CS, Wang HL, Lin PH, Chieh MS, Wu JJ, Yang JC, Chow LP (2016) Combination of OipA, BabA, and SabA as candidate biomarkers for predicting *Helicobacter pylori*-related gastric cancer. Sci Rep 6:36442. https://doi.org/10.1038/srep36442

Sugimoto M, Ohno T, Graham DY, Yamaoka Y (2011) *Helicobacter pylori* outer membrane proteins on gastric mucosal interleukin 6 and 11 expression in *Mongolian gerbils*. J Gastroenterol Hepatol 26:1677–1684. https://doi.org/10.1111/j.1440-1746.2011.06817.x

Ta LH, Hansen LM, Sause WE, Shiva O, Millstein A, Ottemann KM, Castillo AR, Solnick JV (2012) Conserved transcriptional unit organization of the cag pathogenicity island among *Helicobacter pylori* strains. Front Cell Infect Microbiol 2:46. https://doi.org/10.3389/fcimb.2012.00046

Tabassam FH, Graham DY, Yamaoka Y (2008) OipA plays a role in *Helicobacter pylori*-induced focal adhesion kinase activation and cytoskeletal re-organization. Cell Microbiol 10:1008–1020. https://doi.org/10.1111/j.1462-5822.2007.01104.x

Tabassam FH, Graham DY, Yamaoka Y (2011) Paxillin is a novel cellular target for converging *Helicobacter pylori*-induced cellular signaling. Am J Physiol Gastrointest Liver Physiol 301:G601–611. https://doi.org/10.1152/ajpgi.00375.2010

Talarico S, Whitefield SE, Fero J, Haas R, Salama NR (2012) Regulation of *Helicobacter pylori* adherence by gene conversion. Mol Microbiol 84:1050–1061. https://doi.org/10.1111/j.1365-2958.2012.08073.x

Talebi Bezmin Abadi A, Rafiei A, Ajami A, Hosseini V, Taghvaei T, Jones KR, Merrell DS (2011) *Helicobacter pylori* homB, but not cagA, is associated with gastric cancer in Iran. J Clin Microbiol 49:3191–3197. https://doi.org/10.1128/jcm.00947-11

Tammer I, Brandt S, Hartig R, König W, Backert S (2007) Activation of Abl by *Helicobacter pylori*: a novel kinase for CagA and crucial mediator of host cell scattering. Gastroenterology 132(4):1309–19

Tanaka J, Suzuki T, Mimuro H, Sasakawa C (2003) Structural definition on the surface of *Helicobacter pylori* type IV secretion apparatus. Cell Microbiol 5:395–404. https://doi.org/10.1046/j.1462-5822.2003.00286.x

Tegtmeyer N, Harrer A, Schmitt V, Singer BB, Backert S (2019) Expression of CEACAM1 or CEACAM5 in AZ-521 cells restores the type IV secretion deficiency for translocation of CagA by *Helicobacter pylori*. Cell Microbiol 21:e12965. https://doi.org/10.1111/cmi.12965

Tegtmeyer N, Neddermann M, Asche CI, Backert S (2017a) Subversion of host kinases: a key network in cellular signaling hijacked by *Helicobacter pylori* CagA. Mol Microbiol 105:358–372. https://doi.org/10.1111/mmi.13707

Tegtmeyer N, Wessler S, Necchi V, Rohde M, Harrer A, Rau TT, Asche CI, Boehm M, Loessner H, Figueiredo C, Naumann M, Palmisano R, Solcia E, Ricci V, Backert S (2017b) *Helicobacter pylori* employs a unique basolateral type IV secretion mechanism for CagA delivery. Cell Host Microbe 22:552–560.e555. https://doi.org/10.1016/j.chom.2017.09.005

Telford JL, Ghiara P, Dell'Orco M, Comanducci M, Burroni D, Bugnoli M, Tecce MF, Censini S, Covacci A, Xiang Z et al (1994) Gene structure of the *Helicobacter pylori* cytotoxin and evidence of its key role in gastric disease. J Exp Med 179:1653–1658. https://doi.org/10.1084/jem.179.5.1653

Varga MG, Shaffer CL, Sierra JC, Suarez G, Piazuelo MB, Whitaker ME, Romero-Gallo J, Krishna US, Delgado A, Gomez MA, Good JA, Almqvist F, Skaar EP, Correa P, Wilson KT, Hadjifrangiskou M, Peek RM (2016) Pathogenic *Helicobacter pylori* strains translocate DNA and activate TLR9 via the cancer-associated cag type IV secretion system. Oncogene 35:6262–6269. https://doi.org/10.1038/onc.2016.158

Viala J, Chaput C, Boneca IG, Cardona A, Girardin SE, Moran AP, Athman R, Memet S, Huerre MR, Coyle AJ, DiStefano PS, Sansonetti PJ, Labigne A, Bertin J, Philpott DJ, Ferrero RL (2004) Nod1 responds to peptidoglycan delivered by the *Helicobacter pylori* cag pathogenicity island. Nat Immunol 5:1166–1174. https://doi.org/10.1038/ni1131

Winter JA, Letley DP, Cook KW, Rhead JL, Zaitoun AA, Ingram RJ, Amilon KR, Croxall NJ, Kaye PV, Robinson K, Atherton JC (2014) A role for the vacuolating cytotoxin, VacA, in colonization and *Helicobacter pylori*-induced metaplasia in the stomach. J Infect Dis 210:954–963. https://doi.org/10.1093/infdis/jiu154

Wirth HP, Beins MH, Yang M, Tham KT, Blaser MJ (1998) Experimental infection of Mongolian gerbils with wild-type and mutant *Helicobacter pylori* strains. Infect Immun 66:4856–4866. https://doi.org/10.1053/gast.1997.v113.pm9322503

Xia Y, Yamaoka Y, Zhu Q, Matha I, Gao X (2009) A comprehensive sequence and disease correlation analyses for the C-terminal region of CagA protein of *Helicobacter pylori*. PLoS ONE 4:e7736. https://doi.org/10.1371/journal.pone.0007736

Yahiro K, Satoh M, Nakano M, Hisatsune J, Isomoto H, Sap J, Suzuki H, Nomura F, Noda M, Moss J, Hirayama T (2012) Low-density lipoprotein receptor-related protein-1 (LRP1) mediates autophagy and apoptosis caused by *Helicobacter pylori* VacA. J Biol Chem 287:31104–31115. https://doi.org/10.1074/jbc.m112.387498

Yahiro K, Wada A, Nakayama M, Kimura T, Ogushi K, Niidome T, Aoyagi H, Yoshino K, Yonezawa K, Moss J, Hirayama T (2003) Protein-tyrosine phosphatase alpha, RPTP alpha, is a *Helicobacter pylori* VacA receptor. J Biol Chem 278:19183–19189. https://doi.org/10.1074/jbc.m300117200

Yakoob J, Abbas Z, Khan R, Salim SA, Awan S, Abrar A, Jafri W (2016) *Helicobacter pylori* outer membrane protein Q allele distribution is associated with distinct pathologies in Pakistan. Infect Genet Evol 37:57–62. https://doi.org/10.1016/j.meegid.2015.10.027

Yamaoka Y, Kwon DH, Graham DY (2000) A M(r) 34,000 proinflammatory outer membrane protein (oipA) of *Helicobacter pylori*. Proc Natl Acad Sci U S A 97:7533–7538. https://doi.org/10.1073/pnas.130079797

Yamaoka Y, Ojo O, Fujimoto S, Odenbreit S, Haas R, Gutierrez O, El-Zimaity HM, Reddy R, Arnqvist A, Graham DY (2006) *Helicobacter pylori* outer membrane proteins and gastroduodenal disease. Gut 55:775–781. https://doi.org/10.1136/gut.2005.083014

Yang I, Woltemate S, Piazuelo MB, Bravo LE, Yepez MC, Romero-Gallo J, Delgado AG, Wilson KT, Peek RM, Correa P, Josenhans C, Fox JG, Suerbaum S (2016) Different gastric microbiota compositions in two human populations with high and low gastric cancer risk in Colombia. Sci Rep 6:18594. https://doi.org/10.1038/srep18594

Yokoyama K, Higashi H, Ishikawa S, Fujii Y, Kondo S, Kato H, Azuma T, Wada A, Hirayama T, Aburatani H, Hatakeyama M (2005) Functional antagonism between *Helicobacter pylori* CagA and vacuolating toxin VacA in control of the NFAT signaling pathway in gastric epithelial cells. Proc Natl Acad Sci U S A 102:9661–9666. https://doi.org/10.1073/pnas.0502529102

Zhao Q, Busch B, Jimenez-Soto LF, Ishikawa-Ankerhold H, Massberg S, Terradot L, Fischer W, Haas R (2018) Integrin but not CEACAM receptors are dispensable for *Helicobacter pylori* CagA translocation. PLoS Pathog 14:e1007359. https://doi.org/10.1371/journal.ppat.1007359

Zhang XS, Tegtmeyer N, Traube L, Jindal S, Perez-Perez G, Sticht H, Backert S, Blaser MJ (2015) A specific A/T polymorphism in Western tyrosine phosphorylation B-motifs regulates *Helicobacter pylori* CagA epithelial cell interaction. PloS Pathog 11:e1004621. https://doi.org/10.1371/journal.ppat.1004621

Zheng PY, Jones NL (2003) *Helicobacter pylori* strains expressing the vacuolating cytotoxin interrupt phagosome maturation in macrophages by recruiting and retaining TACO (coronin 1) protein. Cell Microbiol 5:25–40. https://doi.org/10.1046/j.1462-5822.2003.00250.x

Zhou P, She Y, Dong N, Li P, He H, Borio A, Wu Q, Lu S, Ding X, Cao Y, Xu Y, Gao W, Dong M, Ding J, Wang D-C, Zamyatina A, Shao F (2018) Alpha-kinase 1 is a cytosolic innate immune receptor for bacterial ADP-heptose. Nature 561:122–126. https://doi.org/10.1038/s41586-018-0433-3

Zimmermann S, Pfannkuch L, Al-Zeer MA, Bartfeld S, Koch M, Liu J, Rechner C, Soerensen M, Sokolova O, Zamyatina A, Kosma P, Mäurer AP, Glowinski F, Pleissner KP, Schmid M, Brinkmann V, Karlas A, Naumann M, Rother M, Machuy N, Meyer TF (2017) ALPK1- and TIFA-Dependent Innate Immune Response Triggered by the *Helicobacter pylori* Type IV Secretion System. Cell Rep 20:2384–2395. https://doi.org/10.1101/139998

Genetic Polymorphisms in Inflammatory and Other Regulators in Gastric Cancer: Risks and Clinical Consequences

Karolina Rudnicka, Steffen Backert and Magdalena Chmiela

Contents

Abstract *Helicobacter pylori* infection is associated with the development of a chronic inflammatory response, which may induce peptic ulcers, gastric cancer (GC), and mucosa-associated lymphoid tissue (MALT) lymphoma. Chronic *H. pylori* infection promotes the genetic instability of gastric epithelial cells and interferes with the DNA repair systems in host cells. Colonization of the stomach with *H. pylori* is an important cause of non-cardia GC and gastric MALT lymphoma. The reduction of GC development in patients who underwent anti-*H. pylori* eradication schemes has also been well described. Individual susceptibility to GC development depends on the host's genetic predisposition, *H. pylori* virulence factors, environmental conditions, and geographical determinants. Biological determinants are urgently sought to predict the clinical course of infec-

K. Rudnicka · M. Chmiela (✉)
Laboratory of Gastroimmunology, Department of Immunology and Infectious Biology, Faculty of Biology and Environmental Protection, University of Lodz, 12/16 Banacha St., 90-237 Lodz, Poland
e-mail: magdalena.chmiela@biol.uni.lodz.pl

K. Rudnicka
e-mail: karolina.rudnicka@biol.uni.lodz.pl

S. Backert
Department of Biology, Division of Microbiology, Friedrich Alexander University Erlangen-Nuremberg, Staudtstr. 5, 91058 Erlangen, Germany
e-mail: Steffen.Backert@fau.de

S. Backert (ed.), *Molecular Mechanisms of Inflammation: Induction, Resolution and Escape by Helicobacter pylori*, Current Topics in Microbiology and Immunology 421, https://doi.org/10.1007/978-3-030-15138-6_3

tion in individuals with confirmed *H. pylori* infection. Possible candidates for such biomarkers include genetic aberrations such as single-nucleotide polymorphisms (SNPs) found in various cytokines/growth factors (e.g., IL-1β, IL-2, IL-6, IL-8, IL-10, IL-13, IL-17A/B, IFN-γ, TNF, TGF-β) and their receptors (IL-RN, TGFR), innate immunity receptors (TLR2, TLR4, CD14, NOD1, NOD2), enzymes involved in signal transduction cascades (PLCE1, PKLR, PRKAA1) as well as glycoproteins (MUC1, PSCA), and DNA repair enzymes (ERCC2, XRCC1, XRCC3). Bacterial determinants related to GC development include infection with CagA-positive (particularly with a high number of EPIYA-C phosphorylation motifs) and VacA-positive isolates (in particular s1/m1 allele strains). The combined genotyping of bacterial and host determinants suggests that the accumulation of polymorphisms favoring host and bacterial features increases the risk for precancerous and cancerous lesions in patients.

1 Introduction

Helicobacter pylori infection, which occurs in ca. 50% of the human population, significantly increases the risk of intestinal and diffuse-type gastric cancer (GC) and is responsible for ~90% of non-cardia GC cases (Plummer et al. 2015) due to various factors including the genetic instability of gastric epithelial cells (GECs) during long-term infection (Machado et al. 2009). A relationship between *H. pylori* seroprevalence and the frequency of GC was shown by the EUROGAST Study Group in 1993 (The EUROGAST Study Group 1993). More recently, Lee et al. (2016) in a systematic review and meta-analysis clearly verified the association between *H. pylori* eradication and reduced incidence of GC. Similarly, Tsukamoto et al. (2017) showed that eradication of *H. pylori* prevents GC development. In high-risk areas of GC such as East Asia, official programmes for *H. pylori* screening and treatment, even in asymptomatic individuals, have started to reduce the prevalence of GC (Kim et al. 2013). However, due to an inverse association between *H. pylori* infection and esophageal adenocarcinoma, the idea of worldwide eradication of *H. pylori* has raised much debate (Islami and Kamagar 2008; Kamada et al. 2012). How the infection could prevent esophageal cancer is still not fully understood (Moss 2016). It is also unknown whether eradication guarantees full protection against stomach cancer and whether the conditions of atrophy or intestinal metaplasia and associated molecular changes are irreversible (Wong et al. 2004). Compared to the total number of *H. pylori*-infected people, only a minority (~1–2%) of *H. pylori*-infected individuals develop GC, and this depends on the genetic variability of the host as well as on specific *H. pylori* virulence factors (Basso et al. 2008; Hunt et al. 2015; McClain et al. 2017; Memon et al. 2014; Nogueira et al. 2001; Winter et al. 2014). One of the key strategies to identify individuals at high risk for developing GC is focusing on individual host determinants. *H. pylori* promotes the development of GC through multiple mechanisms, but the inflammatory response seems to be crucial. Polymorphisms in genes involved in regulation of these inflammatory responses, but also others, may be very important for GC development. In this chapter, we discuss results obtained in large-scale epidemiological studies, literature meta-analyses, and findings from clinical and animal model studies focused on searching for host immune system regulators that increase the risk for GC development and challenging these immune system regulators with *H. pylori* virulence factors facilitates the development of precancerous and cancerous lesions, in the context of environmental and geographical determinants.

2 Gastric Cancer Development is Related to *H. pylori* Infection

Historically, Rudolf Virchow, Pelayo Correa, Robin Warren, and others were pioneers in understanding carcinogenesis initiated by microbes such as *H. pylori*. More than 150 years ago, Rudolf Virchow suggested that cancer in general is derived from normal cells responding to certain damaging or inflammatory factors (Balkwill and Mantovani 2001). However, it was recently shown that specific gastrointestinal stem cells, that undergo genetic/epigenetic mutations and can transform into cancer cells, are involved in the development of precancerous lesions such as metaplasia and dysplasia, and GC (Hayakawa et al. 2017). This stem cell research included basic studies in mouse models and in human patients, and showed enhanced cell proliferation in the isthmus and gastric stem cell region (Hayakawa et al. 2017) and a critical role of the Notch signalling pathway in regulating gastric stem cell proliferation, differentiation, gland fission, tissue expansion, and tumour formation (Demitrack and Samuelson 2017; Gifford et al. 2017). The model of initiation and progression of GC proposed by Correa and co-workers was based on atrophic changes developing from chronic gastritis, which over time increases the risk of intestinal metaplasia, and of GC within 30–50 years of development (Correa et al. 1975; Malaty et al. 2002). Studies by Warren and Marshall (1983) enabled further investigations, that demonstrated the role of *H. pylori* infection as the initial step in the gastric carcinogenesis cascade and led to the categorization of *H. pylori* as a class 1 carcinogen (IARC 1994). Lee et al. (2016) confirmed the association between *H. pylori* eradication and diminished GC incidence.

3 *H. pylori* Virulence Factors and Gastric Cancer Risk

It was shown that *H. pylori* virulence factors differ among strains and contribute to variation in clinical manifestations of the infection (Torres et al. 2009; Zambon et al. 2003; Posselt et al. 2013). In general, genes encoding pathogenic *H. pylori* factors that are involved in GC development include adhesion and colonization properties and/or the capability to induce GEC damage and inflammation (Abdi et al. 2017; Chen et al. 2016b; Cherati et al. 2017; see also Chapter "Impact of *Helicobacter pylori* Virulence Factors on the Host immune Response and Gastric Pathology" of this book). The best-described virulence factors are the cytotoxin-associated gene pathogenicity island (*cag*PAI), effector protein CagA, and cytotoxin VacA (Backert et al. 2017; McClain et al. 2017), as discussed below. Among the genes that determine *H. pylori* colonization, a high frequency of gene mutations has been described in the flagellins, *flaA* and *flaB* (Franco et al. 2009; Ye et al. 2007). Another group of genetic determinants involved in the pathogenicity of *H. pylori* includes the outer membrane proteins (Hop) and Hop-related proteins (Hor), among which the BabA (HopS), encoded by the *babA2* gene, and SabA (HopP) adhesins are best characterized (Oleastro and Ménard 2013). From the four *H. pylori* genotypes, (1) *cagA*(−), *vacA*(s2m2), *babA2*(−); (2) *cagA*(+), *vacA*(*s1m1*) *babA2*(+); (3) *cagA*(+), *vacA*(*s1m2*), *babA2*(+), and (4) *cagA*(+), *vacA*(*s1m2*), *babA2*(−), the second genotype revealed the strongest association with inducing gastric metaplasia (Torres et al. 2009; Zambon et al. 2003). In addition, HopH encoded by the *HP0638/hopH* gene (also known as OipA) is related to *cagA*, *vacA*(*s1m1*), and *babA2*-positive *H. pylori* strains and thus determines gastric disease outcome (Dossumbekova et al. 2006).

Blood-group antigen-binding adhesin (BabA) is a key *H. pylori* component involved in bacterial colonization. BabA binds to Lewis B (Le^b), a fucosylated blood group antigen, that is expressed in the gastric mucosa, or to ABO blood group antigenic determinants (Aspholm et al. 2006; Yamaoka 2008). In addition, BabA is involved in the upregulation of inflammatory responses and is linked with a higher risk for the development of gastroduodenal disorders, including GC (Chen et al. 2016b). By comparison, the sialic acid-binding adhesin A (SabA) of *H. pylori* recognizes sialylated Lewis X (Le^X) on GECs. The existence of polymorphic variants of the *sabA* gene is hypothesized to provide a mechanism for avoiding of the host immune responses (Aspholm et al. 2006). It has been shown that *cag*PAI-positive *H. pylori* strains can upregulate their adhesion capabilities in association with β3*N*-acetylglucosamine T5 and *N*-acetylglucosamine (GlcNAc) transferase, which promote Le^X synthesis (Goodwin et al. 2008).

As mentioned above, CagA, VacA, and various outer membrane proteins (OMPs) have been implicated in *H. pylori*-driven gastric precancerous and cancer lesions. A meta-analysis of the relationship between CagA seropositivity and GC performed by Huang et al. (2003) showed that infection with *H. pylori* CagA-positive strains increased the risk for GC. Similarly, Plummer et al. (2007) showed a strong relationship between the presence of CagA-positive *H. pylori* strains and the severity of precancerous lesions in a large epidemiological study. In fact, CagA has been demonstrated to be translocated into GECs by the *cag*PAI-encoded type IV secretion system (T4SS). After delivery into the cells, CagA undergoes tyrosine phosphorylation and induces morphological changes in GECs due to rearrangement of cytoskeletal factors (Backert et al. 2015; Huang and Tang 2010). The C-terminal region of *cagA* encodes various tyrosine phosphorylation sites in the so-called Glu-Pro-Ile-Tyr-Ala (EPIYA) sequence motifs (called A, B, C, or D), which are involved in the interaction of CagA with various host cell proteins (Alzahrani et al. 2014; Lind et al. 2014, 2016). In Western countries, *H. pylori* strains with EPIYA-A, B, and C-motifs are dominant, whereas Asian strains possess mainly EPIYA-A, B, and D-motifs, and they are associated with the development of GC (Backert et al. 2008; Kawai et al. 2011; Olbermann et al. 2010). Different numbers, configurations, and polymorphisms in the EPIYA motifs have been found and up to seven EPIYA repeats were observed in single CagA variants (Backert et al. 2010). The type and number of EPIYA motifs influences the virulence of *H. pylori* strains and the transcription of host genes involved in carcinogenesis (Vaziri et al. 2015). Several studies have shown that infection with *H. pylori* strains carrying more than one EPIYA-C motif was associated with higher levels of GEC damage, with atrophic gastritis, intestinal metaplasia, and dysplasia, and were more frequently found in GC biopsies (Basso et al. 2008; Sicinschi et al. 2010; Batista et al. 2011; Ferreira et al. 2012). In Western countries, EPIYA-B motifs are also polymorphic and may impact gastric disease development (Zhang et al. 2015). Other components of the T4SS, such as the pilus-associated CagL protein, can also affect the course of disease (Kwok et al. 2007; Barden et al. 2013). Cherati et al. (2017) showed that the presence of amino acids D, I, Q, and N at CagL positions 58, 60, 62, and 122, respectively, was related to an increased risk of peptic ulcer disease (PUD). However, amino acids N, M, Q, and N at the same position alongside V134 increased the risk of GC.

VacA is a typical pore-forming toxin composed of four polymorphic regions—the signal (s) region in variants s1 and s2, which is responsible for anion channel formation; the mid (m) region affecting host cell tropism, which is subdivided into m1 and m2; the intermediate (i) region, subdivided into i1, i2, and i3, which determines the vacuolation and carcinogenic properties of VacA; and the deletion region (d) of 81bp located between the i- and m-regions (Bridge and Merell 2013; Jones et al. 2011; Latifi-Navid et al. 2013). *H. pylori* carrying the *vacA*(s1/m1) allele showed the strongest vacuolation activity in infected host cells; which has been associated with an increased risk for the development of PUD and GC (Jones et al. 2011). In addition, certain polymorphisms in the *vacA* s1, m1, i1, and d1 (no deletion) alleles further increase the risk of GC development (Ogiwara et al. 2009). Recently, Abdi et al. (2017) showed that in Ardabil, a region in north-western Iran with a very high risk of GC, the *vacA* i1, and d1 genotypes significantly increased the risk of cardia and non-cardia GC. Moreover, infection of strains with the *vacA* i1 genotype was significantly associated with an increased risk of intestinal type GC, whereas infection with strains of the *vacA* d1 genotype was associated with the diffuse-type GC (Abdi et al. 2017). Accordingly, the authors suggested that i1 and d1 sequences might be molecular candidates for the prediction of the risk for these cancer subtypes. Considering that both CagA and VacA proteins, affect the formation of precancerous and cancerous lesions in *H. pylori*-induced GC, the co-expression of both virulence factors may constitute an indicator of a higher risk for *H. pylori*-related GC development (Basso et al. 2008; Memon et al. 2014; Rhead et al. 2007). A meta-analysis, performed by Matos et al. (2013), showed that individuals infected with *cagA*$^+$ and *vacA* s1/m1 *H. pylori* strains had an increased risk for GC. González et al. (2011) performed a long-term follow-up study in a high-risk area for GC in Spain and showed that *H. pylori cagA* and *vacA* genotypes may be useful for the identification of patients at high risk of progression of gastric pre-neoplastic lesions.

4 Host Genetic Susceptibility to Gastric Cancer

4.1 Immuno-Regulatory Cytokines

Various cytokines secreted by immune cells infiltrating the gastric mucosa during *H. pylori*-related gastritis have been identified. Due to the broad and pleiotropic effects of cytokines on immune cells, as well as epithelial and endothelial cells, these soluble mediators are considered to be additional risk factors for GC development. Recent research has focused on identifying a panel of biomarkers, including polymorphisms in genes controlling the inflammatory host responses, which would become indicators of individual susceptibility to and consequences of infection, including GC (McLean and El-Omar 2014). The role of cytokines in regulating the severity of gastritis, cell activation and turnover, and neoplastic transformation has been intensively studied using mouse and human models (Bockerstett and DiPaolo 2017; Epplein et al. 2013). Research based on genetic data indicated the involvement of several crucial cytokines, growth factors, and their receptors in GC progression. Additionally, protein-based studies have provided additional data on cytokines, which may potentially drive GC development (Al-Sammak et al. 2013; Buzzeli et al. 2015; Fukui et al. 2014; Howlett et al. 2012; Petersen et al. 2018; Tanaka et al. 2017; Tsai et al. 2014; Wu et al. 2003). Similarly, using mouse models, the cytokines implicated in gastric disorders such as parietal cell atrophy (IL-4, IL-6, IL-11, IL-33, interferon (IFN)-γ, and gp130 receptor), neck cell hyperplasia (IL-1, IL-6, IFN-γ, and gp130 receptor), and metaplasia (IL-1β, IL-6, IL-11, IL-13, IL-33, IFN-γ, and gp130 receptor), and transforming growth factor (TGF)-β have been classified (Bockerstett and DiPaolo 2017). In the last two decades, host genetic polymorphisms of several cytokine genes and innate immune response genes have been found to be involved in various stages of the neoplastic process in gastric carcinoma. Corresponding meta-analysis studies have revealed both consistent and contradictory results, which will be discussed below.

4.1.1 IL-1β and IL-1R

Transgenic mice overproducing interleukin (IL)-1β in the stomach have been shown to develop inflammation and cancer (Tu et al. 2008). In addition, IL-1β$^{-/-}$ knockout mice exhibited a decreased capability to develop gastric tumours compared to wild-type mice (Shigematsu et al. 2013). Thus, IL-1β plays a central role in the development of GC. In fact, the IL-1 family of cytokines comprises 11 members, including 7 pro-inflammatory agonists (IL-1α, IL-1β, IL-18, IL-33, IL-36α, IL-36β, and IL-36γ) and 4 proven or putative antagonists (IL-1R antagonist IL-1Ra, IL-36Ra, IL-37, and IL-38). IL-1 (IL-1α and IL-1β) is a pro-inflammatory cytokine, whereas IL-1Ra possesses natural anti-inflammatory activity (Palomo et al. 2015). Thus, an imbalance between agonist and antagonist levels can lead to exaggerated inflammatory responses. The biological activity of IL-1β favours enhancing the host's inflammatory response mainly by the induced expression of many other pro-inflammatory cytokines such as tumour necrosis factor (TNF)-α, IL-2, IL-6, IL-12, IFN-α, β, and γ and, pro-inflammatory mediators such as cyclooxygenase-2 and C-reactive protein (El-Omar 2001). On the other hand, IL-1β also acts as a powerful gastric acid inhibitor.

In a series of landmark papers, El-Omar and co-workers reported that the levels of IL-1β are upregulated in gastric tissues of patients with *H. pylori* infection (El-Omar et al. 1997, 2000; El-Omar 2001). The *IL-1* gene cluster is located on the long arm of human chromosome 2, comprising *IL-1A*, *IL-1B,* and *IL-1RN* (El-Omar et al. 2000). *IL-1RN* encodes IL-1rα, which is the endogenous receptor antagonist of IL-1α and IL-1β. Several genetic polymorphisms in the *IL-1* locus are associated with dysregulated IL-1 activity and development of gastric diseases in humans. Remarkably, the presence of two pro-inflammatory genotypes, *IL-1β*-31T+ and *IL-1RN*2/*2*, increased the likelihood of chronic hypochlorhydria due to elevated levels of IL-1, which initiate spontaneous inflammation followed by dysplasia and GC through activation of the pro-inflammatory transcription factor NF-κB during *H. pylori* infection, while in the absence of the pathogen, this relationship does not exist (El-Omar et al. 1997, 2000; Hwang et al. 2002). Machado et al. (2001) provided further pro-inflammatory genotype evidence showing increased GC risk associated with an *IL-1β*-511T allele, which substantially increased in individuals homozygous for the *IL-1RN**2 allele (Machado et al. 2001). It was also shown that patients carrying allele 2 (*IL-1RN**2) are more likely to develop adenocarcinoma compared to carriers of the *IL-1RN1/1* variant (Melo Barbosa et al. 2009).

After these important initial findings in the *IL-1* gene locus, a series of meta-analyses followed. For example, one study evaluated the association between *IL-1B-51* and *IL-1RN* polymorphisms and gastric precancerous lesions showing that individuals carrying the *IL-1RN**22 genotype had an increased risk of gastric precancerous lesions (22 versus LL) and no such association was found for *IL-1B*-511 (TT versus CC) (Peleteiro et al. 2010). One of the largest meta-analyses published, included patient populations from various geographic locations, and supported the idea of increased cancer risk for *IL-1RN2* carriers in non-Asian populations and was observed for intestinal and diffuse cancers, distal cancers and, to a lesser extent, cardia cancers (Ma et al. 2017). Reduced risk in association with the *IL-1B*-31C carrier status was observed in Asian and Mexican populations, respectively (Persson et al. 2011; Garza-González et al. 2005). Recently, the results of another meta-analysis revealed that the *IL-1B*-31-C/T polymorphism increased the risk for *H. pylori* infection, whereas *IL-1B*-511-C/T was a risk factor for *H. pylori*-related GC or PUD (Ma et al. 2017). Additionally, Ying et al. (2016) showed that the *IL-1B*-31-T/C polymorphism might confer susceptibility to GC in the ongoing *H. pylori* infection, which may suggest a gene–environment interaction during the process of gastric carcinogenesis.

In addition to the host susceptibility associated with *IL-1B* and *IL-1RN* polymorphisms, specific bacterial factors can affect the course of *H. pylori*-dependent development of GC as discussed above. Thus, combined genotyping analyses have been performed to help define high-risk individuals. For example, in a study of Portuguese patients, the risk of developing GC was the highest in *IL-1B*-511*T carriers or *IL-1RN*2* homozygotes infected with *H. pylori vacA* s1/m1 and *cagA*-positive strains, demonstrating the combinations of both bacterial and high-risk host genotypes (Figueiredo et al. 2002). A similar approach was followed to study the combination of bacterial/host genotypes and gastric histological abnormalities in patients from Germany. The highest prevalence of severe gastric inflammation and intestinal metaplasia was found in patients carrying both the bacterial and host high-risk genotypes *cagA*(+)/*vacAs1*(+) and *IL-1B* −511T/*IL-1RN*2 (Rad et al. 2003). Sicinschi et al. (2006) showed a significant correlation between *IL-1B*-31 polymorphism and CagA status for the risk of intestinal type GC incidence in a Mexican population. Among CagA-positive individuals, those with *IL-1B*-31-CC genotype displayed a higher risk of intestinal GC type than the carriers of *IL-1B*-31-TT genotype. None of these polymorphisms were significantly related to the risk of diffuse-type GC (Sicinschi et al. 2006).

4.1.2 TNF-α

Similar to IL-1β, TNF-α is pro-inflammatory cytokine and inhibits gastric acid production. TNF-α is upregulated in the gastric mucosa in response to *H. pylori* infection and is considered a potential biomarker of *H. pylori*-related gastric diseases (Brenner et al. 2015; Shibata et al. 1999). In a case–control study, including patients with chronic gastritis and GC, Machado et al. (2003) showed that carriers of the TNF-α 308*A allele polymorphism were at increased risk for GC development. Zambon et al. (2005) evaluated five SNPs in the *TNF-A* gene (at −1031 C/T, −857 C/T, −376 A/G, −308 A/G, and −238 A/G), which were related to *H. pylori* infection and gastroduodenal diseases together with genetic polymorphisms of *IL-1B* (−31 C/T), *IL-1RN* (intron 2 VNTR), *IFN-G* (+874 A/T), and *IL-10* (−1082 A/G, −819 C/T, −592 A/C). They showed that the *TNF-A* −857 TT genotype is related to duodenal ulcer disease, and that the *IFN-G* +874 AA genotype promotes $CagA^{+}$ *H. pylori* infections, whereas the *IL-10*-819 TT genotype is related to gastric metaplasia. Another meta-analysis performed by Sun et al. (2016) demonstrated the protective role of *TNF-A* −308G/A and −1031 T/C polymorphisms against *H. pylori* infection and suggested a −863C/A polymorphism as a risk factor for infection. The presence of the TNF-A-G308A (rs1800629) SNP in association with GC has been confirmed in a Chinese population, and haplotypes of TNF-A-308/-238 GA/GG, AA/GG, and AA/GA appeared to increase the susceptibility for GC development (Xu et al. 2017).

4.1.3 Other Cytokines

In addition to the important polymorphisms found initially in IL-1 and TNF-α gene clusters, further studies screened for disease-associated SNPs in other cytokines. For instance, the first stages of acute inflammatory responses by *H. pylori* infection involve the release of chemotactic IL-8 from the epithelium, which attracts the granulocytes to infiltrate the lamina propria to resolve the infection (Ramis et al. 2015). Variation in IL-8 genetic diversity was addressed by Caleman Neto et al. (2014) who showed that the elevated production of IL-8 and the intensity of the inflammatory response are related to the presence of the A allele in the promoter region of the *IL-8* gene (−251 position) and that the *IL-8* −251TT genotype protects, whereas the *IL-8* −251TA genotype promotes *H. pylori* infection. Another study by Ohyachi et al. (2005) indicated that expression of the *IL-8* −251A allele was linked with gastric ulcers, gastric atrophy, and eventually GC development. This study also showed that the *IL-8* −251T and *IL-8* −251A variants are associated with active gastritis accompanied by strong neutrophil infiltration. However, no link between the IL-8 polymorphism at −251 and increased risk of gastric carcinoma was found in other studies (Kamangar et al. 2006; Savage et al. 2006; Canedo et al. 2008; Ramis et al. 2015).

Zhang et al. (2017) investigated the association of genetic variations in the *IL-6* gene [rs6949149, rs1800796, rs10499563, and its receptor (IL-6R) rs2228145] with GC risk in a Chinese population. The study was based on 473 patients with GC and 474 healthy individuals with or without *H. pylori* infection. In the uninfected group, *IL-6R* rs2228145 AC and AC/CC genotypes were associated with a decreased risk of GC, similar to *H. pylori*-positive individuals with *IL-6* rs10499563 TC, CC and TC/CC genotypes. Specifically in males, *IL-6* rs1800796 CG and CG/GG genotypes were associated with diminished risk of GC (Zhang et al. 2017). Furthermore, Kim et al. (2012) performed genotyping of three promoter polymorphisms in the *IL-10* gene (−1082A/G, −819T/C; −592 A/C) and showed that they were associated with an increased risk of the intestinal type of non-cardia GC, specifically in *H. pylori*-infected individuals and active smokers. These results are consistent with another meta-analysis demonstrating that an *IL-10*-1082 GG-plus-GE genotype versus AA genotype was associated with increased risk of intestinal type GC (Ni et al. 2012). In addition, IL-2 is a key pro-inflammatory cytokine, which is mainly synthesized by immuno-regulatory T helper lymphocytes that promote T cell proliferation and activation of other immunocompetent cells, which intensify inflammation and potentiate gastric disorders. Recently, Melchiades et al. (2017) investigated polymorphisms in the *IL-2* gene. It was shown that among *H. pylori*-infected Brazilian patients, the *IL-2*-330 G/G genotype and the G allele, as well as the *IL-2* +114 T/T genotype and the T allele or the −330G/+114T haplotype, increased the risk of GC development.

Finally, several studies reported the association between gene polymorphisms of IL-17 and gastroduodenal diseases. IL-17 is induced in Th17 lymphocytes by *H. pylori* and bridges the innate and adaptive immune responses (Khamri et al. 2010; Mizuno et al. 2005; Tanaka et al. 2017). Arisawa et al. (2012) showed the association between GC and genetic polymorphisms in the *IL-17A* rs2275913 (−197 G/A), rs3748067 (*1249 C/T), and pri-miR-938, rs2505901 (T/C), targeting *IL-17A* 3'-UTR. Six SNPs in the *IL-17A* (rs2275913, rs3748067, and rs3819025) and *IL-17F* (763780, rs9382084, and rs12203582) genes were associated with *H. pylori* infection, and smoking significantly enhanced the risk of GC in a Chinese population (Qinghai et al. 2014). In addition, Liu et al. (2015) analyzed genetic variants in a large cohort of patients and found that the AA genotype of *IL-17A* rs2275913 and *IL-17F* rs763780 polymorphisms were associated with increased GC risk in Asian populations. Another meta-analysis by Jiang et al. (2015) confirmed that the rs2275913 variant was significantly associated with gastrointestinal malignancy, whereas the rs763780 variant was not. Together, this multitude of studies demonstrates that various cytokines contain genetic polymorphisms, that affect their expression levels and play a crucial role in gastric disease development induced by *H. pylori* infection.

4.2 Pattern Recognition Receptors (PRRs)

4.2.1 Toll-Like Receptors (TLRs)

Toll-like receptors (TLRs) are critical for pathogen detection and downstream signalling to stimulate effective immunity (Smith 2014; Varga and Peek 2017). The early detection of pathogen-associated molecular patterns (PAMPs) by TLRs and other PRRs is believed to produce regulatory cytokine or chemokine profiles through activation of NF-κB, MAP kinases, and interferon regulatory factor (IRF) signalling cascades, which finally control *H. pylori* infection (Backert and Naumann 2010; Brisslert et al. 2005; Pachathundikandi et al. 2015). *H. pylori* factors such as lipopolysaccharide (LPS), NapA, HSP-60, RNA, and DNA were described in various reports to be recognized by specific TLRs such as TLR1, TLR2, TLR4, TLR6, TLR9, and TLR10, while *H. pylori* flagellin was found to evade recognition by TLR5 (see Chapter "Importance of Toll-like Receptors in Pro-inflammatory and Anti-inflammatory Responses by *Helicobacter pylori* Infection" of this book). Furthermore, various TLRs have been found to be overexpressed in precancerous and cancerous conditions in the gastric tissues (Pimentel-Nunes et al. 2011). Genetic polymorphisms have been reported in various *TLR* genes, most notably *TLR4* rs4986790 (Asp299Gly), *TLR4* rs4986791 (Thr399Ile), *TLR4* rs10116253, *TLR4* rs10983755, *TLR4* rs11536889 (C3725G/C), and *TLR4* rs1927911 were identified to be associated with GC (Castaño-Rodríguez et al. 2013; Bagheri et al. 2014; Zhou et al. 2014). In addition, CD14 (cluster of differentiation 14) is a factor that co-operates with TLR4 in the recognition of LPS. Interestingly, two single-nucleotide polymorphisms in the promoter region of *CD14* (namely, −260C/T rs2569190 and −561 rs5744455) have been found to be associated with *H. pylori*-related risk of GC development (Castaño-Rodríguez et al. 2013; Wang et al. 2014). Additional polymorphisms in *TLR2*, such as the deletion of nucleotides at position −196 to −174, have also been described (Castaño-Rodríguez et al. 2013). Another well-documented SNP has been reported within the *TLR9* promoter (*TLR9* −1237T/C) and is associated with a variety of inflammatory disorders, including allergic asthma, inflammatory bowel disease, and atopy. The *TLR9* −1237 C allele is also significantly associated with the development of *H. pylori*-induced premalignant gastric changes (Ng et al. 2010). Functional analysis of this SNP showed that carriage of the C allele increases TLR9 transcriptional activity driven by activation of NF-κB. Therefore, the interplay of TLRs and *H. pylori* factors highlights the complexity of innate immune recognition of the pathogen as well as signalling processes in the development of gastric pathology.

4.2.2 NOD-Like Receptors (NLRs)

Nucleotide-binding oligomerization domain (NOD)-like receptors (NLRs) are conserved intracellular sensors of bacterially-released peptidoglycan (Mukherjee et al. 2018; see also Chapter "Role of NOD1 and ALPK1/TIFA Signalling in Innate Immunity Against *Helicobacter pylori* Infection" of this book). Two important NLR members, NOD1 and NOD2, play fundamental and pleiotropic roles in host defence against bacterial infections and in the control of inflammation. NOD1/NOD2 protein expression levels were found to be significantly upregulated in the gastric epithelium of *H. pylori*-positive patients (Rosenstiel et al. 2006). However, *NOD1/NOD2* gene mutations were not found to be linked with gastritis or gastric ulcer development. Nevertheless, a mutation in the leucine-rich region of *NOD2* at R702W was significantly associated with gastric lymphoma. Patients carrying the rare T allele had a more than doubled risk of developing lymphoma compared to controls (Rosenstiel et al. 2006). Thus, *H. pylori*-induced upregulation of NOD1 and NOD2 *in vivo* may play a critical role in the recognition of the pathogen and the R702W mutation of NOD1 is associated with gastric lymphoma. It was further shown that GC risk increases due to SNPs in *NOD1*, namely the rs 7789045 TT genotype (Wang et al. 2012). In addition, Li et al. (2015) showed that patients carrying the *NOD1* rs2709800 TG genotype had a decreased risk of intestinal metaplasia, while the presence of the *NOD2* rs718226 G allele (AG/GG) showed an increased risk of dysplasia and GC development. Moreover, the *NOD2* rs2111235 C allele and rs7205423 G allele were found to be associated with decreased risk of GC progression in *H. pylori*-infected subjects (Li et al. 2015). Finally, Castaño-Rodríguez et al. (2014) investigated genetic variance in the NLR signalling cascade of 310 patients from China in relation to *H. pylori* infection and GC development. They detected 51 polymorphisms in 6 genes comprising *NLRX1, NLRP3, NLRP12, CASP1, CARD8,* and *ASC*. Newly discovered SNPs were located in *NLRP12* (rs2866112) and in *CARD8* (rs11672725) and were found to be associated with an increased risk for *H. pylori* infection and GC formation. Together, the above studies considerably advance our understanding of the roles of TLR and NLR signal transduction cascades and their SNPs during *H. pylori* infection and highlight the importance of these pathways in gastric pathogenesis.

4.3 DNA Repair Enzymes

During *H. pylori* infection, the gastric milieu is heavily infiltrated by various immune cell types, which become activated (see also Chapter "Inflammation, Immunity and Vaccine Development for the Gastric Pathogen *Helicobacter pylori*" of this book). However, the host immune response is not efficient enough to eradicate the pathogen, and as a consequence, the inflammatory response becomes chronic (Chmiela et al. 2014; Grebowska et al. 2008, 2010; Mnich et al. 2015; Paziak-Domanska et al. 2000; Rudnicka et al. 2015). These infection-induced inflammatory reactions in gastric tissue combined with the induction of oxidative stress such as the release of reactive oxygen species (ROS) and reactive nitrogen species (RNS), lead to DNA damage, which also plays a crucial role in tumour development (Bhattacharyya et al. 2014; Butcher et al. 2017; Naumann et al. 2017). It was shown in cell culture and mouse models deficient in DNA repair enzymes that *H. pylori* infection increased the susceptibility to oxidative stress and DNA damage (Koeppel et al. 2015; Machado et al. 2009; Meira et al. 2008). In addition, experimental infection of transgenic BigBlue® mice with *Helicobacter felis* was found to be associated with inactivation of the *p53* tumor suppressor gene, enhancement of gastric mutations, infiltration of inflammatory cells and hyperplasia (Jenks et al. 2003). Similarly, it was demonstrated that *H. pylori* infection of cultured GECs induced DNA double-strand breaks in the chromosome (Hartung et al. 2015) and mutations in the mitochondrial DNA (Machado et al. 2013). Chronic colonization of mice with *H. pylori* induced mutations in the gastric epithelium and the mutagenic potential of the bacteria was determined by the genotype and duration of infection (Sheh et al. 2010). Long-term *H. pylori* infection of mice has been shown to result in more severe gastric lesions and pro-inflammatory, Th1-biased responses promote mutagenesis, presumably by exposing the stomach to prolonged oxidative stress (Sheh et al. 2010). A nested case–control study, including 246 gastric adenocarcinomas, 1175 matched controls, and 91 cases with chronic atrophic gastritis, studied 12 polymorphisms in DNA repair genes (*MLH1, MSH2, OGG1, ERCC2,* and *XRCC1*), combined with seropositivity for *H. pylori* (Capellá et al. 2008). No association was observed for any of the polymorphisms with a risk of GC. However, the *ERCC2* K751Q polymorphism was associated with an increased risk for non-cardia neoplasm, and *ERCC2* D312 N and K751Q alleles were associated with diffuse-type GC. Furthermore, *ERCC2* D312 N and K751Q polymorphisms as well as the *XRCC1* R399Q allele were associated with an increased risk for severe chronic atrophic gastritis (Capellá et al. 2008). Other reports suggested that genetic polymorphisms in *XRCC1* R194W, *XRCC1* R399Q, and *OGG1* S326C may play important roles in the evolution of *H. pylori*-associated gastric lesions in a high-risk population in China (Li et al. 2009). Furthermore, a meta-analysis of published case–control and cohort examinations, which included 18 studies with 3915 GC cases and 6759 controls, revealed that the *XRCC1* R194W homozygous mutant genotype (W/W) was associated with increased GC risk (Chen et al. 2016a). Recently, another meta-analysis was performed in China, including nine case–control studies focusing on the *XRCC3* gene, which is responsible for the maintenance of genome integrity and protection against mutations (Quin et al. 2014). This study revealed a T241M gene polymorphism in *XRCC3*, representing a new risk factor for non-cardia GC development but not for cardia GC, and among Asian, but not Caucasian populations. This finding highlights the importance of genotype differences for specific GC development in various ethnic groups.

4.4 Genetic Polymorphisms in Other Factors

Gastric disease-associated genetic polymorphisms have also been detected in a multitude of other host factors. We will highlight below some representative findings on mucins (MUC), caspases (CASP), prostate stem cell antigen (PSCA), pyruvate kinase (PKLR), catalytic α-subunit of AMP-activated protein kinase (PRKAA1), phospholipase-C epsilon-1 (PLCE1), and cyclooxygenase-2 (Cox2). For example, a recent large-scale meta-analysis identified 11 variants significantly associated with risk for different types of GC (Mocellin et al. 2015). Gene variants associated with cardia, non-cardia, diffuse, intestinal, and mixed gastric cancer, encoding cytokines and kinases as well as glycolipids, mucins and caspases were identified. Notably, *MUC1* rs2070803 at 1q22 and *PKLR* rs3762272 1q22 variants were associated with diffuse-type *PSCA* rs2976392 was linked with the intestinal group, *PSCA* rs2294008 at 8q24.2 and *PRKAA1* rs13361707 5p13 were associated with non-cardia GC, whereas the *PLCE1* rs2274223 10q23 variant was linked to cardia GC. *CASP8* rs3834129 2q33 and *TNF* rs1799724 6p21.3 were associated with the mixed type of GC (Mocellin et al. 2015). Finally, in the inflammatory milieu, cytokines increase the production of Cox2, which activates angiogenesis by inhibiting apoptosis and promotes metastasis and influences GC development (Cheng and Fan 2013). A polymorphism in the *cox2* gene at position 1195AA was found and proposed as a potential new biomarker of GC development (Li et al. 2012), whereas other investigators suggested that the −765G/C rs20417 polymorphism is an indicator of GC risk in Asians and Indians (Zhao et al. 2014). Compelling evidence has been provided in recent years showing that *H. pylori*-induced gastric diseases also strongly depend on the susceptibility of the host, which is associated with various genetic polymorphisms in immuno-regulatory and other genes.

5 Concluding Remarks

H. pylori infection is still one of the most frequent infections that occur in humans worldwide. Apart from chronic inflammation and peptic ulcers, *H. pylori* colonization significantly increases the risk of intestinal and diffuse types of GC, and is responsible for ~90% of non-cardia GC cases. Studies in humans and mouse models have demonstrated that the inflammatory milieu, especially multiple cytokines, contributes to an increased risk for gastric atrophy, intestinal metaplasia, and GC. Nearly 60% of the intestinal type GCs are associated with *H. pylori* infections. Cancer risk increases if the *H. pylori* strain possesses the well-known virulence factors—CagA and VacA. These bacteria promote gastric carcinogenesis, for instance by inducing increased DNA damage in the nucleus and mitochondria and impairing the corresponding DNA repair processes. In addition, the host cell receptors and cytokines, that are involved in the propagation of the immune and inflammatory responses during infection, may also play a role in the development of GC.

More information on cytokines and other host targets as well as bacterial factors that influence GC development is urgently needed to better understand how inflammatory responses influence GC risk. Numerous studies have focused on searching for a panel of biomarkers, including polymorphisms of genes involved in host inflammatory responses to predict individual susceptibility and consequences (including GC) of *H. pylori* infection. The relationship between the risk for GC development and polymorphisms of host genes involved with the maintenance of the genome integrity and protection against mutations has also been reported. Recently, non-coding microRNAs (miRNAs) were also suggested to be key regulators in signalling pathways during oncogenesis (see also Chapter "Mechanisms of Inflammasome Signaling, microRNA Induction and Resolution of Inflammation by *Helicobacter pylori*" of this book). miRNA dysregulation affects GC growth, cell cycle progression, and metastatic processes due to inhibition of tumour suppressor genes and other responses (Wu et al. 2010, 2013). Further studies describing the role of these newly described regulators and possible new SNPs in the development of GC are required to fully address their prognostic value.

Acknowledgements The preparation of the chapter was financed by the National Science Center grant SONATA no. 2016/23/D/NZ6/02553. The work of SB was supported by the German Science Foundation (project A04 in CRC-1181).

References

Abdi E, Latifi-Navid S, Zahri S, Yazdanbod A, Safaralizadeh R (2017) *Helicobacter pylori* genotypes determine risk of non-cardia gastric cancer and intestinal- or diffuse-type GC in Ardabil: A very high-risk area in Northwestern Iran. Microb Pathog 107:287–292. https://doi.org/10.1016/j.micpath.2017.04.007

Al-Sammak F, Kalinski T, Weinert S, Link A, Wex T, Malfertheiner P (2013) Gastric epithelial expression of IL-12 cytokine family in *Helicobacter pylori* infection in humans: is it head or tail of the coin? PLoS One 17 8(9):e75192. https://doi.org/10.1371/journal.pone.0075192

Alzahrani S, Lina TT, Gonzalez J, Pinchuk IV, Beswick EJ, Reyes VE (2014) Effect of *Helicobacter pylori* on gastric epithelial cells. World J Gastroenterol 20:12767–12780. https://doi.org/10.3748/wjg.v20.i36.12767

Arisawa T, Tahara T, Shiroeda H, Matsue Y, Minato T, Nomura T, Yamada H, Hayashi R, Saito T, Matsunaga K, Fukuyama T, Hayashi N, Otsuka T, Fukumura A, Nakamura M, Shibata T (2012) Genetic polymorphisms of *IL17A* and pri-microRNA-938, targeting IL17A3'-UTR, influence susceptibility to gastric cancer. Human Immunol 73:747–752. https://doi.org/10.1016/j.humimm.2012.04.011

Aspholm M, Olfat F, Nordén J, Sondén B, Lundberg C, Sjöström R, Altraja S, Odenbreit S, Haas R, Wadström T, Engstrand L, Semino-Mora C, Liu H, Dubois A, Teneberg S, Arnqvist A, Borén T (2006) SabA is the *H. pylori* hemagglutinin and is polymorphic in binding to sialylated glycans. PLoS Pathog 2(10):e110. https://doi.org/10.1371/journal.ppat.0020110

Backert S, Naumann M (2010) What a disorder: proinflammatory signaling pathways induced by *Helicobacter pylori*. Trends Microbiol 18(11):479–486. https://doi.org/10.1016/j.tim.2010.08.003

Backert S, Feller SM, Wessler S (2008) Emerging roles of Abl family tyrosine kinases in microbial pathogenesis. Trends Biochem Sci 33(2):80–90. https://doi.org/10.1016/j.tibs.2007.10.006

Backert S, Tegtmeyer N, Selbach M (2010) The versatility of *Helicobacter pylori* CagA effector protein functions: the master key hypothesis. Helicobacter 15(3):163–176. https://doi.org/10.1111/j.1523-5378.2010.00759.x

Backert S, Tegtmeyer N, Fischer W (2015) Composition, structure and function of the *Helicobacter pylori* cag pathogenicity island encoded type IV secretion system. Future Microbiol 10:955–965. https://doi.org/10.2217/fmb.15.32

Backert S, Haas R, Gerhard M, Naumann M (2017) The *Helicobacter pylori* Type IV secretion system encoded by the *cag* pathogenicity island: architecture, function, and signaling. Curr Top Microbiol Immunol 413:187–220. https://doi.org/10.1007/978-3-319-75241-9_8

Bagheri N, Azadegan-Dehkordi F, Sanei H, Taghikhani A, Rahimian G, Salimzadeh L, Hashemzadeh-Chaleshtori M, Rafieian-kopaei M, Shirzad M, Shirzad H (2014) Association of TLR4 single-nucleotide polymorphism with *H. pylori* associated gastric diseases in Iranian patients. Clin Res Hepatol Gastroenterol 38:366–371. https://doi.org/10.1016/j.clinre.2013.12.004

Balkwill F, Mantovani A (2001) Inflammation and cancer: back to Virchow? Lancet 357:539–545. https://doi.org/10.1016/S0140-6736(00)04046-0

Barden S, Lange S, Tegtmeyer N, Conradi J, Sewald N, Backert S, Niemann HH (2013) A helical RGD motif promoting cell adhesion: crystal structures of the *Helicobacter pylori* type IV secretion system pilus protein CagL. Structure 21(11):1931–1941. https://doi.org/10.1016/j.str.2013.08.018

Basso D, Zambon CF, Letley DP, Stranges A, Marchet A, Rhead JL, Schiavon S, Guariso G, Ceroti M, Nitti D, Rugge M, Plebani M, Atherton JC (2008) Clinical relevance of *Helicobacter pylori cagA* and *vacA* gene polymorphisms. Gastroenterology 135(1):91–99. https://doi.org/10.1053/j.gastro.2008.03.041

Batista SA, Rocha GA, Rocha AM, Saraiva IE, Cabral MM, Oliveira RC, Queiroz DM (2011) Higher number of *Helicobacter pylori* CagA EPIYA C phosphorylation sites increases the risk of gastric cancer, but not duodenal ulcer. BMC Microbiol 11:61. https://doi.org/10.1186/1471-2180-11-61

Bhattacharyya A, Chattopadhyay R, Mitra S, Crowe SE (2014) Oxidative stress: an essential factor in the pathogenesis of gastrointestinal mucosal diseases. Physiol Rev 94:329–354. https://doi.org/10.1152/physrev.00040.2012

Bockerstett KA, DiPaolo RJ (2017) Regulation of gastric carcinogenesis by inflammatory cytokines. Cell Mol Gastroenterol Hepatol 4:47–53. https://doi.org/10.1016/j.jcmgh.2017.03.005

Brenner D, Blaser H, Mak TW (2015) Regulation of tumour necrosis factor signalling: live or let die. Nat Rev Immunol 15:362–374. https://doi.org/10.1038/nri3834

Bridge DR, Merrell DS (2013) Polymorphism in the *Helicobacter pylori* CagA and VacA toxins and disease. Gut Microbes 4:101–117. https://doi.org/10.4161/gmic.23797

Brisslert M, Enarsson K, Lundin S, Karlsson A, Kusters JG, Svennerholm AM, Backert S, Quiding-Järbrink M (2005) *Helicobacter pylori* induce neutrophil transendothelial migration: role of the bacterial HP-NAP. FEMS Microbiol Lett 249(1):95–103

Butcher LD, den Hartog G, Ernst PB, Crowe SE (2017) Oxidative stress resulting from *Helicobacter pylori* infection contributes to gastric carcinogenesis. Cell Mol Gastroenterol Hepatol 3:316–322. https://doi.org/10.1016/j.jcmgh.2017.02.002

Buzzeli JN, Chalinor HV, Pavlic DI, Sutton P, Menheniott TR, Giraud AS, Judd LM (2015) IL-33 is a stomach alarmin that initiates a skewed Th2 response to injury and infection. Cell Mol Gastroenterol Hepatol 1:203–221. https://doi.org/10.1016/j.jcmgh.2014.12.003

Caleman Neto A, Rasmussen LT, de Labio RW, de Queiroz VF, Smith Mde A, Viani GA, Payão SL (2014) Gene polymorphism of interleukin 1 and 8 in chronic gastritis patients infected with *Helicobacter pylori*. J Venom Anim Toxins Incl Trop Dis 20:17. https://doi.org/10.1186/1678-9199-20-17

Canedo P, Castanheira-Vale AJ, Lunet N, Pereira F, Figueiredo C, Gioia-Patricola L, Canzian F, Moreira H, Suriano G, Barros H, Carneiro F, Seruca R, Machado JC (2008) The interleukin-

8-251*T/*A polymorphism is not associated with risk for gastric carcinoma development in a Portuguese population. Eur J Cancer Prev 17(1):28–32

Capellá G, Pera G, Sala N, Agudo A, Rico F, Del Giudicce G, Plebani M, Palli D, Boeing H, Bueno-de-Mesquita HB, Carneiro F, Berrino F, Vineis P, Tumino R, Panico S, Berglund G, Simán H, Nyrén O, Hallmans G, Martinez C, Dorronsoro M, Barricarte A, Navarro C, Quirós JR, Allen N, Key T, Bingham S, Caldas C, Linseisen J, Nagel G, Overvad K, Tjonneland A, Boshuizen HC, Peeters PH, Numans ME, Clavel-Chapelon F, Trichopoulou A, Lund E, Jenab M, Kaaks R, Riboli E, González CA (2008) DNA repair polymorphisms and the risk of stomach adenocarcinoma and severe chronic gastritis in the EPIC-EURGAST study. Int J Epidemiol 37(6):1316–1325. https://doi.org/10.1093/ije/dyn145

Castaño-Rodríguez N, Kaakoush NO, Goh KL, Fock KM, Mitchell HM (2013) The role of *TLR2*, *TLR4* and *CD14* genetic polymorphisms in gastric carcinogenesis: a case-control study and meta analysis. PLoS ONE 8:e60327. https://doi.org/10.1371/journal.pone.0060327

Castaño-Rodríguez N, Kaakoush NO, Mitchell HM (2014) Pattern-recognition receptors and gastric cancer. Front Immunol 5:336. https://doi.org/10.3389/fimmu.2014.00336

Chen S, Zhu XC, Liu YL, Wang C, Zhang KG (2016a) Investigating the association between *XRCC1* gene polymorphisms and susceptibility to gastric cancer. Genet Mol Res. 15(3). https://doi.org/10.4238/gmr.15038342

Chen YL, Mo XQ, Huang GR, Huang YQ, Xiao J, Zhao LJ, Wei HY (2016b) Gene polymorphisms of pathogenic *Helicobacter pylori* in patients with different types of gastrointestinal disease. World J Gastroenterol 22(44):9718–9726. https://doi.org/10.3748/wjg.v22.i44.9718

Cheng J, Fan XM (2013) Role of cyclooxygenase-2 in gastric cancer development and progression. World J Gastroenterol 19:7361–7368. https://doi.org/10.3748/wjg.v19.i42.7361

Cherati MR, Shokri-Shirvani J, Karkhah A, Rajabnia R, Nouri HR (2017) *Helicobacter pylori cagL* amino acid polymorphism D58E59 pave the way toward peptic ulcer disease while N58E59 is associated with gastric cancer in north of Iran. Microb Pathog 107:413–418. https://doi.org/10.1016/j.micpath.2017.04.025

Chmiela M, Miszczyk K, Rudnicka K (2014) Structural modifications of *Helicobacter pylori* lipopolysaccharide: an idea how to live in peace. World J Gastroenterol 20:9882–9897. https://doi.org/10.3748/wjg.v20.i29.9882

Correa P, Haenszel W, Cuello C, Tannenbaum S, Archer M (1975) A model for gastric cancer epidemiology. Lancet 2:58–60

Demitrack ES, Samuelson LC (2017) Notch as a driver of gastric epithelial cell proliferation. Cell Mol Gastroenterol Hepatol 3:323–330. https://doi.org/10.1016/j.jcmgh.2017.01.012

Dossumbekova A, Prinz C, Mages J, Lang R, Kusters JG, Van Vliet AH, Reindl W, Backert S, Saur D, Schmid RM, Rad R (2006) *Helicobacter pylori* HopH (OipA) and bacterial pathogenicity: genetic and functional genomic analysis of hopH gene polymorphisms. J Infect Dis 194:1346–1355. https://doi.org/10.1086/508426

El-Omar EM (2001) The importance of interleukin 1β in *Helicobacter pylori* associated disease. Gut 48:743–747

El-Omar EM, Oien K, El-Nujumi A, Gillen D, Wirz A, Dahill S, Williams C, Ardill JE, McColl KE (1997) *Helicobacter pylori* infection and chronic gastric acid hyposecretion. Gastroenterology 113:15–24. https://doi.org/10.1016/S0016-5085(97)70075-1

El-Omar EM, Carrington M, Chow WH, McColl KE, Bream JH, Young HA, Herrera J, Lissowska J, Yuan CC, Rothman N, Lanyon G, Martin M, Fraumeni JF Jr, Rabkin CS (2000) Interleukin-1 polymorphisms associated with increased risk of gastric cancer. Nature 404(6776):398–402. https://doi.org/10.1038/35006081

Epplein M, Xiang YB, Cai Q, Peek RM Jr, Li H, Correa P, Gao J, Wu J, Michel A, Pawlita M, Zheng W, Shu XO (2013) Circulating cytokines and gastric cancer risk. Cancer Causes Control 24:2245–2250. https://doi.org/10.1007/s10552-013-0284-z

Ferreira RM, Machado JC, Leite M, Carneiro F, Figueiredo C (2012) The number of *Helicobacter pylori* CagA EPIYA C tyrosine phosphorylation motifs influences the pattern of gastritis and the development of gastric carcinoma. Histopathology 60:992–998. https://doi.org/10.1111/j.1365-2559.2012.04190.x

Figueiredo C, Machado JC, Pharoah P, Seruca R, Sousa S, Carvalho R, Capelinha AF, Quint W, Caldas C, van Doorn LJ, Carneiro F, Sobrinho-Simões M (2002) *Helicobacter pylori* and interleukin 1 genotyping: an opportunity to identify high-risk individuals for gastric carcinoma. J Natl Cancer Inst 94(22):1680–1687

Franco AT, Friedman DB, Nagy TA, Romero-Gallo J, Krishna U, Kendall A, Israel DA, Tegtmeyer N, Washington MK, Peek RM Jr (2009) Delineation of a carcinogenic *Helicobacter pylori* proteome. Mol Cell Proteomics 8:1947–1958. https://doi.org/10.1074/mcp.M900139-MCP200

Fukui H, Zhang X, Sun C, Hara K, Kikuchi S, Yamasaki T, Kondo T, Tomita T, Oshima T, Watari J, Imura J, Fujimori T, Sasako M, Miwa H (2014) IL-22 produced by cancer-associated fibroblasts promotes gastric cancer cell invasion via STAT3 and ERK signalling. Br J Cancer 111:763–771. https://doi.org/10.1038/bjc.2014.336

Garza-González E, Bosques-Padilla FJ, El-Omar E, Hold G, Tijerina-Menchaca R, Maldonado-Garza HJ, Pérez-Pérez GI (2005) Role of the polymorphic *IL-1B*, *IL-1RN* and *TNF-A* genes in distal gastric cancer in Mexico. Int J Cancer 114(2):237–241. https://doi.org/10.1002/ijc.20718

Gifford GB, Demitrack ES, Keeley TM, Tam A, La Cunza N, Dedhia PH, Spence JR, Simeone DM, Saotome I, Louvi A, Siebel CW, Samuelson LC (2017) Notch 1 and Notch 2 receptors regulate mouse and human gastric antral epithelial cell homeostasis. Gut 66:1001–1011. https://doi.org/10.1136/gutjnl-2015-310811

González CA, Figueiredo C, Lic CB, Ferreira RM, Pardo ML, Ruiz Liso JM, Alonso P, Sala N, Capella G, Sanz-Anquela JM (2011) *Helicobacter pylori cagA* and *vacA* genotypes as predictors of progression of gastric preneoplastic lesions: a long-term follow-up in a high-risk area in Spain. Am J Gastroenterol 106(5):867–874. https://doi.org/10.1038/ajg.2011.1

Goodwin AC, Weinberger DM, Ford CB, Nelson JC, Snider JD, Hall JD, Paules CI, Peek RM Jr, Forsyth MH (2008) Expression of the *Helicobacter pylori* adhesin SabA is controlled via phase variation and the ArsRS signal transduction system. Microbiology 154:2231–2240. https://doi.org/10.1099/mic.0.2007/016055-0

Grebowska A, Moran AP, Matusiak A, Bak-Romaniszyn L, Czkwianianc E, Rechciński T, Walencka M, Płaneta-Małecka I, Rudnicka W, Chmiela M (2008) Anti-phagocytic activity of *Helicobacter pylori* lipopolysaccharide (LPS)-possible modulation of the innate immune response to these bacteria. Pol J Microbiol 57:185–192

Grebowska A, Moran AP, Bielanski W, Matusiak A, Rechcinski T, Rudnicka K, Baranowska A, Rudnicka W, Chmiela M (2010) *Helicobacter pylori* lipopolysaccharide activity in human peripheral blood mononuclear leukocyte cultures. J Physiol Pharmacol 61:437–442

Hartung ML, Gruber DC, Koch KN, Grüter L, Rehrauer H, Tegtmeyer N, Backert S, Müller A (2015) *Helicobacter pylori*-induced DNA strand breaks are introduced by nucleotide excision repair endonucleases and promote NF-κB target gene expression. Cell Rep 13(1):70–79. https://doi.org/10.1016/j.celrep.2015.08.074

Hayakawa Y, Fox JG, Wang TC (2017) Isthmus stem cells are the origins of metaplasia in the gastric corpus. Cell Mol Gastroenterol Hepatol 4:89–94. https://doi.org/10.1016/j.jcmgh.2017.02.009

Howlett M, Chalinor HV, Buzzelli JN, Nguyen N, van Driel IR, Bell KM, Fox JG, Dimitriadis E, Menheniott TR, Giraud AS, Judd LM (2012) IL-11 is a parietal cell cytokine that induces atrophic gastritis. Gut 61:1398–1409. https://doi.org/10.1136/gutjnl-2011-300539

Huang ZG, Tang GC (2010) Influence of *Helicobacter pylori* cagA gene of gastric mucosa epithelialcell tumor associated protein expression. Zhongguo Gonggongweishen Zazhi 26:881–883

Huang JQ, Zheng GF, Sumanac K, Irvine EJ, Hunt RH (2003) Meta-analysis of the relationship between cagA seropositivity and gastric cancer. Gastroenterology 125:1636–1644

Hunt RH, Camilleri M, Crowe SE, El-Omar EM, Fox JG, Kuipers EJ, Malfertheiner P, McColl KE, Pritchard DM, Rugge M, Sonnenberg A, Sugano K, Tack J (2015) The stomach in health and disease. Gut 64:1650–1668. https://doi.org/10.1136/gutjnl-2014-307595

Hwang IR, Kodama T, Kikuchi S, Sakai K, Peterson LE, Graham DY, Yamaoka Y (2002) Effect of interleukin 1 polymorphisms on gastric mucosal interleukin 1beta production in *Helicobacter pylori* infection. Gastroenterology 123(6):1793–1803. https://doi.org/10.1053/gast.2002.37043

IARC Working Group on the Evaluation of Carcinogenic Risks to humans (1994) Schistosomes, liver flukes and *Helicobacter pylori*. IIARC Monogr Eval Carcinog Risks Hum 61:1–241

Islami F, Kamangar F (2008) *Helicobacter pylori* and esophageal cancer risk: a meta-analysis. Cancer Prev Res (Phila) 1(5):329–338. https://doi.org/10.1158/1940-6207.capr-08-0109

Jenks PJ, Jeremy AH, Robinson PA, Walker MM, Crabtree JE (2003) Long term infection with *Helicobacter felis* and inactivation of the tumor suppressor gene *p53* cumulatively enhance the gastrin mutation frequency in Big Blue® transgenic mice. J Pathol 201:596–602. https://doi.org/10.1002/path.1488

Jiang YX, Li GM, Yi D, Yu PW (2015) A meta-analysis: The association between interleukin-17 pathway gene polymorphism and gastrointestinal diseases. Gene 572:243–251. https://doi.org/10.1016/j.gene.2015.07.018

Jones KR, Jang S, Chang JY, Kim J, Chung IS, Olsen CH, Merrell DS, Cha JH (2011) Polymorphisms in the intermediate region of VacA impact *Helicobacter-pylori*-induced disease development. J Clin Microbiol 49:101–110. https://doi.org/10.1128/JCM.01782-10

Kamada T, Kurose H, Yamanaka Y, Manabe N, Kusunoki H, Shiotani A, Inoue K, Hata J, Matsumoto H, Akiyama T, Hirai T, Sadahira Y, Haruma K (2012) Relationship between gastroesophageal junction adenocarcinoma and *Helicobacter pylori* infection in Japan. Digestion 85(4):256–260. https://doi.org/10.1159/000336352

Kamangar F, Abnet CC, Hutchinson AA, Newschaffer CJ, Helzlsouer K, Shugart YY, Pietinen P, Dawsey SM, Albanes D, Virtamo J, Taylor PR (2006) Polymorphisms in inflammation-related genes and risk of gastric cancer (Finland). Cancer Causes Control 17(1):117–125

Kawai M, Furuta Y, Yahara K, Tsuru T, Oshima K, Handa N, Takahashi N, Yoshida M, Azuma T, Hattori M, Uchiyama I, Kobayashi I (2011) Evolution in an oncogenic bacterial species with extreme genome plasticity: *Helicobacter pylori* East Asian genomes. BCM Microbiol 11:104. https://doi.org/10.1186/1471-2180-11-104

Khamri W, Walker MM, Clark P, Atherton JC, Thursz MR, Bamford KB, Lechler RI, Lombardi G (2010) *Helicobacter pylori* stimulates dendritic cells to induce interleukin-17 expression from CD4+ T lymphocytes. Infect Immun 78:845–853. https://doi.org/10.1128/IAI.00524-09

Kim J, Cho YA, Choi IJ, Lee YS, Kim SY, Shin A, Cho SJ, Kook MC, Nam JH, Ryu KW, Lee JH, Kim YW (2012) Effects of *Interleukin-10* polymorphisms, *Helicobacter pylori* infection, and smoking on the risk of noncardia gastric cancer. PLoS ONE 7:e29643. https://doi.org/10.1371/journal.pone.0029643

Kim SG, Jung HK, Lee HL, Jang JY, Lee H, Kim CG, Shin WG, Shin ES, Lee YC (2013) Guidelines for the diagnosis and treatment of *Helicobacter pylori* infection in Korea, 2013 revised edition. Korean J Gastroenterol 62:3–26. https://doi.org/10.4166/kjg.2013.62.1.3

Koeppel M, Garcia-Alcalde F, Glowinski F, Schlaermann P, Meyer TF (2015) *Helicobacter pylori* infection causes characteristic DNA damage patterns in human cells. Cell Rep 11:1703–1713. https://doi.org/10.1016/j.celrep.2015.05.030

Kwok T, Zabler D, Urman S, Rohde M, Hartig R, Wessler S, Misselwitz R, Berger J, Sewald N, Konig W, Backert S (2007) *Helicobacter* exploits integrin for type IV secretion and kinase activation. Nature 449(7164):862–866. https://doi.org/10.1038/nature06187

Latifi-Navid S, Mohammadi S, Maleki P, Zahri S, Yazdanbod A, Siavoshi F, Massarrat S (2013) *Helicobacter pylori vacA* d1/-il genotypes and geographic differentiation between high and low incidence areas of gastric cancer in Iran. Arch Iran Med 16:330–337. doi:013166/AIM.005

Lee YC, Chiang TH, Chou CK, Tu YK, Liao WC, Wu MS, Graham DY (2016) Association between *Helicobacter pylori* eradication and gastric cancer incidence: a systemic review and meta-analysis. Gastroenterology 150:1113–1124. https://doi.org/10.1053/j.gastro.2016.01.028

Li WQ, Zhang L, Ma JL, Zhang Y, Li JY, Pan KF, You WC (2009) Association between genetic polymorphisms of DNA base excision repair genes and evolution of precancerous gastric lesions in a Chinese population. Carcinogenesis 30(3):500–505. https://doi.org/10.1093/carcin/bgp018

Li Y, Dai L, Zhang J, Wang P, Chai Y, Ye H, Zhang J, Wang K (2012) *Cyclooxygenase-2* polymorphisms and the risk of gastric cancer in various degrees of relationship in the Chinese Han population. Oncol Lett 3:107–112. https://doi.org/10.3892/ol.2011.426

Li ZX, Wang YM, Tang FB, Zhang L, Zhang, Ma JL, Zhou T, You WC, Pan KF (2015) *NOD1* and *NOD2* genetic variants in association with risk of gastric cancer and its precursors in a Chinese population. PLoS One 1 10(5):e0124949. https://doi.org/10.1371/journal.pone.0124949

Lind J, Backert S, Pfleiderer K, Berg DE, Yamaoka Y, Sticht H, Tegtmeyer N (2014) Systematic analysis of phosphotyrosine antibodies recognizing single phosphorylated EPIYA-motifs in CagA of Western-type *Helicobacter pylori* strains. PLoS ONE 9(5):e96488. https://doi.org/10.1371/journal.pone.0096488

Lind J, Backert S, Hoffmann R, Eichler J, Yamaoka Y, Perez-Perez GI, Torres J, Sticht H, Tegtmeyer N (2016) Systematic analysis of phosphotyrosine antibodies recognizing single phosphorylated EPIYA-motifs in CagA of East Asian-type *Helicobacter pylori* strains. BMC Microbiol 16(1):201. https://doi.org/10.1186/s12866-016-0820-6

Liu J, Xu Q, Yuan Q, Wang Z, Xing C, Yuan Y (2015) Association of *IL-17A* and *IL-17F* polymorphisms with gastric cancer risk in Asians: a meta-analysis. Human Immunl 76:6–12. https://doi.org/10.1016/j.humimm.2014.12.011

Ma J, Wu D, Hu X, Li J, Cao M, Dong W (2017) Associations between cytokine gene polymorphisms and susceptibility to *Helicobacter pylori* infection and *Helicobacter pylori* related gastric cancer, peptic ulcer disease: A meta-analysis. PLoS ONE 4:e0176463. https://doi.org/10.1371/journal.pone.0176463

Machado JC, Pharoah P, Sousa S, Carvalho R, Oliveira C, Figueiredo C, Amorim A, Seruca R, Caldas C, Carneiro F, Sobrinho-Simões M (2001) *Interleukin 1B* and *interleukin 1RN* polymorphisms are associated with increased risk of gastric carcinoma. Gastroenterology 121(4):823–829

Machado JC, Figueiredo C, Canedo P, Pharoah P, Carvalho R, Nabais S, Castro Alves C, Campos ML, Van Doorn LJ, Caldas C, Seruca R, Carneiro F, Sobrinho-Simões M (2003) A proinflammatory genetic profile increases the risk for chronic atrophic gastritis and gastric carcinoma. Gastroenterology 125(2):364–371. https://doi.org/10.1016/S0016-5085(03)00899-0

Machado AM, Figueiredo C, Touati E, Máximo V, Sousa S, Michel V, Carneiro F, Nielsen FC, Seruca R, Rasmussen LJ (2009) *Helicobacter pylori* infection induces genetic instability of nuclear and mitochondrial DNA in gastric cells. Clin Cancer Res 15:2995–3002. https://doi.org/10.1158/1078-0432.CCR-08-2686

Machado AM, Desler C, Bøggild S, Strickertsson JA, Friis-Hansen L, Figueiredo C, Seruca R, Rasmussen LJ (2013) *Helicobacter pylori* infection affects mitochondrial function and DNA repair, thus, mediating genetic instability in gastric cells. Mech Ageing Dev 134(10):460–466. https://doi.org/10.1016/j.mad.2013.08.004

Malaty HM, El-Kasabany A, Graham DY, Miller CC, Reddy SG, Srinivasan SR, Yamaoka Y, Berenson GS (2002) Age at acquisition of *Helicobacter pylori* infection: a follow up study from infancy to adulthood. Lancet 359:931–988. https://doi.org/10.1016/S0140-6736(02)08025-X

Matos JI, de Sousa HA, Marcos-Pinto R, Dinis-Ribeiro M (2013) *Helicobacter pylori* CagA and VacA genotypes and gastric phenotype: a meta-analysis. Eur J Gastroenterol Hepatol 25(12):1431–1441. https://doi.org/10.1097/MEG.0b013e328364b53e

McClain MS, Beckett AC, Cover TL (2017) *Helicobacter pylori* vacuolating toxin and gastric cancer. Toxins 9(10):E316. https://doi.org/10.3390/toxins9100316

McLean MH, El-Omar EM (2014) Genetics of gastric cancer. Nat Rev Gastroenterol Hepatol 11:664–674. https://doi.org/10.1038/nrgastro.2014.143

Meira LB, Bugni JM, Green SL, Lee CW, Pang B, Borenshtein D, Rickman BH, Rogers AB, Moroski-Erkul CA, McFaline JL, Schauer DB, Dedon PC, Fox JG, Samson LD (2008) DNA damage induced by chronic inflammation contributes to colon carcinogenesis in mice. J Clin Invest 118:2516–2525. https://doi.org/10.1172/JCI35073

Melchiades JL, Zabaglia LM, Sallas ML, Orcini WA, Chen E, Smith MAC, Payão SLM, Rasmussen LT (2017) Polymorphisms and haplotypes of the *interleukin 2* gene are associated with an increased risk of gastric cancer. The possible involvement of *Helicobacter pylori*. Cytokine 96:203–207. https://doi.org/10.1016/j.cyto.2017.04.020

Melo Barbosa HP, Martins LC, Dos Santos SE, Demachki S, Assumpção MB, Aragão CD, de Oliveira Corvelo TC (2009) *Interleukin-1* and *TNF-alpha* polymorphisms and *Helicobacter pylori* in a Brazilian Amazon population. World J Gastroenterol 15:1465–1471

Memon AA, Hussein NR, Miendje Deyi VY, Burette A, Atherton JC (2014) Vacuolating cytotoxin genotypes are strong markers of gastric cancer and duodenal ulcer-associated *Helicobacter pylori* strains: a matched case-control study. AJ Clin Microbiol. 52(8):2984–2989. https://doi.org/10.1128/JCM.00551-14

Mizuno T, Ando T, Nobata K, Tsuzuki T, Maeda O, Watanabe O, Minami M, Ina K, Kusugami K, Peek RM, Goto H (2005) Interleukin-17 levels in *Helicobacter pylori*-infected gastric mucosa and pathologic sequelae of colonization. World J Gastroenterol 11:6305–6311

Mnich E, Gajewski A, Rudnicka K, Gonciarz W, Stawerski P, Hinc K, Obuchowski M, Chmiela M (2015) Immunoregulation of antigen presenting and secretory functions of monocytic cells by *Helicobacter pylori* antigens in relation to impairment of lymphocyte expansion. Acta Biochim Pol 62:641–650. https://doi.org/10.18388/abp.2015_1045

Mocellin S, Verdi D, Pooley KA, Nitti D (2015) Genetic variation and gastric cancer risk: a field synopsis and meta-analysis. Gut 64(8):1209–1219. https://doi.org/10.1136/gutjnl-2015-309168

Moss SF (2016) The clinical evidence linking *Helicobacter pylori* to gastric cancer. Cell Mol Gastroenterol Hepatol 3(2):183–191. https://doi.org/10.1016/j.jcmgh.2016.12.001

Mukherjee T, Hovingh ES, Foerster EG, Abdel-Nour M, Philpott DJ, Girardin SE (2018) NOD1 and NOD2 in inflammation, immunity and disease. Arch Biochem Biophys pii S0003–9861(18):30937–30938. https://doi.org/10.1016/j.abb.2018.12.022

Naumann M, Sokolova O, Tegtmeyer N, Backert S (2017) *Helicobacter pylori*: a paradigm pathogen for subverting host cell signal transmission. Trends Microbiol 25:316–328. https://doi.org/10.1016/j.tim.2016.12.004

Ng MT, Van't Hof R, Crockett JC, Hope ME, Berry S, Thomson J, McLean MH, McColl KE, El-Omar EM, Hold GL (2010) Increase in NF-kappaB binding affinity of the variant C allele of the *toll-like receptor 9*–1237T/C polymorphism is associated with *Helicobacter pylori*-induced gastric disease. Infect Immun 78(3):1345–1352. https://doi.org/10.1128/IAI.01226-09

Ni P, Xu H, Xue H, Lin B, Lu Y (2012) A Meta-analysis on *Interleukin-10*-1082 promoter polymorphism associated with gastric cancer risk. DNA and Cell Biol 31:582–590. https://doi.org/10.1089/dna.2011.1440

Nogueira C, Figueiredo C, Carneiro F, Gomes AT, Barreira R, Figueira P, Salgado C, Belo L, Peixoto A, Bravo JC, Bravo LE, Realpe JL, Plaisier AP, Quint WG, Ruiz B, Correa P, van Doorn LJ (2001) *Helicobacter pylori* genotypes may determine gastric histopathology. Am J Pathol 158(2):647–654

Ogiwara H, Sugimoto M, Ohno T, Vilaichone RK, Mahachai V, Graham DY, Yamaoka Y (2009) Role of deletion located between the intermediate and middle regions of the *Helicobacter pylori vacA* gene in cases gastroduodenal diseases. J Clin Microbiol 47:3493–3500. https://doi.org/10.1128/JCM.00887-09

Ohyauchi M, Imatani A, Yonechi M, Asano N, Miura A, Iijima K, Koike T, Sekine H, Ohara S, Shimosegawa T (2005) The polymorphism *interleukin 8*–251 A/T influences the susceptibility of *Helicobacter pylori* related gastric diseases in the Japanese population. Gut 54:330–335. https://doi.org/10.1136/gut.2003.033050

Olbermann P, Josenhans C, Moodley Y, Uhr M, Stamer C, Vauterin M, Suerbaum S, Achtman M, Linz B (2010) A global overview of the genetic and functional diversity in the *Helicobacter pylori* cag pathogenicity island. PloS Genet 6:e1001069. https://doi.org/10.1371/journal.pgen.1001069

Oleastro M, Ménard A (2013) The role of *Helicobacter pylori* outer membrane proteins in adherence and pathogenesis. Biology (Basel) 2:1110–1134. https://doi.org/10.3390/biology2031110

Pachathundikandi SK, Lind J, Tegtmeyer N, El-Omar EM, Backert S (2015) Interplay of the Gastric Pathogen *Helicobacter pylori* with Toll-Like Receptors. Biomed Res Int 2015:192420. https://doi.org/10.1155/2015/192420

Palomo J, Dietrich D, Martin P, Palmer G, Gabay C (2015) The interleukin (IL)-1 cytokine family–Balance between agonists and antagonists in inflammatory diseases. Cytokine 76(1):25–37. https://doi.org/10.1016/j.cyto.2015.06.017

Paziak-Domanska B, Chmiela M, Jarosińska A, Rudnicka W (2000) Potential role of CagA in the inhibition of T cell reactivity in *Helicobacter pylori* infections. Cell Immunol 202:136–139. https://doi.org/10.1006/cimm.2000.1654

Peleteiro B, Lunet N, Carrilho C, Durães C, Machado JC, La Vecchia C, Barros H (2010) Association between cytokine gene polymorphisms and gastric precancerous lesions: systematic review and meta-analysis. Cancer Epidemiol Biomarkers Prev 19(3):762–776. https://doi.org/10.1158/1055-9965.EPI-09-0917

Persson C, Canedo P, Machado JC, El-Omar EM, Forman D (2011) Polymorphisms in inflammatory response genes and their association with gastric cancer: a HuGE systematic review and meta-analyses. Am J Epidemiol 173(3):259–270. https://doi.org/10.1093/aje/kwq370

Petersen CP, Meyer AR, De Salvo C, Choi E, Schlegel C, Petersen A, Engevik AC, Prasad N, Levy SE, Peebles RS, Pizarro TT, Goldenring JR (2018) A signalling cascade of IL-33 to IL-13 regulates metaplasia in the mouse stomach. Gut 67:805–817. https://doi.org/10.1136/gutjnl-2016-312779

Pimentel-Nunes P, Afonso L, Lopes P, Roncon-Albuquerque R Jr, Gonçalves N, Henrique R, Moreira-Dias L, Leite-Moreira AF, Dinis-Ribeiro M (2011) Increased expression of toll-like receptors (TLR) 2, 4, 5 in gastric dysplasia. Pathol Oncol Res 17:677–683. https://doi.org/10.1007/s12253-011-9368-9

Plummer M, van Doorn LJ, Franceschi S, Kleter B, Canzian F, Vivas J, Lopez G, Colin D, Muñoz N, Kato I (2007) *Helicobacter pylori* cytotoxin-associated genotype and gastric precancerous lesions. J Natl Cancer Inst 99:1328–1334. https://doi.org/10.1093/jnci/djm120

Plummer M, Franceschi S, Vignat J, Forman D, de Martel C (2015) Global burden of gastric cancer attributable to *Helicobacter pylori*. Int J Cancer 136:487–490. https://doi.org/10.1002/ijc.28999

Posselt G, Backert S, Wessler S (2013) The functional interplay of *Helicobacter pylori* factors with gastric epithelial cells induces a multi-step process in pathogenesis. Cell Commun Signal 11:77. https://doi.org/10.1186/1478-811X-11-77

Qinghai Z, Yanying W, Yunfang C, Xukui Z, Xiaoqiao Z (2014) Effect of interleukin-17A and *interleukin-17F* gene polymorphisms on the risk of gastric cancer in Chinese population. Gene 537:328–332. https://doi.org/10.1016/j.gene.2013.11.007

Quin XP, Zhou Y, Chen Y, Li NN, Wu XT (2014) *XRCC3* Thr241Met polymorphism and gastric cancer susceptibility: a meta-analysis. Clin Res Hepatol Gastroenterol 38:226–234. https://doi.org/10.1016/j.clinre.2013.10.011

Rad R, Prinz C, Neu B, Neuhofer M, Zeitner M, Voland P, Becker I, Schepp W, Gerhard M (2003) Synergistic effect of *Helicobacter pylori* virulence factors and *interleukin-1* polymorphisms for the development of severe histological changes in the gastric mucosa. J Infect Dis 188(2):272–281. https://doi.org/10.1086/376458

Ramis IB, Vianna JS, Gonçalves CV, von Groll A, Dellagostin OA, da Silva PEA (2015) Polymorphisms of the *IL-6*, *IL-8* and *IL-10* genes and the risk of gastric pathology in patients infected with *Helicobacter pylori*. J Microbiol Immunol Infect 50:153–159. https://doi.org/10.1016/j.jmii.2015.03.002

Rhead JL, Letley DP, Mohammadi M, Hussein N, Mohagheghi MA, Eshagh Hosseini M, Atherton JC (2007) A new *Helicobacter pylori* vacuolating cytotoxin determinant, the intermediate region, is associated with gastric cancer. Gastroenterology 133(3):926–936. https://doi.org/10.1053/j.gastro.2007.06.056

Rosenstiel P, Hellmig S, Hampe J, Ott S, Till A, Fischbach W, Sahly H, Lucius R, Fölsch UR, Philpott D, Schreiber S (2006) Influence of polymorphisms in the NOD1/CARD4 and NOD2/CARD15 genes on the clinical outcome of *Helicobacter pylori* infection. Cell Microbiol 8(7):1188–1198

Rudnicka K, Miszczyk E, Matusiak A, Walencka M, Moran AP, Rudnicka W, Chmiela M (2015) *Helicobacter pylori*-driven modulation of NK cell expansion, intracellular cytokine expression and cytotoxic activity. Innate Immun 21:127–139. https://doi.org/10.1177/1753425913518225

Savage SA, Hou L, Lissowska J, Chow WH, Zatonski W, Chanock SJ, Yeager M (2006) *Interleukin-8* polymorphisms are not associated with gastric cancer risk in a Polish population. Cancer Epidemiol Biomarkers Prev 15(3):589–591

Sheh A, Lee CW, Masumura K, Rickman BH, Nohmi T, Wogan GN, Fox JG, Schauer DB (2010) Mutagenic potency of *Helicobacter pylori* in the gastric mucosa of mice is determined by sex and duration of infection. Proc Natl Acad Sci USA 107:15217–15222. https://doi.org/10.1073/pnas.1009017107

Shibata J, Goto H, Arisawa T, Niwa Y, Hayakawa T, Nakayama A, Mori N (1999) Regulation of tumour necrosis factor (TNF) induced apoptosis by soluble TNF receptors in *Helicobacter pylori* infection. Gut 45:24–31

Shigematsu Y, Niwa T, Rehnberg E, Toyoda T, Yoshida S, Mori A, Wakabayashi M, Iwakura Y, Ichinose M, Kim YJ, Ushijima T (2013) Interleukin-1beta induced by *Helicobacter pylori* infection enhances mouse gastric carcinogenesis. Cancer Lett 340(1):141–147. https://doi.org/10.1016/j.canlet.2013.07.034

Sicinschi LA, Lopez-Carrillo L, Camargo MC, Correa P, Sierra RA, Henry RR, Chen J, Zabaleta J, Piazuelo MB, Schneider BG (2006) Gastric cancer risk in a Mexican population: role of *Helicobacter pylori* CagA positive infection and polymorphisms in *interleukin-1* and *-10* genes. Int J Cancer 118(3):649–657. https://doi.org/10.1002/ijc.21364

Sicinschi LA, Correa P, Peek RM, Camargo MC, Piazuelo MB, Romero-Gallo J, Hobbs SS, Krishna U, Delgado A, Mera R, Bravo LE, Schneider BG (2010) CagA C-terminal variations in *Helicobacter pylori* strains from Colombian patients with gastric precancerous lesions. Clin Microbiol Infect 2010:369–378. https://doi.org/10.1111/j.1469-0691.2009.02811.x

Smith SM (2014) Role of Toll-like receptors in *Helicobacter pylori* infection and immunity. World J Gastrointest Pathophysiol 5(3):133–146. https://doi.org/10.4291/wjgp.v5.i3.133

Sun X, Xu Y, Wang L, Zhang F, Zhang J, Fu X, Jing T, Han J (2016) Association between *TNF A* gene polymorphisms and *Helicobacter pylori* infection: a meta-analysis. PLoS ONE 54:703–706. https://doi.org/10.1371/journal.pone.0147410

Tanaka S, Nagashima H, Cruz M, Uchida T, Uotani T, Jiménez Abreu JA, Mahachai V, Vilaichone RK, Ratanachu-Ek T, Tshering L, Graham DY, Yamaoka Y (2017) Interleukin 17C in human *Helicobacter pylori* gastritis. InfectImmun 85:e00389–17. https://doi.org/10.1128/IAI.00389-17

The EUROGAST Study Group (1993) An international association between *Helicobacter pylori* infection and gastric cancer. Lancet 341:1359–1362

Torres LE, Melián K, Moreno A, Alonso J, Sabatier CA, Hernández M, Bermúdez L, Rodríguez BL (2009) Prevalence of *vacA*, *cagA* and *babA2* genes in Cuban *Helicobacter pylori* isolates. World J Gastroenterol 15:204–210

Tsai CY, Wang CS, Tsai MM, Chi HC, Cheng WL, Tseng YH, Chen CY, Lin CD, Wu JI, Wang LH, Lin KH (2014) Interleukin-32 increases human gastric cancer cell invasion associated with tumor progression and metastasis. Clin Cancer Res 20:2276–2288. https://doi.org/10.1158/1078-0432.CCR-13-1221

Tsukamoto T, Nakagawa M, Kiriyama Y, Toyoda T, Cao X (2017) Prevention of gastric cancer: eradication of *Helicobacter pylori* and beyond. Int J Mol Sci 18:1699. https://doi.org/10.3390/ijms18081699

Tu S, Bhagat G, Cui G, Takaishi S, Kurt-Jones EA, Rickman B, Betz KS, Penz-Oesterreicher M, Bjorkdahl O, Fox JG, Wang TC (2008) Overexpression of interleukin-1beta induces gastric inflammation and cancer and mobilizes myeloid-derived suppressor cells in mice. Cancer Cell 14(5):408–419. https://doi.org/10.1016/j.ccr.2008.10.011

Varga MG, Peek RM (2017) DNA Transfer and Toll-like receptor modulation by *Helicobacter pylori*. Curr Top Microbiol Immunol 400:169–193. https://doi.org/10.1007/978-3-319-50520-6_8

Vaziri F, Peerayeh SN, Alebouyeh M, Maghsoudi N, Azimzadeh P, Siadat SD, Zali MR (2015) Novel effects of *Helicobacter pylori* CagA on key genes of gastric cancer signal transduction: a comparative transfection study. Pathog Dis 73(3):ftu021. https://doi.org/10.1093/femspd/ftu021

Wang P, Zhang L, Jiang JM, Ma D, Tao HX, Yuan SL, Wang YC, Wang LC, Liang H, Zhang ZS, Liu CJ (2012) Association of *NOD1* and *NOD2* genes polymorphisms with *Helicobacter pylori* related gastric cancer risk in Chinese population. Worl J Gastroenterol 18:2112–2120. https://doi.org/10.3748/wjg.v18.i17.2112

Wang J, Guo X, Yu S, Song J, Zhang J, Cao Z, Wang J, Liu M, Dong W (2014) Association between *CD14* gene polymorphisms and cancer risk: a meta analysis. PLoS ONE 9(6):e100122. https://doi.org/10.1371/journal.pone.0100122

Warren JR, Marshall BJ (1983) Unidentified curved bacilli on gastric epithelium in active chronic gastritis. Lancet 1:1273–1275

Winter JA, Letley DP, Cook KW, Rhead JL, Zaitoun AA, Ingram RJ, Amilon KR, Croxall NJ, Kaye PV, Robinson K, Atherton JC (2014) A role for the vacuolating cytotoxin, VacA, in colonization and *Helicobacter pylori*-induced metaplasia in the stomach. J Infect Dis 210(6):954–963. https://doi.org/10.1093/infdis/jiu154

Wong BC, Lam SK, Wong WM, Chen JS, Zheng TT, Feng RE, Lai KC, Hu WH, Yuen ST, Leung SY, Fong DY, Ho J, Ching CK, Chen JS, China Gastric Cancer Study Group (2004) *Helicobacter pylori* eradication to prevent gastric cancer in a high risk region of China: a randomized controlled trial. JAMA 291:187–194. https://doi.org/10.1001/jama.291.2.187

Wu MS, Wu CY, Chen CJ, Lin MT, Shun CT, Lin JT (2003) *Interleukin-10* genotypes associate with the risk of gastric carcinoma in Taiwanese Chinese. Int J Cancer 104(5):617–623. https://doi.org/10.1002/ijc.10987

Wu WK, Lee CW, Cho CH, Fan D, Wu K, Yu J, Sung JJ (2010) MicroRNA dysregulation in gastric cancer: a new player enters the game. Oncogene 29:5761–5771. https://doi.org/10.1038/onc.2010.352

Wu HH, Lin WC, Tsai KW (2013) Advances in molecular biomarkers for gastric cancer: miRNAs as emerging novel cancer markers. Expert Rev Mol Med 16:e1. https://doi.org/10.1017/erm.2013.16

Xu Y, Cao X, Jiang J, Chen Y, Wang K (2017) TNF-α-308/-238 polymorphisms are associated with gastric cancer: a case-control family study in China. Clin Res Hepatol Gastroenterol 41(1):103–109. https://doi.org/10.1016/j.clinre.2016.05.014

Yamaoka Y (2008) Roles of *Helicobacter pylori* BabA in gastroduodenal pathogenesis. World J Gastroenterol 14:4265–4272

Ye F, Brauer T, Niehus E, Drlica K, Josenhans C, Suerbaum S (2007) Flagellar and global gene regulation in *Helicobacter pylori* modulated by changes in DNA supercoiling. Int J Med Microbiol 297:65–81. https://doi.org/10.1016/j.ijmm.2006.11.006

Ying HY, Yu BW, Yang Z, Yang SS, Bo LH, Shan XY, Wang HJ, Zhu YJ, Wu XS (2016) Interleukin-1B 31 C > T polymorphism combined with *Helicobacter pylori*-modified gastric cancer susceptibility: evidence from 37 studies. J Cell Mol Med 20:526–536. https://doi.org/10.1111/jcmm.12737

Zambon CF, Navaglia F, Basso D, Rugge M, Plebani M (2003) *Helicobacter pylori babA2*, *cagA*, and s1 *vacA* genes work synergistically in causing intestinal metaplasia. J Clin Pathol 56:287–291

Zambon CF, Basso D, Navaglia F, Belluco C, Falda A, Fogar P, Greco E, Gallo N, Rugge M, Di Mario F, Plebani M (2005) Pro- and anti-inflammatory cytokines gene polymorphisms and *Helicobacter pylori* infection: interactions influence outcome. Cytokine 29:141–152. https://doi.org/10.1016/j.cyto.2004.10.013

Zhang XS, Tegtmeyer N, Traube L, Jindal S, Perez-Perez G, Sticht H, Backert S, Blaser MJ (2015) A specific A/T polymorphism in Western tyrosine phosphorylation B-motifs regulates *Helicobacter pylori* CagA epithelial cell interaction. PLoS Pathog 11:e1004621. https://doi.org/10.1371/journal.ppat.1004621

Zhang JZ, Liu CM, Peng HP, Zhang Y (2017) Association of genetic variations in *IL-6/IL6R* pathway genes with gastric cancer risk in a Chinese population. Gene 623:1–4. https://doi.org/10.1016/j.gene.2017.04.038

Zhao F, Zhu H, Huang M, Yi C, Huang Y (2014) The 765 G > C polymorphism in the *cyclooxygenase-2* gene and gastric cancer risk: an update by meta-analysis. Asian Pac J Cancer Prev 15:2863–2868

Zhou Q, Wang C, Wang X, Wu X, Zhu Z, Liu B, Su L (2014) Association between *TLR4* (+896A/G and +1196C/T) polymorphisms and gastric cancer risk: an updated meta-analysis. PLoS ONE 9:e109605. https://doi.org/10.1371/journal.pone.0109605

MALT Lymphoma as a Model of Chronic Inflammation-Induced Gastric Tumor Development

Lukas Marcelis, Thomas Tousseyn and Xavier Sagaert

Contents

Abstract Mucosa-associated lymphoid tissue (MALT) lymphoma, or extranodal marginal zone lymphoma of MALT, is an indolent B-cell non-Hodgkin lymphoma linked with preexisting chronic inflammation. The stomach is the most commonly affected organ and the MALT lymphoma pathogenesis is clearly associated with *Helicobacter pylori* gastroduodenitis. Inflammation induces the lymphoid infiltrates in extranodal sites, where the lymphoma then subsequently develops. Genetic aberrations arise through the release of reactive oxygen species (ROS), *H. pylori*-induced endonucleases, and other effects. The involvement of nuclear factor kappa B (NF-κB) pathway activation, a critical regulator of pro-inflammatory responses, further highlights the role of inflammation in gastric MALT lymphoma. The NF-κB pathway

L. Marcelis · T. Tousseyn · X. Sagaert (✉)
Translational Cell and Tissue Research Lab, Department of Imaging and Pathology, KU Leuven, Louvain, Belgium
e-mail: xavier.sagaert@uzleuven.be

L. Marcelis
e-mail: lukas.marcelis@uzleuven.be

T. Tousseyn
e-mail: thomas.tousseyn@uzleuven.be

T. Tousseyn · X. Sagaert
Department of Pathology, UZ Leuven, University Hospitals, Louvain, Belgium

L. Marcelis · T. Tousseyn · X. Sagaert
O&N IV Herestraat 49 - bus 7003 24, 3000 Louvain, Belgium

S. Backert (ed.), *Molecular Mechanisms of Inflammation: Induction, Resolution and Escape by Helicobacter pylori*, Current Topics in Microbiology and Immunology 421, https://doi.org/10.1007/978-3-030-15138-6_4

regulates key elements of normal lymphocyte function, including the transcription of proliferation-promoting and anti-apoptotic genes. Aberrant constitutive activation of NF-κB signaling can lead to autoimmunity and malignancy. NF-κB pathway activation can happen through both the canonical and non-canonical pathways and can be caused by multiple genetic aberrations such as t(11;18)(q12;q21), t(1;14)(p22;q32), and t(14;18)(q32;q21) translocations, chronic inflammation and even directly by *H. pylori*-associated mechanisms. Gastric MALT lymphoma is considered one of the best models of how inflammation initiates genetic events that lead to oncogenesis, determines tumor biology, dictates clinical behavior and leads to viable therapeutic targets. The purpose of this review is to present gastric MALT lymphoma as an outstanding example of the close pathogenetic link between chronic inflammation and tumor development and to describe how this information can be integrated into daily clinical practice.

1 Introduction

Marginal zone lymphomas are the third most frequent type of B-cell non-Hodgkin lymphoma after diffuse large B-cell lymphoma (DLBCL) (~30%) and follicular lymphoma (~20%) (Swerdlow et al. 2008). They develop in the marginal zone of B-cell follicles, which are most prominently developed in organs that are chronically exposed to antigenic stimulation. These organs include spleen, mesenteric lymph nodes, and mucosa-associated lymphoid tissue (MALT) (Martin and Kearney 2002). Marginal zone B (MZB) cells are crucial in T-cell-independent responses to various antigens such as polysaccharides derived from encapsulated bacteria (including *Haemophilus influenzae*, *Neisseria meningitidis*, and *Streptococcus pneumoniae*) (Pillai et al. 2005). The current WHO classification of tumors of hematopoietic and lymphoid tissues subdivides marginal zone lymphomas on the basis of their anatomical location in splenic zone, nodal marginal zone, and extranodal marginal zone or MALT lymphomas (Swerdlow et al. 2008). Extranodal marginal zone lymphoma of MALT, or MALT lymphoma, differs from its splenic and nodal counterparts in that it arises in organs that normally lack lymphoid tissue, but that have accumulated B-cells in response to either chronic infection or autoimmune processes (Wotherspoon et al. 1991; De Re et al. 2000; Roggero et al. 2000; Zucca et al. 2000; Thieblemont et al. 2002; Lecuit et al. 2004; Ferreri et al. 2009; Stefanovic and Lossos 2009; Coeuret et al. 2014). This review focuses on gastric MALT lymphoma, because it is the most frequent primary lymphoma of the gastrointestinal tract (~50%) and 85% of gastrointestinal MALT lymphomas are located in the stomach (Cook et al. 2017). The small intestine is the other potential gastrointestinal site, extra-gastrointestinal sites are, lung, head and neck, ocular adnexa, skin, salivary glands, thyroid, and breast (Thieblemont et al. 1997). Gastric MALT lymphoma has an indolent course; recurrences are less frequent than in patients with extragastric MALT lymphomas and it is associated with *H. pylori* in >90% of cases (Wotherspoon et al. 1991; Raderer et al. 2005).

The strong association with *H. pylori* infection makes gastric MALT lymphoma an ideal model to examine the link between chronic inflammation and lymphoma pathogenesis. Sustained antigenic stimulation triggers a polyclonal B-cell proliferation and attracts neutrophils to the site of inflammation. The release of reactive oxygen species (ROS) by these neutrophils can cause a wide range of genetic aberrations (Coussens and Werb 2009). Next to ROS-induced genetic defects, *H. pylori* can also activate endonucleases XPF and XPG through the nuclear factor kappa B (NF-κB) pathway, which induce double-strand DNA breaks (DSBs) in chromosomal DNA of the host (Hartung et al. 2015). Furthermore, B-cells are inherently instable during somatic hypermutation and class-switching recombination. Prolonged B-cell proliferation, therefore, increases the risk of genetic aberration accrual such as DSBs and translocation events (Goossens et al. 1998). Identified mutagenic events in MALT lymphomas include trisomy of chromosomes 3, 7, 12, or 18, and the disease-specific chromosomal translocations t(1;14)(p22;q32), t(14;18)(q32;q21), t(11;18)(q21;q21), and t(3;14)(p13;q32). In those MALT lymphomas lacking the aforementioned aberrations, inactivating mutations of *TNFAIP3* (tumor necrosis factor alpha-induced protein 3) are frequent (Lin et al. 2010). Remarkably, the majority of mutagenic events in MALT lymphomas, at least two of the three trisomies, three of the translocations and *TNFAIP3* inactivation, all result in activation of the NF-κB pathway. Moreover, *H. pylori*-specific products such as the CagA protein may also play a direct role in oncogenesis and may modulate NF-κB signaling (Brandt et al. 2005; Posselt et al. 2013; Zhang et al. 2015). NF-κB is a key mediator of the immune response and has been the focus of intense investigation over the past thirty years (Zhang et al. 2017; Yu et al. 2017). NF-κB regulates the expression of over 200 genes including several implicated in survival and proliferation of B-cells (Ruefli-Brasse 2003; Siebenlist et al. 2005; Packham 2008).

In this review, we discuss current insights into the pathogenesis of gastric MALT lymphomas and relate this information to daily clinical practice. This review only briefly mentions non-gastric MALT lymphomas, as they are considered to be outside the scope of this article and is an update of a previous review by our group (Sagaert et al. 2010b). In-depth reviews of non-gastric MALT lymphomas are available (Schreuder et al. 2017; Defrancesco and Arcaini 2018).

2 Clinical and Pathological Features

The gold standard for gastric MALT lymphoma diagnosis is histopathological examination of endoscopic mucosal biopsies. The tumor consists of diffusely spread lymphoma cells surrounding reactive B-cell follicles and invading epithelial structures. The malignant B-cells resemble MZB cells with a limited, pale cytoplasm, and slightly irregular nuclei with inconspicuous nucleoli. The presence of reactive B-cell follicles and lympho-epithelial lesions (Fig. 1) is important for histological diagnosis and the absence of these lesions can make diagnosis difficult. Lympho-epithelial lesions are not always prominent, reactive B-cell follicles can become

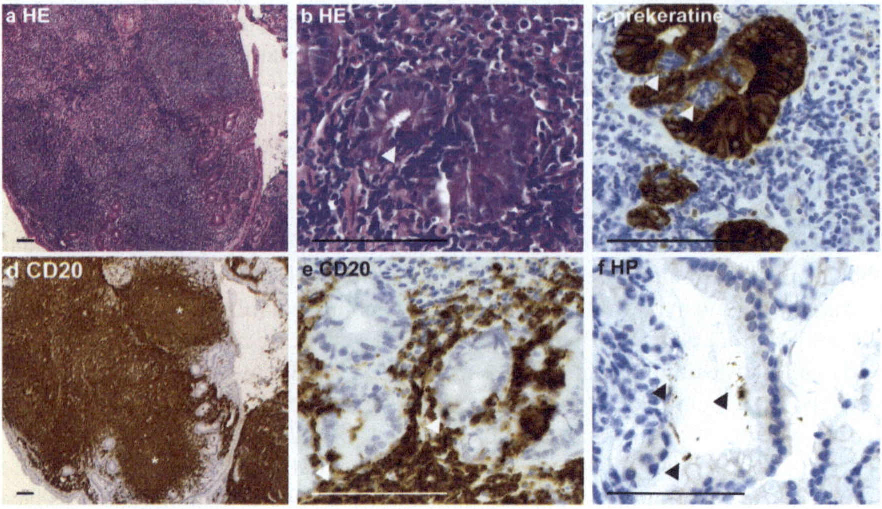

Fig. 1 Morphologic and immunohistochemical features of gastric MALT lymphoma. Gastric MALT lymphoma cells can overgrow reactive B-cell follicles (*) and invade the gastric glandular epithelium, resulting in so-called lympho-epithelial lesions (white arrowheads). **a** Hematoxylin and eosin stain, overview. **b** Hematoxylin and eosin stain, detail of lympho-epithelial lesions. **c** Prekeratin. **d** CD20, overview. **e** CD20, detail of lympho-epithelial lesions. **f** *Helicobacter pylori* staining (black arrowheads). All scale bars are 100 μm Abbreviations: HE, hematoxylin and eosin, HP, *Helicobacter pylori*

unrecognizable due to neoplastic overgrowth and biopsy specimens are sometimes small. Molecular techniques such as polymerase chain reaction (PCR) are, therefore, often used to support MALT lymphoma diagnosis by determining if a clonal B-cell population is present or not (Dreyling et al. 2013; Zucca et al. 2013).

Gastric MALT lymphoma has an indolent behavior and the vast majority of patients diagnosed with MALT lymphoma have a disease limited to the site of origin (Ann Arbor stage IE) at the time of diagnosis. Gastric MALT lymphoma tends to remain localized for several years, nevertheless several prospective studies revealed that up to 15–32% of patients showed involvement of the regional lymph nodes (stage IIE) and around that 11% of patients had secondary tumors in other extranodal sites (Fischbach et al. 2000; Hong et al. 2006; Moleiro et al. 2016). The therapeutic strategy differs greatly for patients with localized (Ann Arbor stage IE/II) versus advanced (Ann Arbor III–IV) disease making thorough tumor staging at primary diagnosis essential and specific attention should be given to the presence of disease in extranodal sites such as Waldeyer's ring, intestine, and spleen (Fischbach et al. 2000; Hong et al. 2006; Moleiro et al. 2016).

Patients with gastric MALT lymphoma have a 10-year survival rate of ~90% and disease-free survival rate of ~70% (Cogliatti et al. 1991; Thieblemont et al. 2000). Transcriptional profiling and existence of patients with diagnosis of synchronous gastric MALT lymphoma and DLBCL diagnosis provided evidence that MALT lymphomas can transform into DLBCL. Transformation into DLBCL is thought to lead to a more aggressive biological behavior, but available clinical data are conflicting. In the early 1990s, two retrospective studies reported 5-year survival of 42 and 56%, respectively (Cogliatti et al. 1991; Radaszkiewicz et al. 1992). However, prospective studies conducted 10 years later showed no or only a marginal difference in overall survival between gastric DLBCL and the more indolent gastric MALT lymphoma (Koch et al. 2001, 2005).

A MALT lymphoma prognostic index (MALT-IPI) for both gastric and non-gastric lymphomas was recently proposed and validated in the largest randomized study conducted in MALT lymphoma to date (Thieblemont et al. 2017). The prognostic index based on three parameters (Ann Arbor stage, Age $\geq$ 70 years and elevated serum lactate dehydrogenase levels) and can help identify patients with increased risk of disease progression and death (Thieblemont et al. 2017).

3 Pathogenesis

MALT lymphomas develop from reactive MZB cells in a multistep process (Fig. 2). According to the current hypothesis, chronic antigen stimulation is the primary insult (Zucca et al. 2003). The antigens can be of exogenous origin as is described for lung and gastric MALT lymphomas both of which come in frequent contact with environmental particles and are linked to various infectious agents (Zucca et al. 2003). Two infectious agents are currently associated with gastric MALT lymphoma: *H. pylori* and *H. heilmannii.* Both ocular and cutaneous MALT lymphomas have also been linked with infectious agents, *Chlamydia psittaci* and *Borrelia burgdorferi,* respectively, but the association is strongly dependent on the geographical area (Roggero et al. 2000; Ferreri et al. 2009). *Campylobacter jejuni* and more recently *Campylobacter coli* have also been detected in patients with small intestinal MALT lymphoma, known as immunoproliferative small intestinal disease (IPSID) (Lecuit et al. 2004; Coeuret et al. 2014). Other MALT lymphomas such as those of the salivary glands and often those of the ocular adnexa are more frequently linked with autogenic antigen stimulation and autoimmune disease (Zucca et al. 2003). It is in this context of chronic immune system activation and continuous lymphocyte proliferation that malignant lymphoid transformation ultimately occurs.

3.1 *First Step: Gastric Acquisition of MALT*

The stomach, unlike the rest of the gastrointestinal tract, lacks MALT under physiological conditions since low pH prevents the survival of lymphocytes in the gastric wall. Microbial agents that thrive in the gastric environment buffer local pH values trough secretion of bacterial urease. The combined presence of bacterial antigens and

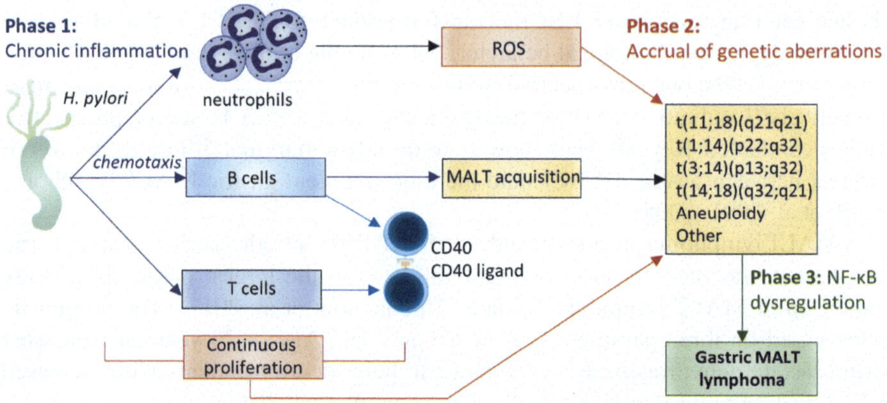

Fig. 2 Hypothetical model of gastric MALT lymphoma pathogenesis. Phase 1: *Helicobacter pylori (H. pylori)* infection attracts B-cells, T-cells, and neutrophils to the gastric mucosa. B-cell proliferation is driven by CD40–CD40 ligand interaction with *H. pylori*-activated, reactive T-cells, as well as by cytokines. Phase 2: The chronic proliferative state of these B-cells, as well as neutrophil–mediated release of reactive oxygen species (ROS) in areas of chronic inflammation, induces additional oncogenic events that eventually make lymphoproliferation independent of antigenic stimulation. Phase 3: The major recurring genetic aberrations all dysregulate NF-κB through various mechanisms

decreased acidity of the stomach triggers lymphoid infiltration and the acquisition of MALT (Zucca et al. 2003).

3.1.1 The Role of Antigen Stimulation

The presence of antigen-driven changes in MALT lymphoma DNA supports the role of antigen stimulation in MALT lymphomagenesis. These changes include somatic hypermutation and intraclonal variation. Sequence analysis of the immunoglobulin heavy chain locus (IGH) in DNA from MALT lymphomas confirmed the presence of somatic hypermutation and about 50% of the tumors also showed intraclonal variation in the IGH locus (Craig et al. 2010).

Next to the DSBs and translocations caused by the inherent genetic instability during somatic hypermutation, the continuous B-cell proliferation and consequentially DNA replication causes an accrual of transcription errors. The DNA mismatch repair (MMR) mechanism normally repairs these errors in physiological conditions. The presence of MMR-defects as illustrated by RT-PCR for 10 microsatellite loci (*BAT25, BAT26, D5S346, D17S250, D2S123, TGFB, BAT40, D18S58, D17S787* and *D18S69*) in 20% of gastric MALT lymphomas further supports the role of antigen stimulation in lymphomagenesis (Degroote et al. 2012).

The role of antigenic stimulation in the late stages of tumor progression is most likely of only marginal importance as illustrated by the reduction of the ongoing mutation rate. Accumulation of ROS-induced genetic anomalies over time resulting in more and more antigen stimulation-independent lymphoproliferation could explain this change.

3.1.2 *H. pylori* Infection

The infectious etiology of gastric MALT lymphoma is the best documented among all marginal zone lymphomas. A thick mucus layer and gastric acid create an environment that is hostile to most bacteria. *H. pylori* is one of the few pathogens, which can survive there and the association with gastric MALT lymphoma, gastric carcinoma, and peptic ulcer disease (PUD) is made clear through several factors (Parsonnet et al. 1994). Epidemiological studies revealed that in patients with MALT lymphoma the prevalence of *H. pylori* infection in gastric mucosa and *H. pylori* seropositivity are well above the population average with a recorded seropositivity in up to 98% of patients (Wotherspoon et al. 1991; Parsonnet et al. 1994; Eck et al. 1997). The incidence of gastric MALT lymphoma is furthermore highest in regions of endemic *H. pylori* infection (Doglioni et al. 1992). Moreover, *H. pylori* eradication therapy leads to complete lymphoma regression in about 80% of patients with early stage disease (Fischbach et al. 2004; Wündisch et al. 2005). Both in vitro studies in cell lines and in vivo studies in mouse models further support a causal role of *H. pylori* gastric MALT lymphomagenesis. First, *H. pylori* triggers T-cell mediated, clonal B-cell expansion by activating the CD40 (cluster of differentiation 40) pathway (Hussell et al. 1993b). Second, chronic *H. pylori* infection induces gastric MALT lymphoma in mice (O'Rourke 2008). Finally, specific bacterial proteins such as CagA (encoded by cytotoxin-associated gene A) with known carcinogenic effects in gastric carcinoma (Backert et al. 2008; Tegtmeyer et al. 2017) have been shown to translocate into B-cells in vitro (Lin et al. 2010) and can affect chronic inflammation through NF-κB activation and interleukin (IL)-8 upregulation (Backert and Naumann 2010; Kuo et al. 2013).

The causal link between *H. pylori* and gastric MALT lymphoma, however, still needs further investigation since not all *H. pylori*-infected patients develop MALT lymphoma and conversely 2–10% of patients with gastric MALT lymphoma are *H. pylori*-negative (Wotherspoon et al. 1991; Parsonnet et al. 1994; Eck et al. 1997). Notably, the antibodies produced by gastric *H. pylori*-positive gastric MALT lymphomas recognize various autoantigens instead of *H. pylori* (Hussell et al. 1993a). One possible explanation for this observation is that gastric MALT lymphoma arises from *H. pylori*-stimulated, but otherwise autoreactive B-cells.

The cause of the imperfect correlation between *H. pylori* infection and gastric MALT lymphoma could lie in the host genetic background, but one study investigating MALT lymphoma translocation protein 1 (*MALT1*) gene polymorphisms in a large cohort of patients failed to identify a risk factor (Hellmig et al. 2009). A second possible explanation could lie in the exact nature and efficacy of the immune response. Specific T-cell subsets such as T helper 1 and T helper 17 cells correlate with the development of gastritis and PUD. On the other hand, T helper 2 and regulatory T-cell responses modulate tissue damage but may lead to persistence of the pathogen and promote malignancy through persisting chronic inflammation (Jafarzadeh et al. 2018). The majority of this data was, however, generated in mouse models and validation in human disease is still needed (Watanabe et al. 2004; Kaparakis et al. 2006).

At least part of *H. pylori*-negative gastric MALT lymphomas are most likely caused by other, not routinely identified, pathogens such as *H. heilmannii*. This is supported by the fact that *H. pylori* eradication therapy can be effective in these cases (Morgner et al. 2000; Asano et al. 2012; Raderer et al. 2015). In other cases, *H. pylori* infection may have simply gone undiagnosed. For this reason, a negative *H. pylori* test in a patient with gastric MALT lymphoma should be routinely followed by a 13 C-urea breath test, serological antibody test, immunostaining of biopsy tissue samples, and/or stool culture. In those cases where a *H. pylori* test is performed on only the biopsy, sample interpretation of a negative result should be done with care since some gastric MALT lymphomas arise in atrophic mucosa where the *H. pylori* bacterial load is low.

3.2 Second Step: Acquisition of Genetic Abnormalities

Specific genetic abnormalities are associated with location and are considered to reflect differences in underlying pathogenesis. MALT lymphomas in distinct anatomical locations have exposure to different inflammatory agents, which could explain the subsequent genetic differences. Trisomy of chromosome 3, 7, 12, and 18, chromosomal translocation t(11;18)(q21;q21), t(1;14)(p22;q32), t(14;18)(q32;q21), and t(3;14)(p13;q32) are all well known to occur with variable frequencies in both gastric and non-gastric MALT lymphomas (Willis et al. 1999; Baens et al. 2000; Streubel et al. 2003; Wlodarska et al. 2005). Translocation detection is possible in both fresh frozen and paraffin-embedded tumor tissue by reverse transcription PCR and fluorescence in situ hybridization (FISH). One or both of these techniques are available in most laboratories as screening tests for various cancers. Aneuploidy can be identified by karyotyping or FISH (using special probes), but these two techniques are only available in more specialized laboratories.

3.2.1 t(11;18)(q21;q21)

The translocation t(11;18)(q21;q21) is the most frequent structural chromosomal abnormality in gastric MALT lymphomas. Different studies have reported the presence of this aberration in 10–50% of gastric MALT lymphomas, but it is infrequent in other non-gastric MALT lymphomas with the exception of pulmonary MALT lymphoma, and it is not found in other lymphoma types (Streubel et al. 2003; Ye et al. 2003; Baens et al. 2006). The presence of t(11;18)(q21;q21) in gastric MALT lymphomas is also strongly associated with infection by *cag*A-positive strains of *H. pylori* (Eck et al. 1997; Ye et al. 2003). The CagA protein can induce various cancer-associated signal transduction pathways and expression of IL-8 (Backert et al. 2017;

Naumann et al. 2017), a powerful chemokine involved in neutrophil activation and subsequent ROS secretion (Ye et al. 2003). A tempting hypothesis, therefore, is that the genetic abnormality t(11;18)(q21;q21) is a consequence of the increased oxidative stress associated with inflammatory responses, specifically in the MALT of organs frequently exposed to exogenous antigens such as the stomach and lung.

The translocation t(11;18)(q21;q21) fuses the amino (N)-terminus of the *BIRC3* gene (Baculoviral IAP repeat-containing 3; formerly termed *API2*, on chromosome 11q21) to the carboxyl (C)-terminus of the *MALT1* gene (located on chromosome 18q21), resulting in the fusion gene *BIRC3–MALT1* (Fig. 3) (Dierlamm et al. 1999). BIRC3 (Baculovirus inhibitor of apoptosis protein repeat-containing protein 3) is a member of the inhibitor of apoptosis family and inhibits the biological activity of certain caspases (Roy et al. 1997). The protein contains three Baculovirus inhibitor of apoptosis repeat (BIR) domains, a CARD (caspase recruitment domain) motif and one RING finger motif. MALT1 (MALT lymphoma translocation protein 1 gene; also known as paracaspase) is a key mediator of the antigen-receptor signaling pathway that leads to NF-κB activation (Uren et al. 2000). This protein contains an N-terminus death domain, two immunoglobulin-like domains, and a C-terminus caspase-like domain (Uren et al. 2000). The break points in both *BIRC3* and *MALT1* are well documented (Dierlamm et al. 1999; Baens et al. 2000; Kalla et al. 2000; Remstein et al. 2000; Motegi et al. 2000; Liu et al. 2001b; Ye et al. 2003). The location of the *MALT1* break point seems to be associated with specific gene expression profiles in t(11;18)(q21;q21)-positive gastric MALT lymphomas (Sagaert et al. 2010a), but all variants result a similar BIRC3–MALT1 fusion protein. BIRC3 is fused in-frame to MALT1 and the resulting protein always contains the three intact BIR domains of the N-terminus BIRC3 portion and the intact caspase-like domain of the C-terminus MALT1 portion. The presence of this structural aberration correlates with the absence of other genetic abnormalities (Starostik et al. 2002; Müller-Hermelink 2003), but more recent high-resolution genome studies revealed the existence of genetically distinct subclones within each lymphoma with a variable amount of other genetic aberrations (Flossbach et al. 2013). Remarkably, there was a correlation between cell size and genetic complexity with the least genetic aberrations in the smaller lymphoma cells (Flossbach et al. 2013). This further validates the finding that a t(11;18)(q21;q21) translocation does not exclude progression to DLBCL (Alpen et al. 2000; Sugiyama et al. 2001; Liu et al. 2001a; Ye et al. 2003; Toracchio et al. 2009).

Multiple studies have revealed that t(11;18)(q21;q21)-positive gastric MALT lymphomas are more often resistant to *H. pylori* eradication treatment than tumors lacking this translocation (Alpen et al. 2000; Sugiyama et al. 2001; Liu et al. 2001a; Ye et al. 2003). Nevertheless, complete lymphoma regression can still be obtained in 20% of patients with t(11;18)(q21;q21)-positive disease after *H. pylori* eradication. Consequently, although FISH for the t(11;18)(q21;q21) translocation can be used to identify patients who are unlikely to respond to *H. pylori* eradication therapy (Zucca et al. 2013), patients who have *H. pylori*-positive gastric MALT lymphoma should always undergo eradication therapy, regardless of their t(11;18) (q21;q21) status (Zullo et al. 2010). Given this, once the diagnosis of MALT lymphoma has

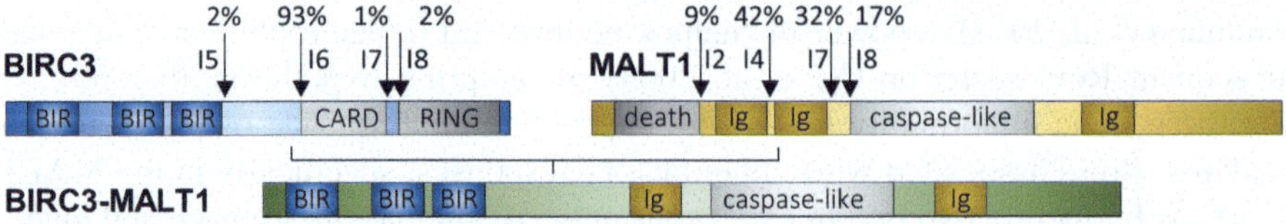

Fig. 3 **BIRC3-MALT1 fusion gene**. Known break points (arrows) in BIRC3 and MALT1 are shown with their respective frequencies. The break points within BIRC3 almost always occur in I6 (according to Ensembl Gene eNSG00000023445), whereas those within MALT1 are located in I2, I4, I7, and I8, which result in four possible versions of the BIRC3–MALT1 fusion gene: BIRC3(I6)–MALT1(I2), BIRC3(I6)–MALT1(I4), BIRC3(I6)–MALT1(I7) and BIRC3(I6)–MALT1(I8). The fusion gene depicted is the BIRC3(I6)–MALT1(I4) version. Abbreviations: BIR, Baculovirus inhibitor of apoptosis repeat; BIRC3, Baculoviral IAP Repeat-Containing 3; CARD, caspase recruitment domain; Death, death domain; I, intron; Ig, immunoglobulin-like; MALT1, MALT lymphoma translocation 1 gene; p20, caspase-like p20 domain; RING, really interesting new gene

been made, FISH for the t(11;18)(q21;q21) translocations is not mandatory (Dreyling et al. 2013), although it can be used in difficult cases to confirm the diagnosis.

3.2.2 t(14;18)(q32;q21)

The translocation t(14;18)(q32;q21), which generates the fusion gene *IGH–MALT1*, is detected in 2–20% of gastric MALT lymphomas (Streubel et al. 2003; Sagaert et al. 2006b). This translocation juxtaposes the *MALT1* gene located at 18q21 to the IGH transcriptional enhancer regions located at 14q32, which results in overexpression of the MALT1 protein. In contrast to t(11;18)(q21;q21)-positive tumors, however, t(14;18) (q32;q21)-positive MALT lymphomas mainly occur in non-gastric tissues and as such are discussed in excellent other review articles (Streubel et al. 2003; Remstein et al. 2004).

3.2.3 t(1;14)(p22;q32) and t(2;3)(p22;p12)

One of either t(1;14)(p22;q32) or its variant t(1;2)(p22;p12) is present in approximately 5% of gastric MALT lymphomas (Thome 2004; Sagaert et al. 2006b). These translocations are again only rarely reported in other extranodal MALT lymphomas, excluding those of the lung (9%) or other lymphoma subtypes (Thome 2004; Sagaert et al. 2006b). These two translocations are grouped together here because they bring the same gene, *BCL10* (B-cell lymphoma 10), in the neighborhood of the IGH gene on chromosome 14 or the IGK gene on chromosome 2, which encode immunoglobulin heavy chains and light chains, respectively. This brings the *BCL10* gene (located at chromosome 1p22) under the control of either the IGH transcriptional enhancer located at 14q32 or the IGK immunoglobulin light-chain κ gene enhancer at 2p12, both of which result in overexpression of BCL10 (Ye et al. 2000). This gene encodes a CARD-containing protein that has a key role in antigen-receptor signaling and NF-κB regulation (Thome 2004; Turvey et al. 2014). Patients with t(1;14)(p22;q32) or t(1;2)(p22;p12) tend to present with more advanced stages of gastric MALT lymphoma and their tumors typically have multiple additional genetic aberrations, such as other structural chromosomal anomalies or alterations in chromosomal number (Willis et al. 1999; Achuthan et al. 2000).

3.2.4 t(3;14)(p12;q32)

The translocation t(3;14)(p13;q32) has a low reported frequency in gastric MALT lymphoma (not more than 10%) and DLBCL of which the majority is extranodal (Streubel et al. 2005; Wlodarska et al. 2005; Sagaert et al. 2006a; Haralambieva et al. 2006). Again the *IGH* enhancer regions on chromosome 14 cause overexpression of a translocated gene, in this case *FOXP1* (forkhead box P1, located at chromosome 3p13). Interestingly, up to 10% of t(3;14)(p13;q32)-negative MALT lymphomas have a strong nuclear expression of FOXP1 (Wlodarska et al. 2005). This suggests other genetic aberrations, such as a trisomy of chromosome 3, which might lead to nuclear FoxP1 overexpression as well. During recent years, multiple studies have revealed more about the potential oncogenic functions of FoxP1 (Banham et al. 2001; van Keimpema et al. 2014; van Keimpema et al. 2015). The *FOXP1* gene encodes a member of the FOX family of transcription factors, which are characterized by a common DNA-binding winged helix or forkhead domain (Banham et al. 2001). FOXP1 has been shown to inhibit pro-apoptotic genes such as Bcl2 homology 3 (BH3)-only genes (*BIK* and *Harakiri*), p53-regulatory genes (*TP63, RASSF6*, and *TP53INP1*, *AIM2* and *EAF2*) as well as transcriptional factors of plasma cell differentiation (*PRDM1, IRF4, XBP1*) (van Keimpema et al. 2014, 2015). Importantly, FOXP1 complements NF-κB activity by further enhancing B-cell survival during constitutive NF-κB activity (van Keimpema et al. 2014). The prognostic implications of nuclear *FOXP1* overexpression in lymphoma remain contested until this day. Previous studies linked *FOXP1* overexpression with reduced survival in patients with (nodal) DLBCL (Barrans et al. 2004; Banham et al. 2005), although another study could not confirm this (Hans et al. 2004). In recent years, one study found an association between FOXP1 positivity and improved survival in extranodal DLBCL of the non-germinal center subtype (non-GCB), but not on the total cohort suggesting the existence of not only a difference in FOXP1 impact between nodal and extranodal DLBCL, but also non-GCB and GCB DLBCL (Petrakis et al. 2017). Expression of *FOXP1* was linked to more adverse overall survival according to one study which looked specifically at gastric MALT lymphoma (27 cases) (Zhai et al. 2011). Another retrospective study of *FOXP1* expression in 115 MALT lymphomas found no impact on survival in the general cohort, but a potential negative influence on prognosis in thyroid cases. The FOXP1-positive group in this cohort had more plasmacytic differentiation and a higher Ki-67 proliferation index. *FOXP1* expression was also higher in MALT lymphoma cases with a large cell component. This is concordant with two previous studies where strong nuclear expression of FOXP1 in gastric MALT lymphoma was only present in a subgroup of lymphomas at risk of transforming into DLBCL (Sagaert et al. 2006a; Han et al. 2009).

3.2.5 Other Translocations

The above-described translocations were all identified in the 1990s using cytogenetic and FISH analyses. More recent described translocation in MALT lymphoma are t(5;14)(q34;q32)/*ODZ2-IGH*, t(9;14)(p24;q32)/*JMJD2C-IGH*, t(1;14)(p21;q32)/*CNN3-IGH* and t(X;14)(p11.4;q32)/*GPR34-IGH*, all of which bring specific genes under influence of the IGH enhancer region (Vinatzer et al. 2008; Baens et al. 2012; Ansell et al. 2012). Due to the low reported incidence of each translocation and absence of these translocations in the few included gastric MALT lymphoma cases, they will not be discussed further in this review.

3.2.6 Other Genetic Aberrations

Over the past decades, a wide array of other genetic and epigenetic aberrations has been reported. First, chromosomal aneuploidies are frequently observed in gastric MALT lymphomas. Returning aneuploidies are trisomy of chromosomes 3, 7, 12, and 18, which all have been reported in t(11;18)(q21;q21)-negative gastric MALT lymphomas. Trisomy of chromosome 3 and 18 are the most frequently observed aneuploidies with an occurrence in one study of 15% and 30%, respectively (Kwee et al. 2011). The exact biological effects of the aneuploidies remain unknown, although both chromosome 3 and 18 each contain an oncogenic gene, also involved in the above-mentioned translocations. These are FOXP1 and MALT1 on chromosomes 3 and 18, respectively.

Next to chromosomal aneuploidies, chromosomal gains and losses are again mainly observed in t(11;18)(q21;q21)-negative gastric MALT lymphomas and not in those harboring the translocation (Barth et al. 2001; Starostik et al. 2002; Zhou et al. 2006; Ferreira et al. 2008). Remarkably, gastric MALT lymphomas have a higher frequency of 8q gains compromising MYC, a proto-oncogene involved in multiple lymphomas such as Burkitt lymphoma, compared to MALT lymphomas of other anatomical sites (Kwee et al. 2011). Chromosomal imbalances (specifically, regions of recurrent gain on chromosomes 3, 1p36.2 and 18q, and regions of recurrent loss on 1p36.3 and 7q31–q3) in t(11;18)(q21;q21)-negative gastric MALT lymphomas may also be linked with non-responsiveness to *H. pylori* eradication therapy (Fukuhara et al. 2007). The observation that the highest amount of aberrations are observed in gastrointestinal MALT lymphomas with both a small and large cell component points toward an accumulation of these errors over time and a possible link with progression to large B-cell lymphoma (Flossbach et al. 2013).

Finally, epigenetic alterations such as DNA methylation are another mechanism that can influence gene expression and have garnered increased interest over the recent years (Arribas and Bertoni 2017). Hypermethylation of CpG islands within gene promoter regions, which lead to silencing of related gene expression is present in gastric MALT lymphoma. Promoter hypermethylation-mediated silencing is reported for *CDKN2A*, *DAPK1*, *APBA1*, *APBA2*, and *MINT31* (a "methylated in tumors" locus mapped to chromosome 17q21) genes (Sinn et al. 2009; Kondo et al. 2009). *INK4A* and *KIP2* (inhibitors of cyclin-dependent kinases) promoter hypermethylation were also implied in the development of gastric MALT lymphoma and accumulation of CpG hypermethylation has been associated with *H. pylori* infection and lymphoma progression (Min et al. 2006; Kondo et al. 2009). Another epigenetic mechanism is gene expression regulation through microRNAs (miRs). miRs are noncoding RNAs that can bind target messenger RNA molecules and induce RNA degradation, which can lead to posttranscriptional gene silencing (Belair et al. 2009). Deregulation of multiple miRs has been implied in the transformation of gastritis to malignancy (Thorns et al. 2012), but only miR-150 upregulation was validated in a follow-up study (Gebauer et al. 2014) and was also observed in primary conjunctival MALT (Cai et al. 2012). MiR-150 targets the transcription factor c-Myb and so influences B-cell differentiation, but its exact role in oncogenesis remains to be investigated (Xiao et al. 2007). Remarkably, miR-150 has been implied as a tumor suppressor, downregulated in DLBCL (Roehle et al. 2008). Increased expression of miR-155 was already observed in other B-cell malignancies (Jiang et al. 2006) and recently found to be upregulated in gastric MALT lymphoma compared to chronic gastritis (Fernández et al. 2017). MiR-155 targets the pro-apoptotic gene *TP53INP1* and could be a factor in lymphoma persistence after *H. pylori* eradication (Saito et al. 2012).

3.3 Third Step: Dysregulation of NF-κB Signaling

The NF-κB signaling pathway is involved in several critical physiological functions including cell proliferation, apoptosis, angiogenesis, and inflammation (Yu et al. 2017). It was first described more than 30 years ago by Sen and Baltimore (1986) as a nuclear factor binding near the κ light-chain gene in B-cells or NF-κB. To date, it is known that NF-κB is not a single transcription factor but rather a family of five-related proteins that share a N-terminal REL homology domain: RelA (p65), RelB, c-Rel, NF-κB 1 (p105/p50) and NF-κB2 (p100/p52). The homology domain is required for dimerization, sequence-specific DNA binding, and inhibitory protein binding. Two independent pathways control NF-κB signaling the canonical pathway and non-canonical pathway (Li and Verma 2002; Pomerantz and Baltimore 2002).

3.3.1 The Canonical Pathway

A broad range of receptors such as the B-cell receptors (BCRs), Toll-like receptors (TLRs), nucleotide oligomerization domain-like receptors (NODs), and TNF family receptors (TNFRs) can activate the canonical pathway. Potential triggers are, therefore, pathogen-associated molecular patterns (PAMPs) from bacteria and viruses, and pro-inflammatory cytokines such as TNF and IL-1. All upstream signaling converges at the IκB kinase (IKK) complex, in unstimulated B-cells, iκB proteins are bound to RelA and p50 to form latent complexes that remain in the cytoplasm. Activation of the canonical pathway induces polyubiquitinylation and activation of "NF-κB essential modifier" (NEMO), which results in the phosphorylation and subsequent proteasomal degradation of iκB by the iKK catalytic subunit iKKα. Because of iκB degradation, p50/RelA and p50/c-Rel heterodimers and p50/p50 homodimers can form and translocate to the nucleus, where they influence gene expression. Upregulated proteins include various cytokines, chemokines, adhesion molecules and anti-apoptotic proteins (Bcl2, Bcl2-related protein A1, Bcl2-like protein 1) as well as proliferation-promoting proteins, such as cyclin-D2. In physiological conditions, canonical signaling is transient due to negative auto-regulators such as IκBα, IκBε, and TNFAIP3/A20 induced by the canonical pathway itself (Bonizzi and Karin 2004; Krappmann and Vincendeau 2016).

3.3.2 The Non-canonical Pathway

A more restricted set of triggers activates the non-canonical pathway, specifically a subset of TNF superfamily members such as B-cell activating factor (BAFF; also known as 13B or BLyS), lymphotoxin-β (LTβ), and CD40 ligand. Non-canonical stimuli lead to the destabilization of TRAF2/3-cIAP E3 ligase complexes, which in unstimulated cells, induce degradation of protein kinase NF-κB inducing kinase (NIK). The stabilized protein kinase NIK phosphorylates and activates IKKα, which in turn phosphorylates the C-terminal serine residues of the precursor protein p100. Ubiquitin-proteasome system-mediated degradation of the C-terminal half of p100 creates p52 and ultimately leads to nuclear translocation of RelB/p52 heterodimers (Krappmann and Vincendeau 2016). Non-canonical signaling is, in physiological conditions, more chronic than canonical pathway activation and has been associated with more long-term effects such as lymphoid organogenesis next to distinct regulatory functions in adaptive immunity, including upregulation of CXCL13 (CXC-motif chemokine ligand 13, formerly known as B lymphocyte chemoattractant) and TNFSF13B (Bonizzi and Karin 2004; Krappmann and Vincendeau 2016).

Many genetic aberrations in MALT lymphomas activate NF-κB signaling through the canonical pathway and this has been studied most intensively (Hussell et al. 1993b). The non-canonical NF-κB signaling pathway is, however, also involved in gastric MALT lymphoma pathogenesis as discussed later. Furthermore, in vitro experiments have shown that *H. pylori* directly activated the non-canonical NF-κB pathway in B lymphocytes (Ohmae et al. 2005). However, more than 16 different and often contradictory signaling cascades have been implicated in *H. pylori*-induced pro-inflammatory signaling, making interpretation difficult (Backert and Naumann 2010).

3.3.3 The CARD11-BCL10-MALT1 (CBM) Signalosome Complex

Both MALT1 and BCL10 are upregulated as a consequence of chromosomal translocations with relatively high frequency in gastric MALT lymphoma, namely t(11;18)(q21;q21) for MALT1 and either t(1;14)(p22;q32) or t(2;3)(p22;p12) for BCL10. Both MALT1 and BCL10 are essential components in the activation of the canonical NF-κB pathway by forming a signalosome complex together with a third "caspase activation and recruitment domain" (CARD) and "membrane-associated guanylate kinase-like" (MAGUK) domain-containing protein (CARMA), known as the CARMA-BCL10-MALT1 (CBM) signalosome complex (Fig. 4) (Ruland et al. 2001, 2003; Ruefli-Brasse 2003). This is illustrated by in vivo data from mouse models with targeted BCL10, MALT1, and CARD11 (=CARMA1) disruptions showing that all three proteins are critical for NF-κB activation after B- and T-cell receptor activation (Rosebeck et al. 2011b; Turvey et al. 2014).

Initial activation can occur in B-cells, for instance, through stimulation of the B-cell receptor (BCR). Antigen binding to the BCR initiates a tyrosine phosphorylation signaling cascade culminating in activation of PKC (protein kinase C) isoforms (Rawlings et al. 2006). PKCβ in turn phosphorylates CARD11, which undergoes a structural change with exposure of its CARD motifs. BCL10 and CARD11 then interact through their respective N-terminal CARD domains (Qiao et al. 2013). Interaction between BCL10 and MALT1 happens through the Ser/Thr-rich C-terminal portion of BCL10 and the immunoglobulin-like domains of MALT1 (Qiao et al. 2013). MALT1-BCL10 heterodimers exist in the absence BCR stimulation and CARD11

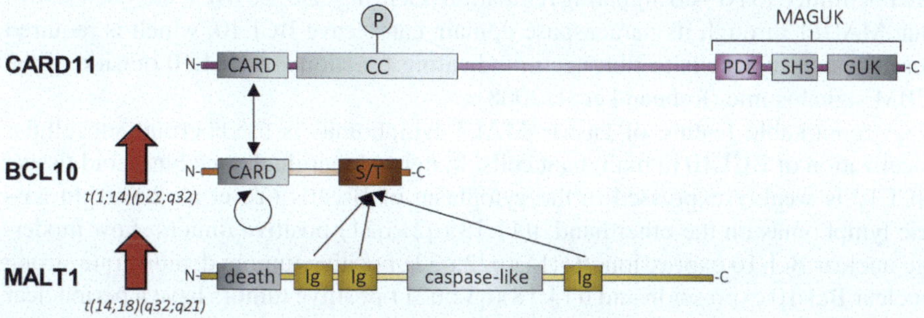

Fig. 4 CARD11-BCL10-MALT1 interaction. The CARD11-BCL10-MALT1 (CBM) signalosome complex forms after upstream signaling lead to phosphorylation of the CARD11 CC-domain resulting in exposure of its CARD motif and CARD–CARD interaction with BCL10. BCL10 also constitutively interacts with MALT1 via its Ser/Thr (S/T)-rich C-terminus. There are no direct interactions between CARD11 and MALT1. BCL10 and MALT1 protein concentration is increased in gastric MALT lymphoma by t(1;14)(p22;q32) and t(14;18)(q32;q21), respectively. Abbreviations: Bcl–10, B–cell lymphoma 10; CARD, caspase activation and recruitment domain; CARD11, caspase recruitment domain family member 11; CC, coiled-coiled domain; death: death domain; GUK, Guanylate kinase; Ig: immunoglobulin-like; MAGUK, membrane-associated guanylate kinase; MALT1, MALT lymphoma translocation protein 1; SH3, SRC Homology 3 domain; S/T: serine/threonine

(Thome et al. 2010). The resulting CARD11–Bcl-10–MALT1 complex interacts with and induces oligomerization of TRAF6 (TNF-receptor associated factor 6) via the C-terminus of MALT1 (Sun et al. 2004). This enables TRAF6 to polyubiquitinylate various target proteins including NEMO and TRAF6 itself on Lys63. This Lys63-linked polyubiquitinylation of TRAF6 is made possible by the E3 ubiquitin ligase activity of the RING E3 domain in TRAF6 and is distinct from Lys-48 linked polyubiquitinylation, which leads to proteasomal degradation. The Lys-63 linked polyubiquitinylation of TRAF6 enables interaction of NEMO and TRAF6 with TAK1 (transforming growth factor β activating kinase). The interaction with TAK1 in turn activates this protein, which subsequently fully activates the IKK complex by phosphorylation of its β subunit. Phosphorylation of IKK then leads to IκB phosphorylation and nuclear translocation of p65-p50 dimers.

MALT1 was for a long time considered to be mainly a scaffolding protein, but discovery and further investigation of its caspase activity by caspase-like or "paracaspase" domain revealed that it can activate NF-κB signaling in an NF-κB-independent manner through cleavage of various substrates including TNFAIP3 (A20) and BCL10 (Afonina et al. 2015). This led to an increased interest for MALT1 inhibitors as a therapeutic tool. However, the development of multi-organ inflammation in knock-out mice and the existence of a link between MALT1 deficiency and combined immunodeficiency disorder (CID) in humans have dramatically tempered initial expectations for this therapy (McKinnon et al. 2014; Bornancin et al. 2015; Demeyer et al. 2016).

BCL10 has been discovered to be a critical regulator of both canonical NF-Kb signaling activation and termination through extensive post-translational regulation involving phosphorylation and ubiquitination. Nonetheless, the functions of BCL10 are not limited to NF-κB signaling regulation (Gehring et al. 2018). It has been shown that MALT1 through its paracaspase domain can cleave BCL10, which is required for adhesion of T-cells to fibronectin indicating functions of BCL10 outside of the CBM signalosome (Rebeaud et al. 2008).

A remarkable feature of gastric MALT lymphomas is the aberrant subcellular localization of BCL10 in malignant cells. In normal marginal zone lymphoid tissue, BCL10 is weakly expressed in the cytoplasm of B-cells (Ye et al. 2000). In gastric lymphomas on the other hand, t(11;18)(q21;q21)-positive tumors show moderate nuclear Bcl-10 expression, t(1;14)(p22;q32)-positive tumors demonstrate strong nuclear Bcl-10 expression and t(14;18)(q32;q21)-positive tumors have a perinuclear staining pattern of BCL10 (Ye et al. 2000, 2005; Sagaert et al. 2010a). The different staining patterns can be explained by the balance between BCL10 and MALT1, which regulates the transport of BCL10 between cytoplasm and nucleus (Nakagawa et al. 2005). In the presence of t(1;14)(p22;q32), there is overexpression of BCL10 resulting in nuclear retention due to a relative shortage of MALT1. In t(14;18)(q32;q21)-positive MALT lymphomas, there is overexpression of MALT1 and BCL10 is mainly cytoplasmic because all nuclear Bcl-10 is exported. A relative shortage of MALT1, due to the loss of one allele in t(11;18)(q21;q21)-positive MALT lymphomas, might be responsible for the nuclear retention of Bcl-10 as the associated BIRC3–MALT1 fusion protein is unable to export Bcl-10. In a diagnostic setting, it is important to note that a nuclear Bcl-10 staining pattern can indirectly indicate the presence of t(1;14)(q32;q21) or t(11;18)(q21;q21).

3.3.4 Aberrant NF-κB Activation

The NF-κB pathway regulates key elements of normal lymphocyte function, including the transcription of proliferation-promoting and anti-apoptotic genes. Aberrant, constitutive, activation of NF-κB signaling can lead to autoimmunity and malignancy. In gastric MALT lymphoma both chronic inflammation, often related to *H. pylori*, and numerous genetic aberrations are linked with dysregulation of NF-κB signaling (Rosebeck et al. 2011b). The phenotypes associated with t(11;18)(q12;q21), t(1;14) (p22;q32), and t(14;18)(q32;q21) all cause aberrant NF-κB activation through *BIRC3–MALT1* expression, *BCL10* overexpression and *MALT1* overexpression, respectively (Table 1). The translocations (1;14)(p22;q32) and t(14;18)(q32;q21) fuse the *IGH* enhancer with *BCL10* and *MALT1*, respectively. Both proteins are key regulators of the canonical NF-κB activation and overexpression caused by the respective translocations induces canonical NF-κB signaling (Uren et al. 2000; Lucas et al. 2001).

BCL10 protein overexpression, the major consequence of the t(1;14)(p22;q32) translocation, leads to auto-oligomerization through N-terminal CARD domain interactions. The oligomerization subsequently causes constitutive BCL10 activation and increased NF-κB activity without a need for upstream signaling activation (Lucas et al. 2001; Du 2016). Next to canonical NF-κB activation, recent data also provided evidence for a role of BCL10 in the regulation of the non-canonical pathway. In vitro experiments with BCL10 deficient B-cells showed a reduced expression of NF-κB2 (p100) and a reduced nuclear accumulation of the p52/RelB complex after BAFF stimulation (Yu et al. 2012).

Table 1 Frequent translocations in gastric MALT lymphomas

Translocation	Fusion transcript	BCL10 staining pattern	Example mechanism of NFκB (hyper)activation
t(11;18)(q21;q21)	BIRC3–MALT1	Moderately nuclear	NEMO polyubiquitinylation
t(14;18)(q32;q21)	IGH–MALT1	Perinuclear	MALT1 oligomerization (via BCL10) caspase activity
t(1;14)(p22;q32)	BCL10–IGH	Strongly nuclear	MALT1 oligomerization
t(3;14)(p13;q32)	FOXP1–IGH	Negative	Inhibition of pro-apoptotic genes

Abbreviations: Bcl10, B-cell lymphoma 10; NEMO, nuclear factor κB essential modulator; MALT1, MALT lymphoma translocation protein 1

MALT1 protein overexpression, the major consequence of the t(14;18)(q32;q21) translocation, and subsequent NF-κB activation through the canonical pathway is partially dependent on the presence of BCL10 since MALT1 does not contain any domains that allow it to oligomerize on its own (Toracchio et al. 2009). This hypothesis is supported by the immunohistochemical staining pattern of BCL10 in t(14;18)(q32;q21), where BCL10 is mainly co-localized with MALT1 in the cytoplasm. Nonetheless, MALT1, through its proteolytic activity, can also initiate NF-κB signalling independent from BCL10. It can cleave and consequentially inactivate several NF-κB regulators including TNFAIP3 (A20), CYLD, and RelB. These substrates are negative regulators of the NF-κB pathway and their inactivation may lead to NF-κB activation (Coornaert et al. 2008; Hailfinger et al. 2011; Staal et al. 2011; Kirchhofer and Vucic 2012; Lim et al. 2017). Just as BCL10, MALT1 may also play a role in non-canonical activation of NF-κB trough degradation of NF-κB2 (p100) as shown in vitro (Tusche et al. 2009). Remarkably, the effect was observed only in B-cells of the marginal zone and not follicular B-cells (Tusche et al. 2009).

The fusion protein BIRC3–MALT1, created by the t(11;18)(q21;q21) translocation is thought to activate NF-κB via both canonical and non-canonical NF-κB pathways. To start, BIRC3 can directly increase canonical NF-κB signaling trough NEMO polyubiquitinylation as has been demonstrated in vitro and in vivo in BIRC3–MALT1 transgenic mice (Zhou et al. 2005; Baens et al. 2006). This process is dependent on the first BIR [baculovirus IAP (inhibitor of apoptosis) repeat] domain of BIRC3 (Zhou et al. 2005; Baens et al. 2006) and the MALT1 C-terminus (Baens et al. 2006). In addition, the BIRC3-MALT1 fusion protein is capable of auto-oligomerization and hetero-oligomerization with proteins involved in downstream signaling (such TRAF2) through heterotypic interaction with again the first BIR domain of BIRC3 (Lucas et al. 2007). This BIR domain has also been tied to association of BIRC3–MALT1 with lipid rafts and subsequent binding of TRAF6 (Baens et al. 2006). All these mentioned oligomer bypass the normal process of antigen-induced oligomerization of MALT1–TRAF2/6 and thus can induce constitutive activation of the canonical NF-κB pathway (Baens et al. 2006; Dong et al. 2006; Lucas et al. 2007).

Just as MALT1, the BIRC3-MALT1 fusion protein has protease activity and can cleave negative regulators of NF-κB signaling such as TNFAIP3 (A20) and CYLD (Coornaert et al. 2008; Staal et al. 2011). Furthermore, non-canonical NF-κB signaling can be activated through recruitment and cleavage of NIK (NF-κB inducing kinase) (Rosebeck et al. 2011a). NIK cleavage requires concert activity from both elements of the BIRC3-MALT1 fusion protein. Specifically, BIRC3 can recruit NIK whereupon MALT1 cleaves it at Arg325. The created C-terminal NIK fragment retains its kinase activity, but is resistant to protein degradation resulting in constitutive NF-κB activation (Rosebeck et al. 2011a). Interestingly, another target of BIRC3-MALT1 cleaving is the tumor suppressor protein LIMA1 (LIM domain and actin-binding protein-1) which generates a novel oncogenic LIM domain-only (LMO) fragment (Nie et al. 2015). LMO expression promotes survival and proliferation of B-cells in vitro and lymphoma generation in a xenograft mice model (Nie et al. 2015). The exact molecular mechanism behind this is not yet fully clear, but this nevertheless suggests that there is also a NF-κB pathway-independent oncogenic effect of t(11;18)(q21;q21).

Finally, it is importantly to note that all the above-mentioned effects are most likely enhanced by a positive feedback loop. The BIRC3–MALT1, namely causes NF-κB pathway activation, which in turn upregulates BIRC3 expression as shown for t(11;18)(q21;q21)-positive MALT lymphomas (Hosokawa et al. 2005).

4 Treatment and Follow-up

First-line treatment for localized gastric MALT lymphoma (stage I–II) is *H. pylori* eradication with antibiotics. This may even be attempted for patients with localized *H. pylori*–negative gastric MALT lymphoma (Morgner et al. 2000; Asano et al. 2012; Dreyling et al. 2013; Raderer et al. 2015). Successful eradication should be verified by urea breath test (or a monoclonal stool antigen test) at least 6 weeks after eradication therapy and 2 weeks after PPI (proton pump inhibitor) discontinuation (Dreyling et al. 2013). For those patients who respond to *H. pylori* eradication therapy, it is important to remember that the time to complete remission can range from about 2 weeks up to more than 1 year. Serial endoscopic follow-up is recommended for a period of 2 years to measure histological regression. A first endoscopic examination is recommended 2–3 months after the cessation of antibiotic treatment, followed by endoscopy every 6 months for 2 years (Dreyling et al. 2013). Correct interpretation of lymphoid infiltrates after treatment is difficult and clinical decisions should result from careful deliberation between clinician and pathologist and scoring of histological specimens should be done according to GELA (Groupe d'Etude des Lymphomes de l'Adulte) scoring system (Copie-Bergman et al. 2003).

For those patients where antibiotic therapy fails to eradicate *H. pylori* or in patients with a localized, *H. pylori*-negative gastric MALT lymphoma, modest doses of involved field radiotherapy (25–40 Gy over 4 weeks) can be applied to the stomach and perigastric nodes with excellent results (Dreyling et al. 2013; Ruskoné-Fourmestraux et al. 2015; Ohkubo et al. 2017). Given the indolent nature of the disease, a conservative "watchful waiting" approach is suggested for patients who demonstrate persistence of histological gastric MALT lymphoma infiltrates despite successful H. pylori eradication and normalization of endoscopic findings (Fischbach et al. 2007; Stathis et al. 2009). These patients are considered to have minimal residual disease in the GELA scoring system (Copie-Bergman et al. 2003).

Disseminated disease (stage III–IV) and/or persisting disease 2 years after radiotherapy should be treated with sequential systemic therapies if symptomatic. Possibilities include anti-CD20 monoclonal antibody (rituximab) therapy only or in combination with chemotherapy. Only a few chemotherapy regimens have been tested specifically in MALT lymphomas. Oral alkylating agents (chlorambucil, cyclophosphamide, bendamustine) can result in a high rate of disease control. One non-randomized, observational trial found rituximab-chlorambucil to be more efficient than rituximab alone but only for patients t(11;18)(q21;q21)-positive gastric MALT lymphoma (Lévy et al. 2013). Purine analogues such as cladribin can be effective in the treatment of extranodal MALT lymphomas (Jäger et al. 2002). Other agents suggested in the literature include immunomodulatory agents and proteasome inhibitors. Lenalidomide, an immunomodulator and a derivative of thalidomide, could be beneficial, but further investigation is required (Kiesewetter et al. 2013). Bortezomib, a proteasome inhibitor is effective in inducing lymphoma regression, but further trials are needed to see if efficacy remains at lower doses, where the side effects such as peripheral neuropathy, hematological toxicity, and diarrhea are less problematic (Troch et al. 2009; Conconi et al. 2011). Surgery no longer has a role in the treatment of gastric MALT lymphomas except for those instances where local complications (such as gastric perforation) occur.

5 Concluding Remarks

Gastric MALT lymphoma pathogenesis is a multistep process initiated by infection with *H. pylori* followed by accumulation of genetic aberration ultimately leading to malignant transformation. In the event of a diagnosis with gastric MALT lymphoma, made by histological examination of an endoscopic mucosal biopsy, the necessary investigations to identify the presence of *H. pylori* need to be performed. For all *H. pylori*-positive patients who have localized disease, triple antibiotic therapy is the sole initial treatment, although the treating physician should bear in mind that the presence of t(11;18)(q21;q21) is associated with a reduced likelihood of successful lymphoma regression. Gene alterations associated with gastric MALT lymphoma, including BIRC3–MALT, IGH–BCL10 and IGH–MALT, all result in constitutive activation of the NF-κB pathway and together with the roles of *H. pylori* and associated chronic inflammation point toward the importance of inflammation in lymphomagenesis and to gastric MALT lymphoma as a model of chronic inflammation-induced gastric tumor development. Pharmaceutical interventions that target members of the NF-κB pathway still represent an attractive theoretical treatment strategy but given the vast influence of the NF-κB pathway in many different cellular processes precision is essential when blocking elements of the NF-κB pathway to prevent side effects. More insight in the exact nature and role of inflammation in gastric MALT lymphoma could help us predict which *H. pylori*-infected individuals have the highest risk for gastric MALT lymphoma development, better predict which lymphomas will respond to *H. pylori* eradication, or which might transform into DLBCL and find more specific therapeutic targets.

Conflicts of Interest The authors have no conflicts of interest to declare.
Financial Support
TT holds a Mandate for Fundamental and Translational Research from the "Stichting tegen Kanker" DNA dama2014-083).
LM is a Ph.D. student, financially supported by KULeuven, Department of Imaging and Pathology, "Stefanie's Rozen fonds", "Fonds Tom Debackere", Stichting Me2You' and "Emmanuel van der Schueren beurs (Kom op tegen Kanker)".

References

Achuthan R, Bell SM, Leek JP et al (2000) Novel translocation of the BCL10 gene in a case of mucosa associated lymphoid tissue lymphoma. Genes Chromosomes Cancer 29:347–349. https://doi.org/10.1002/1098-2264(2000)9999:9999%3c:AID-GCC1048%3e3.0.CO;2-B

Afonina IS, Elton L, Carpentier I, Beyaert R (2015) MALT1—a universal soldier: multiple strategies to ensure NF-κB activation and target gene expression. FEBS J 282:3286–3297. https://doi.org/10.1111/febs.13325

Alpen B, Neubauer A, Dierlamm J et al (2000) Translocation t(11;18) absent in early gastric marginal zone B-cell lymphoma of MALT type responding to eradication of Helicobacter pylori infection. Blood 95:4014–4015. https://doi.org/10.1182/blood.V98.1.250

Ansell SM, Akasaka T, McPhail E et al (2012) t(X;14)(p 11;q32) in MALT lymphoma involving GPR34 reveals a role for GPR34 in tumor cell growth. Blood 120:3949–3957. https://doi.org/10.1182/blood-2011-11-389908

Arribas AJ, Bertoni F (2017) Methylation patterns in marginal zone lymphoma. Best Pract Res Clin Haematol 30:24–31. https://doi.org/10.1016/j.beha.2016.09.003

Asano N, Iijima K, Terai S et al (2012) Eradication therapy is effective for *Helicobacter pylori*-negative gastric mucosa-associated lymphoid tissue lymphoma. Tohoku J Exp Med 228:223–227. https://doi.org/10.1620/tjem.228.223.Correspondence

Backert S, Feller SM, Wessler S (2008) Emerging roles of Abl family tyrosine kinases in microbial pathogenesis. Trends Biochem Sci 33(2):80–90. https://doi.org/10.1016/j.tibs.2007.10.006

Backert S, Naumann M (2010) What a disorder: proinflammatory signaling pathways induced by *Helicobacter pylori*. Trends Microbiol 18:479–486. https://doi.org/10.1016/j.tim.2010.08.003

Backert S, Haas R, Gerhard M, Naumann M (2017) The *Helicobacter pylori* type IV secretion system encoded by the *cag* Pathogenicity Island: architecture, function, and signaling. Curr Top Microbiol Immunol 413:187–220. https://doi.org/10.1007/978-3-319-75241-9_8

Baens M, Ferreiro JF, Tousseyn T et al (2012) t(X;14)(p 11.4;q32.33) is recurrent in marginal zone lymphoma and up-regulates GPR34. Haematologica 97:184–188. https://doi.org/10.3324/haematol.2011.052639

Baens M, Fevery S, Sagaert X et al (2006) Selective expansion of marginal zone B cells in Emicro-API2-MALT1 mice is linked to enhanced IkappaB kinase gamma polyubiquitination. Cancer Res 66:5270–5277. https://doi.org/10.1158/0008-5472.CAN-05-4590

Baens M, Maes B, Steyls A et al (2000) The product of the t(11;18), an API2-MLT fusion, marks nearly half of gastric MALT type lymphomas without large cell proliferation. Am J Pathol 156:1433–1439. https://doi.org/10.1016/S0002-9440(10)65012-2

Banham AH, Beasley N, Campo E et al (2001) The FOXP1 winged helix transcription factor is a novel candidate tumor suppressor gene on chromosome 3p. Cancer Res 61:8820–8829. https://doi.org/10.1038/297365a0

Banham AH, Connors JM, Brown PJ et al (2005) Expression of the FOXP1 transcription factor is strongly associated with inferior survival in patients with diffuse large B-cell lymphoma. Clin Cancer Res 11:1065–1072

Barrans SL, Fenton JAL, Banham A et al (2004) Strong expression of FOXP1 identifies a distinct subset of diffuse large B-cell lymphoma (DLBCL) patients with poor outcome. Blood 104:2933–2935. https://doi.org/10.1182/blood-2004-03-1209

Barth TFE, Bentz M, Leithuser F et al (2001) Molecular-cytogenetic comparison of mucosa-associated marginal zone B-cell lymphoma and large B-cell lymphoma arising in the gastro-intestinal tract. Genes Chromosom Cancer 31:316–325. https://doi.org/10.1002/gcc.1150

Belair C, Belair C, Darfeuille F et al (2009) Helicobacter pylori and gastric cancer: possible role of microRNAs in this intimate relationship. Clin Microbiol Infect 15:806–812. https://doi.org/10.1111/j.1469-0691.2009.02960.x

Bonizzi G, Karin M (2004) The two NF-κB activation pathways and their role in innate and adaptive immunity. Trends Immunol 25:280–288. https://doi.org/10.1016/j.it.2004.03.008

Bornancin F, Renner F, Touil R et al (2015) Deficiency of MALT1 paracaspase activity results in unbalanced regulatory and effector T and B cell responses leading to multiorgan inflammation. J Immunol 194:3723–3734. https://doi.org/10.4049/jimmunol.1402254

Brandt S, Kwok T, Hartig R, König W, Backert S (2005) NF-B activation and potentiation of proinflammatory responses by the *Helicobacter pylori* CagA protein. Proc Natl Acad Sci 102:9300–9305. https://doi.org/10.1073/pnas.0409873102

Cai J, Liu X, Cheng J et al (2012) MicroRNA-200 is commonly repressed in conjunctival MALT lymphoma, and targets cyclin E2. Graefes Arch Clin Exp Ophthalmol 250:523–531. https://doi.org/10.1007/s00417-011-1885-4

Coeuret S, de La Blanchardière A, Saguet-Rysanek V et al (2014) *Campylobacter coli* cultured from the stools of a patient with immunoproliferative small intestinal disease. Clin Microbiol Infect 20:908–911. https://doi.org/10.1111/1469-0691.12545

Cogliatti SB, Schmid U, Schumacher U et al (1991) Primary B-cell gastric lymphoma: a clinicopathological study of 145 patients. Gastroenterology 101:1159–1170. https://doi.org/10.1016/0016-5085(91)90063-Q

Conconi A, Martinelli G, Lopez-Guillermo A et al (2011) Clinical activity of bortezomib in relapsed/refractory MALT lymphomas: results of a phase II study of the international extranodal lymphoma study group (IELSG). Ann Oncol 22:689–695. https://doi.org/10.1093/annonc/mdq416

Cook JR, Müller-Hermelink HK, Isaacson PG et al (2017) Extranodal marginal zone lymphoma of lymphoma of- mucosa-associated lymphoid tissue (MALT) lymphoma. In: Swerdlow SH, Campo E, Harris NL et al (eds) WHO classification of tumours of haematopoietic and lymphoid tissues, Revised 4t. IARC, Lyon, pp 259–262

Coornaert B, Baens M, Heyninck K et al (2008) T cell antigen receptor stimulation induces MALT1 paracaspase—mediated cleavage of the NF-κB inhibitor A20. Nat Immunol 9:263–271. https://doi.org/10.1038/ni1561

Copie-Bergman C, Gaulard P, Lavergne-Slove A et al (2003) Proposal for a new histological grading system for post-treatment evaluation of gastric MALT lymphoma. Gut 52:1656. https://doi.org/10.1136/gut.52.11.1655

Coussens LM, Werb Z (2009) Inflammation and cancer. Nature 420:860–867. https://doi.org/10.1038/nature01322

Craig VJ, Arnold I, Gerke C et al (2010) Gastric MALT lymphoma B cells express polyreactive, somatically mutated immunoglobulins. Blood 115:581–591. https://doi.org/10.1182/blood-2009-06-228015

De Re V, De Vita S, Marzotto A et al (2000) Sequence analysis of the immunoglobulin antigen receptor of hepatitis C virus-associated non-Hodgkin lymphomas suggests that the malignant cells are derived from the rheumatoid factor-producing cells that occur mainly in type II cryoglobulinemia. Blood 96:3578–3584

Defrancesco I, Arcaini L (2018) Overview on the management of non-gastric MALT lymphomas. Best Pract Res Clin Haematol 31:57–64. https://doi.org/10.1016/j.beha.2017.11.001

Degroote A, Knippenberg L, Vander Borght S et al (2012) Analysis of microsatellite instability in gastric mucosa-associated lymphoid tissue lymphoma. Leuk Lymphoma 8194:1–7. https://doi.org/10.3109/10428194.2012.723211

Demeyer A, Staal J, Beyaert R (2016) Targeting MALT1 proteolytic activity in immunity, inflammation and disease: good or bad? Trends Mol Med 22:135–150. https://doi.org/10.1016/j.molmed.2015.12.004

Dierlamm J, Baens M, Wlodarska I et al (1999) The apoptosis inhibitor gene API2 and a novel 18q gene, MLT, are recurrently rearranged in the t(11;18)(q21;q21) associated with mucosa-associated lymphoid tissue lymphomas. Blood 93:3601–3609

Doglioni C, Wotherspoon AC, Moschini A et al (1992) High incidence of primary gastric lymphoma in northeastern Italy. Lancet (London, England) 339:834–835

Dong W, Liu Y, Peng J et al (2006) The IRAK-1-BCL10-MALT1-TRAF6-TAK1 cascade mediates signaling to NF-κB from toll-like receptor 4. J Biol Chem 281:26029–26040. https://doi.org/10.1074/jbc.M513057200

Dreyling M, Thieblemont C, Gallamini A et al (2013) Esmo consensus conferences: guidelines on malignant lymphoma. Part 2: marginal zone lymphoma, mantle cell lymphoma, peripheral T-cell lymphoma. Ann Oncol 24:857–877. https://doi.org/10.1093/annonc/mds643

Du M (2016) MALT lymphoma: a paradigm of NF-κB dysregulation. Semin Cancer Biol 39:49–60. https://doi.org/10.1016/j.semcancer.2016.07.003

Eck M, Schmausser B, Haas R et al (1997) MALT-type lymphoma of the stomach is associated with Helicobacter pylori strains expressing the CagA protein. Gastroenterology 112:1482–1486. https://doi.org/10.1016/S0016-5085(97)70028-3

Fernández C, Bellosillo B, Ferraro M et al (2017) MicroRNAs 142–3p, miR-155 and miR-203 are deregulated in gastric MALT lymphomas compared to chronic gastritis. Cancer Genomics Proteomics 14:75–82. https://doi.org/10.21873/cgp.20020

Ferreira BI, García JF, Suela J et al (2008) Comparative genome profiling across subtypes of low-grade B-cell lymphoma identifies type-specific and common aberrations that target genes with a role in B-cell neoplasia. Haematologica 93:670–679. https://doi.org/10.3324/haematol.12221

Ferreri AJM, Dolcetti R, Magnino S et al (2009) Chlamydial infection: The link with ocular adnexal lymphomas. Nat Rev Clin Oncol 6:658–669. https://doi.org/10.1038/nrclinonc.2009.147

Fischbach W, Dragosics B, Kolve-Goebeler ME et al (2000) Primary gastric B-cell lymphoma: results of a prospective multicenter study. Gastroenterology 119:1191–1202. https://doi.org/10.1053/gast.2000.19579

Fischbach W, Goebeler-Kolve M-E, Dragosics B et al (2004) Long term outcome of patients with gastric marginal zone B cell lymphoma of mucosa associated lymphoid tissue (MALT) following exclusive Helicobacter pylori eradication therapy: experience from a large prospective series. Gut 53:34–37. https://doi.org/10.1136/gut.53.1.34

Fischbach W, Goebeler ME, Ruskone-Fourmestraux A et al (2007) Most patients with minimal histological residuals of gastric MALT lymphoma after successful eradication of *Helicobacter pylori* can be managed safely by a watch and wait strategy: experience from a large international series. Gut 56:1685–1687. https://doi.org/10.1136/gut.2006.096420

Flossbach L, Holzmann K, Mattfeldt T et al (2013) High-resolution genomic profiling reveals clonal evolution and competition in gastrointestinal marginal zone B-cell lymphoma and its large cell variant. Int J Cancer 132:E117–127. https://doi.org/10.1002/ijc.27774

Fukuhara N, Nakamura T, Nakagawa M et al (2007) Chromosomal imbalances are associated with outcome of *Helicobacter pylori* eradication in t(11;18)(q21;q21) negative gastric mucosa-associated lymphoid tissue lymphomas. Genes Chromosomes Cancer 46:784–790. https://doi.org/10.1002/gcc.20464

Gebauer N, Kuba J, Senft A et al (2014) MicroRNA-150 Is up-regulated in extranodal marginal zone lymphoma of MALT type. Cancer Genomics Proteomics 11:51–56

Gehring T, Seeholzer T, Krappmann D (2018) BCL10—Bridging CARDs to immune activation. Front Immunol 9:1–10. https://doi.org/10.3389/fimmu.2018.01539

Goossens T, Klein U, Kuppers R (1998) Frequent occurrence of deletions and duplications during somatic hypermutation: implications for oncogene translocations and heavy chain disease. Proc Natl Acad Sci 95:2463–2468. https://doi.org/10.1073/pnas.95.5.2463

Hailfinger S, Nogai H, Pelzer C et al (2011) Malt1-dependent RelB cleavage promotes canonical NF-B activation in lymphocytes and lymphoma cell lines. Proc Natl Acad Sci 108:14596–14601. https://doi.org/10.1073/pnas.1105020108

Han SL, Wu XL, Wan L et al (2009) FOXP1 expression predicts polymorphic histology and poor prognosis in gastric mucosa-associated lymphoid tissue lymphomas. Dig Surg 26:156–162. https://doi.org/10.1159/000212058

Hans CP, Weisenburger DD, Greiner TC et al (2004) Confirmation of the molecular classification of diffuse large B-cell lymphoma by immunohistochemistry using a tissue microarray. Blood 103:275–282. https://doi.org/10.1182/blood-2003-05-1545

Haralambieva E, Adam P, Ventura R et al (2006) Genetic rearrangement of FOXP1 is predominantly detected in a subset of diffuse large B-cell lymphomas with extranodal presentation. Leukemia 20:1300–1333. https://doi.org/10.1038/sj.leu.2404244

Hartung ML, Gruber DC, Koch KN et al (2015) *H. pylori*-induced DNA strand breaks are introduced by nucleotide excision repair endonucleases and promote NF-κB target gene expression. Cell Rep 13:70–79. https://doi.org/10.1016/j.celrep.2015.08.074

Hellmig S, Bartscht T, Fischbach W et al (2009) Germline variations of the MALT1 gene as risk factors in the development of primary gastric B-cell lymphoma. Eur J Cancer 45:1865–1870. https://doi.org/10.1016/j.ejca.2009.03.010

Hong SS, Jung HY, Choi KD et al (2006) A prospective analysis of low-grade gastric MALT lymphoma after *Helicobacter pylori* eradication. Helicobacter 11:569–573. https://doi.org/10.1111/j.1523-5378.2006.00460.x

Hosokawa Y, Suzuki H, Nakagawa M et al (2005) API2-MALT1 fusion protein induces transcriptional activation of the API2 gene through NF-κB binding elements: Evidence for a positive feed-back loop pathway resulting in unremitting NF-κB activation. Biochem Biophys Res Commun 334:51–60. https://doi.org/10.1016/j.bbrc.2005.06.058

Hussell T, Isaacson PG, Crabtree JE et al (1993a) Immunoglobulin specificity of low grade B cell gastrointestinal lymphoma of mucosa-associated lymphoid tissue (MALT) type. Am J Pathol 142:285–292

Hussell T, Isaacson PG, Spencer J, Crabtree JE (1993b) The response of cells from low-grade B-cell gastric lymphomas of mucosa-associated lymphoid tissue to Helicobacter pylori. Lancet 342:571–574. https://doi.org/10.1016/0140-6736(93)91408-E

Jafarzadeh A, Larussa T, Nemati M, Jalapour S (2018) T cell subsets play an important role in the determination of the clinical outcome of Helicobacter pylori infection. Microb Pathog 116:227–236. https://doi.org/10.1016/j.micpath.2018.01.040

Jäger G, Neumeister P, Brezinschek R et al (2002) Treatment of extranodal marginal zone B-cell lymphoma of mucosa-associated lymphoid tissue type with cladribine: a phase II study. J Clin Oncol 20:3872–3877. https://doi.org/10.1200/JCO.2002.05.117

Jiang J, Eun JL, Schmittgen TD (2006) Increased expression of microRNA-155 in Epstein-Barr virus transformed lymphoblastoid cell lines. Genes Chromosom Cancer 45:103–106. https://doi.org/10.1002/gcc.20264

Kalla J, Stilgenbauer S, Schaffner C et al (2000) Heterogeneity of the API2-MALT1 gene rearrangement in MALT-type lymphoma. Leukemia 14:1967–1974. https://doi.org/10.1038/sj.leu.2401918

Kaparakis M, Laurie KL, Wijburg O et al (2006) CD4+ CD25+ regulatory T cells modulate the T-cell and antibody responses in *Helicobacter*-Infected BALB/c mice. Infect Immun 74:3519–3529. https://doi.org/10.1128/IAI.01314-05

Kiesewetter B, Troch M, Dolak W et al (2013) A phase II study of lenalidomide in patients with extranodal marginal zone B-cell lymphoma of the mucosa associated lymphoid tissue (MALT lymphoma). Haematologica 98:353–356. https://doi.org/10.3324/haematol.2012.065995

Kirchhofer D, Vucic D (2012) Protease activity of MALT1: a mystery unravelled. Biochem J 444:e3–e5. https://doi.org/10.1042/BJ20120414

Koch P, del Valle F, Berdel WE et al (2001) Primary gastrointestinal non-Hodgkin's lymphoma: II. Combined surgical and conservative or conservative management only in localized gastric lymphoma–results of the prospective German Multicenter Study GIT NHL 01/92. J Clin Oncol 19:3874–3883. https://doi.org/10.1200/JCO.2001.19.18.3874

Koch P, Probst A, Berdel WE et al (2005) Treatment results in localized primary gastric lymphoma: data of patients registered within the German multicenter study (GIT NHL 02/96). J Clin Oncol 23:7050–7059. https://doi.org/10.1200/JCO.2005.04.031

Kondo T, Oka T, Sato H et al (2009) Accumulation of aberrant CpG hypermethylation by *Helicobacter pylori* infection promotes development and progression of gastric MALT lymphoma. Int J Oncol 35:547–557. https://doi.org/10.3892/ijo

Krappmann D, Vincendeau M (2016) Mechanisms of NF-κB deregulation in lymphoid malignancies. Semin Cancer Biol 39:3–14. https://doi.org/10.1016/j.semcancer.2016.05.002
Kuo S-H, Chen L-T, Lin C-W et al (2013) Detection of the *Helicobacter pylori* CagA protein in gastric mucosa-associated lymphoid tissue lymphoma cells: clinical and biological significance. Blood Cancer J 3:e125. https://doi.org/10.1038/bcj.2013.22
Kwee I, Rancoita PMV, Rinaldi A et al (2011) Genomic profiles of MALT lymphomas: variability across anatomical sites. Haematologica 96:1064–1066. https://doi.org/10.3324/haematol.2011.040402
Lecuit M, Abachin E, Martin A et al (2004) Immunoproliferative small intestinal disease associated with *Campylobacter jejuni*. N Engl J Med 350:239–248. https://doi.org/10.1056/NEJMoa031887
Lévy M, Copie-Bergman C, Amiot A et al (2013) Rituximab and chlorambucil versus rituximab alone in gastric mucosa-associated lymphoid tissue lymphoma according to t(11;18) status: a monocentric non-randomized observational study. Leuk Lymphoma 54:940–944. https://doi.org/10.3109/10428194.2012.729832
Li Q, Verma IM (2002) NF-κB regulation in the immune system. Nat Rev Immunol 2:725–734. https://doi.org/10.1038/nri910
Lim MCC, Maubach G, Sokolova O, Feige MH, Diezko R, Buchbinder J, Backert S, Schlüter D, Lavrik IN, Naumann M (2017) Pathogen-induced ubiquitin-editing enzyme A20 bifunctionally shuts off NF-κB and caspase-8-dependent apoptotic cell death. Cell Death Differ 24:1621–1631. https://doi.org/10.1038/cdd.2017.89
Lin W-C, Tsai H-F, Kuo S-H et al (2010) Translocation of *Helicobacter pylori* CagA into Human B lymphocytes, the origin of mucosa-associated lymphoid tissue lymphoma. Cancer Res 70:5740–5748. https://doi.org/10.1158/0008-5472.CAN-09-4690
Liu H, Ruskon-Fourmestraux A, Lavergne-Slove A et al (2001a) Resistance of t(11;18) positive gastric mucosa-associated lymphoid tissue lymphoma to *Helicobacter pylori* eradication therapy. Lancet (London, England) 357:39–40. https://doi.org/10.1016/S0140-6736(00)03571-6
Liu H, Ye H, Dogan A et al (2001b) T(11;18)(q21;q21) is associated with advanced mucosa-associated lymphoid tissue lymphoma that expresses nuclear BCL10. Blood 98:1182–1187
Lucas PC, Kuffa P, Gu S et al (2007) A dual role for the API2 moiety in API2-MALT1-dependent NF-κB activation: heterotypic oligomerization and TRAF2 recruitment. Oncogene 26:5643–5654. https://doi.org/10.1038/sj.onc.1210342
Lucas PC, Yonezumi M, Inohara N et al (2001) Bcl10 and MALT1, independent targets of chromosomal translocation in malt lymphoma, cooperate in a novel NF-kappa B signaling pathway. J Biol Chem 276:19012–19019. https://doi.org/10.1074/jbc.M009984200
Martin F, Kearney JF (2002) Marginal-zone B cells. Nat Rev Immunol 2:323–335. https://doi.org/10.1038/nri799
McKinnon ML, Rozmus J, Fung SY et al (2014) Combined immunodeficiency associated with homozygous MALT1 mutations. J Allergy Clin Immunol 133(5):1458–1462. https://doi.org/10.1016/j.jaci.2013.10.045
Min KO, Seo EJ, Kwon HJ et al (2006) Methylation of p16(INK4A) and p57(KIP2) are involved in the development and progression of gastric MALT lymphomas. Mod Pathol 19:141–148. https://doi.org/10.1038/modpathol.3800505
Moleiro J, Ferreira S, Lage P, Dias Pereira A (2016) Gastric malt lymphoma: Analysis of a series of consecutive patients over 20 years. United Eur Gastroenterol J 4:395–402. https://doi.org/10.1177/2050640615612934
Morgner A, Lehn N, Andersen LP et al (2000) Helicobacter heilmannii—associated primary gastric low-grade. Gastroenterology 5:821–828. https://doi.org/10.1053/gg.2000.7032
Motegi M, Yonezumi M, Suzuki H et al (2000) API2-MALT1 chimeric transcripts involved in mucosa-associated lymphoid tissue type lymphoma predict heterogeneous products. Am J Pathol 156:807–812. https://doi.org/10.1016/S0002-9440(10)64948-6
Müller-Hermelink HK (2003) Genetic and molecular genetic studies in the diagnosis of B-cell lymphomas: marginal zone lymphomas. Hum Pathol 34:336–340. https://doi.org/10.1053/hupa.2003.98

Nakagawa M, Hosokawa Y, Yonezumi M et al (2005) MALT1 contains nuclear export signals and regulates cytoplasmic localization of BCL10. Blood 106:4210–4216. https://doi.org/10.1182/blood-2004-12-4785

Naumann M, Sokolova O, Tegtmeyer N, Backert S (2017) *Helicobacter pylori*: a paradigm pathogen for subverting host cell signal transmission. Trends Microbiol 25:316–328. https://doi.org/10.1016/j.tim.2016.12.004

Nie Z, Du MQ, McAllister-Lucas LM et al (2015) Conversion of the LIMA1 tumour suppressor into an oncogenic LMO-like protein by API2-MALT1 in MALT lymphoma. Nat Commun 6:1–13. https://doi.org/10.1038/ncomms6908

O'Rourke JL (2008) Gene expression profiling in Helicobacter-induced MALT lymphoma with reference to antigen drive and protective immunization. J Gastroenterol Hepatol 23:151–156. https://doi.org/10.1111/j.1440-1746.2008.05553.x

Ohkubo Y, Saito Y, Ushijima H et al (2017) Radiotherapy for localized gastric mucosa-associated lymphoid tissue lymphoma: Long-term outcomes over 10 years. J Radiat Res 58:537–542. https://doi.org/10.1093/jrr/rrw044

Ohmae T, Hirata Y, Maeda S et al (2005) Helicobacter pylori activates NF-kappaB via the alternative pathway in B lymphocytes. J Immunol 175:7162–7169. https://doi.org/10.4049/jimmunol.175.11.7162

Packham G (2008) The role of NF-κB in lymphoid malignancies. Br J Haematol 143:3–15. https://doi.org/10.1111/j.1365-2141.2008.07284.x

Parsonnet J, Hansen S, Rodriguez L et al (1994) Helicobacter pylori infection and gastric lymphoma. N Engl J Med 330:1267–1271. https://doi.org/10.1056/NEJM199405053301803

Petrakis G, Kostopoulos I, Venizelos I et al (2017) Expression of the activation markers Blimp1, Foxp1 and pStat3 in extranodal diffuse large B-cell lymphomas. Histol Histopathol 32:825–834. https://doi.org/10.14670/HH-11-852

Pillai S, Cariappa A, Moran ST (2005) Marginal zone B cells. Annu Rev Immunol 23:161–196. https://doi.org/10.1146/annurev.immunol.23.021704.115728

Pomerantz JL, Baltimore D (2002) Two pathways to NF-kappa B. Mol Cell 10:693–695

Posselt G, Backert S, Wessler S (2013) The functional interplay of *Helicobacter pylori* factors with gastric epithelial cells induces a multi-step process in pathogenesis. Cell Commun Signal 11:77. https://doi.org/10.1186/1478-811X-11-77

Qiao Q, Yang C, Zheng C et al (2013) Structural architecture of the CARMA1/Bcl10/MALT1 signalosome: nucleation-induced filamentous assembly. Mol Cell 51:766–779. https://doi.org/10.1016/j.molcel.2013.08.032

Radaszkiewicz T, Dragosics B, Bauer P (1992) Gastrointestinal malignant lymphomas of the mucosa-associated lymphoid tissue: factors relevant to prognosis. Gastroenterology 102:1628–1638. https://doi.org/10.1016/0016-5085(92)91723-H

Raderer M, Streubel B, Woehrer S et al (2005) High relapse rate in patients with MALT lymphoma warrants lifelong follow-up. Clin Cancer Res 11:3349–3352. https://doi.org/10.1158/1078-0432.CCR-04-2282

Raderer M, Wöhrer S, Kiesewetter B et al (2015) Antibiotic treatment as sole management of Helicobacter pylori-negative gastric MALT lymphoma: a single center experience with prolonged follow-up. Ann Hematol 94:969–973. https://doi.org/10.1007/s00277-014-2298-3

Rawlings DJ, Sommer K, Moreno-García ME (2006) The CARMA1 signalosome links the signalling machinery of adaptive and innate immunity in lymphocytes. Nat Rev Immunol 6:799–812. https://doi.org/10.1038/nri1944

Rebeaud F, Hailfinger S, Posevitz-Fejfar A et al (2008) The proteolytic activity of the paracaspase MALT1 is key in T cell activation. Nat Immunol 9:272–281. https://doi.org/10.1038/ni1568

Remstein ED, James CD, Kurtine PJ (2000) Incidence and subtype specificity of API2-MALT1 fusion translocations in extranodal, nodal, and splenic marginal zone lymphomas. Am J Pathol 156:1183–1188. https://doi.org/10.1016/S0002-9440(10)64988-7

Remstein ED, Kurtin PJ, Einerson RR et al (2004) Primary pulmonary MALT lymphomas show frequent and heterogeneous cytogenetic abnormalities, including aneuploidy and translocations

involving API2 and MALT1 and IGH and MALT1. Leukemia 18:156–160. https://doi.org/10.1038/sj.leu.2403185
Roehle A, Hoefig KP, Repsilber D et al (2008) MicroRNA signatures characterize diffuse large B-cell lymphomas and follicular lymphomas. Br J Haematol 142:732–744. https://doi.org/10.1111/j.1365-2141.2008.07237.x
Roggero E, Zucca E, Mainetti C, Bertoni F (2000) Eradication of Borrelia burgdorferi infection in primary marginal zone B-cell lymphoma of the skin. Hum Pathol 31:263–268
Rosebeck S, Madden L, Jin X et al (2011a) Cleavage of NIK by the API2-MALT1 fusion oncoprotein leads to noncanonical NF-kappaB activation. Science 331:468–472. https://doi.org/10.1126/science.1198946
Rosebeck S, Rehman AO, Lucas PC, McAllister-Lucas LM (2011b) From MALT lymphoma to the CBM signalosome: three decades of discovery. Cell Cycle 10:2485–2496. https://doi.org/10.4161/cc.10.15.16923
Roy N, Deveraux QL, Takahashi R et al (1997) The c-IAP-1 and c-IAP-2 proteins are direct inhibitors of specific caspases. EMBO J 16:6914–6925. https://doi.org/10.1093/emboj/16.23.6914
Ruefli-Brasse AA (2003) Regulation of NF-B-dependent lymphocyte activation and development by paracaspase. Science (80-) 302:1581–1584. https://doi.org/10.1126/science.1090769
Ruland J, Duncan GS, Elia A et al (2001) Bcl10 is a positive regulator of antigen receptor-induced activation of NF-κB and neural tube closure. Cell 104:33–42. https://doi.org/10.1016/S0092-8674(01)00189-1
Ruland J, Duncan GS, Wakeham A, Mak TW (2003) Differential requirement for Malt1 in T and B cell antigen receptor signaling. Immunity 19:749–758. https://doi.org/10.1016/S1074-7613(03)00293-0
Ruskoné-Fourmestraux A, Matysiak-Budnik T, Fabiani B et al (2015) Exclusive moderate-dose radiotherapy in gastric marginal zone B-cell MALT lymphoma: results of a prospective study with a long term follow-up. Radiother Oncol 117:178–182. https://doi.org/10.1016/j.radonc.2015.08.029
Sagaert X, De Paepe P, Libbrecht L et al (2006a) Forkhead box protein P1 expression in mucosa-associated lymphoid tissue lymphomas predicts poor prognosis and transformation to diffuse large B-cell lymphoma. J Clin Oncol 24:2490–2497. https://doi.org/10.1200/JCO.2006.05.6150
Sagaert X, Laurent M, Baens M et al (2006b) MALT1 and BCL10 aberrations in MALT lymphomas and their effect on the expression of BCL10 in the tumour cells. Mod Pathol 19:225–232. https://doi.org/10.1038/modpathol.3800523
Sagaert X, Tousseyn T, Vanhentenrijk V et al (2010a) Comparative expressed sequence hybridization studies of t(11;18)(q21;q21)-positive and-negative gastric MALT lymphomas reveal both unique and overlapping gene programs. Mod Pathol 23:458–469. https://doi.org/10.1038/modpathol.2009.182
Sagaert X, Van Cutsem E, De Hertogh G et al (2010b) Gastric MALT lymphoma: a model of chronic inflammation-induced tumor development. Nat Rev Gastroenterol Hepatol 7:336–346. https://doi.org/10.1038/nrgastro.2010.58
Saito Y, Suzuki H, Tsugawa H et al (2012) Overexpression of miR-142-5p and miR-155 in gastric mucosa-associated lymphoid tissue (MALT) lymphoma resistant to Helicobacter pylori eradication. PLoS ONE 7:e47396. https://doi.org/10.1371/journal.pone.0047396
Schreuder MI, van den Brand M, Hebeda KM et al (2017) Novel developments in the pathogenesis and diagnosis of extranodal marginal zone lymphoma. J Hematop 10:91–107. https://doi.org/10.1007/s12308-017-0302-2
Sen R, Baltimore D (1986) Multiple nuclear factors interact with the immunoglobulin enhancer sequences. Cell 46:705–716
Siebenlist U, Brown K, Claudio E (2005) Control of lymphocyte development by nuclear factor-κB. Nat Rev Immunol 5:435–445. https://doi.org/10.1038/nri1629
Sinn DH, Kim YH, Lee EJ et al (2009) Methylation and API2/MALT1 fusion in colorectal extranodal marginal zone lymphoma. Mod Pathol 22:314–320. https://doi.org/10.1038/modpathol.2008.194

Staal J, Driege Y, Bekaert T et al (2011) T-cell receptor-induced JNK activation requires proteolytic inactivation of CYLD by MALT1. EMBO J 30:1742–1752. https://doi.org/10.1038/emboj.2011.85

Starostik P, Patzner J, Greiner A et al (2002) Gastric marginal zone B-cell lymphomas of MALT type develop along 2 distinct pathogenetic pathways. Blood 99:3–9. https://doi.org/10.1182/blood.V99.1.3

Stathis A, Chini C, Bertoni F et al (2009) Long-term outcome following Helicobacter pylori eradication in a retrospective study of 105 patients with localized gastric marginal zone B-cell lymphoma of MALT type. Ann Oncol 20:1086–1093. https://doi.org/10.1093/annonc/mdn760

Stefanovic A, Lossos IS (2009) Extranodal marginal zone lymphoma of the ocular adnexa. Blood 114:501–510. https://doi.org/10.1182/blood-2008-12-195453

Streubel B, Lamprecht A, Dierlamm J et al (2003) T(14;18)(q32;q21) involving IGH and MALT1 is a frequent chromosomal aberration in MALT lymphoma. Blood 101:2335–2339. https://doi.org/10.1182/blood-2002-09-2963

Streubel B, Vinatzer U, Lamprecht A et al (2005) T(3;14)(p14.1;q32) involving IGH and FOXP1 is a novel recurrent chromosomal aberration in MALT lymphoma. Leukemia 19:652–658. https://doi.org/10.1038/sj.leu.2403644

Sugiyama T, Asaka M, Nakamura T et al (2001) API2-MALT1 chimeric transcript is a predictive marker for the responsiveness of H. pylori eradication treatment in low-grade gastric MALT lymphoma [1] (multiple letters). Gastroenterology 120:1884–1885. https://doi.org/10.1053/gast.2001.25305

Sun L, Deng L, Ea CK et al (2004) The TRAF6 ubiquitin ligase and TAK1 kinase mediate IKK activation by BCL10 and MALT1 in T lymphocytes. Mol Cell 14:289–301. https://doi.org/10.1016/S1097-2765(04)00236-9

Swerdlow SH, Campo E, Harris NL et al (2008) WHO classification of tumours of haematopoietic and lymphoid tissues, Revised 4th edn, Revised 4t. World Health Organization

Tegtmeyer N, Neddermann M, Asche CI, Backert S (2017) Subversion of host kinases: a key network in cellular signaling hijacked by *Helicobacter pylori* CagA. Mol Microbiol 105:358–372. https://doi.org/10.1111/mmi.13707

Thieblemont C, Bastion Y, Berger F et al (1997) Mucosa-associated lymphoid tissue gastrointestinal and nongastrointestinal lymphoma behavior: analysis of 108 patients. J Clin Oncol 15:1624–1630. https://doi.org/10.1200/JCO.1997.15.4.1624

Thieblemont C, Berger F, Dumontet C et al (2000) Mucosa-associated lymphoid tissue lymphoma is a disseminated disease in one third of 158 patients analyzed. Blood 95:802–806

Thieblemont C, Cascione L, Conconi A et al (2017) A MALT lymphoma prognostic index. Blood 130:1409–1417. https://doi.org/10.1182/blood-2017-03-771915

Thieblemont C, Mayer A, Dumontet C et al (2002) Primary thyroid lymphoma is a heterogeneous disease. J Clin Endocrinol Metab 87:105–111. https://doi.org/10.1210/jcem.87.1.8156

Thome M (2004) CARMA1, BCL-10 and MALT1 in lymphocyte development and activation. Nat Rev Immunol 4:348–359. https://doi.org/10.1038/nri1352

Thome M, Charton JE, Pelzer C, Hailfinger S (2010) Antigen receptor signaling to NF-kappaB via CARMA1, BCL10, and MALT1. Cold Spring Harb Perspect Biol 2. https://doi.org/10.1101/cshperspect.a003004

Thorns C, Kuba J, Bernard V et al (2012) Deregulation of a distinct set of microRNAs is associated with transformation of gastritis into MALT lymphoma. Virchows Arch 460:371–377. https://doi.org/10.1007/s00428-012-1215-1

Toracchio S, Ota H, de Jong D et al (2009) Translocation t(11;18)(q21;q21) in gastric B-cell lymphomas. Cancer Sci 100:881–887. https://doi.org/10.1111/j.1349-7006.2009.01128.x

Troch M, Jonak C, Müllauer L et al (2009) A phase II study of bortezomib in patients with MALT lymphoma. Haematologica 94:738–742. https://doi.org/10.3324/haematol.2008.001537

Turvey SE, Durandy A, Fischer A et al (2014) The CARD11-BCL10-MALT1 (CBM) signalosome complex: stepping into the limelight of human primary immunodeficiency. J Allergy Clin Immunol 134:276–284. https://doi.org/10.1016/j.jaci.2014.06.015

Tusche MW, Ward LA, Vu F et al (2009) Differential requirement of MALT1 for BAFF-induced outcomes in B cell subsets. J Exp Med 206:2671–2683. https://doi.org/10.1084/jem.20091802
Uren AG, O'Rourke K, Aravind L et al (2000) Identification of paracaspases and metacaspases: two ancient families of caspase-like proteins, one of which plays a key role in MALT lymphoma. Mol Cell 6:961–967. https://doi.org/10.1016/S1097-2765(00)00094-0
van Keimpema M, Gruneberg LJ, Mokry M et al (2014) FOXP1 directly represses transcription of proapoptotic genes and cooperates with NF-B to promote survival of human B cells. Blood 124:3431–3440. https://doi.org/10.1182/blood-2014-01-553412
van Keimpema M, Grüneberg LJ, Mokry M et al (2015) The forkhead transcription factor FOXP1 represses human plasma cell differentiation. Blood 126:2098–2109. https://doi.org/10.1182/blood-2015-02-626176
Vinatzer U, Gollinger M, Müllauer L et al (2008) Mucosa-associated lymphoid tissue lymphoma: novel translocations including rearrangements of ODZ2, JMJD2C, and CNN3. Clin Cancer Res 14:6426–6431. https://doi.org/10.1158/1078-0432.CCR-08-0702
Watanabe K, Murakami K, Sato R et al (2004) CTLA-4 blockade inhibits induction of Helicobacter pylori-associated. Clin Exp Immunol 135(1):29–34. https://doi.org/10.1046/j.1365-2249.2004.02338.x
Willis TG, Jadayel DM, Du MQ et al (1999) Bcl10 is involved in t(1;14)(p22;q32) of MALT B cell lymphoma and mutated in multiple tumor types. Cell 96:35–45. https://doi.org/10.1016/S0092-8674(00)80957-5
Wlodarska I, Veyt E, De Paepe P et al (2005) FOXP1, a gene highly expressed in a subset of diffuse large B-cell lymphoma, is recurrently targeted by genomic aberrations. Leukemia 19:1299–1305. https://doi.org/10.1038/sj.leu.2403813
Wotherspoon AC, Ortiz-Hidalgo C, Falzon MR, Isaacson PG (1991) Helicobacter pylori-associated gastritis and primary B-cell gastric lymphoma. Lancet (London, England) 338:1175–1176
Wündisch T, Thiede C, Morgner A et al (2005) Long-term follow-up of gastric MALT lymphoma after Helicobacter pylori eradication. J Clin Oncol 23:8018–8024. https://doi.org/10.1200/JCO.2005.02.3903
Xiao C, Calado DP, Galler G et al (2007) MiR-150 controls B cell differentiation by targeting the transcription factor c-Myb. Cell 131:146–159. https://doi.org/10.1016/j.cell.2007.07.021
Ye H, Dogan A, Karran L et al (2000) BCL 10 expression in normal and neoplastic lymphoid tissue: nuclear localization in MALT lymphoma. Am J Pathol 157:1147–1154. https://doi.org/10.1016/S0002-9440(10)64630-5
Ye H, Gong L, Liu H et al (2005) MALT lymphoma with t(14;18)(q32;q21)/IGH-MALT1 is characterized by strong cytoplasmic MALT1 and BCL10 expression. J Pathol 205:293–301. https://doi.org/10.1002/path.1715
Ye H, Liu H, Attygalle A et al (2003) Variable frequencies of t(11;18)(q21;q21) in MALT lymphomas of different sites: significant association with CagA strains of H pylori in gastric MALT lymphoma. Blood 102:1012–1018. https://doi.org/10.1182/blood-2002-11-3502
Yu L, Li L, Medeiros LJ, Young KH (2017) NF-κB signaling pathway and its potential as a target for therapy in lymphoid neoplasms. Blood Rev 31:77–92. https://doi.org/10.1016/j.blre.2016.10.001
Yu M, Chen Y, He Y et al (2012) Critical role of B cell lymphoma 10 in BAFF-regulated NF-κB activation and survival of anergic B cells. J Immunol 189:5185–5193. https://doi.org/10.4049/jimmunol.1102952
Zhai L, Zhao Y, Ye S et al (2011) Expression of PIK3CA and FOXP1 in gastric and intestinal non-Hodgkin's lymphoma of mucosa-associated lymphoid tissue type. Tumor Biol 32:913–920. https://doi.org/10.1007/s13277-011-0192-3
Zhang Q, Lenardo MJ, Baltimore D (2017) 30 years of NF-κB: a blossoming of relevance to human pathobiology. Cell 168:37–57. https://doi.org/10.1016/j.cell.2016.12.012
Zhang XS, Tegtmeyer N, Traube L, Jindal S, Perez-Perez G, Sticht H, Backert S, Blaser MJ (2015) A specific A/T polymorphism in Western tyrosine phosphorylation B-motifs regulates Helicobacter

pylori CagA epithelial cell interaction. PloS Pathog 11:e1004621. https://doi.org/10.1371/journal.ppat.1004621
Zhou H, Du MQ, Dixit VM (2005) Constitutive NF-κB activation by the t(11;18)(q21;q21) product in MALT lymphoma is linked to deregulated ubiquitin ligase activity. Cancer Cell 7:425–431. https://doi.org/10.1016/j.ccr.2005.04.012
Zhou Y, Ye H, Martin-Subero JI et al (2006) Distinct comparative genomic hybridisation profiles in gastric mucosa-associated lymphoid tissue lymphomas with and without t(11;18)(q21;q21). Br J Haematol 133:35–42. https://doi.org/10.1111/j.1365-2141.2006.05969.x
Zucca E, Conconi A, Pedrinis E et al (2003) Nongastric marginal zone B-cell lymphoma of mucosa-associated lymphoid tissue. Arbor Cienc Pensam Y Cult 101:2489–2495. https://doi.org/10.1182/blood-2002-04-1279.Supported
Zucca E, Copie-Bergman C, Ricardi U et al (2013) Gastric marginal zone lymphoma of MALT type: ESMO clinical practice guidelines for diagnosis, treatment and follow-up. Ann Oncol 24:145–148. https://doi.org/10.1093/annonc/mdt343
Zucca E, Roggero E, Maggi-Solcà N, Conconi A (2000) Prevalence of Helicobacter pylori and hepatitis C virus infections among non-Hodgkin's lymphoma patients in Southern Switzerland. Haematologica 85:147–153
Zullo A, Hassan C, Cristofari F et al (2010) Effects of helicobacter pylori eradication on early stage gastric mucosa-associated lymphoid tissue lymphoma. Clin Gastroenterol Hepatol 8:105–110. https://doi.org/10.1016/j.cgh.2009.07.017

Crosstalk Between DNA Damage and Inflammation in the Multiple Steps of Gastric Carcinogenesis

Olga Sokolova and Michael Naumann

Contents

Abstract Over the last years, intensive investigations in molecular biology and cell physiology extended tremendously the knowledge about the association of inflammation and cancer. In frame of this paradigm, the human pathogen *Helicobacter pylori* triggers gastritis and gastric ulcer disease, and contributes to the development of gastric cancer. Mechanisms, by which the bacteria-induced inflammation in gastric mucosa leads to intestinal metaplasia and carcinoma, are represented in this review. An altered cell-signaling response and increased production of free radicals by epithelial and immune cells account for the accumulation of DNA damage in gastric mucosa, if infection stays untreated. Host genetics and environmental factors, especially diet, can accelerate the process, which offers the opportunity of intervention based on a balanced nutrition. It is supposed that inflammation might influence stem- or progenitor cells in gastric tissue predisposing for metaplasia or tumor relapse. Herein, DNA is strongly mutated and labile, which restricts therapy options. Thus, the understanding of the mechanisms that underlie gastric carcinogenesis will be of preeminent importance for the development of strategies for screening and early

---

O. Sokolova (✉) · M. Naumann
Institute of Experimental Internal Medicine, Otto von Guericke University,
Leipziger Str. 44, 39120 Magdeburg, Germany
e-mail: olga.sokolova@med.ovgu.de

S. Backert (ed.), *Molecular Mechanisms of Inflammation: Induction, Resolution and Escape by Helicobacter pylori*, Current Topics in Microbiology and Immunology 421, https://doi.org/10.1007/978-3-030-15138-6_5

detection. As most gastric cancer patients face late-stage disease with a poor overall survival, the development of multi-targeted therapeutic intervention strategies is a major challenge for the future.

1 Introduction

An important risk factor in the multifactorial etiology of gastric cancer (GC) is a *Helicobacter pylori* infection. Mechanisms by which *H. pylori* contributes to cancer development are complex and incompletely revealed (Posselt et al. 2013). Persistent inflammatory state and affected secretory function in the stomach combined with a certain host susceptibility genotype and environmental factors trigger the transformation from chronic atrophic gastritis to metaplasia, epithelial dysplasia and, finally, adenocarcinoma (Miftahussurur et al. 2017). During all these steps, changes in DNA integrity and regulation play an extremely important role. Here, we discuss how *H. pylori* supports carcinogenesis via triggering chromosomal DNA modifications in the host.

2 *H. pylori* and Inflammation

Several *H. pylori* virulence factors can induce an inflammatory response in the gastric mucosa, and the bacterial type IV secretion system (T4SS) is the major of them. The T4SS is encoded by a ~40 kb genome fragment carrying about 31 genes known as the *cag* pathogenicity island (*cag*PAI). The *cag*PAI-positive *H. pylori* isolates are highly virulent and associated with gastritis, gastric ulcer, and cancer development (Cover and Blaser 2009; Backert et al. 2017). The bacteria-host cell contact and the T4SS functionality are strongly required for the activation of c-Jun kinase (JNK) and p38 mitogen-activated protein kinase (MAPK) pathways, as well as induction of early response transcription factors activator protein (AP)-1 and nuclear factor kappa B (NF-κB) (Backert and Naumann 2010). This leads to expression of pro-inflammatory chemokines interleukin (IL)-8, IL-6, and tumor necrosis factor (TNF) in epithelial cells (Jüttner et al. 2003; Sokolova et al. 2013, 2014), which attract immune cells, including neutrophils and macrophages, to infiltrate into the infected gastric mucosa. In addition to various structural proteins of the T4SS, the *cag*PAI encodes the 125-135-kDa effector protein cytotoxin-associated gene A (CagA) (Backert et al. 2008; Zhang et al. 2015). Translocated by the T4SS into the host cell, CagA interferes with several signaling pathways, especially with these involving tyrosine kinase receptors epidermal growth factor receptor (EGFR) and c-met protein (c-Met), and kinases cellular and sarcoma (c-Src) and focal adhesion kinase (FAK), thereby inducing cytoskeletal rearrangements and inhibiting apoptosis of infected cells (Tegtmeyer et al. 2017; Naumann et al. 2017). CagA might essentially contribute to the progression toward GC (Hatakeyama 2014). Patients infected with CagA-positive (and *cag*PAI-positive) *H. pylori* revealed a 5.8-fold higher risk of developing diffuse and intestinal gastric adenocarcinoma compared to non-infected patients (Parsonnet et al. 1997). In addition, ectopic expression of CagA triggered the spontaneous development of gastrointestinal and hematopoietic malignancies in some transgenic mice (Ohnishi et al. 2008).

Other important bacterial virulence factors are the secreted vacuolating cytotoxin A (VacA) and γ-glutamyl transpeptidase (GGT), which contribute to the *H. pylori*-related pathology by inducing host cell damage. Processed VacA protein inserts itself into host cell membranes, forms channels, damages mitochondria, triggers vacuolization, autophagy and apoptosis (Terebiznik et al. 2009; Ricci 2016). The cytotoxic VacA effects are eventually antagonized by translocated CagA (Argent et al. 2008; Tegtmeyer et al. 2009). The $cagA^{+}vacA$ s1a^{+} (type I) strains of *H. pylori* are highly pathogenic, and it has been speculated that the CagA–VacA interaction can promote long-term colonization of the stomach by ameliorating the detrimental effects of these virulence factors (Oldani et al. 2009).

Constitutively expressed and highly conserved among different strains, GGT functions to metabolize extracellular glutathione and glutamine to glutamate, which can be used by the bacteria for the tricarboxylic acid cycle (Ling et al. 2013). GGT has been also demonstrated to induce cell cycle arrest at the G1-S phase, triggers activation of caspases 3 and 9, expression of pro-apoptotic Bax and anti-apoptotic Bcl-2, release of cytochrome *c* from mitochondria leading to the epithelial cell apoptosis (Ling et al. 2013). Both VacA and GGT act on antigen-presenting cells and on T cells, and thereby promote immune tolerance and long-lasting infection (Schmees et al. 2007; Larussa et al. 2015).

H. pylori can persist, sometimes asymptomatically, for decades in the human stomach. Epithelial cells as well as macrophages and neutrophils infiltrated into the gastric mucosa release anti-microbial compounds such as reactive oxygen species (ROS) and reactive nitrogen species (RNS) as well as chemokines/cytokines (IL-1β, TNF, IL-6, IL-8), metalloproteinases (MMP-1, MMP-9) and prostaglandins, e.g., prostaglandin E2 (PGE_2), maintaining the inflammatory environment (Serizawa et al. 2015; Sulzbach et al. 2016) (Fig. 1). Remarkably, overexpression of a single pro-inflammatory cytokine, IL-1β, in transgenic mice is sufficient for stepwise progression of gastric dysplasia and cancer formation. IL-1β stimulates myeloid-derived cell recruitment to the stomach at the early stage of gastric inflammation. The cells further support an inflammatory environment at the site of infection by producing IL-6 and TNF and chemokine SDF1 (via NF-κB), and by suppressing T- and B-cell proliferation (Tu et al. 2008). In humans, enhanced IL-1β production in carrier with IL-1B-511T and IL-1B-31C polymorphisms in the IL-1β gene contributes to *H. pylori*-induced gastritis and GC development (Persson et al. 2011).

In the inflammatory environment, host cells are activated, e.g., via cytokine and growth factor receptors; the process involves cellular signaling pathways, such as the Janus kinase (JAK)/signal activator and transducer of transcription (STAT), and modulate widely cellular functions (Sokolova and Naumann 2017) (Fig. 1). The cytokine-induced transcriptional factors, STATs and NF-κB, drive expression of anti-apoptotic factors Bcl-2, Bcl-X_L, BIRCs, and pro-proliferative c-Myc, cyclins, EGFR, vascular

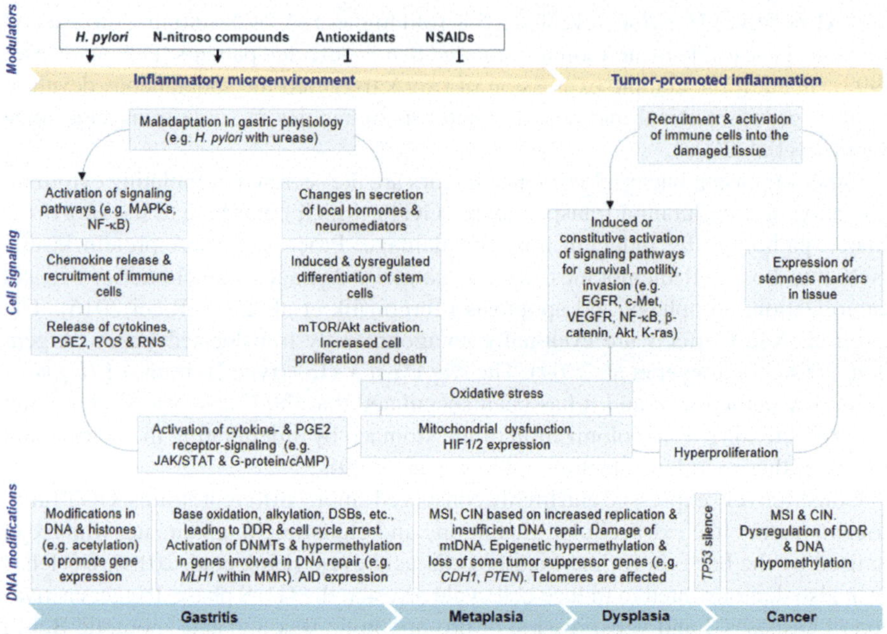

Fig. 1 Model of pathogenic mechanisms of multistep gastric carcinogenesis. *H. pylori* infection together with environmental factors triggers chronic inflammation accompanied with DNA damage. Increased cell turnover supports genetic instability, leading to key mutations and malignization in gastric mucosa. At gastritis, anti-microbial, anti-oxidative, or anti-inflammatory therapies are beneficial to prevent tissue damage. At the late metaplasia-cancer stages, when DNA modifications become pathology drivers, molecular targeted therapy might be beneficial *via* controlling mutated signaling pathways. Please see the main text for details

endothelial growth factors and M-CSF as well as the master enzyme in PGE_2 synthesis cyclooxygenase 2 (COX-2), and could promote thereby gastric carcinogenesis (Aggarwal et al. 2006; Rocha et al. 2015; Echizen et al. 2016).

Importantly, *H. pylori* strains expressing a functional T4SS and CagA are able to interact directly with gastric progenitor or stem cells to accelerate their proliferation in the glands of the mice and human stomachs (Sigal et al. 2015). The continuous inflammatory environment promotes genetic and epigenetic changes in differentiated- and stem cells concomitantly with cell turnover. This leads to genomic instability in cell divisions and enforces tumor formation. A six-month infection of C57Bl/6 mice with the *H. pylori* strain SS1 leads not only to severe gastritis, but also to a fourfold increase of the mutation frequency in the gastric mucosa (Touati et al. 2003). According to a meta-analysis by Wang et al. (2011), eradication of *H. pylori* at stages of gastritis and gastric corpus atrophy improves histological lesions preventing thereby progression to GC. Prolonged damage of the gastric mucosa leads to intestinal metaplasia, and this advanced precancerous stage cannot be reverted by anti-microbial therapy (Wang et al. 2011). On the other hand, additional therapy with non-steroidal anti-inflammatory drugs (NSAIDs) or selective inhibitors of COX-2, such as aspirin or celecoxib and etodolac, might exert cancer-preventive effects in patients with precancerous lesions (Zhang et al. 2009; Yanaoka et al. 2010; Langley and Rothwell 2014).

3 Mechanisms of DNA Damage

Epigenetic and genetic changes introduced during long-term infection with *H. pylori* form the molecular pathological background of carcinogenesis (Kalisperati et al. 2017). Epigenetic regulation includes the expression of non-coding RNAs (miRNAs), DNA methylation events via DNA methyltransferases (DNMTs), and histone modifications (acetylation, de-acetylation, methylation, phosphorylation, etc.), which impact on chromatin folding and gene expression. Aberrant expression of the miRNAs regulating proteins involved in cell cycle, proliferation, cellular transport, and invasion including p21 (CDKN1A), p27 (CDKN1B), FOXP1, HRAS, HER2 (ERBB2), HMGA2, FGFR2 takes place during *H. pylori*-associated inflammation, and correlates with GC progression and survival in humans (Ueda et al. 2010; Matsushima et al. 2011; Wang et al. 2014). Herein, miRNAs could exert anti-oncogenic roles in *H. pylori*-related diseases (Valenzuela et al. 2015).

H. pylori-related gastritis and GC development are both characterized by redistribution in the genomic methylation patterns. DNA methylation generally decreases with progression of gastric lesions along with intestinal metaplasia, gastric adenoma and GC, while promoters of certain genes, especially of tumor suppressors, are hypermethylated (Park et al. 2009; Ding et al. 2010a). *H. pylori* infection initiates methylation within CpG islands in and around promoters of the genes implicated in cell proliferation, apoptosis, autophagy, and DNA repair (Muhammad et al. 2017; Maeda et al. 2018). In tissue samples of infected gastric mucosa, methylation of *LOX*, *ARF* (p14), *RUNX3*, *MLH1*, *BRCA1*, and 6-O-methylguanine-DNA transferase (*MGMT*) has been described (Ushijima et al. 2006; Ding et al. 2010a; Nakamura et al. 2014). Eradication of bacteria decreases the methylation of cancer-related genes *CDH1*, *CDKN2A* (p16), *MLH1*, *PTGS2* (COX-2) in non-neoplastic gastric mucosa (Perri et al. 2007). Despite *H. pylori* infection is closely associated with enhanced CpG methylation in chronic gastritis, further methylation changes in intestinal metaplasia and gastric carcinoma accumulate regardless of *H. pylori* infection (Park et al. 2009). Notably, promoter methylation can be triggered by cytokines. It has been demonstrated that IL-1β is able to induce hypermethylation of the *CDH1* promoter in GC cell lines (Qian et al. 2008). In *H. pylori*-infected gerbils, a repression of inflammatory responses with cyclosporine A, an immunosuppressive agent, suppressed the induction of aberrant DNA methylation (Niwa et al. 2013). Thus, the inflammatory response triggered by *H. pylori* is a major inducer of aberrant DNA methylation (Maeda et al. 2017).

In addition to CpG methylation, histone modifications regulate gene expression. In GC specimens, reduced acetylation of core histones H3 and H4 in *HLTF* and *SLC5A8* promoters was associated with CpG hypermethylation and decreased expression of the genes (Ding et al. 2010a). Hyperacetylation of the histones H3 and H4 in the gene promoter of ataxia-telangiectasia mutated (ATM), a kinase involved in DNA damage response (DDR), activates expression of the ATM gene in *H. pylori*-infected human gastric mucosa (Santos et al. 2017). Similarly, *H. pylori*-induced hyperacetylation of histone H4 and release of histone deacetylase 1 from *CDKN1A* (p21) promoter might upregulate gene expression in gastric epithelial cells (GECs) (Xia et al. 2008). Phosphorylation of histone H2A variant X (γH2AX), attributable to DNA damage repair (Hauer and Gasser 2017), has been described in *H. pylori*-infected GECs, as well as time- and dose-dependent dephosphorylation of histone H3 at serine 10 (Ding et al. 2010b; Toller et al. 2011). Interestingly, these phosphorylation events were independent of *H. pylori* virulence factors VacA and CagA.

Pro-carcinogenic properties of *H. pylori* could be also related to its ability to induce genetic alterations in infected tissue. Each day in every human cell, DNA experiences about 10^5 damage events caused by both endogenous and exogenous factors. DNA lesions then trigger the DDR, which involves recognition, signaling, and repair of the damage. DNA damage can be represented by (1) oxidation, reduction, alkylation, deamination, or methylation of bases, abasic sites [are repaired by the base excision repair pathway (BER)], (2) formation of DNA adducts caused by chemical agents or environmental stress, particularly by ultraviolet [are repaired via nucleotide excision repair (NER) pathway], (3) incorporation of mismatched bases during DNA replication [is repaired via mismatch repair pathway (MMR)], and (4) double-strand breaks (DSBs) [are repaired via the homologous recombination repair (HEJ) and non-homologous end-joining (NHEJ) pathways] (Pollard et al. 2008; van Loon et al. 2010; Hauer and Gasser 2017). *H. pylori* has been described to influence host genetics by three ways: (a) by potentiating oxidative stress, (b) through its effect on DNA-modifying enzymes, and (c) by interfering with DDR signaling pathways.

3.1 Oxidative Stress

Excessive generation and accumulation of oxidative stress is the major DNA-damaging factor in *H. pylori* infection (Katsurahara et al. 2009; Shimizu et al. 2017). ROS and RNS are by-products of oxidative phosphorylation in mitochondria. Due to unpaired electron(s), the free radicals interact with other molecules, including proteins, lipids and nucleic acids. ROS comprise the superoxide radical $O_2^{\bullet-}$, the hydrogen peroxide H_2O_2, and hydroxyl radical $^\bullet OH$ (Davalli et al. 2018). $O_2^{\bullet-}$ is generated via the NADPH-oxidase(NOX)-dependent reduction of molecular oxygen (with electrons leaking during ATP synthesis). $O_2^{\bullet-}$ is further degraded to H_2O_2 mainly with help of the manganese-superoxide dismutase (Mn-SOD). H_2O_2 diffuses easily through cellular membranes and, in the presence of reduced transition metals, e.g., Fe^{2+}, is turned to more hazardous $^\bullet OH$ (Jimenes-Del-Rio and Velez-Pardo 2012). A strong oxidant peroxynitrite $ONOO^-$, a precursor of RNS, is generated from $O_2^{\bullet-}$ and nitric oxide $^\bullet NO$. Nitric oxide production requires L-arginine and NADPH, and is catalyzed by NO synthases (NOS). The constitutive forms of neuronal NOS and epithelial NOS produce relatively small amounts of •NO, which acts as neurotransmitter and vasodilatator. The inducible NOS isoform (iNOS) produces much larger amounts of •NO and can be stimulated by cytokines, including TNF, IFNγ, and IL-1β (Reuter et al. 2010).

Infection with *H. pylori* has been shown to stimulate the expression of TNF, IFNγ, and iNOS in the stomach epithelium of mice and Wistar rats (Sheh et al. 2010; Elseweidy et al. 2010). In Mongolian gerbils, *H. pylori*-induced iNOS expression was promoted by a high-salt diet (Toyoda et al. 2008). In human gastric biopsy specimens, an increase in iNOS mRNA accompanies development of *H. pylori*-associated gastritis (Fu et al. 1999; Naito et al. 2008; Sasaki et al. 2013). Elevated iNOS expression and NO release in response to lipopolysaccharide isolated from *Helicobacter* increase the risk of permanent genotoxic effects in the colon carcinoma cell lines SW480 and LoVo (Cavallo et al. 2011).

Under healthy conditions, ROS and RNS are constantly eliminated via protective antioxidant mechanisms. SODs, H_2O_2-reducing enzyme catalase, glutathione peroxidases and thioredoxin reductases are major proteins maintaining the vital oxidative (redox) balance. ROS and RNS contribute to functions related to mitochondrial biogenesis, angiogenesis, and immunity. Redox regulation of protein tyrosine kinases, phosphatases and the transcription factors AP-1, FOXO and NF-κB has been described (Nathan and Cunningham-Bussel 2013). However, disturbances in the oxidative balance are deleterious for biomolecules. Particularly in the DNA, which is relatively unstable chemically, a variety of potentially mutagenic and carcinogenic adducts can be formed, and strand breaks and aberrant cross-linking may occur (Butcher et al. 2017). Base guanine (G) can be oxidized to 7, 8-dihydro-8-oxo-2′-deoxyguanine (8-oxo-G). During DNA synthesis, 8-oxo-G is recognized as thymine (T), and this leads to formation of adenine (A):8-oxo-G mispair, instead of C:G pair. This mispair often remains uncorrected, thus, one of the daughter cells inherits C:G-to-A:T transversions. In addition, ROS can oxidize cellular pools of desoxyribonucleotides, before they are incorporated in DNA during replication or repair. The 8-oxo-7, 8-dihydro-2′-deoxyguanosine-5′-Triphosphate (8-oxo-dGTP) insertion into DNA has the potential to lead to A:T-to-G:C transversions in mammalian cells (Markkanen 2017). RNS induce 8-nitro-Guanine (8-nitro-G) formation (Kim et al. 2012).

It is not surprising that oxidative stress has been implicated in the pathogenesis of diverse gastrointestinal diseases including *H. pylori*-associated gastritis and GC, and anti-oxidative therapy is beneficial in gastrointestinal disorders (Kim et al. 2012).

H. pylori activates SOD and catalase in the gastric mucosa and itself possess ROS-detoxifying enzymes, which contribute to redox balance in the inflammatory milieu (Bulbuloglu et al. 2005; Lee et al. 2016). Despite that *H. pylori* infection is characterized by increased amount of ROS metabolites in human epithelial cells in vivo and in vitro (Davies et al. 1994; Ding et al. 2007; Jang et al. 2012).

It has been demonstrated that CagA-positive *H. pylori* simulates expression of spermine oxidase (SMOX), which catalyzes back conversion of the polyamine spermine into spermidine and generates H_2O_2. This additional (to the NADPH-dependent reduction of molecular oxygen to $O_2^{\bullet-}$) mechanism of ROS generation has been described by Gobert and Wilson (2017) in gastric tissue samples from *H. pylori*-positive patients as well as in infected gerbils and mice. Notably, the cells with SMOX activation demonstrated high levels of DNA damage and apoptosis resistance (Chaturvedi et al. 2011; Gobert and Wilson 2017). In addition, CagA-positive bacteria induce the expression of Cation transport regulator 1 (CHAC1), which possesses γ-glutamylcyclotransferase activity and degrades glutathione, leading to ROS production in AGS cells. Overexpression of CHAC1 was accompanied by nucleotide alterations in the *TP53*, a gene encoding tumor repressor p53 (Wada et al. 2018).

In gastric tissue, the presence of CagA-expressing *H. pylori* and a high level of 8-oxo-7,8-dihydro-2′-deoxyguanosine (8-OHdG), a biomarker of oxidative DNA damage, correlate with severe gastric inflammation, intestinal-metaplasia- and intestinal-type of GC (Raza et al. 2014). Elevated levels of 8-OHdG have been detected also in serum of dyspeptic patients carrying CagA- and BabA-positive *H. pylori* (Yeniova et al. 2015; Shahi et al. 2016). Bacterial GGT has been also reported to trigger H_2O_2 generation and 8-OHdG increase in AGS and primary GECs (Gong et al. 2010). *H. pylori*-induced DNA oxidation results in typical A:T-to-G:C and G:C-to-T:A mutations in the gastric mucosa, as it has been detected in the *gpt* delta C57BL/6 mice, a model for in vivo genotoxicity assays, six-month post-infection (Sheh et al. 2010). Further, *H. pylori*-infected patients demonstrate increasing levels of lipid peroxidation product malondialdehyde in gastric juice in chronic gastritis and further in gastric intestinal metaplasia. Malondialdehyde is able to react with deoxyadenosine and deoxyguanosine in DNA, forming mutagenic DNA adducts (Wang et al. 2018).

DNA damage and DDR hallmarks, such as expression of poly(ADP-ribose) polymerase-1 (PARP-1), phosphorylation of H2AX and increase in the NHEJ, were found in gastric biopsy specimens from *H. pylori*-infected patients with chronic gastritis and peptic ulcer disease (PUD). It has been proposed that inflammation-mediated overproduction of ROS and PARP-1 cause telomere shortening (Lee et al. 2016).

The high level of ROS induced by *H. pylori* is toxic for mitochondrial DNA (mtDNA) (Fig. 1). Indeed, transitions are the most frequent mtDNA alterations described in gastritis patients infected with *cagA*$^+$*vacA* s1/m1 *H. pylori* (Machado et al. 2010). Human mtDNA is a 16.6-kB double-stranded circular molecule that encodes 13 respiratory enzyme complex polypeptides, 22 transfer RNAs, and 2 ribosomal RNAs required for mitochondrial protein synthesis (Wallace 2012). In vitro, *H. pylori* or bacterial extracts cause mtDNA mutations and decrease mtDNA content (Huang et al. 2011; Machado et al. 2013). In tissue specimens of 32% of *H. pylori*-associated PUD patients, a number of heteroplasmic mtDNA mutations in the 303 polyC, 16184 polyC, and 514 (CA) were found (Lee et al. 2007). Cellular oxidative stress induced by *H. pylori* leads to increased Bax, decreased Bcl-2 expression, and activation of apoptotic pathways, in sum. Interestingly, pre-treatment with vitamin E or lycopene mitigates these effects (Calvino-Fernandez et al. 2008; Jang et al. 2012).

In addition to epithelial cells, *H. pylori* can stimulate the production of ROS and RNS in immune cells. For example, recombinant *H. pylori*'s neutrophil-activating protein (NapA) dose dependently activates NOX in the plasma membrane of isolated human peripheral blood neutrophils, which increases H_2O_2 production in the leukocytes. Signaling molecules involved include G proteins, phosphatidylinositol 3-kinase (PI3K), the Src family of tyrosine kinases and ERKs (Satin et al. 2000; Fu et al. 2014). NapA also induces the transendothelial migration and secretion of myeloperoxidase by neutrophils into the extracellular space, where it catalyzes the conversion of H_2O_2 to hypochlorous acid (HOCl), another potent oxidant (Brisslert et al. 2005; Fu 2014). Myeloperoxidase has been described to contribute to DNA damage, *H. pylori*-related gastric ulceration and cancer development (Mika and Guruvayoorappan 2011). HOCl, in particular hypochlorite ions OCl^-, can react with ammonia (NH_3), whose production from urea is catalyzed by bacterial urease. As a result, cytotoxic NH_2Cl is formed, which could affect cell cycle and induce apoptosis in gastric mucosa (Handa et al. 2011). *H. pylori* infection causes oxidative DNA damage in immune cells in vivo, e.g., in human peripheral lymphocytes, and anti-microbial therapy significantly decreases oxidative stress and its deleterious consequences (Dulger et al. 2011; Tsai et al. 2013).

Further, *H. pylori* infection might potentiate genotoxic effects of some exogenous mutagens and dietary risk factors, including heterocyclic amines (in cell culture, Poplawski et al. 2013), indole-3-acetonitrile (in Mongolian gerbils, Matsubara et al. 2012) and iron (in human stomach tissue, Ward et al. 2012). For example, DNA-damaging agent 1-Nitrosoindole-3-acetonitrile and *H. pylori* infection induced glandular stomach adenocarcinomas in Mongolian gerbils, only when applied in combination, but not alone (Matsubara et al. 2012). Antioxidants, such as vitamin C, carotenoids or melatonin, mitigate the effects of some mutagens (Poplawski et al. 2013; Kim and Kim 2015).

3.2 DNA-Modifying Enzymes

Genome wide analyses demonstrated that not G:C to T:A transversions induced by 8-OHdG, but rather C:G to T:A transitions are prevalent within the total mutations identified in human cancer cells, including inflammation-associated cancers (Greenman et al. 2007). Such kind of transitions can be introduced by the single-strand DNA activation-induced cytidine deaminase (AID). AID deaminates cytosine (C) to produce uracil (U), and therefore generates a DNA C:G pair into a U:G mismatch. This is either further replicated as T:A or is recognized by uracil-DNA-glycosylase or *mutS* homolog (MSH)2/MSH6 heterodimer and cut, resulting in DSBs and mutations at this DNA site (Chiba et al. 2012).

Under physiological conditions, AID induces DNA somatic hypermutation of immunoglobulin gene in B cells, thereby providing antibody diversification (Smith et al. 2012). In epithelial cells, aberrant expression of AID can be induced by inflammation and microbial infection, and results to genetic instability and mutations in proto-oncogenes *KRAS*, *MYC*, or *TP53*. This leads to neoplasia development in epithelial tissues, particularly in the stomach of humans or in the AID transgenic mice model (Morisawa et al. 2008; Takeda et al. 2012; Shimizu et al. 2014; Nishizawa et al. 2015).

AID has been described by several groups as an important molecular link between inflammation, DNA damage, and cancer formation (Touati 2010). *H. pylori*- or cytokine-activated NF-κB can induce aberrant expression of AID in human gastric mucosa (Shimizu et al. 2012). The expression is highest in the antrum and correlates with severity of chronic inflammation, glandular atrophy, and intestinal metaplasia (Goto et al. 2011; Nagata et al. 2014). Eradication of *H. pylori* significantly reduces AID expression (Nagata et al. 2014).

N-alkylnitroso compounds from food, alkylating chemotherapeutic agents, such as nitrosourea or temozolomide, and endogenous S-adenosyl methionine favor the formation of a cytotoxic DNA lesion 6-O-Methylguanine. Increased RNS might contribute to formation of the N-alkylnitroso compounds. *H. pylori* itself has been suspected to secrete an alkylative mutagenic component, which induces DNA damage studied by a Comet assay (Yan et al. 2008) and enhances mutagenicity of the N-methyl-N-nitrosourea and N-methyl-N′-nitro-N-nitrosoguanidine studied by a Ames test (Arimoto-Kobayashi et al. 2015). The alkylated guanine is repaired by MGMT without causing DSBs. The MGMT transfers the methyl/alkyl group from O^6 position of guanine to its own cysteine residue, which irreversibly inhibits the enzyme (Liu and Gerson 2006). Infection with CagA-positive *H. pylori* results in hypermethylation of the *MGTM* promoter and down-regulation of the protein amount in the stomach of adult gastritis patients (Perri et al. 2007; Alvarez et al. 2013). *MGMT* methylation frequency increases further in GC patients (Alvarez et al. 2013). According to the meta-analysis by Ding et al. (2016), *MGMT* promoter methylation was significantly associated with GC risk and substantial heterogeneity, but did not correlate with tumor types, clinical stage, age, and *H. pylori* status.

3.3 DNA Repair System

H. pylori acts also at the level of the host DNA repair machinery and DDR pathways. The BER pathway diminishes the oxidative DNA damage described above. Within this pathway, modified DNA bases are recognized and removed from the DNA sugar-phosphate backbone by a specific damage-recognizing glycosylase. Resulting abasic sites are further processed by certain glycosylases and endonucleases, e.g., by apurinic/apyrimidinic (AP) endonuclease-1 (APE-1), which cleaves the DNA phosphodiester backbone in the AP site and generates thereby ssDNA breaks. Missing sequence is reconstituted by DNA polymerase β using the template DNA strand, and DNA ligase III-Xrrc1 seals the nick (Machado et al. 2010; van Loon et al. 2010).

Some of the DNA's damage-recognizing glycosylases are bi-functional since they not only cleave the base, but also have an associated AP lyase activity. One group includes endonuclease III homolog 1 (NTH1) and 8-oxo-guanine-DNA glycosylase 1 (OGG1), which repair damaged pyrimidines and purine-derived lesions, respectively. Another group comprises of Nei-like (NEIL)1, NEIL2, and NEIL3, which preferentially target ssDNA and also lesions from a DNA bubble, and, thus, fix replication and transcription errors (Sahan et al. 2018). The potentially toxic-nicked DNA intermediates resulting from AP lyase activity of bi-functional glycosylases can be processed by APE-1. For example, APE1 removes 3′-α,β-unsaturated aldehyde and leaves a 1 nucleotide gap with a 3′-OH terminus (Vidal et al. 2001).

Infection with *H. pylori* promotes accumulation of abasic sites in GECs (Kidane et al. 2014) (Fig. 1). Elevated expression of APE1 has been detected in *H. pylori*-infected epithelial cell lines or in gastric biopsies from patients with gastritis (Ding et al. 2004; Futagami et al. 2008). In addition to bacteria, CagA-containing water extracts of *H. pylori* were able to induce APE1 expression in cultured epithelial cells and human peripheral macrophages (Futagami et al. 2008). It has been suggested that APE1 can untraditionally act in cell-signaling pathways: the enzyme interacts through its N-terminal lysine residues with Rac1. In this way, the *H. pylori*-induced APE1 overexpression in human GECs led to Rac1 activation and NOX1 expression, and mitigated thereby ROS generation (den Hartog et al. 2016).

Inconsistently, another group has described the down-regulation of APE1 expression in *H. pylori*-infected GECs (Machado et al. 2010). The APE1 underload together with insufficiency of other DNA repair systems has been suggested to be responsible for accumulation of mtDNA mutations induced by *H. pylori* in GECs (Machado et al. 2013).

Expression of the BER-related enzymes and, therefore susceptibility to oxidative DNA damage and GC, might depend on single nucleotide polymorphisms (SNPs) in the endonuclease genes, such as NEIL2 (rs804270) and APE1 (rs2275008), described in the Chinese population (Elingarami et al. 2015). Indeed, OGG1 1245C/G polymorphism was usual in intestinal metaplasia, atrophy and GC, but very rare in control patients, and correlated even more closely with 8-OHdG levels than did *H. pylori* infection or bacterial CagA status (Farinati et al. 2008).

In the MMR pathway, which deals with DNA replication errors, mismatched bases are recognized by the MSH2/MSH3 or MSH2/MSH6 dimers. The dimer undergoes an ATP-dependent transition to a sliding clamp and recruits a second heterodimer MLH1/PMS1, MLH1/PMS2, or MLH1/MLH3. The complex recruits the exonuclease EXO1 and degrades the newly synthesized DNA back to mismatch (Pollard et al. 2008; Velho et al. 2014). Genetic or epigenetic alterations of genes encoding for the MMR proteins MLH1, MSH2, MSH6, PMS2 were described to result in mutations within short DNA sequence repeats in tumor-suppressor genes. This is a hallmark of several cancers, including GC (Santos and Ribeiro 2015).

Immunohistochemical analysis revealed that MLH1 amount is decreased in biopsy samples of *H. pylori*-positive patients with gastritis; eradication therapy, however, restores levels of the protein (Park et al. 2005; Mirzaee et al. 2008). The down-regulated expression of *MLH1* in the patients with chronic gastritis is frequently caused by hypermethylation of the gene (Alvarez et al. 2013) (Fig. 1).

H. pylori infection is able to impair the MMR pathway via miRNAs. It has been demonstrated that different strains of *H. pylori* modulate expression of miRNAs miR-150-5p, miR-155-5p, and miR-3163, which target POLD3, MSH2, and MSH3, respectively, after 4, 8, and 12 h of infection in GECs (Santos et al. 2017).

DSBs can be repaired via two major pathways: the HEJ that acts mostly during late S and G2 cell cycle phases and uses undamaged sister chromatid as template, and the NHEJ that operates throughout the cell cycle with higher activity in G1 and early S phases (Pollard et al. 2008). Both pathways implicate the MRN complex, composed of meiotic recombination 11 homolog (Mre11), Rad50 and Nijmegen breakage syndrome 1 (Nbs1) proteins, which functions to detect and repair DSBs. The NHEJ implicates an initial recruitment of the Ku70/Ku80 heterodimers and DNA-dependent protein kinase to DSB sites. After aligning, DNA broken ends are ligated through DNA ligase IV/XRCC4 complex (Pollard et al. 2008; Stracker et al. 2013). The NHEJ pathway has been shown to account for repair of DSBs in *H. pylori*-infected epithelial cells. In particular, siRNA-mediated knockdown of the XRCC4 resulted in accumulation of fragmented DNA in AGS cells infected with *H. pylori* strain G27 (Hartung et al. 2015).

Two kinases, ATM and ataxia-telangiectasia and Rad9 related (ATR), are key transducers of the DDR. DNA damage induces phosphorylation, dissociation of ATM dimers to active monomers, and their recruitment to DSBs together with Nbs1 subunit of the MRN complex. ATR together with ATR-interacting protein (ATRIP) binds ssDNA and operates during DNA replication. Activated ATM and ATR phosphorylate several substrates, including checkpoint kinase (Chk)1 and Chk2, p53 and H2AX. Adapter proteins, the p53 binding protein 1 (53BP1) and BRCA1 attract Chk2, γ-H2AX, and other proteins to ATM. Phosphorylated Chk1 triggers destruction of Cdc25A phosphatase and stabilizes p53, which induces G_2 checkpoint genes, such as *CDKN1A* (p21) or *CFN* (14-3-3σ). Finally, these events result in Cdk1-cyclin A and Cdk1-cyclin B inhibition and arrest the G_2 to M transition (Pollard et al. 2008; Stracker et al. 2013). Experiments using Mongolian gerbils demonstrated that inoculation with *H. pylori* resulted, in about seven weeks, in increased levels of DNA repair proteins (Ku70/80, ATM, ATR), cell cycle regulators (Chk1, Chk2) and phosphorylated p53 in gastric mucosa, pointing to the DDR activation by DSBs. These effects were ameliorated by an antioxidant N-acetylcysteine in diet (Bae et al. 2013).

The exposure of cultured epithelial cells to live *H. pylori* has been shown to induce DSBs and DNA repair, represented by accumulation of the mediator of DNA damage checkpoint protein 1 (MDC1), p53BP1 (Toller et al. 2011), ATM (Hanada et al. 2014) and γ-H2AX foci (Jang et al. 2012). It has been suggested that *H. pylori* activates ATM expression through histone H3 and H4 hyperacetylation and promoter hypomethylation (Santos et al. 2017). In AGS cells, *H. pylori*-induced activation of ATM/Chk2 and ATR/Chk1, phosphorylations of p53 and H2AX are concomitant to delay in the cell cycle S phase progression (Jang et al. 2012).

Some experimental data, however, suggest that *H. pylori* mitigates DDR. For example, reduced levels of ATM substrate Nbs1 and impaired phosphorylation of ATRIP were detected in AGS and primary GECs infected with wild-type, but not with *cag*PAI-deficient *H. pylori* (Koeppel et al. 2015). DNA damage should trigger the induction of p53, which mediates cell cycle arrest and down-regulates proliferation to allow DNA repair. Interaction of *H. pylori* with GECs, however, inhibits p53 activity via inducing AP-1-dependent expression of truncated Δ133p53 and

Δ160p53 isoforms of the protein, which leads to survival of the infected cells (Wei et al. 2012).

No bacterial factor directly involved in the host DNA modification has been identified yet. However, the endonucleases XPF and XPG, which operate within the host NER pathway, are recruited to chromatin and might be responsible for DNA fragmentation induced by *H. pylori* (Hartung et al. 2015). Interestingly, XPG expression is increased in gastric tumor tissue and correlates with disease progression from superficial gastritis to atrophic gastritis and to gastric carcinoma (Deng et al. 2014). DSBs and DNA repair occur in a CagA-, VacA- and GGT-independent fashion, but require direct contact of live bacteria with their host cells as well as activity of the bacterial T4SS. DNA fragmentation in infected primary and transformed human and murine epithelial cells might implicate a mechanism independent of ROS (Toller et al. 2011).

In summary, *H. pylori* infection results in increased DNA damage, via oxidation or by influencing DNA-modifying enzymes in host target cells. In such conditions, any deficit in DNA repair and cell cycle control can be harmful. Investigating the link between inflammation and cancer in pancreas in vivo, Kiraly et al. (2015) have demonstrated that inflammation, although associated with ROS, RNS and DNA damage, does not drive sequence rearrangements alone but in combination with active DNA replication and hyperproliferation. DNA damage together with replication can lead to the formation of DSBs. Their repair by homologous recombination can result in loss of heterozygosity (LOH), sequence rearrangements and point mutations, contributing thereby to cancer development (Kiraly et al. 2015). Indeed, the inflammation-linked precancerous stage is mainly caused by mutations in the cell proliferation pathways, so that cells do not experience DNA damage-induced death, but instead survive and accumulate mutations (Guo et al. 2017).

4 Genetic Alterations During Gastric Carcinogenesis

4.1 Mutational Landscape of Gastric Cancer

Proposed in the beginning of the nineteenth century, the mutation hypothesis of cancer might consider experimentally approved nowadays. Persistent inflammation caused by *H. pylori* infection is accompanied by chronic oxidative stress, DNA damage, accumulation of genetic and epigenetic changes in gastric tissue, which leads to genomic instability at the next cell divisions and forces neoplastic transformation (Fig. 1). Besides these genetic and epigenetic events, progress toward GC involves the interplay between epithelial cells, cancer cells and stroma through the growth factor/cytokine receptor system, which modulates cell growth and apoptosis, angiogenesis, tumor progression and metastasis. Recent investigations demonstrated that intestinal metaplasia and GC are associated with high numbers of infiltrated neutrophils independently of tumor type and stage. Neutrophils density in gastric mucosa correlated positively with cell proliferation, CDH1 status, IL-8 level, but not with patient, age, or *H. pylori* persistence in cancer adjacent tissues (Fu et al. 2016). Eradication of *H. pylori* in non-atrophic gastritis stage reduces or reverses infection-associated molecular changes, including AID expression, DSBs, aberrant DNA methylation and deficiency in the MMR, and prevents thereby development of GC. At the atrophic gastritis stage, however, cancer risk remains high despite anti-bacterial therapy, thus additional cancer prevention strategies should be implemented (Shiotani et al. 2013).

The Lauren histological classification recognizes two GC types: intestinal (well differentiated) and diffuse (poorly differentiated) (Lauren 1965). Intestinal tumors are more common and develop through the sequential steps: gastritis-atrophic gastritis-intestinal metaplasia-dysplasia-cancer. The diffuse type has no such preceding steps, originates usually from superficial pangastritis, and is represented by cells with impaired ability to adhere and to form glandular structures, diffusely infiltrated into the gastric stroma (Santos and Ribeiro 2015). Diffuse GC affects younger people, is hereditary (HDGC) in about 1–3% of patients, up to 40% of which carry germline mutation in *CDH1*.

Generally, a range of germline mutations has been described to predispose to GC. 18 from 43 adults with GC of any histology had a mutation in one of the genes *CDH1*, *ATR*, *BRCA2*, *BRIP1*, *FANCC*, *CTNNA1*, *FLCN*, *SBDS* or *GNAS*, as revealed by sequencing of germline DNA samples (Slavin et al. 2017). Many somatic mutations in human GC samples have been described; they modulate cell-cell adhesion, activity of the WNT, Hedgehog and receptor tyrosine kinase (RTK) signaling pathways, and affect protein guardians of the genome integrity and cell cycle (Holbrook et al. 2011).

Both histological cancer types are characterized by a high rate of genomic alterations represented by chromosomal instability (CIN) and microsatellite instability (MSI), depending on the alterations level (Hudler 2012; Maleki and Röcken 2017). CIN, the most common instability in sporadic gastric tumors, comprise copy number gains or loss of whole chromosomes (aneuploidy) or parts of chromosomes (LOH, translocations and amplifications) (Hudler 2012). Gains at 12q and 13q are more common in diffuse-type cancer, and gains at 8q, 17q, and 20q are more often found in intestinal-type cancer (Liu and Meltzer 2017). Most frequent LOHs were at 3p14.2 in early GC, and at 11q24.3-25, 11q23.2-24.1, 11q14.1, 12p11.21-13.33 in advanced GC (Tahara 2004). CIN in gastric precancerous lesions or in GC affects expression of a number of tumor suppressor genes, including *TP53*, *PTEN* and *RB1* (Rb) (Karaman et al. 2010; Nagini 2012). CIN results in amplification of genes *EGFR*, *MET*, *ERBB2* (HER2), *KRAS*, and, thus, in activation of RTK pathways within progression of GC. CIN-related amplification of cell cycle regulators CCNE1 (cyclin E1) and CCND1 (cyclin D1), and pro-oncogenic transcription factors MYC, GATA4 and GATA6 has been described (Liu and Meltzer 2017).

MSI, a high frequency of DNA replication errors resulting in insertions or deletions of nucleotides within microsatellite repeats, is detected in about 15–30% of GC cases. MSI is caused mainly by deficiency in the MMR pathway, especially by genetic or epigenetic inactivation of *MLH1* (Velho et al. 2014). MSI is often represented by mutations in genes encoding tumor suppressors, cell cycle and apoptosis regulators, such as transforming growth factor β (TGF-β) receptor, insulin-like growth factor II receptor (IGFIIR), TCF4, BAX, FAS, APAF1, Bcl10. MSI can trigger mutations in DDR proteins, including members of the MMR (*MSH3*, *MSH6*), ATR and poly(adenosine diphosphate-ribose) polymerases (Kim et al. 2011; Hudler 2012). In GC with MSI, *ATM* intron mutations and ATM protein loss have also been detected (Kim et al. 2013). More examples of MSI in GC are described (Kim et al. 2010; Woerner et al. 2010).

Epi/genetic patterns typical for *H. pylori*-associated chronic superficial gastritis and further the gastric precancerous cascade have been discovered. *H. pylori*-related gastritis is characterized by the hypermethylation of CpG islands in tumor-suppressing genes (*CDH1*, *CDKN2A*) as well as in genes encoding members of the MMR pathway, e.g., in *MLH1* (Perri et al. 2007; Valenzuela et al. 2015). It favors DNA damage accumulation despite induction of the DDR (described above). At stage of gastric intestinal metaplasia, hypermethylation of the gene encoding for the adenomatous polyposis coli (APC), demethylation of *MAGE*, recurrent mutations in certain tumor suppressors (but not yet in *TP53*, *ARID1A*), chromosome 8q amplification, and shortened telomeres are frequent (Fig. 1). Mutated *KRAS* oncogene was found in intestinal metaplasia and intestinal-type cancer, but not in the diffuse-type cancer (Nagini 2012; Huang et al. 2018). Further, mutations in *APC* are prevalent in developed intestinal-type cancer (30–40% of cases) (Nishizawa and Suzuki 2015). Intestinal-type gastric tumors are additionally characterized by *EGFR* (HER1) and *ERBB2* (HER2) amplification, and overexpression of cyclin E (Tahara 2004). These alterations are rare in gastric adenomas and can be used as indicators of their malignant transformation. In 20% of gastric adenomas, *APC* mutations, which can lead to β-catenin stabilization, were described as DNA structural alterations (Tamura 2006; Park et al. 2009), predisposing to the adenoma-to-carcinoma transition (Tahara 2004). On the other hand, the development of poorly differentiated GCs involves, besides *CDH1* silence, LOH at chromosome 17p and mutations in *TP53*, and aberrant amplification within the FGFR2/ErbB3/PI3K pathway. In both intestinal and diffuse GC types, reduced or lost *CDKN1B* (p27) expression has been reported. *MET* is amplified in 39% and 19% of diffuse type and intestinal type of gastric tumors, respectively (Tahara 2004; Yamashita et al. 2011; Nagini 2012). Analysis of the current literature reveals that about four mutations are required for gastrointestinal cancer to develop, and according to a PubMed database search, the most frequently reported genome alteration for GC is the +20q13 mutation of *TP53* (Nishimura 2008).

Promoter methylation or demethylation is relatively early events in the intestinal-type GC progression cascade. In diffuse gastric carcinoma, DNA methylation profile has been also investigated and revealed several groups with different clinical features: (i) hypermethylation of *BRCA1* in association with young age (<45 years old), (ii) hypermethylation of *CDK2AP2* (p14) and *CDKN2A* (p16) together with male predominance and Epstein-Barr virus infection, and (iii) hypermethylation of *FHIT* and antral tumors (Bernal et al. 2008).

Studies of genetic alterations in GC allow the development of therapeutic strategies. Paradoxically, mutations in genes related to the DDR pathways, e.g., in *ARID1A*, can be beneficial in anti-cancer therapy (Ronchetti et al. 2017). The defects in *BRCA1* and *BRCA2* found in 7–12% of GCs mitigate the HEJ DNA repair. It can be advantageous in chemo- and radiotherapy, where DSBs are major triggers of cancer cell death (Alexandrov et al. 2015; De Luca and De Siervi 2016).

Somatic point mutations and decrease in copy number of mtDNA can compromise mitochondrial function toward reduced mitochondrial respiration and contribute to GC progression (Lee et al. 2014). Among the mtDNA mutations identified in about 65% of GCs, 46% are represented by transition mutations (e.g., T-to-C or G-to-A), and 46% result from mononucleotide or dinucleotide MSI (Hung et al. 2010). Mutations in the mtDNA D-loop region affect replication and transcription, thereby decreasing mtDNA copy number (Lee et al. 2014). In addition to ROS, decreased expression in several factors involved in mtDNA replication and maintenance contribute to mtDNA mutagenesis in GC. Suppressed expression of SIRT3, a mitochondrial NAD(+)-dependent histone deacetylase, in human GC specimens correlates with clinicopathological variables, including tumor infiltration, poor cell differentiation, advanced stage and less overall survival of the patients (Yang et al. 2014).

Progress in molecular profiling of GC and intension toward personalized therapy requested development of a GC classification based on epi/genetics. Several classifications were proposed (Liu and Meltzer 2017). According to one of them, The Cancer Genome Atlas classification, four GC molecular subtypes exist: (1) Epstein-Barr virus-positive tumors, which display recurrent *PI3KCA* mutations, hypermethylation, CDKN2A promoter hypermethylation, amplification of *JAK2* and Programmed cell death 1 ligand 2 (*PDCD1LG2*), (2) tumors with MSI, (3) genomic stable tumors, which correlate well with the diffuse histological variant and mutations of *RHOA* or fusions involving RHO-family GTPase-activating protein, and (4) tumors with CIN, which show focal amplification of RTKs, aneuploidy, *TP53* mutations in 71% of cases, and correlate mostly with the intestinal histological type (Cancer Genome Atlas Research Network 2014). In frame of this classification, *MLH1* silencing is associated to MSI; mutated *CDH1* is a feature of genomic stable gastric tumors of diffuse histology; high frequency of mutations, LOH or aneuploidy in *TP53* as well as RTK-RAS activation and intestinal histology are associated with CIN tumors. The four subtypes are distributed as 8.8%:21.7%:19.7%:49.8% (Liu and Meltzer 2017).

4.2 Stem Cells

In *H. pylori*-related carcinogenesis, the hypotheses of (1) de-differentiation of epithelial cells, (2) mutations, (3) stem-or progenitor cells transformation, and (4) bone-marrow-derived sources of cancer are all in work (Sell 2011; Waldum et al. 2018). In frames of the de-differentiation, *H. pylori* stimulates pro-oncogenic cellular pathways (c-Src, NF-κB, EGFR, β-catenin), induces cells renewal and expression of genes characteristic for stem cells (Fig. 1). For example, *H. pylori* infection in both humans and mice results in aberrant activation of the intestine-specific caudal-related homeobox transcription factors CDX1 and CDX2. It is followed by expression of stemness-associated reprogramming factors SALL4 and Krüppel-like factor (KLF)5 and intestinal-differentiation markers in cultured GECs and in intestinal metaplastic regions of the stomach (Stairs et al. 2010; Fujii et al. 2012). The mRNA levels of TGF-β1, Twist, Snail, Slug and vimentin have been found up-regulated in *H. pylori*-positive patients with dysplasia or early GC, and eradication of *H. pylori* reduced the expression of the biomarkers (Choi et al. 2015). Parietal cell atrophy has been suggested to promote de-differentiation of the digestive-enzyme secreting chief cells into the so-called Spasmolytic Polypeptide Expressing Metaplasia (SPEM) cells, which can form intestinal metaplasia (Goldenring et al. 2011).

Further, epi/genetic alterations in oncogenes, tumor suppressors, cell cycle regulators and DNA repair genes, genetic instability and telomerase activation have been implicated in *H. pylori*/inflammation-mediated carcinogenesis (described above).

Cell-of-origin analysis indicates that many of the mutations detectable in tissue were acquired in the long-lived tissue containing stem or progenitor cells (Yadav et al. 2016). In human gastric areas, various populations of gastric stem- or progenitor cells were found, including highly proliferative Lgr5-expressing ($Lgr5^+$) cells which reside in the base of antral, but not fundic glands, $Sox2^+$ stem cells located slightly above the base of pyloric and fundic glands, and $Mist1^+$ and Villin/β-gal-expressing stem cells in the isthmus. They differ in topology, markers, proliferation rate; their hierarchy remains not clear (Hoffmann 2015). In conditions of damage-stimulated cell turnover, epi/genetic changes in stem cells are extremely fatal. Firstly, mutations in growth regulation genes might affect differentiation of dividing stem cells (Waldum et al. 2018). For example, disruption of *KLF4*, a transcriptional factor involved in the differentiation control, in $Villin^+$ stem cells of antral mucosa promotes development of GC in mice (Li et al. 2012). Secondly, the molecular alterations are maintained in stem cells despite *H. pylori* eradication, can pass to the epithelial clonal progenies and give rice to the intestinal metaplasia and GC (McDonald et al. 2008; Ushijima and Hattori 2012). Thus, the field cancerization, a tissue apparently benign on pathological level, might contain affected stem cells, which cause cancer relapse (Yadav et al. 2016).

In a mouse model, $Mist1^+$ isthmus stem cells (but not $Mist1^+$ chief cells) with *KRAS* mutation are an origin of intestinal metaplasia and dysplasia. CDH1 loss in $Mist1^+$ cells is not sufficient alone to initiate cancer. Importantly, chronic inflammation induced by infection with *H. felis* or crossing with H^+/K^+-ATPase-IL-1β transgenic mice promotes diffuse GC in CDH1-deficient mice (at 18 months post-infection). Addition of TP53 mutation in the infection model leads to invasive diffuse GC within nine months. Therefore, development of diffuse-type GC derived from $Mist1^+$ stem cells is promoted by chronic inflammation and could be compromised by anti-inflammatory therapy (Hayakawa et al. 2015).

What kind of effects does *H. pylori* exert on the gastric stem- or progenitor cells? *H. pylori* can and need to directly interact with stem cells in the human and mouse stomachs to induce gene expression and proliferation (Sigal et al. 2015). In mouse stem cells in vitro and in vivo, the cancer-associated (but not atrophic gastritis-associated) *H. pylori* isolate specifically activates expression of ornithine decarboxylase 1, a rate limiting enzyme in polyamine biosynthesis and a biomarker of gastric intestinal metaplasia. The isolate up-regulates also expression of antizyme inhibitor 1, which is induced in human GC (Giannakis et al. 2008). Noto and co-workers have found that the rodent-adapted *H. pylori* strain PMSS1 induced the expansion of $KLF5^{+}$ cells enriched in the stemness marker Lrig1 in C57BL/6 mice (Noto et al. 2013). In non-neoplastic gastric mucosa of patients with GC (but not in non-cancer patients), *H. pylori* infection was associated with elevated level 8-oxi-dG in $Lgr5^{+}$ stem cells and their induced expansion (Uehara et al. 2013). In the macroscopically normal mucosa resected from the stomach with cancer and *H. pylori* infection, most of the nuclei were positive for ssDNA, suggesting DNA damage. Immunohistochemical analysis demonstrated that the neck portions of gastric units were enriched in cells positive for proliferation marker Ki67 and stemness marker CD117 (Kit), in parallel to a number of apoptotic cells, suggesting stem cells proliferation in inflamed field cancerization (Kato et al. 2008).

In addition to tissue-resident stem cells, bone-marrow-derived cells (BMDCs) might be able to migrate to the sites of gastric inflammation and injury (Donnelly et al. 2014). Studies of mice infected with human-derived and mouse-adapted *H. pylori* have demonstrated that BMDCs expressing the green fluorescent protein accumulate among the GECs in parallel to development of dysplastic lesions (Varon et al. 2012). The BMDCs are long-lived, chemoresistant and have the capacity to self-renew and to differentiate along diverse lineages (Houghton and Wang 2005). It has been suggested that resident stem- or progenitor cells or BMDCs can give rise to GC (Singh 2013).

Takaishi et al. (2009) have described another minor subpopulation of gastric cells, which self-regenerate and resist to the cell death induced by chemotherapy or radiation, the CD44-positive cancer stem cells (CSCs). The CSCs express such markers as aldehyde dehydrogenase 1 (ALDH)1, SOX2, OCT-4, CD133, NANOG and can expand toward genetically and phenotypically heterogeneous GC. High levels of some CD44 splicing variants (CD44v) and ALDH1 correlate with metastasis and relapse of the GC (Wakamatsu et al. 2012; Gao et al. 2018). It has been suggested that *H. pylori's* CagA can promote cancerogenesis because it is not destroyed by autophagy in CD44v9 CSCs, accumulates there and can, theoretically, be even transferred to daughter cells (Tsugawa et al. 2012).

It is not known, whether *H. pylori* induces DNA damage in BMDCs or CSCs. At least CD44v9-expressing GC cells are resistant to ROS, provided through increased intracellular glutathione synthesis (Ishimoto et al. 2011). However, stem cells and CSCs are controlled by a range of niche-specific environmental factors. Thus, the inflammation initiated by *H. pylori* and supported by infiltrated immune cells might lead to a dysbalance in proliferation and differentiation of the stem cells, which predisposes to malignant transformation.

5 Concluding Remarks

The hypothesis about association between sustained inflammation and carcinogenesis was proposed more than 150 years ago (Virchow 1863). Basic research has provided much information regarding the molecular mechanisms underlying the inflammation-related neoplastic transformation, in particular, in *H. pylori* infection. The inflammatory environment created by epithelial and immune cells, as well as bacterial compounds, promote oxidative stress and DNA damage, apoptosis and cell proliferation in tissue. Importantly, gastric stem cells can also be influenced. DNA damage together with deficits in DNA repair systems, stimulated DNA replication and cell turnover elevate the mutagenesis rate. Acquired mutations determine progression from gastritis toward GC (see also Chapter "Genetic Polymorphisms in Inflammatory and Other Regulators in Gastric Cancer: Risks and Clinical Consequences" of this book). Overall, the tremendous number of descriptions about the impact of environmental factors including *H. pylori* on DNA damage and inflammation in the multiple steps of gastric carcinogenesis disclosed a complex disease picture. Herein, future work should dissolve in detail the intervening net of certain molecular- and cellular processes to generate promising opportunities to prevent and heal gastric diseases.

Acknowledgements This work was supported by the German Research Foundation to M.N. (CRC-854 A04).

References

Aggarwal BB, Sethi G, Nair A, Ichikawa H (2006) Nuclear factor-κB: a Holy Grail in cancer prevention and therapy. Curr Signal Transduct Ther 1(1):25–52. https://doi.org/10.2174/157436206775269235

Alexandrov LB, Nik-Zainal S, Siu HC, Leung SY, Stratton MR (2015) A mutational signature in gastric cancer suggests therapeutic strategies. Nat Commun 6:8683. https://doi.org/10.1038/ncomms9683

Alvarez MC, Santos JC, Maniezzo N, Ladeira MS, da Silva AL, Scaletsky IC, Pedrazzoli J Jr, Ribeiro ML (2013) MGMT and MLH1 methylation in *Helicobacter pylori*-infected children and adults. World J Gastroenterol 19(20):3043–3051. https://doi.org/10.3748/wjg.v19.i20.3043

Argent RH, Thomas RJ, Letley DP, Rittig MG, Hardie KR, Atherton JC (2008) Functional association between the *Helicobacter pylori* virulence factors VacA and CagA. J Med Microbiol 57(Pt 2):145–150. https://doi.org/10.1099/jmm.0.47465-0

Arimoto-Kobayashi S, Ohta K, Yuhara Y, Ayabe Y, Negishi T, Okamoto K, Nakajima Y, Ishikawa T, Oguma K, Otsuka T (2015) Mutagenicity and clastogenicity of extracts of *Helicobacter pylori* detected by the Ames test and in the micronucleus test using human lymphoblastoid cells. Mutagenesis 30(4):537–544. https://doi.org/10.1093/mutage/gev016

Backert S, Naumann M (2010) What a disorder: Proinflammatory signaling pathways induced by *Helicobacter pylori*. Trends Microbiol 18:479–486. https://doi.org/10.1016/j.tim.2010.08.003

Backert S, Feller SM, Wessler S (2008) Emerging roles of Abl family tyrosine kinases in microbial pathogenesis. Trends Biochem Sci. 33(2):80–90. https://doi.org/10.1016/j.tibs.2007.10.006

Backert S, Haas R, Gerhard M, Naumann M (2017) The *Helicobacter pylori* type IV secretion system encoded by the cag Pathogenicity island: architecture, function, and signaling. Curr Top Microbiol Immunol 413:187–220. https://doi.org/10.1007/978-3-319-75241-9_8

Bae M, Lim JW, Kim H (2013) Oxidative DNA damage response in *Helicobacter pylori*-infected mongolian gerbils. J Cancer Prev 18(3):271–275. https://doi.org/10.15430/JCP.2013.18.3.271

Bernal C, Vargas M, Ossandón F, Santibáñez E, Urrutia J, Luengo V, Zavala LF, Backhouse C, Palma M, Argandoña J, Aguayo F, Corvalán A (2008) DNA methylation profile in diffuse type gastric cancer: evidence for hypermethylation of the BRCA1 promoter region in early-onset gastric carcinogenesis. Biol Res 41(3):303–315. https://doi.org/10.4067/S0716-97602008000300007

Brisslert M, Enarsson K, Lundin S, Karlsson A, Kusters JG, Svennerholm AM, Backert S, Quiding-Järbrink M (2005) *Helicobacter pylori* induce neutrophil transendothelial migration: role of the bacterial HP-NAP. FEMS Microbiol Lett 249(1):95–103

Bulbuloglu E, Inanc F, Bakaris S, Kantarceken B, Cetinkaya A, Cağlar R, Ilhami TK, Kilinc M (2005) Association of adenosine deaminase, superoxide dismutase, and catalase activities with *Helicobacter pylori*. Dig Dis Sci 50(12):2296–2299. https://doi.org/10.1007/s10620-005-3050-6

Butcher LD, den Hartog G, Ernst PB, Crowe SE (2017) Oxidative stress resulting from *Helicobacter pylori* infection contributes to gastric cancerogenesis. Cell Mol Gastroenterol Hepathol 3:316–322

Calvino-Fernández M, Benito-Martínez S, Parra-Cid T (2008) Oxidative stress by *Helicobacter pylori* causes apoptosis through mitochondrial pathway in gastric epithelial cells. Apoptosis 13(10):1267–1280. https://doi.org/10.1007/s10495-008-0255-0

Cancer Genome Atlas Research Network. Collaborators (315) (2014) Comprehensive molecular characterization of gastric adenocarcinoma. Nature 513(7517):202–209. https://doi.org/10.1038/nature13480

Cavallo P, Cianciulli A, Mitolo V, Panaro MA (2011) Lipopolysaccharide (LPS) of helicobacter modulates cellular DNA repair systems in intestinal cells. Clin Exp Med 11(3):171–179. https://doi.org/10.1007/s10238-010-0118-1

Chaturvedi R, Asim M, Romero-Gallo J, Barry DP, Hoge S, de Sablet T, Delgado AG, Wroblewski LE, Piazuelo MB, Yan F, Israel DA, Casero RA Jr, Correa P, Gobert AP, Polk DB, Peek RM Jr, Wilson KT (2011) Spermine oxidase mediates the gastric cancer risk associated with *Helicobacter pylori* CagA. Gastroenterology 141(5):1696–1708. https://doi.org/10.1053/j.gastro.2011.07.045

Chiba T, Marusawa H, Ushijima T (2012) Inflammation-associated cancer development in digestive organs: mechanisms and roles for genetic and epigenetic modulation. Gastroenterology 143(3):550–563. https://doi.org/10.1053/j.gastro.2012.07.009

Choi YJ, Kim N, Chang H, Lee HS, Park SM, Park JH, Shin CM, Kim JM, Kim JS, Lee DH, Jung HC (2015) *Helicobacter pylori*-induced epithelial-mesenchymal transition, a potential role of gastric cancer initiation and an emergence of stem cells. Carcinogenesis 36(5):553–563. https://doi.org/10.1053/j.gastro.2012.07.009

Cover TL, Blaser MJ (2009) *Helicobacter pylori* in health and disease. Gastroenterology 136(6):1863–1873. https://doi.org/10.1053/j.gastro.2009.01.073

Davalli P, Marverti G, Lauriola A, D'Arca D (2018) Targeting oxidatively induced DNA damage response in cancer: opportunities for novel cancer therapies. Oxid Med Cell Longev 2018:2389523. https://doi.org/10.1155/2018/2389523

Davies GR, Simmonds NJ, Stevens TR, Sheaff MT, Banatvala N, Laurenson IF, Blake DR, Rampton DS (1994) *Helicobacter pylori* stimulates antral mucosal reactive oxygen metabolite production in vivo. Gut 35:179–185. https://doi.org/10.1136/gut.35.2.179

De Luca P, De Siervi A (2016) Critical role for BRCA1 expression as a marker of chemosensitivity response and prognosis. Front Biosci (Elite Ed) 8:72–83. https://doi.org/10.2741/752

den Hartog G, Chattopadhyay R, Ablack A, Hall EH, Butcher LD, Bhattacharyya A, Eckmann L, Harris PR, Das S, Ernst PB, Crowe SE (2016) Regulation of Rac1 and reactive oxygen species production in response to infection of gastrointestinal epithelia. PLoS Pathog 12(1):e1005382. https://doi.org/10.1371/journal.ppat.1005382

Deng N, Liu JW, Sun LP, Xu Q, Duan ZP, Dong NN, Yuan Y (2014) Expression of XPG protein in the development, progression and prognosis of gastric cancer. PLoS ONE 9(9):e108704. https://doi.org/10.1371/journal.pone.0108704

Ding SZ, O'Hara AM, Denning TL, Dirden-Kramer B, Mifflin RC, Reyes VE, Ryan KA, Elliott SN, Izumi T, Boldogh I, Mitra S, Ernst PB, Crowe SE (2004) *Helicobacter pylori* and H_2O_2 increase AP endonuclease-1/redox factor-1 expression in human gastric epithelial cells. Gastroenterology 127(3):845–858. https://doi.org/10.1053/j.gastro.2004.06.017

Ding SZ, Minohara Y, Fan XJ, Wang J, Reyes VE, Patel J, Dirden-Kramer B, Boldogh I, Ernst PB, Crowe SE (2007) *Helicobacter pylori* infection induces oxidative stress and programmed cell death in human gastric epithelial cells. Infect Immun 75:4030–4039. https://doi.org/10.1128/IAI.00172-07

Ding SZ, Goldberg JB, Hatakeyama M (2010a) *Helicobacter pylori* infection, oncogenic pathways and epigenetic mechanisms in gastric carcinogenesis. Future Oncol 6(5):851–862. https://doi.org/10.2217/fon.10.37

Ding SZ, Fischer W, Kaparakis-Liaskos M, Liechti G, Merrell DS, Grant PA, Ferrero RL, Crowe SE, Haas R, Hatakeyama M, Goldberg JB (2010b) *Helicobacter pylori*-induced histone modification, associated gene expression in gastric epithelial cells, and its implication in pathogenesis. PLoS ONE 5(4):e9875. https://doi.org/10.1371/journal.pone.0009875

Ding Y, Yang Q, Wang B, Ye G, Tong X (2016) The correlation of MGMT promoter methylation and clinicopathological features in gastric cancer: a systematic review and meta-analysis. PLoS ONE 11(11):e0165509. https://doi.org/10.1371/journal.pone.0165509

Donnelly JM, Engevik AC, Engevik M, Schumacher MA, Xiao C, Yang L, Worrell RT, Zavros Y (2014) Gastritis promotes an activated bone marrow-derived mesenchymal stem cell with a phenotype reminiscent of a cancer-promoting cell. Dig Dis Sci 59(3):569–582. https://doi.org/10.1007/s10620-013-2927-z

Dulger AC, Aslan M, Nazligul Y, Horoz M, Bolukbas C, Bolukbas FF, Celik H, Kocyigit A (2011) Peripheral lymphocyte DNA damage and oxidative status after eradication therapy in patients infected with *Helicobacter pylori*. Pol Arch Med Wewn 121(12):428–432 PMID: 22012003

Echizen K, Hirose O, Maeda Y, Oshima M (2016) Inflammation in gastric cancer: Interplay of the COX-2/prostaglandin E2 and Toll-like receptor/MyD88 pathways. Cancer Sci 107(4):391–397. https://doi.org/10.1111/cas.12901

Elingarami S, Liu H, Kalinjuma AV, Hu W, Li S, He N (2015) Polymorphisms in NEIL-2, APE-1, CYP2E1 and MDM2 genes are independent predictors of gastric cancer risk in a Northern Jiangsu Population (China). J Nanosci Nanotechnol 15(7):4815–4828. https://doi.org/10.1166/jnn.2015.10028

Elseweidy MM, Taha MM, Younis NN, Ibrahim KS, Hamouda HA, Eldosouky MA, Soliman H (2010) Gastritis induced by *Helicobacter pylori* infection in experimental rats. Dig Dis Sci 55(10):2770–2777. https://doi.org/10.1007/s10620-009-1103-y

Farinati F, Cardin R, Bortolami M, Nitti D, Basso D, de Bernard M, Cassaro M, Sergio A, Rugge M (2008) Oxidative DNA damage in gastric cancer: CagA status and OGG1 gene polymorphism. Int J Cancer 123(1):51–55. https://doi.org/10.1159/000320052

Fu HW (2014) *Helicobacter pylori* neutrophil-activating protein: from molecular pathogenesis to clinical applications. World J Gastroenterol 20(18):5294–5301. https://doi.org/10.3748/wjg.v20.i18.5294

Fu S, Ramanujam KS, Wong A, Fantry GT, Drachenberg CB, James SP, Meltzer SJ, Wilson KT (1999) Increased expression and cellular localization of inducible nitric oxide synthase and cyclooxygenase 2 in *Helicobacter pylori* gastritis. Gastroenterology 116(6):1319–1329. https://doi.org/10.1016/S0016-5085(99)70496-8

Fu HW (2014) *Helicobacter pylori* neutrophil-activating protein: from molecular pathogenesis to clinical applications. World J Gastroenterol 20(18):5294–5301. https://doi.org/10.3748/wjg.v20.i18.5294.

Fu H, Ma Y, Yang M, Zhang C, Huang H, Xia Y, Lu L, Jin W, Cui D (2016) Persisting and increasing neutrophil infiltration associates with gastric carcinogenesis and E-cadherin downregulation. Sci Rep 6:29762. https://doi.org/10.1038/srep29762

Fujii Y, Yoshihashi K, Suzuki H, Tsutsumi S, Mutoh H, Maeda S, Yamagata Y, Seto Y, Aburatani H, Hatakeyama M (2012) CDX1 confers intestinal phenotype on gastric epithelial cells via induction of stemness-associated reprogramming factors SALL4 and KLF5. Proc Natl Acad Sci USA 109(50):20584–20589. https://doi.org/10.1073/pnas.1208651109

Futagami S, Hiratsuka T, Shindo T, Horie A, Hamamoto T, Suzuki K, Kusunoki M, Miyake K, Gudis K, Crowe SE, Tsukui T, Sakamoto C (2008) Expression of apurinic/apyrimidinic endonuclease-1 (APE-1) in *Helicobacter pylori*-associated gastritis, gastric adenoma, and gastric cancer. Helicobacter 13(3):209–218. https://doi.org/10.1111/j.1523-5378.2008.00605.x

Gao JP, Xu W, Liu WT, Yan M, Zhu ZG (2018) Tumor heterogeneity of gastric cancer: from the perspective of tumor-initiating cell. World J Gastroenterol 24(24):2567–2581. https://doi.org/10.3748/wjg.v24.i24.2567

Giannakis M, Chen SL, Karam SM, Engstrand L, Gordon JI (2008) *Helicobacter pylori* evolution during progression from chronic atrophic gastritis to gastric cancer and its impact on gastric stem cells. Proc Natl Acad Sci USA 105(11):4358–4363. https://doi.org/10.1073/pnas.0800668105

Gobert AP, Wilson KT (2017) Polyamine- and NADPH-dependent generation of ROS during *Helicobacter pylori* infection: A blessing in disguise. Free Radic Biol Med 105:16–27. https://doi.org/10.1016/j.freeradbiomed.2016.09.024

Goldenring JR, Nam KT, Mills JC (2011) The origin of pre-neoplastic metaplasia in the stomach: chief cells emerge from the Mist. Exp Cell Res 317(19):2759–2764. https://doi.org/10.1016/j.yexcr.2011.08.017

Gong M, Ling SS, Lui SY, Yeoh KG, Ho B (2010) *Helicobacter pylori* gamma-glutamyl transpeptidase is a pathogenic factor in the development of peptic ulcer disease. Gastroenterology 139(2):564–573. https://doi.org/10.1053/j.gastro.2010.03.050

Goto A, Hirahashi M, Osada M, Nakamura K, Yao T, Tsuneyoshi M, Takayanagi R, Oda Y (2011) Aberrant activation-induced cytidine deaminase expression is associated with mucosal intestinalization in the early stage of gastric cancer. Virchows Arch 458(6):717–724. https://doi.org/10.1007/s00428-011-1086-x

Greenman C, Stephens P, Smith R, Dalgliesh GL, Hunter C, Bignell G et al (2007) Patterns of somatic mutation in human cancer genomes. Nature 446(7132):153–158. https://doi.org/10.1038/nature05610

Guo Y, Nie Q, MacLean AL, Li Y, Lei J, Li S (2017) Multiscale modeling of inflammation-induced tumorigenesis reveals competing oncogenic and oncoprotective roles for inflammation. Cancer Res 77(22):6429–6441. https://doi.org/10.1158/0008-5472.CAN-17-1662

Hanada K, Uchida T, Tsukamoto Y, Watada M, Yamaguchi N, Yamamoto K, Shiota S, Moriyama M, Graham DY, Yamaoka Y (2014) *Helicobacter pylori* infection introduces DNA double-strand breaks in host cells. Infect Immun 82(10):4182–4189. https://doi.org/10.1128/IAI.02368-14

Handa O, Naito Y, Yoshikawa T (2011) Redox biology and gastric carcinogenesis: the role of *Helicobacter pylori*. Redox Rep 16(1):1–7. https://doi.org/10.1179/174329211X12968219310756

Hartung ML, Gruber DC, Koch KN, Grüter L, Rehrauer H, Tegtmeyer N, Backert S, Müller A (2015) *Helicobacter pylori*-Induced DNA strand breaks are introduced by nucleotide excision repair endonucleases and promote NF-κB target gene expression. Cell Rep 13(1):70–79. https://doi.org/10.1016/j.celrep.2015.08.074

Hatakeyama M (2014) *Helicobacter pylori* CagA and gastric cancer: a paradigm for hit-and-run carcinogenesis. Cell Host Microbe 15(3):306–316. https://doi.org/10.1016/j.chom.2014.02.008

Hauer MH, Gasser SM (2017) Chromatin and nucleosome dynamics in DNA damage and repair. Genes Dev 31(22):2204–2221. https://doi.org/10.1101/gad.307702.117

Hayakawa Y, Ariyama H, Stancikova J, Sakitani K, Asfaha S, Renz BW, Dubeykovskaya ZA, Shibata W, Wang H, Westphalen CB, Chen X, Takemoto Y, Kim W, Khurana SS, Tailor Y, Nagar K, Tomita H, Hara A, Sepulveda AR, Setlik W, Gershon MD, Saha S, Ding L, Shen Z, Fox JG, Friedman RA, Konieczny SF, Worthley DL, Korinek V, Wang TC (2015) Mist1 expressing gastric stem cells maintain the normal and neoplastic gastric epithelium and are supported by a perivascular stem cell niche. Cancer Cell 28(6):800–814. https://doi.org/10.1016/j.ccell.2015.10.003

Hoffmann W (2015) Current status on stem cells and cancers of the gastric epithelium. Int J Mol Sci 16(8):19153–19169. https://doi.org/10.3390/ijms160819153

Holbrook JD, Parker JS, Gallagher KT, Halsey WS, Hughes AM, Weigman VJ, Lebowitz PF, Kumar R (2011) Deep sequencing of gastric carcinoma reveals somatic mutations relevant to personalized medicine. J Transl Med 9:119. https://doi.org/10.1186/1479-5876-9-119

Houghton J, Wang TC (2005) *Helicobacter pylori* and gastric cancer: a new paradigm for inflammation-associated epithelial cancers. Gastroenterology 128(6):1567–1578. https://doi.org/10.1053/j.gastro.2005.03.037

Huang XW, Luo RH, Zhao Q, Shen ZZ, Huang LL, An XY, Zhao LJ, Wang J, Huang YZ (2011) *Helicobacter pylori* induces mitochondrial DNA mutation and reactive oxygen species level in AGS cells. Int J Med Sci 8(1):56–67. https://doi.org/10.7150/ijms.8.56

Huang KK, Ramnarayanan K, Zhu F, Srivastava S, Xu C, Tan ALK, Lee M, Tay S, Das K, Xing M, Fatehullah A, Alkaff SMF, Lim TKH, Lee J, Ho KY, Rozen SG, Teh BT, Barker N, Chia CK, Khor C, Ooi CJ, Fock KM, So J, Lim WC, Ling KL, Ang TL, Wong A, Rao J, Rajnakova A, Lim LG, Yap WM, Teh M, Yeoh KG, Tan P (2018) Genomic and epigenomic profiling of high-risk intestinal metaplasia reveals molecular determinants of progression to gastric cancer. Cancer Cell 33(1):137–150.e5. https://doi.org/10.1016/j.ccell.2017.11.018

Hudler P (2012) Genetic aspects of gastric cancer instability. Sci World J 2012:761909. https://doi.org/10.1100/2012/761909

Hung WY, Wu CW, Yin PH, Chang CJ, Li AF, Chi CW, Wei YH, Lee HC (2010) Somatic mutations in mitochondrial genome and their potential roles in the progression of human gastric cancer. Biochim Biophys Acta 1800:264–270. https://doi.org/10.1016/j.bbagen.2009.06.006

Ishimoto T, Nagano O, Yae T, Tamada M, Motohara T, Oshima H, Oshima M, Ikeda T, Asaba R, Yagi H, Masuko T, Shimizu T, Ishikawa T, Kai K, Takahashi E, Imamura Y, Baba Y, Ohmura M, Suematsu M, Baba H, Saya H (2011) CD44 variant regulates redox status in cancer cells by stabilizing the xCT subunit of system xc(-) and thereby promotes tumor growth. Cancer Cell 19(3):387–400. https://doi.org/10.1016/j.ccr.2011.01.038

Jang SH, Lim JW, Morio T, Kim H (2012) Lycopene inhibits *Helicobacter pylori*-induced ATM/ATR-dependent DNA damage response in gastric epithelial AGS cells. Free Radic Biol Med 52(3):607–615. https://doi.org/10.1016/j.freeradbiomed.2011.11.010

Jimenez-Del-Rio M, Velez-Pardo C (2012) The bad, the good, and the ugly about oxidative stress. Oxid Med Cell Longev 2012:163913. https://doi.org/10.1155/2012/163913

Jüttner S, Cramer T, Wessler S, Walduck A, Gao F, Schmitz F, Wunder C, Weber M, Fischer SM, Schmidt WE, Wiedenmann B, Meyer TF, Naumann M, Höcker M (2003) *Helicobacter pylori* stimulates host cyclooxygenase-2 gene transcription: critical importance of MEK/ERK-dependent activation of USF1/-2 and CREB transcription factors. Cell Microbiol 5(11):821–834. https://doi.org/10.1046/j.1462-5822.2003.00324.x

Kalisperati P, Spanou E, Pateras IS, Korkolopoulou P, Varvarigou A, Karavokyros I, Gorgoulis VG, Vlachoyiannopoulos PG, Sougioultzis S (2017) Inflammation, DNA damage, *Helicobacter pylori* and gastric tumorigenesis. Front Genet 8:20. https://doi.org/10.1038/oncsis.2014.42

Karaman A, Kabalar ME, Binici DN, Oztürk C, Pirim I (2010) Genetic alterations in gastric precancerous lesions. Genet Couns 21(4):439–450

Kato K, Hasui K, Wang J, Kawano Y, Aikou T, Murata F (2008) Homeostatic mass control in gastric non-neoplastic epithelia under infection of *Helicobacter pylori*: an immunohistochemical analysis of cell growth, stem cells and programmed cell death. Acta Histochem Cytochem 41(3):23–38. https://doi.org/10.1267/ahc.07021

Katsurahara M, Kobayashi Y, Iwasa M, Ma N, Inoue H, Fujita N, Tanaka K, Horiki N, Gabazza EC, Takei Y (2009) Reactive nitrogen species mediate DNA damage in *Helicobacter pylori*-infected gastric mucosa. Helicobacter 14(6):552–558. https://doi.org/10.1111/j.1523-5378.2009.00719.x

Kidane D, Murphy DL, Sweasy JB (2014) Accumulation of abasic sites induces genomic instability in normal human gastric epithelial cells during *Helicobacter pylori* infection. Oncogenesis 3:e128. https://doi.org/10.1038/oncsis.2014.42

Kim MJ, Kim H (2015) Anticancer effect of lycopene in gastric carcinogenesis. J Cancer Prev 20(2):92–96. https://doi.org/10.15430/JCP.2015.20.2.92

Kim YR, Chung NG, Kang MR, Yoo NJ, Lee SH (2010) Novel somatic frameshift mutations of genes related to cell cycle and DNA damage response in gastric and colorectal cancers with microsatellite instability. Tumori 96(6):1004–1009. PMID: 21388066 https://www.ncbi.nlm.nih.gov/pubmed/21388066

Kim MS, An CH, Kim SS, Yoo NJ, Lee SH (2011) Frameshift mutations of poly(adenosine diphosphate-ribose) polymerase genes in gastric and colorectal cancers with microsatellite instability. Hum Pathol 42(9):1289–1296. https://doi.org/10.1016/j.humpath.2010.11.020

Kim YJ, Kim EH, Hahm KB (2012) Oxidative stress in inflammation-based gastrointestinal tract diseases: challenges and opportunities. J Gastroenterol Hepatol 27(6):1004–1010. https://doi.org/10.1111/j.1440-1746.2012.07108.x

Kim HS, Choi SI, Min HL, Kim MA, Kim WH (2013) Mutation at intronic repeats of the ataxia-telangiectasia mutated (ATM) gene and ATM protein loss in primary gastric cancer with microsatellite instability. PLoS ONE 8(12):e82769. https://doi.org/10.1371/journal.pone.0082769

Kiraly O, Gong G, Olipitz W, Muthupalani S, Engelward BP (2015) Inflammation-induced cell proliferation potentiates DNA damage-induced mutations in vivo. PLoS Genet 11(2):e1004901. https://doi.org/10.1371/journal.pgen.1004901

Koeppel M, Garcia-Alcalde F, Glowinski F, Schlaermann P, Meyer TF (2015) *Helicobacter pylori* infection causes characteristic DNA damage patterns in human cells. Cell Rep 11(11):1703–1713. https://doi.org/10.1016/j.celrep.2015.05.030

Langley RE, Rothwell PM (2014) Aspirin in gastrointestinal oncology: new data on an old friend. Curr Opin Oncol 26(4):441–447. https://doi.org/10.1097/CCO.0000000000000098

Larussa T, Leone I, Suraci E, Imeneo M, Luzza F (2015) *Helicobacter pylori* and T helper cells: mechanisms of immune escape and tolerance. J Immunol Res 2015:981328. https://doi.org/10.1155/2015/981328

Lauren P (1965) The two histological main types of gastric carcinoma: diffuse and so-called intestinal-type carcinoma. An attempt at histo-clinical classification. Acta Pathol Microbiol Scand 64:31–49. https://doi.org/10.1111/apm.1965.64.1.31

Lee S, Shin MG, Jo WH, Kim MJ, Kim HR, Lee WS, Park DH, Won JH, Shin JH, Suh SP, Ryang DW (2007) Association between *Helicobacter pylori*-related peptic ulcer tissue and somatic mitochondrial DNA mutations. Clin Chem 53(7):1390–1392. https://doi.org/10.1373/clinchem.2007.088047

Lee HC, Huang KH, Yeh TS, Chi CW (2014) Somatic alterations in mitochondrial DNA and mitochondrial dysfunction in gastric cancer progression. World J Gastroenterol 20(14):3950–3959. https://doi.org/10.3748/wjg.v20.i14.3950

Lee WP, Hou MC, Lan KH, Li CP, Chao Y, Lin HC, Lee SD (2016) *Helicobacter pylori*-induced chronic inflammation causes telomere shortening of gastric mucosa by promoting PARP-1-mediated non-homologous end joining of DNA. Arch Biochem Biophys 606:90–98. https://doi.org/10.1016/j.abb.2016.07.014

Li Q, Jia Z, Wang L, Kong X, Li Q, Guo K, Tan D, Le X, Wei D, Huang S, Mishra L, Xie K (2012) Disruption of Klf4 in villin-positive gastric progenitor cells promotes formation and progression of tumors of the antrum in mice. Gastroenterology 142(3):531–542. https://doi.org/10.1053/j.gastro.2011.11.034

Ling SS, Yeoh KG, Ho B (2013) *Helicobacter pylori* γ-glutamyl transpeptidase: a formidable virulence factor. World J Gastroenterol 19(45):8203–8210. https://doi.org/10.3748/wjg.v19.i45.8203

Liu L, Gerson SL (2006) Targeted modulation of MGMT: clinical implication. Clin Cancer Res 12(2):328–331. https://doi.org/10.1158/1078-0432.CCR-05-2543

Liu X, Meltzer SJ (2017) Gastric cancer in the era of precision medicine. Cell Mol Gastroenterol Hepatol 3(3):348–358. https://doi.org/10.1016/j.jcmgh.2017.02.003

Machado AM, Figueiredo C, Seruca R, Rasmussen RJ (2010) *Helicobacter pylori* infection generates genetic instability in gastric cells. Biochim Biophys Acta 1806(1):58–65. https://doi.org/10.1016/j.bbcan.2010.01.007

Machado AM, Desler C, Bøggild S, Strickertsson JA, Friis-Hansen L, Figueiredo C, Seruca R, Rasmussen LJ (2013) *Helicobacter pylori* infection affects mitochondrial function and DNA repair, thus, mediating genetic instability in gastric cells. Mech Ageing Dev 134(10):460–466. https://doi.org/10.1016/j.mad.2013.08.004

Maeda M, Moro H, Ushijima T (2017) Mechanisms for the induction of gastric cancer by *Helicobacter pylori* infection: aberrant DNA methylation pathway. Gastric Cancer 20(Suppl 1):8–15. https://doi.org/10.1007/s10120-016-0650-0

Maeda M, Yamashita S, Shimazu T, Iida N, Takeshima H, Nakajima T, Oda I, Nanjo S, Kusano C, Mori A, Moro H, Yamada H, Tsugane S, Sugiyama T, Sakai Y, Ushijima T (2018) Novel epigenetic markers for gastric cancer risk stratification in individuals after *Helicobacter pylori* eradication. Gastric Cancer. https://doi.org/10.1007/s10120-018-0803-4

Maleki SS, Röcken C (2017) Chromosomal instability in gastric cancer biology. Neoplasia 19(5):412–420. https://doi.org/10.1016/j.neo.2017.02.012

Markkanen E (2017) Not breathing is not an option: how to deal with oxidative DNA damage. DNA Repair (Amst) 59:82–105. https://doi.org/10.1016/j.dnarep.2017.09.007

Matsubara S, Takasu S, Tsukamoto T, Mutoh M, Masuda S, Sugimura T, Wakabayashi K, Totsuka Y (2012) Induction of glandular stomach cancers in *Helicobacter pylori*-infected Mongolian gerbils by 1-nitrosoindole-3-acetonitrile. Int J Cancer 130(2):259–266. https://doi.org/10.1002/ijc.26020

Matsushima K, Isomoto H, Inoue N, Nakayama T, Hayashi T, Nakayama M, Nakao K, Hirayama T, Kohno S (2011) MicroRNA signatures in *Helicobacter pylori*-infected gastric mucosa. Int J Cancer 128(2):361–370. https://doi.org/10.1002/ijc.25348

McDonald SA, Greaves LC, Gutierrez-Gonzalez L, Rodriguez-Justo M, Deheragoda M, Leedham SJ, Taylor RW, Lee CY, Preston SL, Lovell M, Hunt T, Elia G, Oukrif D, Harrison R, Novelli MR, Mitchell I, Stoker DL, Turnbull DM, Jankowski JA, Wright NA (2008) Mechanisms of field cancerization in the human stomach: the expansion and spread of mutated gastric stem cells. Gastroenterology 134(2):500–510. https://doi.org/10.1053/j.gastro.2007.11.035

Miftahussurur M, Yamaoka Y, Graham DY (2017) *Helicobacter pylori* as an oncogenic pathogen, revisited. Expert Rev Mol Med 19:e4. https://doi.org/10.1017/erm.2017.4

Mika D, Guruvayoorappan C (2011) Myeloperoxidase: the yin and yang in tumour progression. J Exp Ther Oncol 9(2):93–100. PMID: 21699016 http://www.oldcitypublishing.com/pdf/2427

Mirzaee V, Molaei M, Shalmani HM, Zali MR (2008) *Helicobacter pylori* infection and expression of DNA mismatch repair proteins. World J Gastroenterol 14(43):6717–6721. https://doi.org/10.3748/wjg.14.6717

Morisawa T, Marusawa H, Ueda Y, Iwai A, Okazaki IM, Honjo T, Chiba T (2008) Organ-specific profiles of genetic changes in cancers caused by activation-induced cytidine deaminase expression. Int J Cancer 123(12):2735–2740. https://doi.org/10.1002/ijc.23853

Muhammad JS, Nanjo S, Ando T, Yamashita S, Maekita T, Ushijima T, Tabuchi Y, Sugiyama T (2017) Autophagy impairment by *Helicobacter pylori*-induced methylation silencing of MAP1LC3Av1 promotes gastric carcinogenesis. Int J Cancer 140(10):2272–2283. https://doi.org/10.1002/ijc.30657

Nagata N, Akiyama J, Marusawa H, Shimbo T, Liu Y, Igari T, Nakashima R, Watanabe H, Uemura N, Chiba T (2014) Enhanced expression of activation-induced cytidine deaminase in human gastric mucosa infected by *Helicobacter pylori* and its decrease following eradication. J Gastroenterol 49(3):427–435. https://doi.org/10.1007/s00535-013-0808-z

Nagini S (2012) Carcinoma of the stomach: a review of epidemiology, pathogenesis, molecular genetics and chemoprevention. World J Gastrointest Oncol 4(7):156–169. https://doi.org/10.4251/wjgo.v4.i7.156

Naito Y, Takagi T, Okada H, Nukigi Y, Uchiyama K, Kuroda M, Handa O, Kokura S, Yagi N, Kato Y, Osawa T, Yoshikawa T (2008) Expression of inducible nitric oxide synthase and nitric oxide-modified proteins in *Helicobacter pylori*-associated atrophic gastric mucosa. J Gastroenterol Hepatol 23(Suppl 2):S250–S257. https://doi.org/10.1111/j.1440-1746.2008.05412.x

Nakamura J, Tanaka T, Kitajima Y, Noshiro H, Miyazaki K (2014) Methylation-mediated gene silencing as biomarkers of gastric cancer: a review. World J Gastroenterol 20(34):1991–2006. https://doi.org/10.3748/wjg.i34.11991

Nathan C, Cunningham-Bussel A (2013) Beyond oxidative stress: an immunologist's guide to reactive oxygen species. Nat Rev Immunol 13(5):349–361. https://doi.org/10.1038/nri3423

Naumann M, Sokolova O, Tegtmeyer N, Backert S (2017) *Helicobacter pylori*: a paradigm pathogen for subverting host cell signal transmission. Trends Microbiol 25(4):316–328. https://doi.org/10.1016/j.tim.2016.12.004

Nishimura T (2008) Total number of genome alterations in sporadic gastrointestinal cancer inferred from pooled analyses in the literature. Tumour Biol 29(6):343–350. https://doi.org/10.1159/000176044

Nishizawa T, Suzuki H (2015) Gastric carcinogenesis and underlying molecular mechanisms: *Helicobacter pylori* and novel targeted therapy. Biomed Res Int 2015:794378. https://doi.org/10.1155/2015/794378

Niwa T, Toyoda T, Tsukamoto T, Mori A, Tatematsu M, Ushijima T (2013) Prevention of *Helicobacter pylori*-induced gastric cancers in gerbils by a DNA demethylating agent. Cancer Prev Res (Philadelphia, PA) 6(4):263–270. https://doi.org/10.1158/1940-6207.capr-12-0369

Noto JM, Khizanishvili T, Chaturvedi R, Piazuelo MB, Romero-Gallo J, Delgado AG, Khurana SS, Sierra JC, Krishna US, Suarez G, Powell AE, Goldenring JR, Coffey RJ, Yang VW, Correa P, Mills JC, Wilson KT, Peek RM Jr (2013) *Helicobacter pylori* promotes the expression of Krüppel-like factor 5, a mediator of carcinogenesis, in vitro and in vivo. PLoS ONE 8(1):e54344. https://doi.org/10.1371/journal.pone.0054344

Ohnishi N, Yuasa H, Tanaka S, Sawa H, Miura M, Matsui A, Higashi H, Musashi M, Iwabuchi K, Suzuki M, Yamada G, Azuma T, Hatakeyama M (2008) Transgenic expression of *Helicobacter pylori* CagA induces gastrointestinal and hematopoietic neoplasms in mouse. Proc Natl Acad Sci USA 105(3):1003–1008. https://doi.org/10.1073/pnas.0711183105

Oldani A, Cormont M, Hofman V, Chiozzi V, Oregioni O, Canonici A, Sciullo A, Sommi P, Fabbri A, Ricci V, Boquet P (2009) *Helicobacter pylori* counteracts the apoptotic action of its VacA toxin by injecting the CagA protein into gastric epithelial cells. PLoS Pathog 5(10):e1000603. https://doi.org/10.1371/journal.ppat.1000603

Park DI, Park SH, Kim SH, Kim JW, Cho YK, Kim HJ, Sohn CI, Jeon WK, Kim BI, Cho EY, Kim EJ, Chae SW, Sohn JH, Sung IK, Sepulveda AR, Kim JJ (2005) Effect of *Helicobacter pylori* infection on the expression of DNA mismatch repair protein. Helicobacter 10(3):179–184. https://doi.org/10.1111/j.1523-5378.2005.00309.x

Park SY, Yoo EJ, Cho NY, Kim N, Kang GH (2009) Comparison of CpG island hypermethylation and repetitive DNA hypomethylation in premalignant stages of gastric cancer, stratified for *Helicobacter pylori* infection. J Pathol 219(4):410–416. https://doi.org/10.1002/path.2596

Parsonnet J, Friedman GD, Orentreich N, Vogelman H (1997) Risk for gastric cancer in people with CagA positive or CagA negative *Helicobacter pylori* infection. Gut 40(3):297–301. https://www.ncbi.nlm.nih.gov/pubmed/9135515

Perri F, Cotugno R, Piepoli A, Merla A, Quitadamo M, Gentile A, Pilotto A, Annese V, Andriulli A (2007) Aberrant DNA methylation in non-neoplastic gastric mucosa of *Helicobacter pylori* infected patients and effect of eradication. Am J Gastroenterol 102(7):1361–1371. https://doi.org/10.1111/j.1572-0241.2007.01284.x

Persson C, Canedo P, Machado JC, El-Omar EM, Forman D (2011) Polymorphisms in inflammatory response genes and their association with gastric cancer: a HuGE systematic review and meta-analyses. Am J Epidemiol 173(3):259–270. https://doi.org/10.1093/aje/kwq370

Pollard TD, Earnshaw WC, Lippincott-Schwartz J (2008) Cell Biology, 2nd edn. Saunders/Elsevier, Philadelphia, p 905

Poplawski T, Chojnacki C, Czubatka A, Klupinska G, Chojnacki J, Blasiak J (2013) *Helicobacter pylori* infection and antioxidants can modulate the genotoxic effects of heterocyclic amines in gastric mucosa cells. Mol Biol Rep 40(8):5205–5212. https://doi.org/10.1007/s11033-013-2622-3

Posselt G, Backert S, Wessler S (2013) The functional interplay of *Helicobacter pylori* factors with gastric epithelial cells induces a multi-step process in pathogenesis. Cell Commun Signal 11:77. https://doi.org/10.1186/1478-811X-11-77

Qian X, Huang C, Cho CH, Hui WM, Rashid A, Chan AO (2008) E-cadherin promoter hypermethylation induced by interleukin-1beta treatment or *Helicobacter pylori* infection in human gastric cancer cell lines. Cancer Lett 263(1):107–113. https://doi.org/10.1016/j.canlet.2007.12.023

Raza Y, Khan A, Farooqui A, Mubarak M, Facista A, Akhtar SS, Khan S, Kazi JI, Bernstein C, Kazmi SU (2014) Oxidative DNA damage as a potential early biomarker of *Helicobacter pylori* associated carcinogenesis. Pathol Oncol Res 20(4):839–846. https://doi.org/10.1007/s12253-014-9762-1

Reuter S, Gupta SC, Chaturvedi MM, Aggarwal BB (2010) Oxidative stress, inflammation, and cancer: how are they linked? Free Radic Biol Med 49(11):1603–1616. https://doi.org/10.1016/j.freeradbiomed.2010.09.006

Ricci V (2016) Relationship between VacA toxin and host cell autophagy in *Helicobacter pylori* infection of the human stomach: a few answers, many questions. Toxins (Basel) 8(7): pii E203. https://doi.org/10.3390/toxins8070203

Rocha GA, Rocha AM, Gomes AD, Faria CL Jr, Melo FF, Batista SA, Fernandes VC, Almeida NB, Teixeira KN, Brito KS, Queiroz DM (2015) STAT3 polymorphism and *Helicobacter pylori* CagA strains with higher number of EPIYA-C segments independently increase the risk of gastric cancer. BMC Cancer 15:528. https://doi.org/10.1186/s12885-015-1533-1

Ronchetti L, Melucci E, De Nicola F, Goeman F, Casini B, Sperati F, Pallocca M, Terrenato I, Pizzuti L, Vici P, Sergi D, Di Lauro L, Amoreo CA, Gallo E, Diodoro MG, Pescarmona E, Vitale I, Barba M, Buglioni S, Mottolese M, Fanciulli M, De Maria R, Maugeri-Saccà M (2017) DNA damage repair and survival outcomes in advanced gastric cancer patients treated with first-line chemotherapy. Int J Cancer 140(11):2587–2595. https://doi.org/10.1002/ijc.30668

Sahan AZ, Hazra TK, Das S (2018) The pivotal role of DNA repair in infection mediated-inflammation and cancer. Front Microbiol 9:663. https://doi.org/10.3389/fmicb.2018.00663

Santos JC, Ribeiro ML (2015) Epigenetic regulation of DNA repair machinery in *Helicobacter pylori*-induced gastric carcinogenesis. World J Gastroenterol 21(30):9021–9037. https://doi.org/10.3748/wjg.v21.i30.9021

Santos JC, Brianti MT, Almeida VR, Ortega MM, Fischer W, Haas R, Matheu A, Ribeiro ML (2017) *Helicobacter pylori* infection modulates the expression of miRNAs associated with DNA mismatch repair pathway. Mol Carcinog 56(4):1372–1379. https://doi.org/10.1002/mc.22590

Sasaki T, Kuniyasu H, Luo Y, Kitayoshi M, Tanabe E, Kato D, Shinya S, Fujii K, Ohmori H, Yamashita Y (2013) Increased phosphorylation of AKT in high-risk gastric mucosa. Anticancer Res 33(8):3295–3300 PMID: 23898095

Satin B, Del Giudice G, Della Bianca V, Dusi S, Laudanna C, Tonello F, Kelleher D, Rappuoli R, Montecucco C, Rossi F (2000) The neutrophil-activating protein (HP-NAP) of *Helicobacter pylori* is a protective antigen and a major virulence factor. J Exp Med 191:1467–1476. https://doi.org/10.1084/jem.191.9.1467

Schmees C, Prinz C, Treptau T, Rad R, Hengst L, Voland P, Bauer S, Brenner L, Schmid RM, Gerhard M (2007) Inhibition of T-cell proliferation by *Helicobacter pylori* gamma-glutamyltranspeptidase. Gastroenterology 132(5):1820–1833. https://doi.org/10.1053/j.gastro.2007.02.031

Sell S (2011) Infection, stem cells and cancer signals. Curr Pharm Biotechnol 12(2):182–188. https://doi.org/10.2174/138920111794295675

Serizawa T, Hirata Y, Hayakawa Y, Suzuki N, Sakitani K, Hikiba Y, Ihara S, Kinoshita H, Nakagawa H, Tateishi K, Koike K (2015) Gastric metaplasia induced by *Helicobacter pylori* is associated with enhanced SOX9 expression via Interleukin-1 signaling. Infect Immun 84(2):562–572. https://doi.org/10.1128/IAI.01437-15

Shahi H, Bahreiny R, Reiisi S (2016) *Helicobacter pylori* and its virulence factors' effect on serum oxidative DNA damages in adults with dyspepsia. Acta Med Iran 54(11):704–708 PMID: 28033692

Sheh A, Lee CW, Masumura K, Rickman BH, Nohmi T, Wogan GN, Fox JG, Schauer DB (2010) Mutagenic potency of *Helicobacter pylori* in the gastric mucosa of mice is determined by sex and duration of infection. Proc Natl Acad Sci USA 107(34):15217–15222. https://doi.org/10.1073/pnas.1009017107

Shimizu T, Marusawa H, Endo Y, Chiba T (2012) Inflammation-mediated genomic instability: roles of activation-induced cytidine deaminase in carcinogenesis. Cancer Sci 103(7):1201–1206. https://doi.org/10.1111/j.1349-7006.2012.02293.x

Shimizu T, Marusawa H, Matsumoto Y, Inuzuka T, Ikeda A, Fujii Y, Minamiguchi S, Miyamoto S, Kou T, Sakai Y, Crabtree JE, Chiba T (2014) Accumulation of somatic mutations in TP53 in gastric epithelium with *Helicobacter pylori* infection. Gastroenterology 147(2):407–417.e3. https://doi.org/10.1053/j.gastro.2014.04.036

Shimizu T, Chiba T, Marusawa H (2017) *Helicobacter pylori*-mediated genetic instability and gastric carcinogenesis. Curr Top Microbiol Immunol 400:305–323. https://doi.org/10.1007/978-3-319-50520-6_13

Shiotani A, Cen P, Graham DY (2013) Eradication of gastric cancer is now both possible and practical. Semin Cancer Biol 23(6 Pt B):492–501. https://doi.org/10.1016/j.semcancer.2013.07.004

Sigal M, Rothenberg ME, Logan CY, Lee JY, Honaker RW, Cooper RL, Passarelli B, Camorlinga M, Bouley DM, Alvarez G, Nusse R, Torres J, Amieva MR (2015) *Helicobacter pylori* activates and expands Lgr5(+) stem cells through direct colonization of the gastric glands. Gastroenterology 148(7):1392–1404. https://doi.org/10.1053/j.gastro.2015.02.049

Singh SR (2013) Gastric cancer stem cells: a novel therapeutic target. Cancer Lett 338(1):110–119. https://doi.org/10.1016/j.canlet.2013.03.035

Slavin T, Neuhausen SL, Rybak C, Solomon I, Nehoray B, Blazer K, Niell-Swiller M, Adamson AW, Yuan YC, Yang K, Sand S, Castillo D, Herzog J, Wu X, Tao S, Chavez T, Woo Y, Chao J, Mora P, Horcasitas D, Weitzel J (2017) Genetic gastric cancer susceptibility in the international clinical cancer genomics community research network. Cancer Genet 216–217:111–119. https://doi.org/10.1016/j.cancergen.2017.08.001

Smith HC, Bennett RP, Kizilyer A, McDougall WM, Prohaska KM (2012) Functions and regulation of the APOBEC family of proteins. Semin Cell Dev Biol 23(3):258–268. https://doi.org/10.1016/j.semcdb.2011.10.004

Sokolova O, Naumann M (2017) NF-κB signaling in gastric cancer. Toxins (Basel) 9(4):pii E119. https://doi.org/10.3390/toxins9040119

Sokolova O, Borgmann M, Rieke C, Schweitzer K, Rothkötter HJ, Naumann M (2013) *Helicobacter pylori* induces type 4 secretion system-dependent, but CagA-independent activation of IκBs and NF-κB/RelA at early time points. Int J Med Microbiol 303(8):548–552. https://doi.org/10.1016/j.ijmm.2013.07.008

Sokolova O, Maubach G, Naumann M (2014) MEKK3 and TAK1 synergize to activate IKK complex in *Helicobacter pylori* infection. Biochim Biophys Acta 1843(4):715–724. https://doi.org/10.1016/j.bbamcr.2014.01.006

Stairs DB, Kong J, Lynch JP (2010) Cdx genes, inflammation, and the pathogenesis of intestinal metaplasia. Prog Mol Biol Transl Sci 96:231–270. https://doi.org/10.1016/B978-0-12-381280-3.00010-5

Stracker TH, Roig I, Knobel PA, Marjanović M (2013) The ATM signaling network in development and disease. Front Genet 4:37. https://doi.org/10.3389/fgene.2013.00037

Sulzbach DE, Oliveira HS, Biolchi V, Richardt Medeiros HR, Bizerra Gandor Jantsch DB, Oliveira Knabben DE, Becker Delving LK, Reckziegel R, Goettert MI, Brum IS, Pozzobon A (2016) Effect of *Helicobacter pylori* on *NFKB1*, *p38α* and *TNF-α* mRNA expression levels in human gastric mucosa. Exp Ther Med 11(6):2365–2372. https://doi.org/10.3892/etm.2016.3213

Tahara E (2004) Genetic pathways of two types of gastric cancer. IARC Sci Publ 157:327–349 PMID: 15055305

Takaishi S, Okumura T, Tu S, Wang SS, Shibata W, Vigneshwaran R, Gordon SA, Shimada Y, Wang TC (2009) Identification of gastric cancer stem cells using the cell surface marker CD44. Stem Cells 27(5):1006–1020. https://doi.org/10.1002/stem.30

Takeda Y, Yashima K, Hayashi A, Sasaki S, Kawaguchi K, Harada K, Murawaki Y, Ito H (2012) Expression of AID, P53, and Mlh1 proteins in endoscopically resected differentiated-type early gastric cancer. World J Gastrointest Oncol 4(6):131–137. https://doi.org/10.4251/wjgo.v4.i6.131

Tamura G (2006) Alterations of tumor suppressor and tumor-related genes in the development and progression of gastric cancer. World J Gastroenterol 12(2):192–198. https://doi.org/10.3748/wjg.v12.i2.192

Tegtmeyer N, Zabler D, Schmidt D, Hartig R, Brandt S, Backert S (2009) Importance of EGF receptor, HER2/Neu and Erk1/2 kinase signalling for host cell elongation and scattering induced by the *Helicobacter pylori* CagA protein: antagonistic effects of the vacuolating cytotoxin VacA. Cell Microbiol 11(3):488–505. https://doi.org/10.1111/j.1462-5822.2008.01269.x

Tegtmeyer N, Neddermann M, Asche CI, Backert S (2017) Subversion of host kinases: a key network in cellular signaling hijacked by *Helicobacter pylori* CagA. Mol Microbiol 105(3):358–372. https://doi.org/10.1111/mmi.13707

Terebiznik MR, Raju D, Vázquez CL, Torbricki K, Kulkarni R, Blanke SR, Yoshimori T, Colombo MI, Jones NL (2009) Effect of *Helicobacter pylori*'s vacuolating cytotoxin on the autophagy pathway in gastric epithelial cells. Autophagy 5(3):370–379. https://doi.org/10.4161/auto.5.3.7663

Toller IM, Neelsen KJ, Steger M, Hartung ML, Hottiger MO, Stucki M, Kalali B, Gerhard M, Sartori AA, Lopes M, Müller A (2011) Carcinogenic bacterial pathogen *Helicobacter pylori* triggers DNA double-strand breaks and a DNA damage response in its host cells. Proc Natl Acad Sci USA 108(36):14944–14949. https://doi.org/10.1073/pnas.1100959108

Touati E (2010) When bacteria become mutagenic and carcinogenic: lessons from *Helicobacter pylori*. Mutat Res 703(1):66–70. https://doi.org/10.1016/j.mrgentox.2010.07.014

Touati E, Michel V, Thiberge JM, Wuscher N, Huerre M, Labigne A (2003) Chronic *Helicobacter pylori* infections induce gastric mutations in mice. Gastroenterology 124(5):1408–1419. https://doi.org/10.1016/S0016-5085(03)00266-X

Toyoda T, Tsukamoto T, Hirano N, Mizoshita T, Kato S, Takasu S, Ban H, Tatematsu M (2008) Synergistic upregulation of inducible nitric oxide synthase and cyclooxygenase-2 in gastric mucosa of Mongolian gerbils by a high-salt diet and *Helicobacter pylori* infection. Histol Histopathol 23:593–599. https://doi.org/10.14670/HH-23.593

Tsai YC, Li PY, Chen CC, Liu YC (2013) Is the oxidative DNA damage level of human lymphocyte correlated with the antioxidant capacity of serum or the base excision repair activity of lymphocyte? Oxid Med Cell Longev 2013:237583. https://doi.org/10.1155/2013/237583

Tsugawa H, Suzuki H, Saya H, Hatakeyama M, Hirayama T, Hirata K, Nagano O, Matsuzaki J, Hibi T (2012) Reactive oxygen species-induced autophagic degradation of *Helicobacter pylori* CagA is specifically suppressed in cancer stem-like cells. Cell Host & Microbe 12(6):764–777. https://doi.org/10.1016/j.chom.2012.10.014

Tu S, Bhagat G, Cui G, Takaishi S, Kurt-Jones EA, Rickman B, Betz KS, Penz-Oesterreicher M, Bjorkdahl O, Fox JG, Wang TC (2008) Overexpression of interleukin-1beta induces gastric inflammation and cancer and mobilizes myeloid-derived suppressor cells in mice. Cancer Cell 14(5):408–419. https://doi.org/10.1016/j.ccr.2008.10.011

Ueda T, Volinia S, Okumura H, Shimizu M, Taccioli C, Rossi S, Alder H, Liu CG, Oue N, Yasui W, Yoshida K, Sasaki H, Nomura S, Seto Y, Kaminishi M, Calin GA, Croce CM (2010) Relation between microRNA expression and progression and prognosis of gastric cancer: a microRNA expression analysis. Lancet Oncol 11(2):136–146. https://doi.org/10.1016/S1470-2045(09)70343-2

Uehara T, Ma D, Yao Y, Lynch JP, Morales K, Ziober A, Feldman M, Ota H, Sepulveda AR (2013) *Helicobacter pylori* infection is associated with DNA damage of Lgr5-positive epithelial stem cells in the stomach of patients with gastric cancer. Dig Dis Sci 58(1):140–149. https://doi.org/10.1007/s10620-012-2360-8

Ushijima T, Hattori N (2012) Molecular pathways: involvement of *Helicobacter pylori*-triggered inflammation in the formation of an epigenetic field defect, and its usefulness as cancer risk and exposure markers. Clin Cancer Res 18(4):923–929. https://doi.org/10.1158/1078-0432.CCR-11-2011

Ushijima T, Nakajima T, Maekita T (2006) DNA methylation as a marker for the past and future. J Gastroenterol 41(5):401–407. https://doi.org/10.1007/s00535-006-1846-6

Valenzuela MA, Canales J, Corvalán AH, Quest AF (2015) *Helicobacter pylori*-induced inflammation and epigenetic changes during gastric carcinogenesis. World J Gastroenterol 21(45):12742–12756. https://doi.org/10.3748/wjg.v21.i45.12742

van Loon B, Markkanen E, Hübscher U (2010) Oxygen as a friend and enemy: how to combat the mutational potential of 8-oxo-guanine. DNA Repair (Amst) 9(6):604–616. https://doi.org/10.1016/j.dnarep.2010.03.004

Varon C, Dubus P, Mazurier F, Asencio C, Chambonnier L, Ferrand J, Giese A, Senant-Dugot N, Carlotti M, Mégraud F (2012) *Helicobacter pylori* infection recruits bone marrow-derived cells that participate in gastric preneoplasia in mice. Gastroenterology 142(2):281–291. https://doi.org/10.1053/j.gastro.2011.10.036

Velho S, Fernandes MS, Leite M, Figueiredo C, Seruca R (2014) Causes and consequences of microsatellite instability in gastric carcinogenesis. World J Gastroenterol 20(44):16433–16442. https://doi.org/10.3748/wjg.v20.i44.16433

Vidal AE, Hickson ID, Boiteux S, Radicella JP (2001) Mechanism of stimulation of the DNA glycosylase activity of hOGG1 by the major human AP endonuclease: bypass of the AP lyase activity step. Nucleic Acids Res 29(6):1285–1292. https://www.ncbi.nlm.nih.gov/pmc/articles/PMC29755/

Virchow R (1863) Die Krankhaften Geschwülste. Vorlesungen über Pathologie. Verlag von August Hirschwald, Berlin, p 543

Wada Y, Takemura K, Tummala P, Uchida K, Kitagaki K, Furukawa A, Ishige Y, Ito T, Hara Y, Suzuki T, Mimuro H, Board PG, Eishi Y (2018) *Helicobacter pylori* induces somatic mutations in *TP53* via overexpression of CHAC1 in infected gastric epithelial cells. FEBS Open Bio 8(4):671–679. https://doi.org/10.1002/2211-5463.12402

Wakamatsu Y, Sakamoto N, Oo HZ, Naito Y, Uraoka N, Anami K, Sentani K, Oue N, Yasui W (2012) Expression of cancer stem cell markers ALDH1, CD44 and CD133 in primary tumor and lymph node metastasis of gastric cancer. Pathol Int 62(2):112–119. https://doi.org/10.1111/j.1440-1827.2011.02760.x

Waldum HL, Öberg K, Sørdal ØF, Sandvik AK, Gustafsson BI, Mjønes P, Fossmark R (2018) Not only stem cells, but also mature cells, particularly neuroendocrine cells, may develop into tumours: time for a paradigm shift. Therap Adv Gastroenterol 11:1756284818775054. https://doi.org/10.1177/1756284818775054

Wallace DC (2012) Mitochondria and cancer. Nat Rev Cancer 12:685–698. https://doi.org/10.1038/nrc3365

Wang J, Xu L, Shi R, Huang X, Li SW, Huang Z, Zhang G (2011) Gastric atrophy and intestinal metaplasia before and after *Helicobacter pylori* eradication: a meta-analysis. Digestion 83(4):253–260. https://doi.org/10.1159/000280318

Wang XW, Wu Y, Wang D, Qin ZF (2014) MicroRNA network analysis identifies key microRNAs and genes associated with precancerous lesions of gastric cancer. Genet Mol Res 13(4):8695–8703. https://doi.org/10.4238/2014.October.27.10

Wang YK, Chiang WC, Kuo FC, Wu MC, Shih HY, Wang SSW, Liu CJ, Chen YH, Wu DC, Su WW, Huang YL (2018) Levels of malondialdehyde in the gastric juice: its association with *Helicobacter pylori* infection and stomach diseases. Helicobacter 23(2):e12460. https://doi.org/10.1111/hel.12460

Ward MH, Cross AJ, Abnet CC, Sinha R, Markin RS, Weisenburger DD (2012) Heme iron from meat and risk of adenocarcinoma of the esophagus and stomach. Eur J Cancer Prev 21(2):134–138. https://doi.org/10.1097/CEJ.0b013e32834c9b6c

Wei J, Noto J, Zaika E, Romero-Gallo J, Correa P, El-Rifai W, Peek RM, Zaika A (2012) Pathogenic bacterium *Helicobacter pylori* alters the expression profile of p53 protein isoforms and p53 response to cellular stresses. Proc Natl Acad Sci USA 109(38):E2543–E2550. https://doi.org/10.1073/pnas.1205664109

Woerner SM, Yuan YP, Benner A, Korff S, von Knebel Doeberitz M, Bork P (2010) SelTarbase, a database of human mononucleotide-microsatellite mutations and their potential impact to tumorigenesis and immunology. Nucleic Acids Res 38:D682–D689. https://doi.org/10.1093/nar/gkp839

Xia G, Schneider-Stock R, Diestel A, Habold C, Krueger S, Roessner A, Naumann M, Lendeckel U (2008) *Helicobacter pylori* regulates p21(WAF1) by histone H4 acetylation. Biochem Biophys Res Commun 369(2):526–531. https://doi.org/10.1016/j.bbrc.2008.02.073

Yadav VK, DeGregori J, De S (2016) The landscape of somatic mutations in protein coding genes in apparently benign human tissues carries signatures of relaxed purifying selection. Nucleic Acids Res 44(5):2075–2084. https://doi.org/10.1093/nar/gkw086

Yamashita K, Sakuramoto S, Watanabe M (2011) Genomic and epigenetic profiles of gastric cancer: potential diagnostic and therapeutic applications. Surg Today 41(1):24–38. https://doi.org/10.1007/s00595-010-4370-5

Yan YG, Zhao G, Ma JP, Cai SR, Zhan WH (2008) Effects of different *Helicobacter pylori* culture filtrates on growth of gastric epithelial cells. World J Gastroenterol 14(23):3745–3749. https://doi.org/10.3748/wjg.14.3745

Yanaoka K, Oka M, Yoshimura N, Deguchi H, Mukoubayashi C, Enomoto S, Maekita T, Inoue I, Ueda K, Utsunomiya H, Iguchi M, Tamai H, Fujishiro M, Nakamura Y, Tsukamoto T, Inada K, Takeshita T, Ichinose M (2010) Preventive effects of etodolac, a selective cyclooxygenase-2 inhibitor, on cancer development in extensive metaplastic gastritis, a *Helicobacter pylori*-negative precancerous lesion. Int J Cancer 126(6):1467–1473. https://doi.org/10.1002/ijc.24862

Yang B, Fu X, Shao L, Ding Y, Zeng D (2014) Aberrant expression of SIRT3 is conversely correlated with the progression and prognosis of human gastric cancer. Biochem Biophys Res Commun 443(1):156–160. https://doi.org/10.1016/j.bbrc.2013.11.068

Yeniova AO, Uzman M, Kefeli A, Basyigit S, Ata N, Dal K, Guresci S, Nazligul Y (2015) Serum 8 Hydroxydeoxyguanosine and Cytotoxin Associated Gene A as Markers for *Helicobacter pylori* Infection. Asian Pac J Cancer Prev 16(13):5199–5203 PMID: 26225653

Zhang LJ, Wang SY, Huo XH, Zhu ZL, Chu JK, Ma JC, Cui DS, Gu P, Zhao ZR, Wang MW, Yu J (2009) Anti-*Helicobacter pylori* therapy followed by celecoxib on progression of gastric precancerous lesions. World J Gastroenterol 15(22):2731–2738. https://doi.org/10.3748/wjg.15.2731

Zhang XS, Tegtmeyer N, Traube L, Jindal S, Perez-Perez G, Sticht H, Backert S, Blaser MJ (2015) A specific A/T polymorphism in western tyrosine phosphorylation B-motifs regulates *Helicobacter pylori* CagA epithelial cell interaction. PloS Pathog 11:e1004621. https://doi.org/10.1371/journal.ppat.1004621

Importance of Toll-like Receptors in Pro-inflammatory and Anti-inflammatory Responses by *Helicobacter pylori* Infection

Hiroyuki Nagashima and Yoshio Yamaoka

Contents

Abstract Infectious diseases have been paramount among the threats to human health and survival throughout evolutionary history. Bacterial cell-surface molecules are key factors in the microorganism–host crosstalk, as they can interact with host pattern-recognition receptors (PRRs) of the gastrointestinal mucosa. The best-studied PRRs are toll-like receptors (TLRs). Because TLRs play an important key role in host defense, they have received increasing interest in the evolutionary and population genetics literature, and their variation represents a potential target of adaptive evolution. *Helicobacter pylori* is one of the commensal bacteria in our body and can have pathogenic properties in a subset of infected people. The history of *H. pylori* research indicated that humans and bacteria co-evolved during evolution. A genome-wide association study (GWAS) has opened the way for investigating the genomic evolution of bacterial pathogens during the colonization and infection of humans. Recent GWAS research emphasized the importance of TLRs, especially

H. Nagashima
Department of Gastroenterology, Hokkaido Cancer Center, Sapporo, Hokkaido, Japan

H. Nagashima · Y. Yamaoka (✉)
Department of Environmental and Preventive Medicine, Faculty of Medicine, Oita University, 1-1 Idaigaoka, Hasama-machi, Yufu-City, Oita 879-5593, Japan
e-mail: yyamaoka@oita-u.ac.jp

Y. Yamaoka
Department of Gastroenterology and Hepatology, Baylor College of Medicine, Houston, TX, USA

S. Backert (ed.), *Molecular Mechanisms of Inflammation: Induction, Resolution and Escape by Helicobacter pylori*, Current Topics in Microbiology and Immunology 421, https://doi.org/10.1007/978-3-030-15138-6_6

TLR10 during pathogenesis in *H. pylori* infection. We demonstrated that TLR10, whose ligand was unknown for a long time, can recognize *H. pylori* LPS. Our results of *H. pylori* research suggest that TLR10 might play an important role to also recognize other commensal bacteria. In this review, we discuss the importance of TLRs in pro-inflammatory and anti-inflammatory responses by *H. pylori* infection. Especially, we highlight the TLR10 interaction with *H. pylori* infection, providing new insights about TLR10 signaling.

1 Introduction

The gastric epithelium forms the first innate immune defense barrier against *Helicobacter pylori* (Naumann et al. 2017). The direct interaction of *H. pylori* with epithelial cells occurs via different adhesins and a type IV secretion system (T4SS) with subsequent exposure to several virulence factors encoded in the *cag* pathogenicity island (PAI) (Backert et al. 2015). The *cag*PAI-positive strains are clearly associated with the development of more severe gastric diseases and higher gastric cancer (GC) risk compared to *cag*PAI-negative strains (Censini et al. 1996; Posselt et al. 2013).

Whereas *cag*PAI-induced responses are best characterized and well understood in *H. pylori* infection, recently it has been revealed that the most important host factor in *H. pylori* infection is the innate immune system (Mayerle et al. 2013; et al. 2015; Zhang et al. 2015). The system of innate immunity recognizing *H. pylori* is very unique as it predominantly results in anti-inflammatory responses, rather than pro-inflammatory responses (Arnold et al. 2011). Several microorganism-associated molecular patterns (MAMPs) of *H. pylori* have evolved to evade excessive inflammation.

Genetic variation among pathogenic isolates and in the human hosts can explain variation in disease outcomes, and that interactions between the bacterial genome and host genome should be incorporated into genetic models of disease caused by infectious agents (see also Chapter "Genetic Polymorphisms in Inflammatory and Other Regulators in Gastric Cancer: Risks and Clinical Consequences" of this book). Genetic epidemiological studies, that fail to take both the pathogen and host into account, can lead to false and misleading conclusions about disease etiology.

Innate immunity is characterized by an immediate response against pathogens and is paramount in the initial control of infection. Toll-like receptors (TLRs) are thought to be among the most ancient pathogen recognition systems in mammals (Kawai and Akira 2010; Pachathundikandi et al. 2013). TLRs, which can be expressed on the cell surface or in intracellular compartments, detect various microorganisms by sensing viral nucleic acids and several bacterial MAMPs, and then signal through two major pathways, which are named according to their adaptor proteins: myeloid differentiation primary response protein 88 (MyD88) and TIR domain-containing adaptor, respectively (Schnare et al. 2001). Bacteria are selected by environmental pressures, resulting in quantitative or qualitative changes in their envelope structures that often promote evasion of host immune responses. Microbes continuously developed to escape from the host immune system. This evolutionary race has also led to the development and improvement of the innate immune system by the host. In general, the innate immune system clearly distinguishes beneficial and harmful microbes. During the evolutionary history of our species, infections shaped the genetics and function of the immune apparatus, leading to variation of host defense responses in human populations as a result of local infectious pressures and migration events (Kodaman et al. 2014a, b). Recent studies have shown that slight, adaptive changes in MAMPs in the bacterial cell wall may result in their ability to induce the secretion not only of pro-inflammatory cytokines, but also of anti-inflammatory cytokines (Bach 2018; see also Chapter "Mechanisms of Inflammasome Signaling, microRNA Induction and Resolution of Inflammation by *Helicobacter pylori*" of this book).

The evolution of TLRs has been studied by many research laboratories (Roach et al. 2005; Jann et al. 2008; Ferrer-Admetlla et al. 2008; Barreiro et al. 2009; Barreiro and Quintana-Murci 2010; Huang et al. 2011). Phylogenetic analyses showed that the vertebrate TLRs evolved independently by gene duplication prior to the divergence of protostomes and deuterostomes (Hughes and Piontkivska 2008). The phylogeny of each major vertebrate TLR family recapitulates the phylogeny of vertebrate species, and sequence analyses show that all vertebrate TLRs evolved at about the same slow rate, suggesting strong selection for maintenance of specific functions (Roach et al. 2005). This high conservation relates to the fact that microorganisms cannot easily mutate their MAMPs, which are recognized by TLRs.

H. pylori lipopolysaccharide (LPS) is an example of evading the recognition by the host's immune system through modifying its lipid A component. Despite a robust host immune response, *H. pylori* persists within the gastric niche and promotes a chronic infection for the duration of the host's life. LPS modification of *H. pylori* is considered as a commensal bacterial strategy to escape from host immune system. In this review, we discuss recent work on TLR research and *H. pylori* infection, especially focused on *H. pylori* LPS and TLR10.

2 Overview of TLRs

Individual persons display variable abilities to fight infections, as well as variable susceptibilities to inflammatory and auto-immune diseases. Accumulating evidence suggests that such heterogeneity reflects differences in the genetic development of a given host (Barreiro and Quintana-Murci 2010). For some decades, genetic studies have provided numerous examples of genes accounting for differences in the susceptibility to rare or common infectious diseases (Alcais et al. 2009; Chapman and Hill 2012). The advent of new exhaustive technologies, such as DNA microarrays and next-generation sequencing, has greatly accelerated the field of human genetics, made it possible to evaluate the contribution of genetic diversity to differences in immunity to infection at the level of the entire genome. For example, current reports have demonstrated the power of whole-genome sequencing for dissecting the immunological mechanisms that underlie the pathogenesis of severe, rare infectious diseases (Quintana-Murci and Clark 2013). Although these studies have provided the proof-of-concept, increasing susceptibility to infectious diseases may result from various types of inborn errors of immunity.

Innate immunity is at the front line of host defense against pathogens and is also important for controlling the microbiome. Thus, innate immunity-associated genes provide an excellent model for studies on the selective pressure that is exerted by microorganisms on the host genome. Infectious diseases have been paramount among the threats to human health and survival throughout evolutionary history. Natural selection is therefore expected to act strongly on host defense genes, particularly on innate immunity genes, whose products mediate the direct interaction between the host and the microbial environment. In insects and mammals, the TLRs appear to play a major role in initiating innate immune responses against microbes. In humans, however, it has been speculated that different TLRs are being under specific positive selection (Quintana-Murci and Clark 2013).

Population genetics studies have provided important insights in terms of the evolution and function of TLRs (Barreiro et al. 2009; Casanova et al. 2011; Mukherjee et al. 2009). In humans, TLRs are broadly subdivided into two groups: those primarily expressed at the cell surface (TLR1, TLR2, TLR4, TLR5, TLR6, and TLR10), which detect predominantly MAMPs from bacteria, fungi, and protozoa, and those expressed within endosomes (TLR3, TLR7, TLR8, and TLR9), which primarily recognize nucleic acids from viruses and bacteria (Beutler et al. 2006; Kawai and Akira 2011). Cell-surface TLRs display higher evolutionary flexibility compared to endosomes TLRs (i.e., lesser selective constraint on non-synonymous mutations) (Barreiro et al. 2009). The higher evolutionary flexibility of the cell-surface TLRs can result in an increased number of potential targets for positive selection. Barreiro et al. (2009) showed how variation in human TLRs may cause a different outcome of infectious diseases.

3 Lipopolysaccharide and TLR Interaction

Gram-negative bacteria exhibit outer membranes, which include lipopolysaccharide (LPS, also called endotoxin) anchored in the outer leaflet. LPS is generally composed of three domains: the lipid A, the core oligosaccharide, and the O antigen (Alexander and Rietschel 2001). LPS is one of the most conserved structures within all Gram-negative bacterial species. Therefore, this structure makes LPS as an important MAMP to be recognized by the mammalian innate immune system, which can subsequently initiate the clearance of a given bacterial infection (Triantafilou et al. 2007). For example, recognition and sensing of LPS of many invasive Gram-negative bacteria accounts for strong host immune system activation.

Lipid A is commonly recognized by the so-called TLR4–MD2 complex. Myeloid differentiation factor 2 (MD2) binds to the concave surface of the N-terminal and central domains (Kim et al. 2007). Binding of TLR4–MD2 to lipid A with the help of cluster of differentiation 14 (CD14) and lipopolysaccharide-binding protein (LBP) triggers a signal cascade that leads to cytokine production, inflammation, and eventually clearance of the bacteria through recruitment of effector cells, phagocytosis, cytotoxicity, and activation of the complement system (Needham and Trent 2013; Raetz et al. 2007; Sassi et al. 2010). Excessive activation of these pathways, however, can lead to sepsis, and endotoxin is generally considered as major driver of septic shock.

Usually, lipid A is hexa-acetylated, meaning that six acyl chains with the core oligosaccharide backbone (Needham and Trent 2013). Primary acyl chains are directly esterified with the sugar moiety while so-called secondary acyl chains form ester bonds with hydroxyl groups of primary acyl chains. Acetylated lipid A means that each glucosamine moiety carries the same number of acyl chains. *Escherichia coli* LPS is an example for containing an asymmetrically acetylated lipid A, since four of its six acyl chains are carried by the first glucosamine. Lipid A is embedded in the outer leaflet of the bacterial outer membrane through electrostatic and mainly hydrophobic interactions. The glucosamine part of the lipid A is oriented toward the exterior environment, while the lipid A acyl chains point to the hydrophilic interior of the membrane (Raetz and Whitfield 2002).

Modification of lipid A equips Gram-negative bacteria to evade immune recognition and survive within a host (Montminy et al. 2006). First, by changing the overall charge of the bacterial surface through the addition of chemical groups to lipid A, such as the addition of phosphate group, resistance to innate immune effectors increases. Second, changes in the structure of lipid A are the most important with regard to bacterial pathogenesis, because they directly affect recognition by the TLR4–MD2 receptor (Steimle et al. 2016). Bioinformatics-based studies revealed that many Gram-negative bacteria share a cascade of enzymes that are used by *E. coli* for LPS biosynthesis. However, since lipid A structures differ between distinct bacteria, alterations in the structure are supposed to be due to genetic and/or environmental differences or influences.

Lipid A structures widely differ among the various bacterial species (Needham and Trent 2013; Raetz et al. 2007). Lipid A modification enzymes were subjected to both transcriptional and post-translational regulation. Modification systems are variable and often regulated by environmental conditions. For example, *E. coli* lipid A contains six acyl chains and two phosphate groups, and is the strongest known TLR4 ligand. In contrast, structurally different lipid A can also cause weak inflammatory host responses. Hexa-acylated lipid A seems to promote the strongest pro-inflammatory immune reactions after binding to TLR4 (Hankins et al. 2012; Kim et al. 2007). Therefore, TLR4-mediated signaling usually is drastically reduced if these six acyl chains are missing (Steimle et al. 2016; Tan et al. 2015). However, due to the already mentioned differences in the structures of the lipid A-binding cavity of the TLR4 receptor between different mammal species, distinct lipid A structures can provide a different immunogenic potential between these species. Immune evasion strategies based on lipid A modification help to promote the survival of Gram-negative bacteria by evading host immune responses that are directed against bacterial invasion, chronic persistence, or susceptibility to cationic antimicrobial peptides (AMPs) (Cullen et al. 2011; Needham and Trent 2013). Bacteria can remove acyl chains and phosphate groups to escape from detection by pattern-recognition receptors (PRRs) or to change the type of immune response (Needham and Trent

2013). Finally, certain modifications (such as those regulated by PhoPQ) alter the properties of the outer-membrane permeability barrier, which provides resistance to the environment such as pH and antibiotics and other stresses. There are bacterial benefits associated with each modification (Cullen et al. 2011; Hankins et al. 2012; Needham and Trent 2013; Raetz et al. 2007).

TLR2 is also reported as direct mediator of signaling by LPS in some bacteria (Yang et al. 1998). The glycolipid anchor often contains two acyl chains that may be responsible for binding to TLR2 (Morath et al. 2002). Atypical LPS with tetra- or penta-acyl lipid A is recognized by TLR2, e.g., that of *Leptospira interrogans* (Werts et al. 2001), *Bacteroides fragilis* (Erridge et al. 2004), *Chlamydia trachomatis* (Erridge et al. 2004), and *Pseudomonas aeruginosa* (Erridge et al. 2004). *H. pylori* LPS is also recognized by TLR2 as reported by various researchers (Mandell et al. 2004; Smith et al. 2003; Torok et al. 2005; Yokota et al. 2007). Furthermore, numerous non-*pylori Helicobacter* species such as *Helicobacter felis* and *Helicobacter hepaticus* are also recognized via TLR2 (Mandell et al. 2004). These ligands are considered as weak agonists.

Signaling through TLR2 and TLR4 is initiated by their ligand-induced dimerization. TLR4 is known to utilize two distinct signaling pathways (MyD88-dependent and MyD88-independent pathways), mediated by different TIR domain-containing adaptor molecules, and leading to the induction of pro-inflammatory cytokines through mitogen-activated protein (MAP) kinase and transcription factor NF-κB activation. In contrast to TLR4, TLR2 mostly depends on the adaptors MyD88 and TIRAP/MAL for signaling (Horng et al. 2002; Yamamoto et al. 2002). TLR2 has been considered a factor potentially determining the ensuing both pro- or anti-inflammatory responses (Chau et al. 2009).

4 TLRs and *H. pylori* Research

Gastric epithelial cells (GECs) play an important role in innate immune signaling to *H. pylori* (Backert and Naumann 2010). The interaction between the human host and pathogen ancestries completely accounted for the difference in the severity of gastric lesions in two regions of Colombia with mostly African ancestry in the coastal region, and mostly Amerindian ancestry in the mountain region, respectively (Kodaman et al. 2014a, b). They considered that co-evolution likely modulated gastric disease risk, and the difference between co-evolved human and *H. pylori* genomes can explain the high incidence of gastric disease in the mountain population. These reports showed that *H. pylori* infection severity is consisted of bacteria and host factors combination.

TLRs are one of the most important host factors in *H. pylori* infection (Mayerle et al. 2013; Pachathundikandi et al. 2013). Many of the early studies on innate immune responses to *H. pylori* in GECs have focused on TLR4, because it is a specific PRR of Gram-negative LPS. TLR4 expression was detected in human GECs such as AGS, MKN45, and NCI-N87 (Smith et al. 2003). In contrast, GECs were non-responsive to *H. pylori* LPS, even when relatively high concentrations (1 μg/mL) of this endotoxin were added to the cells (Backhed et al. 2003). Consistent with this result, a TLR4-neutralizing antibody did not block *H. pylori*-induced secretion of the pro-inflammatory cytokine interleukin-8 (IL-8) in AGS cells (Smith et al. 2003). The TLR involved in the residual detection of *H. pylori* LPS remained a matter of debate, even now, whereas several studies using purified LPS have implicated the classical LPS sensor TLR4 (Ishihara et al. 2004; Kawahara et al. 2001).

Furthermore, *Tlr2*$^{-/-}$ mice that were infected with *H. felis*, a close relative of *H. pylori*, were better able to control experimental infections than wild-type mice and developed stronger T cell responses and T cell-driven immunopathology (Sayi et al. 2011). The effects of *tlr2* gene deletion are phenocopied by Myd88$^{-/-}$ mice, indicating that the absence of anti-inflammatory signals induced by *Helicobacter* spp. is phenotypically dominant over the simultaneous lack of MyD88-dependent pro-inflammatory signals that are induced by other TLRs (Sayi et al. 2011). Kim et al. (2013) identified the bacterial *cag*PAI and the cooperative interaction among host innate receptors TLR2, NOD2, and NLRP3 as important regulators of IL-1β production in *H. pylori*-infected dendritic cells. TLR2 activation by *H. pylori* has been confirmed in various subsequent studies, and its role in inflammasome signaling was established (Koch et al. 2015; Pachathundikandi and Backert 2016).

Another putative *H. pylori* MAMP, flagellin, escapes recognition by TLR5 owing to modifications in the N-terminal TLR5 recognition domain (Andersen-Nissen et al. 2005). However, TLR5 can be activated by *H. pylori* through a yet unknown bacterial factor using HEK293 reporter cells (Kumar Pachathundikandi et al. 2011). Expression of numerous TLRs (i.e., TLR2, TLR4, and TLR5) has been confirmed in many GECs, including AGS, MKN28, MKN45, NUGC3, and KATOIII (Chochi et al. 2008; Gewirtz et al. 2004; Ishihara et al. 2004; Maeda et al. 2001; Takenaka et al. 2004).

Considering the proposed *H. pylori* escape from TLR4 and TLR5 recognition, *H. pylori* applies a very smart strategy in order to colonize the human stomach. In addition, it has been reported that TLR2 recognized *H. pylori* LPS by various researchers as mentioned above (Mandell et al. 2004; Smith et al. 2003; Torok et al. 2005; Yokota et al. 2007).

5 Microbes Drive TLR Evolution

The known repertoire of TLRs mainly occurred by gene duplication events (Rast et al. 2006). Gene duplication is an important mechanism for acquiring new genes and creating genetic novelty in organisms, and is believed to play an important role in evolution by providing a mechanism to generate new gene functions (Innan and Kondrashov 2010). A duplicated gene provides a greater, less-constrained chance for natural selection to shape a novel function (Long et al. 2003) (Fig. 1a). During evolution, continuous diversification of a given TLR has resulted in a family of structurally distinct TLRs (Matsushima et al. 2007). In order to survive, microbes must escape from host defenses, whereas hosts have to prepare various TLRs to detect them. For example, humans exhibit 10 TLRs, mice 11 TLRs, while sea urchin has altogether 253 TLRs in order to recognize a multitude of microbes. Interestingly,

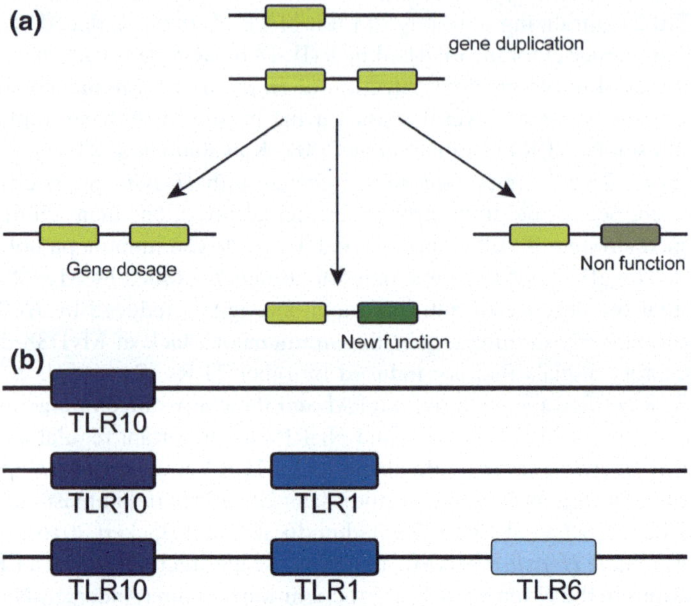

Fig. 1 Duplication mechanism for genes and evolution of the TLR10 gene cluster. **a** A duplicated gene provides a greater, less-constrained chance for natural selection to shape a novel function. Gene duplications and their divergence play an important role in the evolution of novel gene functions. **b** Phylogeny studies support the theory that TLR10 existed before the gene duplication event, which generated TLR1 and TLR6 (Roach et al. 2005; Zhou et al. 2007). Several reports underline the notion that positive selection has targeted the TLR10–TLR1–TLR6 gene cluster in humans (Barreiro et al. 2009). Positive selection increases the frequency of an advantageous mutation in the population. Advantageous mutations can be either fixed or polymorphic in the population

vertebrates contain almost similar numbers of TLR genes (from 10 TLRs in humans to 21 TLRs in amphibians). The large TLR repertoire in the sea urchin is caused by gene duplication (Rast et al. 2006). It is considered that these animals have compensated for their lack of adaptive immune system functions by increasing the number of their TLRs.

TLR10 was identified as a receptor with characteristic leucin-rich repeats (LRRs) and Toll/IL-1 receptor TIR domains shared by all TLRs; however, a microbial ligand was unknown for long time (Chuang and Ulevitch 2001). Importantly, this study found that TLR10 was highly homologous to TLR1 and TLR6 with an overall amino acid identity of 50 and 49%, respectively, whereas it was only 30% identical to TLR2 and less than 25% for the other TLRs. These genes were previously thought to have arisen by a series of independent gene duplications (Huang et al. 2011). Phylogenetic studies on TLR evolution, however, indicate that TLR10 predates TLR1 and TLR6, suggesting that TLR1 and TLR6 arose from gene duplication, consistent with the observation that TLR10, TLR1, and TLR6 are located in tandem on human chromosome 4p14 (Roach et al. 2005; Xu et al. 2016). These phylogenetic and structural analyses of TLR10, TLR1, and TLR6 have clarified their evolutionary origins and predict amino acid residues likely to be important in the host's defense against invading pathogens (Huang et al. 2011) (Fig. 1b).

6 TLR2/TLR10 and *H. pylori* LPS

H. pylori has been proposed as a model for investigating and understanding the dynamics of bacterial persistence and parasitism in chronic infections (Blaser and Kirschner 1999). In general, induction of the antibacterial defense by inflammatory host responses is a crucial function of the innate immune system in response to invading pathogens. This reaction prevents the spreading of the microbe and suppresses its growth. In many infections with pathogens, the innate immune responses slow the progress of infection and allow for the adaptive immune responses to develop (Montminy et al. 2006). Evading the host immune responses is a successful strategy of pathogenic bacteria, known as immune evasion. Immune evasion strategies based on lipid A modification may help to promote the survival of some Gram-negative bacteria by evading host immune responses that are directed against bacterial invasion, chronic persistence, or susceptibility to cationic AMPs (Cullen et al. 2011).

H. pylori LPS is mainly tetra-acylated lipid A and is 1000-fold less biologically active than the hexa-acylated LPS of *E. coli*. This striking structural difference between the originally synthesized lipid A is due to the actions of several enzymes, including dephosphorylation by LpxE and LpxF, addition of phosphate by EptA and deacylation by LpxR (Cullen et al. 2011). Therefore, *H. pylori* is able to evade from host recognition via TLR4. When these modification systems are inactivated through mutation, *E. coli* displays hexa-acylated lipid A, which is a strong stimulator of TLR4 (Cullen et al. 2011). The constitutive lipid A modifications to the acyl chains and phosphate groups are adaptations that allow this bacterium to persist in the harsh gastric environment amidst the several antibacterial components of the innate immune response. It was shown that *H. pylori* modifies LPS, which is a primary surface component, making the bacterium undetectable by components of the innate immune system and highly resistant to antimicrobial compounds secreted by host cells (Cullen et al. 2011).

As outlined above, it has been reported by several researchers that *H. pylori* LPS can be recognized by TLR2, but not TLR4 (Smith et al. 2003, 2011; Pimentel-Nunes et al. 2011). A first report demonstrated that purified *H. pylori* LPS is a TLR2 agonist, but not for TLR4 (Smith et al. 2003). Furthermore, Yokota et al. (2007) reported that *H. pylori* LPS preparations significantly induced an inflammatory reaction via the receptor complex containing TLR2–TLR1 or TLR2–TLR6, but not that containing TLR4.

Unlike TLR4, which signals as a homodimer, TLR2-dependent MAMP recognition and signaling require the formation of TLR2 heterodimers (Jin et al. 2007; Jin and Lee 2008). TLR2 is thought to exist in preformed low-affinity complexes associated with TLR1 and TLR6 under basal conditions and dimerizes upon ligand binding, heterodimerizing with TLR1 or TLR6 upon recognition of tri-acylated and di-acylated lipopeptides, respectively. Almost all studies on TLR2 ligands were demonstrated by using the HEK293 cell model (Smith et al. 2003, 2011; Yokota et al. 2007). However, HEK293 cells express low endogenous levels of TLR1, TLR6, and TLR10 (Regan et al. 2013). Therefore, potential TLR1, TLR6, or TLR10 signaling should be considered when TLR2 ligand research is performed.

Various studies have shown that *H. pylori* regulates and signals through TLR10 (Nagashima et al. 2015; Ram et al. 2015; Pachathundikandi and Backert 2016). We have recently performed microarray studies to compare *H. pylori*-positive subjects and negative subjects to determine host factors, which might influence *H. pylori* infection (Nagashima et al. 2015). TLR10 mRNA levels were upregulated approximately 15-fold in infected subjects; these findings were confirmed by real-time quantitative PCR analysis. Immunohistochemical investigation showed increased TLR10 expression in the GEC biopsies of infected individuals. When *H. pylori* was co-cultured with NCI-N87 GECs, both TLR10 and TLR2 mRNA levels were also upregulated (Nagashima et al. 2015). We used NCI-N87 cells because these GECs represent a naturally polarized cell line that forms a tight monolayer and exhibits typical characteristics like ZO-1 and E-cadherin expression. They also express TLR2, while AGS cells, which are widely used for *H. pylori* infection, do not (Smith et al. 2011). Immunohistochemical investigation showed increased TLR10 expression in GECs of infected individuals. We compared the ability of TLR combinations to mediate NF-κB activation. We identified that *H. pylori* LPS was recognized by expression of the TLR2/TLR10 heterodimer in HEK293 reporter cells. We concluded that TLR10 is a functional receptor involved in the innate immune response to *H. pylori* infection and that the TLR2/TLR10 heterodimer functions in *H. pylori* LPS recognition (Nagashima et al. 2015).

This is the first report that indicates a TLR10 ligand structure (Fig. 2). As mentioned above, *H. pylori* LPS has a tetra-acylated lipid A. TLR2 forms a heterodimer with TLR1 or TLR6 and can recognize tri-acyl or di-acyl lipopeptides. The TLR10-TLR1-TLR6 locus was reported that it is usually present in tandem in mammalian genomes, and these tandems are likely to be the product of successive rounds of gene duplication events from an ancestral gene (Huang et al. 2011). Our results might suggest that tetra-acylated lipid A is a ligand of the TLR2 and TLR10 heterodimer. Not only our findings suggest that *H. pylori* LPS is recognized by TLR10, but also several other researchers reported the *H. pylori*–TLR10 interaction by genetic analysis such as single-nucleotide polymorphisms (SNPs). A highly remarkable report by Mayerle et al. (2013) was performed using GWAS. Of 10,938 participants, 6160 (56.3%) were seropositive for *H. pylori*. GWAS identified the TLR locus to play a role. This report showed that rs10004195 ("rs" means reference SNP ID number, which is an identification tag assigned by NCBI) is the most interactive SNP and belongs to TLR10 in the database. Therefore, TLR10 should be responsible at least for some of the individual risks during *H. pylori* infection. The authors discussed that TLR10 was one of the most important host factors in *H. pylori* infection, rather than other TLRs such as TLR2 and TLR4 (Mayerle et al. 2013).

Tang et al. (2015) reported that genetic polymorphisms of TLR1 and TLR10 might influence the host susceptibility for *H. pylori*. They showed that TLR1 and TLR10

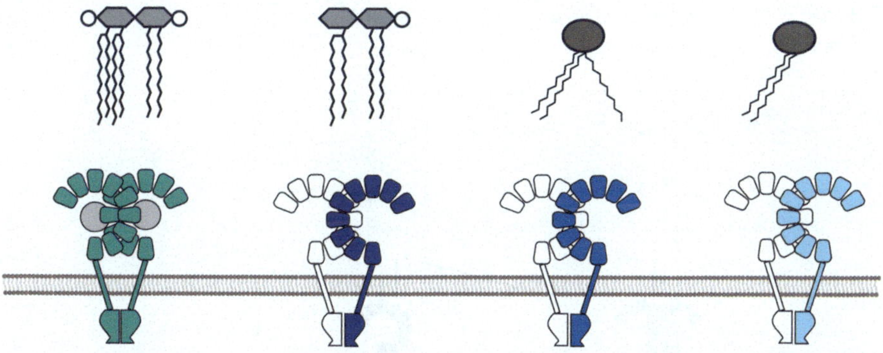

Fig. 2 *H. pylori* LPS is recognized via TLR2/TLR10, and it was suggested that the tetra-acyl lipid A portion is a ligand of TLR2/TLR10. Our data suggest TLR2/TLR10 may recognize tetra-acylated lipid A. TLR10, TLR1, and TLR6 may distinguish the acylation number to recognize various bacterial envelope structure patterns

were associated with *H. pylori* infection and precancerous gastric lesions. Sung et al. (2015) reported variants at the chromosomal locus 4p14, carrying the *TLR10, TLR1,* and *TLR6* genes, which were annotated as expression quantitative trait loci (eQTL) associated with TLR6/10. Their findings suggest that 4p14 polymorphisms are linked to the host immune response during *H. pylori* infection. In addition, it was reported that the TLR1 (rs4833095) C allele and the TLR10 (rs10004195) A allele frequencies significantly increased the risk in the *H. pylori* infection group (Tongtawee et al. 2017, 2018). The TLR1 (rs4833095) C allele and TLR10 (rs10004195) A allele were susceptible TLRs polymorphisms. GC is also associated with rs10004195 (Ram et al. 2015). It was shown that this TLR10 SNP makes a difference in *H. pylori* infection, although these studies were performed in geographically different populations.

7 TLR10 and Anti-inflammatory Responses

H. pylori can be recognized by several TLRs. Rad et al. (2009) demonstrated that activation of TLR2 triggers the MyD88-dependent expression of several anti-inflammatory genes, most notably IL-10, via transcription factor NF-κB (Fig. 3). The DNA of the bacterium as well as a currently uncharacterized MAMP (possibly *H. pylori* LPS) is detected by TLR9 and TLR2, respectively. These TLRs predominantly activate anti-inflammatory signaling pathways and anti-inflammatory IL-10 expression (Rad et al. 2009). Intracellular delivery of *H. pylori* DNA is recognized by endosomal TLR9; however, the effect of this activation is mainly anti-inflammatory response rather than pro-inflammatory response (Varga et al. 2016). TLR9 signaling has anti-inflammatory consequences in the early stages of infection in a mouse model. The effects of *TLR2* gene deletion are phenocopied by $Myd88^{-/-}$ mice, indi-

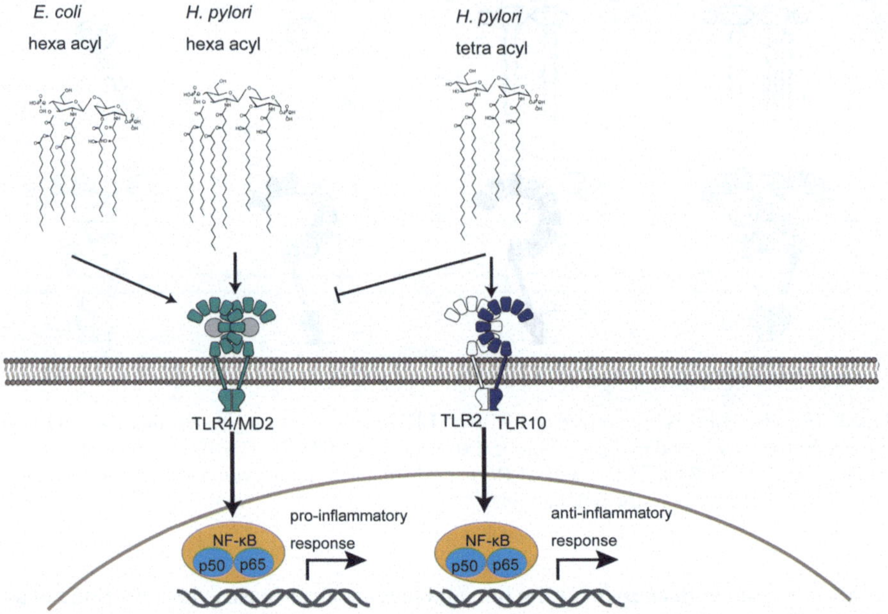

Fig. 3 *H. pylori* and pro-inflammatory and anti-inflammatory responses. *H. pylori* lipid A consists of a tetra-acylated form that lacks the 4′-phosphate group. Lipid A modification mutants present a hexa-acylated, bis-phosphorylated form that is a TLR4 ligand, but *H. pylori* LPS with tetra-acylated is not a ligand of TLR4, but recognized by the TLR2/TLR10 heterodimer (Cullen et al. 2011; Needham and Trent 2013)

cating that the absence of anti-inflammatory signals induced by *Helicobacter* spp. is phenotypically dominant over the simultaneous lack of MyD88-dependent pro-inflammatory signals that are induced by other TLRs (Sayi et al. 2011).

There are two reasons why TLR10 is of major interest in *H. pylori* research. First reason is, TLR10 is reported as an anti-inflammatory receptor. Second reason is that the TLR10-TLR1-TLR6 locus affects the susceptibility for various infectious and allergic diseases (Table 1). It was revealed by Pachathundikandi and Backert (2016) that there is a significant link between TLR10 expression and signaling during *H. pylori* infection. It was demonstrated that blocking of TLR10 by antagonistic antibodies enhanced pro-inflammatory cytokine production, including IL-1β an important mediator of inflammation and gastric acid secretion, specifically after exposure with TLR2 ligands (Oosting et al. 2014). The results showed that TLR10 is a modulatory PRR with mainly inhibitory properties. Furthermore, Jiang et al. (2016) also reported that TLR10 suppressed the production of an array of cytokines in stably transfected human myelomonocytic U937 cells in response to other TLR agonists. Compared with non-transgenic controls, monocytes of TLR10 transgenic mice exhibited blunted IL-6 production following ex vivo blood stimulation with other TLR agonists (Jiang et al. 2016). TLR10 is a suppressor of TLR2/1-induced responses, and TLR10 inhibits TLR-induced inflammatory responses. They also showed that TLR10 suppresses Myd88 and TRIF signaling. These are in vitro data, but two other studies (Lee et al. 2014; Regan et al. 2013) have proposed opposing results, as a pro-inflammatory function for TLR10. One is that siRNA knockdown of endogenous TLR10 in HT-29 colonic epithelial cells blunted the inflammatory response to *Listeria monocytogenes* (Regan et al. 2013). The other study showed that knockdown of TLR10 in the THP-1 myelomonocytic cell line inhibited cellular responses to the H1N1 and H5N1 flu virus strains (Lee et al. 2014). Limitation of these reports is that the ligand of TLR10 was unknown at the time. Our data suggest that *H. pylori* LPS is a ligand of TLR10 as discussed above (Nagashima et al. 2015).

Most data about the possible function of TLR10 are reported by using SNP analysis (Ram et al. 2015; Tang et al. 2015). During the evolutionary history of *Homo sapiens*, infections shaped the genetics and function of the immune system, leading to variation of host defense responses in different geographic populations as a result of local infectious pressures and migration events. Investigation of the mechanisms mediating these changes is crucial for a comprehensive understanding of immune responses in health and disease and for adapting our immune-based treatments to the characteristics of each population. From this perspective, an important new insight of the evolution of the human immune system has been recently provided by two studies (Dannemann et al. 2016; Deschamps et al. 2016). They described introgression of archaic immune alleles in the modern innate immune genes and discussed a special impact on the TLR10–TLR1–TLR6 locus. The biological importance of these findings is obvious: (1) Positive selection in European populations, but not African, has been previously reported at the TLR10–TLR1–TLR6 locus (Ferrer-Admetlla et al. 2008); (2) this locus has been reported to represent a common target of convergent evolution in European, and Sinti and Romany people populations living in Europe. Genes in the immune system were highly represented among those under strong evolutionary pressure in Europeans, and infections are likely to have played an important role. Europeans and Gypsies have lived in the same geographic area and have been exposed to similar environmental hazards, including infections. Laayouni and co-workers (2014) identified convergent evolution signals in genes from different two human populations with different origins; (3) and it has been identified to be under selection in ancient European populations (Mathieson et al. 2015). Furthermore, these findings might also indicate functional consequences of genetic variation in this locus. Indeed, genetic variants in the TLR10–TLR1–TLR6 locus influence the expression of all three TLRs encoded at the locus, and they affect the susceptibility to immune-mediated diseases and allergies (Dannemann et al. 2016). The authors suggest that selection is mediated by the increased expression of the receptors by the archaic haplotypes and association with *H. pylori* infection and allergies (Dannemann et al. 2016). In addition, it is difficult to indicate with certainty, which one of the three TLR receptors is the target of selection, although some studies suggest that TLR10 is the main target (Ferrer-Admetlla et al. 2008).

Strikingly, the TLR10–TLR1–TLR6 gene cluster has been proposed to be a hot spot of positive selection, as it appears to have also been targeted by positive selection

Table 1 TLR10 gene variants and their potential role in various disease outcomes

Disease	Race	Number	Reference SNP ID number	References
H. pylori infection	Chinese	2553	rs10004195, rs4129009	Tang et al. (2015)
Chronic gastritis	Thai	400	rs10004195	Tongtawee et al. (2017)
Gastric carcinoma	Malaysian	95	rs10004195	Ram et al. 2015
H. pylori infection	European	10,938	rs10004195, rs12233670, rs7653908	Mayerle et al. (2013)
Crohn's disease	New Zealanders	1044	rs6841698, rs10024216, rs4274855, rs7658893	Morgan et al. (2012)
Asthma	Finnish	953	rs4129009, rs1109657	Renkonen et al. (2010)
Asthma	Canadian and Australian	5565	rs11096957	Daley et al. (2009)
Allergic rhinitis	Swedish	288	Not mentioned in the paper	Henmyr et al. (2017)
Childhood asthma	Germany	1872	rs11096956, rs4129009	Kormann et al. (2008)
Infant bronchiolitis	Finnish	103	rs4129009	Lauhkonen et al. (2016)
Tuberculosis	Tibetan, Chinese Han	2401	rs11466617 and rs4129009	Wang et al. (2018)
Allergic disease	American	53,862	rs2101521	Hinds et al. (2013)
Severe RSV-associated diseases	Germany	356	rs4129009, rs11466657, rs11096955	Mailaparambil et al. (2008)
Complicated skin and skin structure infections	Belgian	646	rs4129009, rs11096955, rs11096957	Stappers et al. (2015)

in ape genomes such as chimpanzee and orangutan (Enard et al. 2010). These results provide a considerable example of a selective advantage provided by variation in the TLRs to both humans and primates. Functional analysis of this genomic region in non-human primates is now needed to elucidate whether such a shared hot spot of positive selection reflects a case of parallel or convergent evolution, or adaptations toward different phenotypic directions involving the same locus.

TLR10 is reported that it interacts with various illnesses such as asthma and skin disease (Table 1). Koch et al. (2015) demonstrated that TLR2/NLRP3/CASP1/IL-18 axis was critical to *H. pylori*-specific immune regulation. These diseases are increasing recently, and they are reported to be caused by disruption of the microbiome (Bach 2018). Pachathundikandi and Backert (2018) reported that inflammasome-forming NLRP3, an important innate immunity component, is crucial for the manipulation of pro- and anti-inflammatory cytokines in *H. pylori* infection. Thus, further studies are required to investigate this interesting phenomenon.

8 Concluding Remarks

While the pro-inflammatory responses activated upon detection of Gram-negative bacteria by TLRs have been well characterized, the anti-inflammatory responses of TLRs are just beginning to be elucidated. The wide variety of lipid A modifications equips Gram-negative bacteria for survival in the host and other environments. Other functions of lipid A are highlighted by the interplay of lipid A modifications with a diverse set of cellular processes, and both of these facets of lipid A biology emphasize the importance of a continued effort to understand the regulation, benefits, and host response to lipid A in its numerous modified forms. Atypical LPS with tetra- or penta-acyl lipid A was reported that they are recognized by TLR2. It was reported that the TLR2 signal pathway establishes colonization by commensal bacteria (Round et al. 2011). Our data might suggest TLR2 and TLR10 recognize atypical LPS.

Acknowledgements The author acknowledges support from the National Institute of Diabetes and Digestive and Kidney Diseases of the United States (grant DK62813); the Grants-in-Aid for Scientific Research from the Ministry of Education, Culture, Sports, Science, and Technology of Japan (grant 16H05191); and from the Japan Society for the Promotion of Science (Core-to-Core Program).

References

Alcais A, Abel L, Casanova JL (2009) Human genetics of infectious diseases: between proof of principle and paradigm. J Clin Invest 119:2506–2514. https://doi.org/10.1172/JCI38111

Alexander C, Rietschel ET (2001) Bacterial lipopolysaccharides and innate immunity. J Endotoxin Res 7:167–202

Andersen-Nissen E, Smith KD, Strobe KL, Barrett SL, Cookson BT, Logan SM, Aderem A (2005) Evasion of Toll-like receptor 5 by flagellated bacteria. Proc Natl Acad Sci USA 102:9247–9252. https://doi.org/10.1073/pnas.0502040102

Arnold IC, Dehzad N, Reuter S, Martin H, Becher B, Taube C, Muller A (2011) *Helicobacter pylori* infection prevents allergic asthma in mouse models through the induction of regulatory T cells. J Clin Invest 121:3088–3093. https://doi.org/10.1172/JCI45041

Bach JF (2018) The hygiene hypothesis in autoimmunity: the role of pathogens and commensals. Nat Rev Immunol 18:105–120. https://doi.org/10.1038/nri.2017.111

Backhed F et al (2003) Gastric mucosal recognition of Helicobacter pylori is independent of Toll-like receptor 4. J Infect Dis 187:829–836. https://doi.org/10.1086/367896

Backert S, Naumann M (2010) What a disorder: proinflammatory signaling pathways induced by *Helicobacter pylori*. Trends Microbiol 18:479–486. https://doi.org/10.1016/j.tim.2010.08.003

Backert S, Tegtmeyer N, Fischer W (2015) Composition, structure and function of the *Helicobacter pylori cag* pathogenicity island encoded type IV secretion system. Future Microbiol 10:955–965. https://doi.org/10.2217/fmb.15.32

Barreiro LB, Quintana-Murci L (2010) From evolutionary genetics to human immunology: how selection shapes host defence genes. Nat Rev Genet 11:17–30. https://doi.org/10.1038/nrg2698
Barreiro LB et al (2009) Evolutionary dynamics of human toll-like receptors and their different contributions to host defense. PLoS Genet 5:e1000562. https://doi.org/10.1371/journal.pgen.1000562
Beutler B et al (2006) Genetic analysis of host resistance: toll-like receptor signaling and immunity at large. Ann Rev Immunol 24:353–389. https://doi.org/10.1146/annurev.immunol.24.021605.090552
Blaser MJ, Kirschner D (1999) Dynamics of Helicobacter pylori colonization in relation to the host response. Proc Natl Acad Sci USA 96:8359–8364
Casanova JL, Abel L, Quintana-Murci L (2011) Human TLRs and IL-1Rs in host defense: natural insights from evolutionary, epidemiological, and clinical genetics. Annu Rev Immunol 29:447–491. https://doi.org/10.1146/annurev-immunol-030409-101335
Censini S et al (1996) Cag, a pathogenicity island of *Helicobacter pylori*, encodes type I-specific and disease-associated virulence factors. Proc Nat Acad Sci USA 93:14648–14653
Chapman SJ, Hill AV (2012) Human genetic susceptibility to infectious disease. Nat Rev Genet 13:175–188. https://doi.org/10.1038/nrg3114
Chau TA et al (2009) Toll-like receptor 2 ligands on the *staphylococcal* cell wall downregulate superantigen-induced T cell activation and prevent toxic shock syndrome. Nat Med 15:641–648. https://doi.org/10.1038/nm.1965
Chochi K et al (2008) *Helicobacter pylori* augments growth of gastric cancers via the lipopolysaccharide-toll-like receptor 4 pathway whereas its lipopolysaccharide attenuates antitumor activities of human mononuclear cells. Clin Cancer Res: An Official J Am Assoc Cancer Res 14:2909–2917. https://doi.org/10.1158/1078-0432.ccr-07-4467
Chuang T, Ulevitch RJ (2001) Identification of hTLR10: a novel human toll-like receptor preferentially expressed in immune cells. Biochimica et biophysica acta 1518:157–161
Cullen TW, Giles DK, Wolf LN, Ecobichon C, Boneca IG, Trent MS (2011) *Helicobacter pylori* versus the host: remodeling of the bacterial outer membrane is required for survival in the gastric mucosa. PLoS Pathog 7:e1002454. https://doi.org/10.1371/journal.ppat.1002454
Daley D et al (2009) Analyses of associations with asthma in four asthma population samples from Canada and Australia. Hum Genet 125:445–459. https://doi.org/10.1007/s00439-009-0643-8
Dannemann M, Andres AM, Kelso J (2016) Introgression of Neandertal- and Denisovan-like haplotypes contributes to adaptive variation in human toll-like receptors. Am J Hum Genet 98:22–33. https://doi.org/10.1016/j.ajhg.2015.11.015
Deschamps M et al (2016) Genomic signatures of selective pressures and introgression from archaic hominins at human innate immunity genes. Am J Hum Genet 98:5–21. https://doi.org/10.1016/j.ajhg.2015.11.014
Enard D, Depaulis F, Roest Crollius H (2010) Human and non-human primate genomes share hotspots of positive selection. PLoS Genet 6:e1000840. https://doi.org/10.1371/journal.pgen.1000840
Erridge C, Pridmore A, Eley A, Stewart J, Poxton IR (2004) Lipopolysaccharides of *Bacteroides fragilis*, *Chlamydia trachomatis* and *Pseudomonas aeruginosa* signal via toll-like receptor 2. J Med Microbiol 53:735–740. https://doi.org/10.1099/jmm.0.45598-0
Ferrer-Admetlla A et al (2008) Balancing selection is the main force shaping the evolution of innate immunity genes. J Immunol 181:1315–1322
Gewirtz AT, Yu Y, Krishna US, Israel DA, Lyons SL, Peek RM Jr (2004) *Helicobacter pylori* flagellin evades toll-like receptor 5-mediated innate immunity. J Infect Dis 189:1914–1920. https://doi.org/10.1086/386289
Hankins JV, Madsen JA, Giles DK, Brodbelt JS, Trent MS (2012) Amino acid addition to Vibrio cholerae LPS establishes a link between surface remodeling in gram-positive and gram-negative bacteria. Proc Natl Acad Sci USA 109:8722–8727. https://doi.org/10.1073/pnas.1201313109
Henmyr V, Carlberg D, Manderstedt E, Lind-Hallden C, Sall T, Cardell LO, Hallden C (2017) Genetic variation of the toll-like receptors in a Swedish allergic rhinitis case population. BMC Med Genet 18:18. https://doi.org/10.1186/s12881-017-0379-6

Hinds DA et al (2013) A genome-wide association meta-analysis of self-reported allergy identifies shared and allergy-specific susceptibility loci. Nat Genet 45:907–911. https://doi.org/10.1038/ng.2686

Horng T, Barton GM, Flavell RA, Medzhitov R (2002) The adaptor molecule TIRAP provides signalling specificity for toll-like receptors. Nature 420:329–333. https://doi.org/10.1038/nature01180

Huang YH, Temperley ND, Ren LM, Smith J, Li N, Burt DW (2011) Molecular evolution of the vertebrate TLR1 gene family—a complex history of gene duplication, gene conversion, positive selection and co-evolution. BMC Evol Biol 11:149. https://doi.org/10.1186/1471-2148-11-149

Hughes AL, Piontkivska H (2008) Functional diversification of the toll-like receptor gene family. Immunogenetics 60:249–256. https://doi.org/10.1007/s00251-008-0283-5

Innan H, Kondrashov F (2010) The evolution of gene duplications: classifying and distinguishing between models. Nat Rev Genet 11:97–108. https://doi.org/10.1038/nrg2689

Ishihara S et al (2004) Essential role of MD-2 in TLR4-dependent signaling during *Helicobacter pylori*-associated gastritis. J Immunol 173:1406–1416

Jann OC, Werling D, Chang JS, Haig D, Glass EJ (2008) Molecular evolution of bovine Toll-like receptor 2 suggests substitutions of functional relevance. BMC Evol Biol 8:288. https://doi.org/10.1186/1471-2148-8-288

Jiang S, Li X, Hess NJ, Guan Y, Tapping RI (2016) TLR10 Is a Negative Regulator of Both MyD88-Dependent and -Independent TLR Signaling. J Immunol 196:3834–3841. https://doi.org/10.4049/jimmunol.1502599

Jin MS et al (2007) Crystal structure of the TLR1-TLR2 heterodimer induced by binding of a tri-acylated lipopeptide. Cell 130:1071–1082. https://doi.org/10.1016/j.cell.2007.09.008

Jin MS, Lee JO (2008) Structures of the toll-like receptor family and its ligand complexes. Immunity 29:182–191. https://doi.org/10.1016/j.immuni.2008.07.007

Kawahara T, Teshima S, Oka A, Sugiyama T, Kishi K, Rokutan K (2001) Type I *Helicobacter pylori* lipopolysaccharide stimulates toll-like receptor 4 and activates mitogen oxidase 1 in gastric pit cells. Infect Immun 69:4382–4389. https://doi.org/10.1128/iai.69.7.4382-4389.2001

Kawai T, Akira S (2010) The role of pattern-recognition receptors in innate immunity: update on Toll-like receptors. Nat Immunol 11:373–384. https://doi.org/10.1038/ni.1863

Kawai T, Akira S (2011) Toll-like receptors and their crosstalk with other innate receptors in infection and immunity. Immunity 34:637–650. https://doi.org/10.1016/j.immuni.2011.05.006

Kim HM et al (2007) Crystal structure of the TLR4-MD-2 complex with bound endotoxin antagonist. Eritoran Cell 130:906–917. https://doi.org/10.1016/j.cell.2007.08.002

Kim DJ, Park JH, Franchi L, Backert S, Nunez G (2013) The Cag pathogenicity island and interaction between TLR2/NOD2 and NLRP3 regulate IL-1beta production in *Helicobacter pylori* infected dendritic cells. Eur J Immunol 43:2650–2658. https://doi.org/10.1002/eji.201243281

Koch KN, Hartung ML, Urban S, Kyburz A, Bahlmann AS, Lind J, Backert S, Taube C, Müller A (2015) *Helicobacter* urease-induced activation of the TLR2/NLRP3/IL-18 axis protects against asthma. J Clin Invest 125:3297–3302. https://doi.org/10.1172/jci79337

Kodaman N et al (2014a) Human and *Helicobacter pylori* coevolution shapes the risk of gastric disease. Proc Natl Acad Sci USA 111:1455–1460. https://doi.org/10.1073/pnas.1318093111

Kodaman N, Sobota RS, Mera R, Schneider BG, Williams SM (2014b) Disrupted human-pathogen co-evolution: a model for disease. Front Genet 5:290. https://doi.org/10.3389/fgene.2014.00290

Kormann MS et al (2008) Toll-like receptor heterodimer variants protect from childhood asthma. J Allergy Clin Immunol 122:86–92, e81–88. https://doi.org/10.1016/j.jaci.2008.04.039

Kumar Pachathundikandi S, Brandt S, Madassery J, Backert S (2011) Induction of TLR-2 and TLR-5 expression by *Helicobacter pylori* switches *cag*PAI-dependent signalling leading to the secretion of IL-8 and TNF-α. PLoS ONE 6:e19614. https://doi.org/10.1371/journal.pone.0019614

Laayouni H et al (2014) Convergent evolution in European and Rroma populations reveals pressure exerted by plague on Toll-like receptors. Proc Natl Acad Sci USA 111:2668–2673. https://doi.org/10.1073/pnas.1317723111

Lauhkonen E et al (2016) Gene polymorphism of toll-like receptors and lung function at five to seven years of age after infant bronchiolitis. PloS one 11:e0146526. https://doi.org/10.1371/journal.pone.0146526

Lee SM et al (2014) Toll-like receptor 10 is involved in induction of innate immune responses to *influenza virus* infection. Proc Nat Acad Sci USA 111(10):3793–3798. https://doi.org/10.1073/pnas.1324266111

Long M, Betran E, Thornton K, Wang W (2003) The origin of new genes: glimpses from the young and old. Nat Rev Genet 4:865–875. https://doi.org/10.1038/nrg1204

Maeda S et al (2001) Distinct mechanism of *Helicobacter pylori*-mediated NF-kappa B activation between gastric cancer cells and monocytic cells. J Biol Chem 276:44856–44864. https://doi.org/10.1074/jbc.M105381200

Mailaparambil B, Krueger M, Heinze J, Forster J, Heinzmann A (2008) Polymorphisms of toll like receptors in the genetics of severe RSV associated diseases. Dis Markers 25:59–65

Mandell L et al (2004) Intact gram-negative *Helicobacter pylori*, *Helicobacter felis*, and *Helicobacter hepaticus* bacteria activate innate immunity via toll-like receptor 2 but not toll-like receptor 4. Infect Immun 72:6446–6454 https://doi.org/10.1128/iai.72.11.6446-6454.2004

Mathieson I et al (2015) Genome-wide patterns of selection in 230 ancient. Eurasians Nat 528:499–503. https://doi.org/10.1038/nature16152

Matsushima N, Tanaka T, Enkhbayar P, Mikami T, Taga M, Yamada K, Kuroki Y (2007) Comparative sequence analysis of leucine-rich repeats (LRRs) within vertebrate toll-like receptors. BMC Genomics 8:124. https://doi.org/10.1186/1471-2164-8-124

Mayerle J et al (2013) Identification of genetic loci associated with *Helicobacter pylori* serologic status. JAMA, J Am Med Assoc 309:1912–1920. https://doi.org/10.1001/jama.2013.4350

Montminy SW et al (2006) Virulence factors of Yersinia pestis are overcome by a strong lipopolysaccharide response. Nat Immunol 7:1066–1073. https://doi.org/10.1038/ni1386

Morath S, Stadelmaier A, Geyer A, Schmidt RR, Hartung T (2002) Synthetic lipoteichoic acid from *Staphylococcus aureus* is a potent stimulus of cytokine release. J Exp Med 195:1635–1640

Morgan AR, Lam WJ, Han DY, Fraser AG, Ferguson LR (2012) Genetic variation within TLR10 is associated with Crohn's disease in a New Zealand population. Hum Immunol 73:416–420. https://doi.org/10.1016/j.humimm.2012.01.015

Mukherjee S, Sarkar-Roy N, Wagener DK, Majumder PP (2009) Signatures of natural selection are not uniform across genes of innate immune system, but purifying selection is the dominant signature. Proc Nat Acad Sci USA 106:7073–7078. https://doi.org/10.1073/pnas.0811357106

Nagashima H et al (2015) Toll-like receptor 10 in *Helicobacter pylori* infection. The Journal of Infectious Diseases. 212:1666. https://doi.org/10.1093/infdis/jiv270

Naumann M, Sokolova O, Tegtmeyer N, Backert S (2017) *Helicobacter pylori*: a paradigm pathogen for subverting host cell signal transmission. Trends Microbiol 25:316–328. https://doi.org/10.1016/j.tim.2016.12.004

Needham BD, Trent MS (2013) Fortifying the barrier: the impact of lipid A remodelling on bacterial pathogenesis. Nat Rev Microbiol 11:467–481. https://doi.org/10.1038/nrmicro3047

Oosting M et al (2014) Human TLR10 is an anti-inflammatory pattern-recognition receptor. Proc Nat Acad Sci USA 111:E4478–E4484. https://doi.org/10.1073/pnas.1410293111

Pachathundikandi SK, Backert S (2016) Differential expression of interleukin 1β during *Helicobacter pylori* infection of toll-like receptor 2 (TLR2)- and TLR10-expressing HEK293 cell lines. J Infect Dis 214:166–167. https://doi.org/10.1093/infdis/jiw154

Pachathundikandi SK, Backert S (2018) *Helicobacter pylori* controls NLRP3 expression by regulating hsa-miR-223-3p and IL-10 in cultured and primary human immune cells. Innate Immun 24:11–23. https://doi.org/10.1177/1753425917738043

Pachathundikandi SK, Tegtmeyer N, Backert S (2013) Signal transduction of *Helicobacter pylori* during interaction with host cell protein receptors of epithelial and immune cells. Gut Microbes 4:454–474. https://doi.org/10.4161/gmic.27001

Pachathundikandi SK, Lind J, Tegtmeyer N, El-Omar EM, Backert S (2015) Interplay of the gastric pathogen *Helicobacter pylori* with toll-like receptors. Biomed Res Int 2015:192420. https://doi.org/10.1155/2015/192420

Posselt G, Backert S, Wessler S (2013) The functional interplay of *Helicobacter pylori* factors with gastric epithelial cells induces a multi-step process in pathogenesis. Cell Commun Signal 11:77. https://doi.org/10.1186/1478-811X-11-77

Quintana-Murci L, Clark AG (2013) Population genetic tools for dissecting innate immunity in humans. Nat Rev Immunol 13:280–293. https://doi.org/10.1038/nri3421

Rad R et al (2009) Extracellular and intracellular pattern recognition receptors cooperate in the recognition of *Helicobacter pylori*. Gastroenterology 136:2247–2257. https://doi.org/10.1053/j.gastro.2009.02.066

Raetz CR, Whitfield C (2002) Lipopolysaccharide endotoxins. Ann Rev Biochem 71:635–700. https://doi.org/10.1146/annurev.biochem.71.110601.135414

Raetz CR, Reynolds CM, Trent MS, Bishop RE (2007) Lipid A modification systems in gram-negative bacteria. Annu Rev Biochem 76:295–329. https://doi.org/10.1146/annurev.biochem.76.010307.145803

Ram MR et al (2015) Polymorphisms at locus 4p14 of toll-like receptors TLR-1 and TLR-10 confer susceptibility to gastric carcinoma in *Helicobacter pylori* infection. PloS One 10:e0141865. https://doi.org/10.1371/journal.pone.0141865

Rast JP, Smith LC, Loza-Coll M, Hibino T, Litman GW (2006) Genomic insights into the immune system of the *sea urchin*. Science 314:952–956. https://doi.org/10.1126/science.1134301

Regan T, Nally K, Carmody R, Houston A, Shanahan F, Macsharry J, Brint E (2013) Identification of TLR10 as a key mediator of the inflammatory response to *Listeria monocytogenes* in intestinal epithelial cells and macrophages. Journal of immunology 191:6084–6092. https://doi.org/10.4049/jimmunol.1203245

Renkonen J, Joenvaara S, Parviainen V, Mattila P, Renkonen R (2010) Network analysis of single nucleotide polymorphisms in asthma. J Asthma Allergy 3:177–186. https://doi.org/10.2147/JAA.S14459

Roach JC et al (2005) The evolution of vertebrate toll-like receptors. Proc Natl Acad Sci USA 102:9577–9582. https://doi.org/10.1073/pnas.0502272102

Round JL, Lee SM, Li J, Tran G, Jabri B, Chatila TA, Mazmanian SK (2011) The toll-like receptor 2 pathway establishes colonization by a commensal of the human microbiota. Science 332:974–977. https://doi.org/10.1126/science.1206095

Sassi N, Paul C, Martin A, Bettaieb A, Jeannin JF (2010) Lipid A-induced responses in vivo. Adv Exp Med Biol 667:69–80. https://doi.org/10.1007/978-1-4419-1603-7_7

Sayi A, Kohler E, Toller IM, Flavell RA, Muller W, Roers A, Muller A (2011) TLR-2-activated B cells suppress *Helicobacter*-induced preneoplastic gastric immunopathology by inducing T regulatory-1 cells. J Immunol 186:878–890. https://doi.org/10.4049/jimmunol.1002269

Schnare M, Barton GM, Holt AC, Takeda K, Akira S, Medzhitov R (2001) Toll-like receptors control activation of adaptive immune responses. Nat Immunol 2:947–950. https://doi.org/10.1038/ni712

Smith MF Jr et al (2003) Toll-like receptor (TLR) 2 and TLR5, but not TLR4, are required for *Helicobacter pylori*-induced NF-kappa B activation and chemokine expression by epithelial cells. J Biol Chem 278:32552–32560. https://doi.org/10.1074/jbc.m305536200

Smith SM et al (2011) Tribbles 3: a novel regulator of TLR2-mediated signaling in response to *Helicobacter pylori* lipopolysaccharide. J Immunol 186:2462–2471. https://doi.org/10.4049/jimmunol.1000864

Stappers MH et al (2015) Genetic variation in TLR10, an inhibitory toll-like receptor, influences susceptibility to complicated skin and skin structure infections. J Infect Dis 212:1491–1499. https://doi.org/10.1093/infdis/jiv229

Steimle A, Autenrieth IB, Frick JS (2016) Structure and function: lipid A modifications in commensals and pathogens. Int J Med Microbiol IJMM 306:290–301. https://doi.org/10.1016/j.ijmm.2016.03.001

Sung H, Camargo MC, Yu K, Weinstein SJ, Morgan DR, Albanes D, Rabkin CS (2015) Association of 4p14 TLR locus with antibodies to *Helicobacter pylori*. Genes Immun 16(8):567–570. https://doi.org/10.1038/gene.2015.33

Takenaka R et al (2004) *Helicobacter pylori* heat-shock protein 60 induces inflammatory responses through the toll-like receptor-triggered pathway in cultured human gastric epithelial cells. Microbiology 150:3913–3922. https://doi.org/10.1099/mic.0.27527-0

Tan Y, Zanoni I, Cullen TW, Goodman AL, Kagan JC (2015) Mechanisms of toll-like receptor 4 endocytosis reveal a common immune-evasion strategy used by pathogenic and commensal bacteria. Immunity 43:909–922. https://doi.org/10.1016/j.immuni.2015.10.008

Tang FB et al (2015) Toll-like receptor 1 and 10 polymorphisms, *Helicobacter pylori* susceptibility and risk of gastric lesions in a high-risk Chinese population. Infect Genet Evol 31:263–269. https://doi.org/10.1016/j.meegid.2015.02.005

Tongtawee T, Bartpho T, Wattanawongdon W, Dechsukhum C, Leeanansaksiri W, Matrakool L, Panpimanmas S (2017) Role of toll-like receptor 10 gene polymorphism and gastric mucosal pattern in patients with chronic gastritis. Turk J Gastroenterol 28:243–247. https://doi.org/10.5152/tjg.2017.16673

Tongtawee T et al (2018) Genetic polymorphisms in TLR1, TLR2, TLR4, and TLR10 of *Helicobacter pylori*-associated gastritis: a prospective cross-sectional study in Thailand. Eur J Cancer Prev 27:118–123. https://doi.org/10.1097/cej.0000000000000347

Torok AM, Bouton AH, Goldberg JB (2005) *Helicobacter pylori* induces interleukin-8 secretion by toll-like receptor 2- and toll-like receptor 5-dependent and -independent pathways. Infect Immun 73:1523–1531. https://doi.org/10.1128/iai.73.3.1523-1531.2005

Triantafilou M, Uddin A, Maher S, Charalambous N, Hamm TS, Alsumaiti A, Triantafilou K (2007) Anthrax toxin evades Toll-like receptor recognition, whereas its cell wall components trigger activation via TLR2/6 heterodimers. Cell Microbiol 9:2880–2892. https://doi.org/10.1111/j.1462-5822.2007.01003.x

Varga MG et al (2016) Pathogenic *Helicobacter pylori* strains translocate DNA and activate TLR9 via the cancer-associated cag type IV secretion system. Oncogene 35:6262–6269. https://doi.org/10.1038/onc.2016.158

Wang Y et al (2018) Polymorphisms in toll-Like receptor 10 and tuberculosis susceptibility: evidence from three independent series. Front Immunol 9:309. https://doi.org/10.3389/fimmu.2018.00309

Werts C et al (2001) *Leptospiral* lipopolysaccharide activates cells through a TLR2-dependent mechanism. Nat Immunol 2:346–352. https://doi.org/10.1038/86354

Xu TJ, Wang YJ, Li JR, Shu C, Han JJ, Chu Q (2016) Comparative genomic evidence for duplication of TLR1 subfamily and miiuy croaker TLR1 perceives LPS stimulation via MyD88 and TIRAP. Fish Shellfish Immun 56:336–348. https://doi.org/10.1016/j.fsi.2016.07.024

Yamamoto M et al (2002) Essential role for TIRAP in activation of the signalling cascade shared by TLR2 and TLR4. Nature 420:324–329. https://doi.org/10.1038/nature01182

Yang RB et al (1998) Toll-like receptor-2 mediates lipopolysaccharide-induced cellular signalling. Nature 395:284–288. https://doi.org/10.1038/26239

Yokota S, Ohnishi T, Muroi M, Tanamoto K, Fujii N, Amano K (2007) Highly-purified *Helicobacter pylori* LPS preparations induce weak inflammatory reactions and utilize toll-like receptor 2 complex but not toll-like receptor 4 complex. FEMS Immunol Med Microbiol 51:140–148. https://doi.org/10.1111/j.1574-695x.2007.00288.x

Zhang XS, Tegtmeyer N, Traube L, Jindal S, Perez-Perez G, Sticht H, Backert S, Blaser MJ (2015) A specific A/T polymorphism in western tyrosine phosphorylation B-motifs regulates *Helicobacter pylori* CagA epithelial cell interaction. PloS Pathog 11:e1004621. https://doi.org/10.1371/journal.ppat.1004621

Zhou H, Gu J, Lamont SJ, Gu X (2007) Evolutionary analysis for functional divergence of the toll-like receptor gene family and altered functional constraints. J Mol Evol 65:119–123. https://doi.org/10.1007/s00239-005-0008-4

Role of NOD1 and ALPK1/TIFA Signalling in Innate Immunity Against *Helicobacter pylori* Infection

Le Ying and Richard L. Ferrero

Contents

Abstract The human pathogen *Helicobacter pylori* interacts intimately with gastric epithelial cells to induce inflammatory responses that are a hallmark of the infection. This inflammation is a critical precursor to the development of peptic ulcer disease and gastric cancer. A major driver of this inflammation is a type IV secretion system (T4SS) encoded by the cag pathogenicity island (*cag*PAI), present in a subpopulation of more virulent *H. pylori* strains. The *cag*PAI T4SS specifically activates signalling pathways in gastric epithelial cells that converge on the transcription factor, nuclear factor-κB (NF-κB), which in turn upregulates key immune and inflammatory genes, resulting in various host responses. It is now clear that *H. pylori* possesses several mechanisms to activate NF-κB in gastric epithelial cells and, moreover, that multiple

L. Ying · R. L. Ferrero (✉)
Centre for Innate Immunity and Infectious Diseases, Hudson Institute of Medical Research, Monash University, Clayton, VIC, Australia
e-mail: richard.ferrero@hudson.org.au

L. Ying
e-mail: le.ying@hudson.org.au

R. L. Ferrero
Department of Molecular and Translational Medicine, Monash University, Clayton, VIC, Australia

Department of Microbiology, Biomedicine Discovery Institute, Monash University, Clayton, VIC, Australia

S. Backert (ed.), *Molecular Mechanisms of Inflammation: Induction, Resolution and Escape by Helicobacter pylori*, Current Topics in Microbiology and Immunology 421, https://doi.org/10.1007/978-3-030-15138-6_7

signalling pathways are involved in these responses. Two of the dominant signalling pathways implicated in NF-κB-dependent responses in epithelial cells are nucleotide-binding oligomerisation domain 1 (NOD1) and a newly described pathway involving alpha-kinase 1 (ALPK1) and tumour necrosis factor (TNF) receptor-associated factor (TRAF)-interacting protein with forkhead-associated domain (TIFA). Although the relative roles of these two pathways in regulating NF-κB-dependent responses still need to be clearly defined, it is likely that they work cooperatively and non-redundantly. This chapter will give an overview of the various mechanisms and pathways involved in *H. pylori* induction of NF-κB-dependent responses in gastric epithelial cells, including a 'state-of-the-art' review on the respective roles of NOD1 and ALPK1/TIFA pathways in these responses.

1 Introduction

Helicobacter pylori is a Gram-negative bacterial pathogen widely found in the stomach of half of the global population (Salama et al. 2013). Infection with *H. pylori* is a risk factor in the development of gastric cancer (GC), which is responsible for 5% of all malignancies worldwide (Parkin et al. 2005). GC is a multifactorial disease that depends on a variety of bacterial, environmental and host genetic factors (El–Omar et al. 2001, see also Chapter "Genetic Polymorphisms in Inflammatory and Other Regulators in Gastric Cancer: Risks and Clinical Consequences" of this book). *H. pylori* promotes gastric carcinogenesis by the induction of inflammation, which results from increased expression of mucosal cytokines, including interleukin-1β (IL-1β) (Gööz et al. 2003), IL-6 (Lu et al. 2005), IL-8 (Rieder et al. 2001), IL-10 (Bodger et al. 2001) and tumour necrosis factor-α (TNF-α) (Tanahashi et al. 2000). Nuclear factor-κB (NF-κB) is a master regulator of these pro-inflammatory responses induced by *H. pylori* infection (Aihara et al. 1997; Glocker et al. 1998; Keates et al. 1997; Sharma et al. 1998). Indeed, activated NF-κB complexes were detected within epithelial and mononuclear cells of the gastric mucosa in response to the infection (Ferrero et al. 2008; Isomoto et al. 2000; Kudo et al. 2007; van Den Brink et al. 2000).

An important but often overlooked fact is that *H. pylori* induces NF-κB activation in different cell types via distinct mechanisms. In human gastric epithelial cells (GECs), *H. pylori* activates a classic NF-κB pathway involving the translocation of classical NF-κB p50/p65 complex to the nucleus, leading to the transcription and secretion of pro-inflammatory factors, including cytokines and chemokines (Hirata et al. 2006a; Maeda et al. 2001) (Fig. 1a).This was reported to occur via the signalling molecules TNF receptor-associated factor 2 (TRAF2)/TRAF6, transforming growth factor, beta-activated kinase 1 (TAK1) and IκB kinase (IKK) complex (Hirata et al. 2006a; Maeda et al. 2001). The induction of NF-κB signalling in human GECs is strictly dependent upon the presence in *H. pylori* bacteria of a functional type IV secretion system (T4SS) encoded by the *cag* pathogenicity island

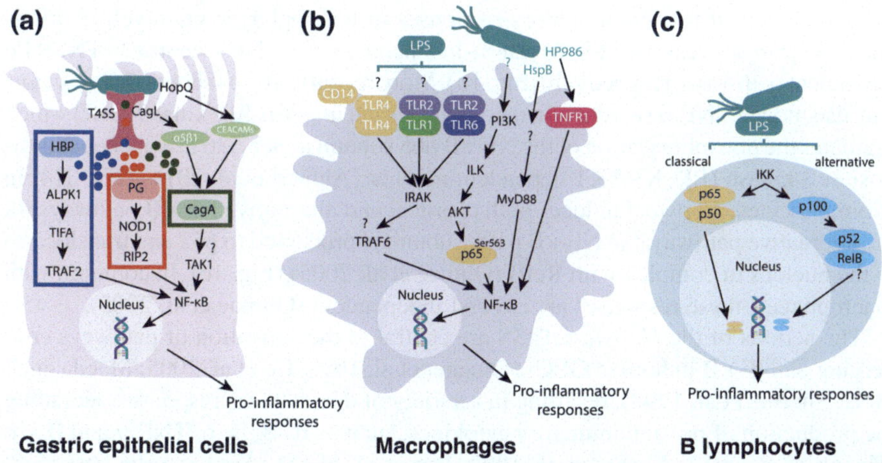

Fig. 1 *H. pylori* induction of NF-κB signalling in different host cell types. **a** NF-κB signalling cascades are activated sequentially following contact between the *H. pylori* T4SS and human gastric epithelial cells (Gall et al. 2017). The initial activation is mediated by the ALPK1/TIFA pathway (in blue), followed by a later response which is mediated by the NOD1/RIPK2 pathway (in red) and, finally, by the translocation of CagA into cells via the cell surface receptors, α5β1 and CEACAMs (in green). The combined activation of NF-κB signalling by these three pathways promotes maximal production of cytokine responses in gastric epithelial cells. **b** NF-κB signalling in macrophages is activated in response to several bacterial factors, independently of the *H. pylori* T4SS, and occurs via multiple pathways all converging on this transcription factor. (See text for details.) **c** NF-κB signalling in B lymphocytes is mediated by both classical and alternative pathways in response to *H. pylori* LPS. Aspects of the pathways that are unknown or require further investigation are indicated by the '?' symbol

(*cag*PAI) (Boonyanugomol et al. 2013; Fischer et al. 2001; Shibata et al. 2005; Viala et al. 2004). Interestingly, *H. pylori* T4SS is not essential for the induction of NF-κB-dependent chemokine responses in mouse GECs (Ferrero et al. 2008). The *cag*PAI T4SS shares features of the prototypical T4SS structure described in the plant pathogen, *Agrobacterium tumefaciens* (Backert et al. 2015; Merino et al. 2017; Olbermann et al. 2010). *H. pylori* encodes several other T4SSs, but only the *cag*PAI T4SS is known to be important for host–pathogen interactions (Fernandez-Gonzalez and Backert 2014).

In contrast to *H. pylori* interactions with human GECs, it was shown that *H. pylori* activates NF-κB signalling in macrophages independently of both the *cag*PAI and direct contact between the bacterium and host cells (Maeda et al. 2001). Macrophage responses were reported to be mediated by the Toll-like receptor 4 (TLR4) signalling pathway, involving CD14/TLR4 → IL-1 receptor-associated kinase (IRAK) → TRAF6 (Fig. 1b). Subsequent work suggested that TLR2 and not TLR4 is required (Mandell et al. 2004); however, this question is still largely unresolved. Several *H. pylori* products, including the heat-shock protein B (HspB) (Gobert et al. 2004) and HP986 (Alvi et al. 2011), have been reported to upregulate the production of pro-inflammatory responses in macrophages via myeloid differentiation primary response 88 (MyD88)-independent and TNF receptor 1 (TNFR1)-dependent pathways, respectively (Fig. 1b). More recently, it was shown that *H. pylori* can also induce TNF-α secretion in macrophages via integrin-like kinase (ILK) which regulates the phosphorylation of the NF-κB p65 subunit at Ser-536 in a phosphatidylinositol 3-kinase (PI3 K)/AKT-dependent manner (Ahmed et al. 2014) (Fig. 1b). In B lymphocytes, *H. pylori* induces both classical and alternative NF-κB pathways. In the alternative pathway, the NF-κB p100 subunit is processed to p52 and translocates to the nucleus in complex with RelB (Ohmae et al. 2005) (Fig. 1c). In common with macrophages, these responses are *cag*PAI independent (Ohmae et al. 2005).

The actions of the *H. pylori* T4SS are central to the activation of pathways converging on NF-κB in human GECs (Glocker et al. 1998; Lu et al. 2005; Maeda et al. 2001; Sharma et al. 1998), resulting in a variety of downstream responses, including the production of pro-inflammatory cytokines, such as IL-8, IL-6, TNF-α and IL-1β (Huang et al. 2012; Keates et al. 1997; Lin et al. 2015). Amongst the best characterised of these pathways is that regulated by nucleotide-binding oligomerisation domain 1 (NOD1), a member of the NOD-like receptors (NLRs) family of innate immune signalling molecules (Viala et al. 2004). More recently, it has emerged that TRAF-interacting protein with forkhead-associated domain (TIFA) is a critical host factor mediating NF-κB-dependent responses to *H. pylori* (Gall et al. 2017; Stein et al. 2017; Zimmermann et al. 2017). This review will broadly discuss the various mechanisms by which *H. pylori* mediates pro-inflammatory responses in human GECs, as well as the signalling pathways involved. It will particularly focus on the respective roles of NOD1 and TIFA in GEC responses to *H. pylori* infection.

2 *H. pylori* Induction of NF-κB Signalling in Human Epithelial Cells

2.1 *The* H. pylori cag*PAI Is Required for Induction of NF-κB Activation in Epithelial Cells*

The screening of an *H. pylori* expression library with serum from an infected individual identified a major high molecular weight protein associated with cytotoxic effects of the toxin VacA on cells (Tummuru et al. 1993). The corresponding gene was thus assigned the name cytotoxin-associated gene A (*cagA*) (Tummuru et al. 1993). Later studies, however, demonstrated that CagA was not directly responsible for this activity (Tummuru et al. 1993). Instead, *cagA* was reported to be a marker for more virulent *H. pylori* strains and to be part of a 40-kilobase DNA region exhibiting features consistent with those of bacterial PAIs (Censini et al. 1996). The *H. pylori cag*PAI was subsequently shown to encode a T4SS that forms a pilus-like structure (Kwok et al. 2007). This structure mediates the translocation of CagA into epithelial cells (Backert et al. 2000; Odenbreit et al. 2000; Segal et al. 1999; Stein et al. 2000), leading to the rearrangement of the host cell actin cytoskeleton, also known as the 'hummingbird' phenotype (Segal et al. 1996). Importantly, several proteins encoded by the *H. pylori cag*PAI were reported to be required for NF-κB activation in KATO-III cells, a GEC line (Glocker et al. 1998). Furthermore, it was shown that *H. pylori* induction of IL-8 production in KATO-III cells was strictly dependent upon NF-κB activation (Münzenmaier et al. 1997), with a lesser role identified for the transcription factor, activator protein 1 (AP-1) (Aihara et al. 1997; Allison et al. 2009).

In addition to IL-8, *H. pylori* induction of active NF-κB complexes leads to the production of IL-6, TNF-α and metalloproteinases in GECs (Caruso et al. 2007; Lin et al. 2015; Lu et al. 2005). In the case of IL-6, maximal responses were induced by *H. pylori* strains carrying both the *cag*PAI and an outer membrane protein, outer inflammatory protein A (OipA) (Lu et al. 2005). In contrast to *H. pylori*-induced IL-8 responses, however, IL-6 responses in different human GECs required activation of the transcription factors NF-κB and AP-1, as well as cAMP response element (CRE) and CCAAT/enhancer-binding protein (C/EBP) (Lu et al. 2005). This suggests that there may be some differences in the mechanisms by which *H. pylori* upregulates the production of different pro-inflammatory factors.

2.2 cag*PAI-Encoded Proteins Associated with NF-κB Activation*

The first detailed study on the mechanism(s) by which *H. pylori* induces NF-κB activation identified the following *cag*PAI-encoded proteins: CagE, CagG, CagH, CagI, CagL and CagM, as being indispensable for NF-κB activation in KATO-III cells (Glocker et al. 1998). Furthermore, it was reported that a secreted product, which was not released by an *H. pylori cagE* mutant, was responsible for the induction of NF-κB activation and IL-8 production in KATO-III cells (Münzenmaier et al. 1997). Interestingly, purified *H. pylori* lipopolysaccharide (LPS) did not appear to mediate NF-κB activation in MKN45 and TMK-1 cells, both of which were derived from gastric adenocarcinomas (Maeda et al. 2001; Münzenmaier et al. 1997). Although another study reported that highly purified *H. pylori* LPS was able to induce IL-8 secretion via TLR2-TLR1 or TLR2-TLR6 complexes, it was approximately 500-fold less active than smooth-form *Escherichia coli* LPS (Yokota et al. 2007). Taken together, the findings suggested that a non-LPS-secreted factor, associated with the *H. pylori cag*PAI, was required for the induction of NF-κB-dependent responses in GECs.

As reported above, the CagA protein is translocated into host cells via the *H. pylori* T4SS (Backert et al. 2008; Tegtmeyer et al. 2017). Studies using isogenic *H. pylori cagA* bacteria, however, showed that CagA was dispensable for *H. pylori* induction of IL-8 production in GECs (Crabtree et al. 1995; Fischer et al. 2001; Zhang et al. 2015). Conversely, one study found that *H. pylori cagA* mutant bacteria induced reduced levels of IL-8 promoter activity, when compared with the parental strain, although this finding could not be explained (Sharma et al. 1998). It was suggested that CagA may physically associate with TAK1 to mediate NF-κB activation (Lamb et al. 2009), however, subsequent work showed that the anti-TAK1 antibody used in that study was able to partially immunoprecipitate CagA protein in *H. pylori* lysates (Sokolova et al. 2014). Thus, the roles of TAK1 or other host proteins as interacting partner(s) involved in CagA-mediated induction of the NF-κB pathway remain to be elucidated.

It was reported that CagA induces IL-8 expression in AGS cells in a time-dependent manner, with maximal IL-8 responses observed at the late stages of infection in vitro, i.e. at 24 and 36 h (Brandt et al. 2005). This IL-8inducing activity of CagA was dependent on signalling through the classical mitogen-activated protein kinase (MAPK)/extracellular signal regulated kinase (ERK) pathway, culminating in NF-κB activation and IL-8 production (Brandt et al. 2005). Importantly, the effect of CagA on IL-8 responses also varied between different *H. pylori* strains (Brandt et al. 2005).

CagA translocation into host cells requires interactions between the CagL protein, present on the *H. pylori* T4SS pilus, and integrins expressed on the surfaces of GECs, most notably α5β1 integrin (Jiménez-Soto et al. 2009; Kwok et al. 2007; Fig. 1a). *H. pylori* CagL–integrin interactions were shown to trigger the induction of IL-8 responses in AGS cells independently of CagA translocation (Gorrell et al. 2013). These responses were dependent on a signalling pathway involving Src kinase, ERK, JNK and NF-κB (Gorrell et al. 2013). It was proposed that CagL–integrin interactions represent an additional pathway whereby the *H. pylori* T4SS mediates NF-κB-dependent pro-inflammatory responses in AGS cells (Gorrell et al. 2013). A recent study, however, reported that carcinoembryonic antigen-related cell adhesion molecules (CEACAMs), and not integrins, are essential for CagA translocation by the *H. pylori* T4SS in KATO-III cells (Fig. 1a). (Zhao et al. 2018). *H. pylori* interacts with CEACAMs via the adhesin, HopQ (Zhao et al. 2018).

2.3 H. pylori *Activates the NF-κB Pathway in Epithelial Cells via NOD1*

Although translocated CagA appeared to be one mediator of *H. pylori* T4SS-dependent effects on NF-κB signalling in epithelial cells (Brandt et al. 2005; Lamb et al. 2009; Sharma et al. 1998), it seemed likely that (an)other T4SS-secreted factor(s) may be involved. One such factor was revealed by studies on the invasive Gram-negative pathogen *Shigella flexneri* (Girardin et al. 2001). This pathogen was found to activate NF-κB signalling in epithelial cells via the innate immune molecule, NOD1 (Girardin et al. 2001).

NOD1 is an intracellular pattern recognition receptor from the NLR family and is characterised by an N-terminal caspase recruitment domain (CARD), a centrally located NOD and a C-terminal sensing domain, consisting of a leucine-rich repeat

(LRR) domain that is involved in recognition of conserved microbial patterns (Sorbara and Philpott 2011; Strober et al. 2006). NOD1 shares similarities with plant pathogen resistance proteins and is an intracellular sensor of bacteria via its recognition of a molecular structure, Tri-DAP, present in the breakdown products (muropeptides) of Gram-negative-type peptidoglycan (PG) (Murray 2005). NOD1 has been reported to mediate NF-κB-dependent responses to a range of pathogens, including *H. pylori* (Viala et al. 2004), *Escherichia coli* (Tan et al. 2015), *Haemophilus parasuis* (Ma et al. 2018), *S. flexneri* (Girardin et al. 2001; Fukazawa et al. 2008), *Chlamydophila pneumoniae* (Opitz et al. 2005) and *Legionella pneumophila* (Shin et al. 2008). Importantly, *H. pylori* strains with an intact *cag*PAI were shown to selectively mediate NOD1 signalling via the T4SS-dependent delivery of PG muropeptides, resulting in NF-κB signalling in AGS cells (Viala et al. 2004). Moreover, NOD1 recognition of *H. pylori* PG and downstream NF-κB activation were dependent on T4SS interactions with α5β1 integrin, associated with cholesterol-rich microdomains in host cell membranes (Hutton et al. 2010). Additionally, *H. pylori* and other Gram-negative bacteria can deliver PG to NOD1 via outer membrane vesicles (OMVs), which enter AGS cells, leading to the activation of NOD1 → NF-κB signalling and downstream IL-8 and human beta-defensin (hBD) responses (Kaparakis et al. 2010). It was shown that PG-containing OMVs traffic to early endosomes and associate with NOD1 and its adapter protein, receptor-interacting serine/threonine kinase 2 (RIPK2), resulting in autophagy in epithelial but not myeloid cells (Irving et al. 2014).

In contrast to *S. flexneri*, which mediates NOD1-dependent c-Jun N-terminal kinase (JNK) signalling in epithelial cells (Girardin et al. 2001), *H. pylori* activation of the NOD1 signalling pathway was reported to induce activation of the MAPKs p38 and ERK, as well as AP-1, but not JNK (Allison et al. 2009). This difference is likely to reflect the diverse lifestyles (invasive versus extracellular) of these bacteria (Allison et al. 2009). Indeed, *H. pylori* activation of the NOD1 pathway was shown to promote Tyr701/Ser727 phosphorylation of the transcription factor, signal transducer and activator of transcription 1 (STAT1), and expression of interferon regulatory factor 1 (IRF1), resulting in enhanced production of the NOD1 and IFN-γ-regulated chemokines, IL-8 and C-X-C motif chemokine ligand 10 (CXCL10), respectively (Allison et al. 2013). *H. pylori* induction of the NOD1 pathway was also reported to mediate type I interferon (Watanabe et al. 2010a) and antimicrobial peptide responses (Boughan et al. 2006; Grubman et al. 2010), which may be involved in controlling the infection. Furthermore, it was reported that NOD1 may regulate NF-κB-dependent repression of the H,K-ATPase α subunit (HKα), leading to inhibition of gastric acid secretion (Hammond et al. 2015). Nevertheless, NOD1 seems to play a minor role in mediating NF-κB p50 homodimer binding to the HKα promoter (Saha et al. 2010).

3 Role of NOD1 in Host Immune Responses to *H. pylori* In Vivo

NOD1 has been reported to mediate host defence responses against infection by a broad range of bacterial pathogens (Sorbara and Philpott 2011). This concept was first established in the mouse model of *H. pylori* infection, in which *Nod1*$^{-/-}$ mice exhibited higher bacterial loads than wild-type animals (Viala et al. 2004). The bacterial loads between animal groups differed by 1 log colony-forming units (CFUs) at day 7 after oral gavage, and as much as 2–3 log CFUs, at 30 days post-gavage (Viala et al. 2004). Similar findings were reported by another group (Watanabe et al. 2010a). Moreover, NOD1 had a protective effect against infection by *cag*PAI-positive *H. pylori* with a functional T4SS, but had no effect on colonisation levels by *cag*PAI-negative bacteria (Viala et al. 2004). Conversely, it was reported that the glycoprotein olfactomedin 4, which is upregulated in the gastric mucosa of *H. pylori*-infected subjects, negatively regulated NOD1-mediated NF-κB signalling, thereby inhibiting host immune responses and thus contributing to bacterial persistence (Liu et al. 2010).

Several studies have sought to identify the mechanism(s) by which NOD1 mediates host defence responses against *H. pylori* infection. NOD1 was shown to regulate hBD-2 gene expression in response to *H. pylori* via a *cag*PAI-dependent mechanism (Boughan et al. 2006; Grubman et al. 2010). Furthermore, *H. pylori* activation of the NOD1 signalling pathway upregulated the production of hBD-2 by epithelial cells in vitro, resulting in direct killing of the bacterium (Grubman et al. 2010). These findings were consistent with those from a primate infection model in which *cag*PAI-positive *H. pylori* bacteria induced gene expression levels for several antimicrobial peptides, including monkey homologues of hBD-2 (Hornsby et al. 2008). It was suggested that these antimicrobial responses may provide an advantage to *H. pylori*, by allowing the bacterium to establish its niche within the stomach, and may even be beneficial to the host, by ensuring a form of colonisation resistance against overgrowth by more harmful bacteria (Hornsby et al. 2008). NOD1 was also reported to activate the IFN-stimulated gene factor 3 (ISGF3) pathway against *H. pylori* infection leading the authors to propose that a type I IFN response is associated with NOD1 host defence functions (Watanabe et al. 2010a).

Consistent with its host defence role, NOD1 has been reported to regulate lymphocyte responses against *H. pylori* infection. Specifically, it was shown that NOD1 signalling was required for priming of antigen-specific responses against *H. pylori*, with *Nod1*$^{-/-}$ mice exhibiting reduced levels of antigen-specific IgG2c serum antibodies when compared with wild-type animals (Fritz et al. 2007). Furthermore, *Nod1*$^{-/-}$ mice exhibited significantly reduced levels of serum antibody titres in response to oral gavage with *H. pylori* OMVs harbouring the NOD1 ligand, PG (Kaparakis et al. 2010).

In contrast to the host protective roles of NOD1 described above, it was also reported that NOD1 recognition of bacterial PG by GECs upregulates expression of *CD274*, encoding programmed cell death ligand 1 (PD-L1), also known as B7-H1

(Lina et al. 2015). This protein downregulates T-cell activation through its interaction with its cognate receptor, PD-1 (Lina et al. 2015). It was suggested that T-regulatory cell expansion may, in turn, decrease inflammation and promote *H. pylori* persistence (Lina et al. 2015). Although the authors did not directly demonstrate that *H. pylori* activation of NOD1 signalling in GECs was able to mediate PD-1-dependent expansion of T-regulatory responses, they did show that *cag*PAI-positive *H. pylori* specifically induced *Cd274* expression in the mouse gastric mucosa (Lina et al. 2015). Furthermore, infection studies in *Cd274*$^{-/-}$ mice suggested that *H. pylori* uses its T4SS to mediate T-regulatory cell-dependent anti-inflammatory responses, thereby promoting bacterial persistence (Lina et al. 2015).

It is plausible that *cag*PAI-positive *H. pylori* bacteria exploit the NOD1 signalling pathway to modulate host immune responses and thus promote their persistence in the host. Indeed, recent data from our laboratory showed that, similar to previous reports (Viala et al. 2004; Watanabe et al. 2010b), a *cag*PAI-positive *H. pylori* strain colonised *Nod1*$^{-/-}$ mice to significantly increase bacterial loads than their wild-type counterparts in the early phase of infection ($\leq$1 month post-gavage), yet colonisation levels were similar, if not reduced, in the *Nod1*$^{-/-}$ animals later in infection, i.e. at 2 months post-gavage (D'Costa and Ferrero; unpublished data). Consistent with these data, NOD1 and its adaptor protein RIPK2 were reported to play a protective role against excessive inflammation by promoting IL-33 production, leading to the development of T helper 2-type responses in vivo (Tran et al. 2018). *H. pylori* bacteria regulated IL-33 responses in AGS cells in a T4SS- and CagA-dependent manner (Tran et al. 2018). Taken together, the data reveal a complex role for NOD1 in host immune responses against infection by *cag*PAI-positive *H. pylori* bacteria.

NOD1 has been reported to protect against tumorigenesis in different models (Chen et al. 2008; da Silva Correia et al. 2006; Suarez et al. 2015; Zhan et al. 2016). NOD1 induced apoptosis in MCF-7 and SK-BR3 breast cancer cell lines (da Silva Correia et al. 2007). Furthermore, in a breast cancer xenograft model, mice injected with NOD1-deficient MCF-7 cells displayed more tumour formation and heightened sensitivity to oestrogen-induced cell proliferation (da Silva Correia et al. 2006). Conversely, NOD1 overexpression had an inhibitory effect on oestrogen-dependent tumour growth in vivo (da Silva Correia et al. 2006). NOD1 also seemed to play a protective role in tumorigenesis in different mouse colitis models (Chen et al. 2008; Zhan et al. 2016). Consistent with these findings, pre-treatment of Mongolian gerbils with C12-iE-DAP, a commercial NOD1 agonist, significantly reduced the levels of bacterial colonisation, inflammation and GC development in response to *H. pylori* challenge (Suarez et al. 2015).

A critical step in the development of gastric adenocarcinoma is the change of the epithelial cells in the stomach to the intestinal type, a process known as intestinal metaplasia (Salama et al. 2013). Interestingly, expression of the intestinal epithelial-specific transcription factor, caudal-type homeobox 2 (CDX2), was reported to be downregulated by NOD1 in AGS and KATO-III cells stimulated with *H. pylori* bacteria (Asano et al. 2016). This was dependent upon NOD1-mediated activation of TRAF3, a negative regulator of NF-κB signalling (Asano et al. 2016). Consistent with these findings, *Nod1*$^{-/-}$ mice with chronic *H. pylori* infection exhibited increased

Cdx2 expression and intestinal metaplasia (Asano et al. 2016). Furthermore, *CDX2* expression levels in human gastric biopsy samples were significantly increased and *TRAF3* levels decreased in tumour when compared with non-tumour tissues (Suarez et al. 2015).

Analysis of human gastric biopsies from subjects with *H. pylori* infection revealed significantly increased gene expression levels of *NOD1*, *CXCL8*, *CXCL10* and *IRF1* in tumour-affected tissues, as well as those from subjects with severe gastritis, when compared with non-tumour and normal tissues, respectively (Allison et al. 2013). NOD1 activation was reported to enhance IFN-γ signalling via STAT1 and IRF1 in response to *H. pylori* infection. It was proposed that cross-talk between NOD1 and IFN-γ signalling pathways promote the development of more severe pathology in *H. pylori* infection (Allison et al. 2013). In agreement with this proposal, one group reported the upregulation of *NOD1* gene expression in gastric biopsies from *H. pylori*-infected subjects with gastritis (Rosenstiel et al. 2006). Conversely, other workers reported NOD1 to be significantly upregulated in non-tumour as opposed to tumour samples, as determined by qPCR and immunohistochemistry (Suarez et al. 2015). It was suggested that during the late stages of gastric carcinogenesis, *NOD1* expression is reduced in intestinal-type gastric adenocarcinomas compared with uninvolved gastric tissue (Suarez et al. 2015). The differences in the findings of these studies are unclear but warrant further investigation.

Several studies have been undertaken to determine the potential impact of *NOD1* gene polymorphisms on the severity of *H. pylori*-related disease, but these have been largely inconclusive. In one study, *NOD1* gene polymorphisms were not linked with susceptibility to *H. pylori* infection nor with gastric carcinogenesis in a Caucasian population (Kupcinskas et al. 2011). Therefore, it was suggested that it was unlikely to be a potential biomarker for advanced GC in this population (Kupcinskas et al. 2011). Similarly, no associations could be found between *NOD1* polymorphisms and either gastric ulcer or mucosa-associated lymphoid tissue lymphoma, associated with chronic *H. pylori* infection, in a German population (Rosenstiel et al. 2006). Contrary to those two studies, *NOD1* gene polymorphisms were associated with an elevated GC risk and gastric mucosal inflammation in Chinese (Wang et al. 2012) and South Korean populations (Kim et al. 2013), respectively. Taken together, it is plausible that host genetic factors influence the impact of NOD1 signalling on the severity of disease outcomes associated with *H. pylori* infection.

4 TIFA, A New Host Factor Mediating NF-κB-Dependent Responses to *H. pylori* Infection

The original report on NOD1-dependent signalling to *H. pylori* infection in AGS cells (Viala et al. 2004) has been confirmed in multiple studies, as well as independently by different research groups (Allison et al. 2009, 2013; Boughan et al. 2006; Cook et al. 2014; Gall et al. 2017; Grubman et al. 2010; Hutton et al. 2010; Kim et al.

2015; Patel et al. 2013; Rosenstiel et al. 2006; Suarez et al. 2015; Watanabe et al. 2010a. Nevertheless, this finding has been questioned by some authors (Backert and Naumann 2010; Hirata et al. 2006b; Zimmermann et al. 2017). Notably, Hirata et al. (2006b) reported that *NOD1* knockdown by siRNA had no effect on IL-8 responses to *H. pylori*; however, these authors did not provide any evidence for the efficacy of the *NOD1* knockdown in their study. Another group performed an RNAi screen using a GFP-expressing AGS reporter cell line to identify factors required for *H. pylori* regulation of NF-κB responses (Zimmermann et al. 2017). They did not observe any significant 'hits' for NOD1 or downstream factors, but did observe that Tri-DAP added to permeabilised AGS cells induced upregulation of *CXCL8* transcription albeit at lower levels than for *H. pylori* lysates (Zimmermann et al. 2017). These authors concluded that NOD1 was clearly stimulated, but noted that no evidence could be found to show that peptidoglycan muropeptides were responsible for *H. pylori*-induced NF-κB responses, although this question was not directly addressed in the study.

The effects of *NOD1* knockdown in cell lines have tended to be variable and not complete, resulting in reductions of approximately 50–90% for NF-κB-dependent responses to *H. pylori*, depending on the study (Boughan et al. 2006; Grubman et al. 2010; Kim et al. 2015; Suarez et al. 2015; Viala et al. 2004). This can be due to a variety of technical reasons, such as the type of *H. pylori* strain used, the procedures and times used to seed and co-culture cells with the bacteria (R. L. F., unpublished data). More significantly, all of these studies used either dominant-negative constructs, siRNA or shRNA to knock down *NOD1* expression either transiently or stably. These techniques suffer from both the variable levels of knockdown efficacy and the potential for off-target effects. The CRISPR/Cas9 gene-editing technology limits these issues and, moreover, has made it possible, for the first time, to readily and specifically delete genes of interest in human cell lines. Thus, *NOD1*-deficient AGS cells generated by the CRISPR technique exhibited reduced, but not completely abrogated IL-8 production in response to *H. pylori* bacteria (Gall et al. 2017; Tran et al. 2018). This finding unambiguously confirmed that at least one other signalling pathway is important in mediating NF-κB-dependent responses to *H. pylori* infection.

Support for this observation emerged from three independent and complementary studies which reported that *H. pylori* promotes NF-κB and IL-8 responses in GECs via the activation of a new innate immune signalling pathway, dependent on the adapter protein, TIFA (Gall et al. 2017; Stein et al. 2017; Zimmermann et al. 2017). TIFA had previously been identified by a genomewide RNA interference screen, in which TIFA was shown to be part of a signalling axis involving the ubiquitin ligase, TRAF6 (Gaudet et al. 2015). This pathway mediated NF-κB responses to monosaccharide heptose-1,7-bisphosphate (HBP), a metabolic intermediate in the LPS biosynthesis pathway of Gram-negative bacteria (Gaudet et al. 2015). Subsequently, it was reported that the induction of TIFA oligomerisation and IL-8 responses by Gram-negative bacteria (*S. flexneri, S. Typhimurium* and *Neisseria meningitidis*) is dependent on the functions of alpha-kinase 1 (ALPK1) (Milivojevic et al. 2017). It was recently established that ALPK1 is a cytosolic innate immune receptor for the LPS metabolite, ADP-β-D-manno-heptose (ADP-Hep) (Zhou et al. 2018). Furthermore, ALPK1 mediates TIFA phosphorylation and 'TIFAsome' formation in response to *H. pylori* (Zimmermann et al. 2017). Gene mutagenesis studies showed that *H. pylori* activation of the ALPK1-TIFA signalling axis occurs via a T4SS-dependent mechanism (Stein et al. 2017; Zimmermann et al. 2017).

H. pylori mutant bacteria with a deletion in the gene encoding the HldE enzyme, which is responsible for HBP synthesis (Pachathundikandi and Backert 2018), induced very low levels of IL-8 responses when compared with the parental strain (Stein et al. 2017; Zimmermann et al. 2017). Furthermore, lysates of *H. pylori* wild-type and *cag*PAI mutant bacteria had no effect on NF-κB responses in HEK293 cells, unless these preparations were transfected into cells (Stein et al. 2017). Cells also did not respond to lysates prepared from *H. pylori hldE* mutant bacteria (Stein et al. 2017). Importantly, *H. pylori* HldE mediates the synthesis of the respective TIFA and ALPK1 agonists, HBP and ADP-Hep. In contrast, two distinct enzymes mediate the biosynthesis of these metabolites in other bacteria (Stein et al. 2017). The absence of the ADP-Hep metabolite in *H. pylori hldE* lysates may explain their lack of activity when added externally to cells. Interestingly, it was reported that ADP-Hep is involved in type III secretion system (T3SS)-dependent NF-κB activation and cytokine secretion, but unlike HBP, is also able to act from the extracellular compartment (Zhou et al. 2018). ADP-Hep has therefore been proposed to act as a trigger for innate immune responses to both extracellular bacteria (e.g. *N. meningitidis*, diffuse-adhering *E. coli*, enterotoxigenic *E. coli*) those with secretion systems (Zhou et al. 2018). Nevertheless, if ADP-Hep can enter cells freely, why then do some bacteria also require secretion systems to activate ALPK1 in cells? Another outstanding question is the in vivo role for ALPK1-TIFA signalling in *H. pylori* and other bacterial infections. Unfortunately, *H. pylori hldE* mutants grow poorly and are more sensitive to external stresses, thus making it problematic to undertake infection studies in animal models (Stein et al. 2017).

5 Concluding Remarks

It is now well established that *H. pylori* exploits multiple T4SS-dependent mechanisms to upregulate pro-inflammatory signalling in human GECs via the critical regulator, NF-κB. Moreover, it appears that *H. pylori* T4SS-dependent activation of NF-κB signalling in these cells in vitro follows a sequential process in which NF-κB is initially activated by the ALPK1/TIFA pathway, followed by the NOD1/RIPK2 pathway, and finally, by translocated CagA (Gall et al. 2017; Fig. 1a). Interestingly, this is the inverse to the situation in the *Shigella* infection model, in which NOD1 activation of NF-κB precedes that mediated by the ALPK1/TIFA pathway (Gall et al. 2017; Gaudet et al. 2015). This is likely to be due to the different infection processes of these two pathogens. Nevertheless, the relevance of sequential NF-κB activation in an in vivo context remains unclear. Future studies will be required to clarify the relative roles of NOD1 and ALPK1-TIFA pathways in epithelial cell responses to the *H. pylori* T4SS, including those previously attributed to NOD1 signalling. It also needs to be determined whether there is any cross-talk and/or co-regulation between NOD1, ALPK1-TIFA, and CagA-dependent pathways in NF-κB responses to *H. pylori* infection in GECs. Indeed, it has been suggested that the convergence of ALPK1/TIFA and NOD1 pathways on TRAF6 may potentiate CagA interactions with TAK1 (Lamb et al. 2009), thereby potentiating NF-κB responses in cells (Gall et al. 2017). Finally, it remains to be determined whether these signalling pathways play other roles in *H. pylori* infection beyond their known functions as mediators of host immune and inflammatory responses.

Acknowledgements The authors thank Christian Ferrero for the preparation of Fig. 1. Research in RLF's laboratory is supported by a project grant (APP1107930) and a Senior Research Fellowship (APP1079904) from the National Health and Medical Research Council of Australia. Dr. L. Ying's position is supported by funding from the U. S. Department of Defense (Award No. W81XWH-17-1-0606). Research at the Hudson Institute of Medical Research is supported by the Victorian Government's Operational Infrastructure Support Program.

References

Ahmed AU, Sarvestani ST, Gantier MP, Williams BRG, Hannigan GE (2014) Integrin-linked kinase modulates lipopolysaccharide- and *Helicobacter pylori*-induced nuclear factor κB-activated tumor necrosis factor-α production via regulation of p 65 serine 536 phosphorylation. J Biol Chem 289(40):27776–27793. https://doi.org/10.1074/jbc.M114.574541

Aihara M, Tsuchimoto D, Takizawa H, Azuma A, Wakebe H, Ohmoto Y, Imagawa K Kikuchi M, Mukaida N, Matsushima K (1997) Mechanisms involved in *Helicobacter pylori*-induced interleukin-8 production by a gastric cancer cell line, MKN45. Infect Immun 65(8):3218–3224

Allison CC, Kufer TA, Kremmer E, Kaparakis M, Ferrero RL (2009) Helicobacter pylori induces MAPK phosphorylation and AP-1 activation via a NOD1-dependent mechanism. J Immunol 183(12):8099–8109. https://doi.org/10.4049/jimmunol.0900664

Allison CC, Ferrand J, McLeod L, Hassan M, Kaparakis-Liaskos M, Grubman A, Bhathal PS, Dev A, Sievert W, Jenkins BJ, Ferrero RL (2013) Nucleotide oligomerization domain 1 enhances IFN-γ signaling in gastric epithelial cells during *Helicobacter pylori* infection and exacerbates disease severity. J Immunol 190(7):3706–3715. https://doi.org/10.4049/jimmunol.1200591

Alvi A, Ansari SA, Ehtesham NZ, Rizwan M, Devi S, Sechi LA, Qureshi IA, Hasnain SE, Ahmed N (2011) Concurrent proinflammatory and apoptotic activity of a *Helicobacter pylori* protein (HP986) points to its role in chronic persistence. PLoS One. 6(7):e22530. https://doi.org/10.1371/journal.pone.0022530

Asano N, Imatani A, Watanabe T, Fushiya J, Kondo Y, Jin X, Ara N, Uno K, Iijima K, Koike T, Strober W, Shimosegawa T (2016) Cdx2 expression and intestinal metaplasia induced by *H. pylori* infection of gastric cells is regulated by NOD1-mediated innate immune responses. Cancer Res 76(5):1135–1145. https://doi.org/10.1158/0008-5472.CAN-15-2272

Backert S, Naumann M (2010) What a disorder: proinflammatory signaling pathways induced by *Helicobacter pylori*. Trends Microbiol 18(11):479–486. https://doi.org/10.1016/j.tim.2010.08.003

Backert S, Ziska E, Brinkmann V, Zimny-Arndt U, Fauconnier A, Jungblut PR, Naumann M, Meyer TF (2000) Translocation of the *Helicobacter pylori* CagA protein in gastric epithelial cells by a type IV secretion apparatus. Cell Microbiol 2(2):155–164. https://doi.org/10.1046/j.1462-5822.2000.00043.x

Backert S, Feller SM, Wessler S (2008) Emerging roles of Abl family tyrosine kinases in microbial pathogenesis. Trends Biochem Sci 33(2):80–90. https://doi.org/10.1016/j.tibs.2007.10.006

Backert S, Tegtmeyer N, Fischer W (2015) Composition, structure and function of the *Helicobacter pylori* cag pathogenicity island encoded type IV secretion system. Future Microbiol 10:955–965
Bodger K, Bromelow K, Wyatt JI, Heatley RV (2001) Interleukin 10 in *Helicobacter pylori* associated gastritis: immunohistochemical localisation and *in vitro* effects on cytokine secretion. J Clin Pathol 54(4):285–292. https://doi.org/10.1136/JCP.54.4.285
Boonyanugomol W, Chomvarin C, Hahnvajanawong C, Sripa B, Kaparakis-Liaskos M, Ferrero RL (2013) *Helicobacter pylori cag* pathogenicity island (*cag*PAI) involved in bacterial internalization and IL-8 induced responses via NOD1- and MyD88-dependent mechanisms in human biliary epithelial cells. PLoS ONE 8(10):e77358. https://doi.org/10.1371/journal.pone.0077358
Boughan PK, Argent RH, Body-Malapel M, Park J-H, Ewings KE, Bowie AG, Ong SJ, Cook SJ, Sorensen OE, Manzo BA, Inohara N, Klein NJ, Nuñez G, Atherton JC, Bajaj-Elliott M (2006) Nucleotide-binding oligomerization domain-1 and epidermal growth factor receptor. J Biol Chem 281(17):11637–11648. https://doi.org/10.1074/jbc.M510275200
Brandt S, Kwok T, Hartig R, König W, Backert S (2005) NF-kappaB activation and potentiation of proinflammatory responses by the *Helicobacter pylori* CagA protein. Proc Natl Acad Sci U S A 102(26):9300–9305. https://doi.org/10.1073/pnas.0409873102
Caruso R, Fina D, Peluso I, Fantini MC, Tosti C, Del Vecchio Blanco G, Paoluzi OA, Caprioli F, Andrei F, Stolfi C, Romano M, Ricci V, MacDonald TT, Pallone F, Monteleone G (2007) IL-21 is highly produced in *Helicobacter pylori*-infected gastric mucosa and promotes gelatinases synthesis. J Immunol 178(9):5957–5965. https://doi.org/10.4049/jimmunol.178.9.5957
Censini S, Lange C, Xiang Z, Crabtree JE, Ghiara P, Borodovsky M, Rappuoli R, Covacci A (1996) *cag*, a pathogenicity island of *Helicobacter pylori*, encodes type I-specific and disease-associated virulence factors. Proc Natl Acad Sci U S A 93(25):14648–14653. https://doi.org/10.1073/PNAS.93.25.14648
Chen GY, Shaw MH, Redondo G, Nunez G (2008) The innate immune receptor Nod1 protects the intestine from inflammation-induced tumorigenesis. Cancer Res 68(24):10060–10067. https://doi.org/10.1158/0008-5472.CAN-08-2061
Cook KW, Letley DP, Ingram RJM, Staples E, Skjoldmose H, Atherton JC, Robinson K (2014) CCL20/CCR6-mediated migration of regulatory T cells to the *Helicobacter pylori*-infected human gastric mucosa. Gut 63(10):1550–1559. https://doi.org/10.1136/gutjnl-2013-306253
Crabtree JE, Xiang Z, Lindley IJ, Tompkins DS, Rappuoli R, Covacci A (1995) Induction of interleukin-8 secretion from gastric epithelial cells by a cagA negative isogenic mutant of *Helicobacter pylori*. J Clin Pathol 48(10):967–969. https://doi.org/10.1136/jcp.48.10.967
da Silva Correia J, Miranda Y, Austin-Brown N, Hsu J, Mathison J, Xiang R, Zhou H, Li Q, Han J, Ulevitch RJ (2006) Nod1-dependent control of tumor growth. Proc Natl Acad Sci 103(6):1840–1845. https://doi.org/10.1073/pnas.0509228103
da Silva Correia J, Miranda Y, Leonard N, Hsu J, Ulevitch RJ (2007) Regulation of Nod1-mediated signaling pathways. Cell Death Differ 14(4):830–839. https://doi.org/10.1038/sj.cdd.4402070
El-Omar EM, Chow W, Rabkin CS (2001) Gastric cancer and *H. pylori*: host genetics open the way. Gastroenterology 121(4):1002–1004. https://doi.org/10.1053/gast.2001.28739
Fernandez-Gonzalez E, Backert S (2014) DNA transfer in the gastric pathogen *Helicobacter pylori*. J Gastroenterol 49(4):594–604. https://doi.org/10.1007/s00535-014-0938-y
Ferrero RL, Ave P, Ndiaye D, Bambou J-C, Huerre MR, Philpott DJ, Memet S (2008) NF-kappaB activation during acute *Helicobacter pylori* infection in mice. Infect Immun 76(2):551–561. https://doi.org/10.1128/IAI.01107-07
Fischer W, Püls J, Buhrdorf R, Gebert B, Odenbreit S, Haas R (2001) Systematic mutagenesis of the *Helicobacter pylori* cag pathogenicity island: essential genes for CagA translocation in host cells and induction of interleukin-8. Mol Microbiol 42(5):1337–1348. https://doi.org/10.1046/j.1365-2958.2001.02714.x
Fritz JH, Le Bourhis L, Sellge G, Magalhaes JG, Fsihi H, Kufer TA, Collins C, Viala J, Ferrero RL, Girardin SE, Philpott DJ (2007) Nod1-mediated innate immune recognition of peptidoglycan contributes to the onset of adaptive immunity. Immunity 26(4):445–459. https://doi.org/10.1016/j.immuni.2007.03.009

Fukazawa A, Alonso C, Kurachi K, Gupta S, Lesser CF, McCormick BA, Reinecker H-C (2008) GEF-H1 mediated control of NOD1 dependent NF-κB activation by *Shigella* effectors. PLoS Pathog 4(11):e1000228. https://doi.org/10.1371/journal.ppat.1000228

Gall A, Gaudet RG, Gray-Owen SD, Salama NR (2017) TIFA signaling in gastric epithelial cells initiates the *cag* type 4 secretion system-dependent innate immune response to *Helicobacter pylori* infection. MBio 8(4):e01168–17. https://doi.org/10.1128/mBio.01168-17

Gaudet RG, Sintsova A, Buckwalter CM, Leung N, Cochrane A, Li J, Cox AD, Moffat J, Gray-Owen SD (2015) Cytosolic detection of the bacterial metabolite HBP activates TIFA-dependent innate immunity. Science 348(6240):1251–1255. https://doi.org/10.1126/science.aaa4921

Girardin SE, Tournebize R, Mavris M, Page A-L, Li X, Stark GR, Bertin J, DiStefano PS, Yaniv M, Sansonetti PJ, Philpott DJ (2001) CARD4/Nod1 mediates NF-κB and JNK activation by invasive *Shigella flexneri*. EMBO Rep 2(8):736–742. https://doi.org/10.1093/embo-reports/kve155

Glocker E, Lange C, Covacci A, Bereswill S, Kist M, Pahl HL (1998) Proteins encoded by the cag pathogenicity island of *Helicobacter pylori* are required for NF-kappaB activation. Infect Immun 66(5):2346–2348

Gobert AP, Bambou JC, Werts C, Balloy V, Chignard M, Moran AP, Ferrero RL (2004) *Helicobacter pylori* heat shock protein 60 mediates interleukin-6 production by macrophages via a toll-like receptor (TLR)-2-, TLR-4-, and myeloid differentiation factor 88-independent mechanism. J Biol Chem 279(1):245–250

Gööz M, Shaker M, Gööz P, Smolka AJ (2003) Interleukin 1beta induces gastric epithelial cell matrix metalloproteinase secretion and activation during *Helicobacter pylori* infection. Gut 52(9):1250–1256. https://doi.org/10.1136/gut.52.9.1250

Gorrell RJ, Guan J, Xin Y, Tafreshi MA, Hutton ML, McGuckin MA, Ferrero RL, Kwok T (2013) A novel NOD1- and CagA-independent pathway of interleukin-8 induction mediated by the *Helicobacter pylori* type IV secretion system. Cell Microbiol 15(4):554–570. https://doi.org/10.1111/cmi.12055

Grubman A, Kaparakis M, Viala J, Allison C, Badea L, Karrar A, Boneca IG, Le Bourhis L, Reeve S, Smith IA, Hartland EL, Philpott DJ, Ferrero RL (2010) The innate immune molecule, NOD1, regulates direct killing of *Helicobacter pylori* by antimicrobial peptides. Cell Microbiol 12(5):626–639. https://doi.org/10.1111/j.1462-5822.2009.01421.x

Hammond CE, Beeson C, Suarez G, Peek RM, Backert S, Smolka AJ, Smolka AJ (2015) *Helicobacter pylori* virulence factors affecting gastric proton pump expression and acid secretion. Am J Physiol Gastrointest Liver Physiol 309(3):G193–G201. https://doi.org/10.1152/ajpgi.00099.2015

Hirata Y, Maeda S, Ohmae T, Shibata W, Yanai A, Ogura K, Yoshida H, Kawabe T, Omata M (2006a) *Helicobacter pylori* induces IkappaB kinase alpha nuclear translocation and chemokine production in gastric epithelial cells. Infect Immun 74(3):1452–1461. https://doi.org/10.1128/IAI.74.3.1452-1461.2006

Hirata Y, Ohmae T, Shibata W, Maeda S, Ogura K, Yoshida H, Kawabe T, Omata M (2006b) MyD88 and TNF receptor-associated factor 6 are critical signal transducers in *Helicobacter pylori*-infected human epithelial cells. J Immunol 176(6):3796–3803. https://doi.org/10.4049/JIMMUNOL.176.6.3796

Hornsby MJ, Huff JL, Kays RJ, Canfield DR, Bevins CL, Solnick JV (2008) *Helicobacter pylori* induces an antimicrobial response in rhesus macaques in a cag pathogenicity island-dependent manner. Gastroenterology 134(4):1049–1057. https://doi.org/10.1053/j.gastro.2008.01.018

Huang FY, Chan AOO, Rashid A, Wong DKH, Cho CH, Yuen MF (2012) *Helicobacter pylori* induces promoter methylation of E-cadherin via interleukin-1β activation of nitric oxide production in gastric cancer cells. Cancer 118(20):4969–4980. https://doi.org/10.1002/cncr.27519

Hutton ML, Kaparakis-Liaskos M, Turner L, Cardona A, Kwok T, Ferrero RL (2010) *Helicobacter pylori* exploits cholesterol-rich microdomains for induction of NF-kappaB-dependent responses and peptidoglycan delivery in epithelial cells. Infect Immun 78(11):4523–4531. https://doi.org/10.1128/IAI.00439-10

Irving AT, Mimuro H, Kufer TA, Lo C, Wheeler R, Turner LJ, Thomas BJ, Malosse C, Gantier MP, Casillas LN, Votta BJ, Bertin J, Boneca IG, Sasakawa C, Philpott DJ, Ferrero RL, Kaparakis-

Liaskos M (2014) The immune receptor NOD1 and kinase RIP2 interact with bacterial peptidoglycan on early endosomes to promote autophagy and inflammatory signaling. Cell Host Microbe 15(5):623–635. https://doi.org/10.1016/j.chom.2014.04.001

Isomoto H, Miyazaki M, Mizuta Y, Takeshima F, Murase K, Inoue K, Yamasaki K, Murata I, Koji T, Kohno S (2000) Expression of nuclear factor-kappaB in *Helicobacter pylori*-infected gastric mucosa detected with southwestern histochemistry. Scand J Gastroenterol 35(3):247–254. https://doi.org/10.1080/003655200750024092

Jiménez-Soto LF, Kutter S, Sewald X, Ertl C, Weiss E, Kapp U, Rohde M, Pirch T, Jung K, Retta SF, Terradot L, Fischer W, Haas R (2009) Helicobacter pylori type IV secretion apparatus exploits beta1 integrin in a novel RGD-independent manner. PLoS Pathog 5(12):e1000684. https://doi.org/10.1371/journal.ppat.1000684

Kaparakis M, Turnbull L, Carneiro L, Firth S, Coleman HA, Parkington HC, Le Bourhis L, Karrar A, Viala J, Mak J, Hutton ML, Davies JK, Crack PJ, Hertzog PJ, Philpott DJ, Girardin SE, Whitchurch CB, Ferrero RL (2010) Bacterial membrane vesicles deliver peptidoglycan to NOD1 in epithelial cells. Cell Microbiol 12(3):372–385. https://doi.org/10.1111/j.1462-5822.2009.01404.x

Keates S, Hitti YS, Upton M, Kelly CP (1997) *Helicobacter pylori* infection activates NF-kappa B in gastric epithelial cells. Gastroenterology 113(4):1099–1109. https://doi.org/10.1053/gast.1997.v113.pm9322504

Kim EJ, Lee JR, Chung WC, Jung SH, Sung HJ, Lee YW, Oh YS, Kim SB, Paik CN, Lee K-M, Noh SJ (2013) Association between genetic polymorphisms of NOD1 and *Helicobacter pylori*-induced gastric mucosal inflammation in healthy Korean population. Helicobacter 18(2):143–150. https://doi.org/10.1111/hel.12020

Kim BJ, Kim JY, Hwang ES, Kim JG (2015) Nucleotide binding oligomerization domain 1 is an essential signal transducer in human epithelial cells infected with *Helicobacter pylori* that induces the transepithelial migration of neutrophils. Gut Liver 9(3):358–369. https://doi.org/10.5009/gnl13218

Kudo T, Lu H, Wu J, Ohno T, Wu MJ, Genta RM, Graham DY, Yamaoka Y (2007) Pattern of transcription factor activation in *Helicobacter pylori*–infected Mongolian gerbils. Gastroenterology 132(3):1024–1038. https://doi.org/10.1053/j.gastro.2007.01.009

Kupcinskas J, Wex T, Bornschein J, Selgrad M, Leja M, Juozaityte E, Kiudelis G, Jonaitis L, Malfertheiner P (2011) Lack of association between gene polymorphisms of Angiotensin converting enzyme, Nod-like receptor 1, Toll-like receptor 4, FAS/FASL and the presence of *Helicobacter pylori*-induced premalignant gastric lesions and gastric cancer in Caucasians. BMC Med Genet 12(1):112. https://doi.org/10.1186/1471-2350-12-112

Kwok T, Zabler D, Urman S, Rohde M, Hartig R, Wessler S, Misselwitz R, Berger J, Sewald N, König W, Backert S (2007) *Helicobacter* exploits integrin for type IV secretion and kinase activation. Nature 449(7164):862–866. https://doi.org/10.1038/nature06187

Lamb A, Yang XD, Tsang YHN, Li JD, Higashi H, Hatakeyama M, Peek RM, Blanke SR, Chen LF (2009) *Helicobacter pylori* CagA activates NF-kappaB by targeting TAK1 for TRAF6-mediated Lys 63 ubiquitination. EMBO Rep 10(11):1242–1249. https://doi.org/10.1038/embor.2009.210

Lin Q, Xu H, Xi Chen, Tang G, Gu L, Wang Y (2015) *Helicobacter pylori* cytotoxin-associated gene A activates tumor necrosis factor-α and interleukin-6 in gastric epithelial cells through P300/CBP-associated factor-mediated nuclear factor-κB p65 acetylation. Mol Med Rep 12(4):6337–6345. https://doi.org/10.3892/mmr.2015.4143

Lina TT, Alzahrani S, House J, Yamaoka Y, Sharpe AH, Rampy BA, Pinchuk IV, Reyes VE (2015) *Helicobacter pylori cag* pathogenicity island's role in B7-H1 induction and immune evasion. PLoS ONE 10(3):e0121841. https://doi.org/10.1371/journal.pone.0121841

Liu W, Yan M, Liu Y, Wang R, Li C, Deng C, Singh A, Coleman WG, Rodgers GP (2010) Olfactomedin 4 down-regulates innate immunity against *Helicobacter pylori* infection. Proc Natl Acad Sci 107(24):11056–11061. https://doi.org/10.1073/pnas.1001269107

Lu H, Wu JY, Kudo T, Ohno T, Graham DY, Yamaoka Y (2005) Regulation of interleukin-6 promoter activation in gastric epithelial cells infected with *Helicobacter pylori*. Mol Biol Cell 16(10):4954–4966. https://doi.org/10.1091/mbc.e05-05-0426

Ma B, Hua K, Zhou S, Zhou H, Chen Y, Luo R, Bi D, Zhou R, He Q, Jin H (2018) *Haemophilus parasuis* infection activates NOD1/2-RIP2 signaling pathway in PK-15 cells. Dev Comp Immunol 79:158–165. https://doi.org/10.1016/j.dci.2017.10.021

Maeda S, Akanuma M, Mitsuno Y, Hirata Y, Ogura K, Yoshida H, Shiratori Y, Omata M (2001) Distinct mechanism of *Helicobacter pylori*-mediated NF-kappa B activation between gastric cancer cells and monocytic cells. J Biol Chem 276(48):44856–44864. https://doi.org/10.1074/jbc.M105381200

Mandell L, Moran AP, Cocchiarella A, Houghton J, Taylor N, Fox JG, Wang TC, Kurt-Jones EA (2004) Intact gram-negative *Helicobacter pylori*, *Helicobacter felis*, and *Helicobacter hepaticus* bacteria activate innate immunity via toll-like receptor 2 but not toll-like receptor 4. Infect Immun 72(11):6446–6454. https://doi.org/10.1128/IAI.72.11.6446-6454.2004

Merino E, Flores-Encarnación M, Aguilar-Gutiérrez GR (2017) Functional interaction and structural characteristics of unique components of *Helicobacter pylori* T4SS. FEBS J 284(21):3540–3549. https://doi.org/10.1111/febs.14092

Milivojevic M, Dangeard A-S, Kasper CA, Tschon T, Emmenlauer M, Pique C, Schnupf P, Guignot J, Arrieumerlou C (2017) ALPK1 controls TIFA/TRAF6-dependent innate immunity against heptose-1,7-bisphosphate of gram-negative bacteria. PLoS Pathog 13(2):e1006224. https://doi.org/10.1371/journal.ppat.1006224

Münzenmaier A, Lange C, Glocker E, Covacci A, Moran A, Bereswill S, Baeuerle PA, Kist M, Pahl HL (1997) A secreted/shed product of *Helicobacter pylori* activates transcription factor nuclear factor-kappa B. J Immunol 159(12):6140–6147

Murray PJ (2005) NOD proteins: an intracellular pathogen-recognition system or signal transduction modifiers? Curr Opin Immunol 17(4):352–358. https://doi.org/10.1016/j.coi.2005.05.006

Odenbreit S, Püls J, Sedlmaier B, Gerland E, Fischer W, Haas R (2000) Translocation of *Helicobacter pylori* CagA into gastric epithelial cells by type IV secretion. Science 287(5457):1497–1500. https://doi.org/10.1126/science.287.5457.1497

Ohmae T, Hirata Y, Maeda S, Shibata W, Yanai A, Ogura K, Yoshida H, Kawabe T, Omata M (2005) *Helicobacter pylori* activates NF-kappaB via the alternative pathway in B lymphocytes. J Immunol 175(11):7162–7169. https://doi.org/10.4049/JIMMUNOL.175.11.7162

Olbermann P, Josenhans C, Moodley Y, Uhr M, Stamer C, Vauterin M, Suerbaum S, Achtman M, Linz B (2010) A global overview of the genetic and functional diversity in the *Helicobacter pylori cag* pathogenicity island. PLoS Genet 6(8):e1001069. https://doi.org/10.1371/journal.pgen.1001069

Opitz B, Förster S, Hocke AC, Maass M, Schmeck B, Hippenstiel S, Suttorp N, Krüll M (2005) Nod1-mediated endothelial cell activation by *Chlamydophila pneumoniae*. Circ Res 96(3):319–326. https://doi.org/10.1161/01.RES.0000155721.83594.2c

Pachathundikandi K, Backert S (2018) Heptose 1,7-Bisphosphate directed TIFA oligomerization: a novel PAMP-recognizing signaling platform in the control of bacterial infections. Gastroenterology 154:778–783. https://doi.org/10.1053/j.gastro.2018.01.009

Parkin DM, Bray F, Ferlay J, Pisani P (2005) Global cancer statistics, 2002. CA Cancer J Clin 55(2):74–108. https://doi.org/10.3322/canjclin.55.2.74

Patel SR, Smith K, Letley DP, Cook KW, Memon AA, Ingram RJM, Staples E, Backert S, Zaitoun AM, Atherton JC, Robinson K (2013) *Helicobacter pylori* downregulates expression of human β-defensin 1 in the gastric mucosa in a type IV secretion-dependent fashion. Cell Microbiol 15(12):2080–2092. https://doi.org/10.1111/cmi.12174

Rieder G, Einsiedl W, Hatz RA, Stolte M, Enders GA, Walz A (2001) Comparison of CXC chemokines ENA-78 and interleukin-8 expression in *Helicobacter pylori*-associated gastritis. Infect Immun 69(1):81–88. https://doi.org/10.1128/IAI.69.1.81-88.2001

Rosenstiel P, Hellmig S, Hampe J, Ott S, Till A, Fischbach W, Sahly H, Lucius R, Folsch UR, Philpott D, Schreiber S (2006) Influence of polymorphisms in the NOD1/CARD4 and NOD2/CARD15 genes on the clinical outcome of *Helicobacter pylori* infection. Cell Microbiol 8(7):1188–1198. https://doi.org/10.1111/j.1462-5822.2006.00701.x

Saha A, Hammond CE, Beeson C, Peek RM, Smolka AJ, Smolka AJ (2010) *Helicobacter pylori* represses proton pump expression and inhibits acid secretion in human gastric mucosa. Gut 59(7):874–881. https://doi.org/10.1136/gut.2009.194795

Salama NR, Hartung ML, Müller A (2013) Life in the human stomach: persistence strategies of the bacterial pathogen *Helicobacter pylori*. Nat Rev Microbiol 11(6):385–399. https://doi.org/10.1038/nrmicro3016

Segal ED, Falkow S, Tompkins LS (1996) *Helicobacter pylori* attachment to gastric cells induces cytoskeletal rearrangements and tyrosine phosphorylation of host cell proteins. Proc Natl Acad Sci U S A 93(3):1259–1264. https://doi.org/10.1073/pnas.93.3.1259

Segal ED, Cha J, Lo J, Falkow S, Tompkins LS (1999) Altered states: involvement of phosphorylated CagA in the induction of host cellular growth changes by *Helicobacter pylori*. Proc Natl Acad Sci U S A 96(25):14559–14564. https://doi.org/10.1073/pnas.96.25.14559

Sharma SA, Tummuru MK, Blaser MJ, Kerr LD (1998) Activation of IL-8 gene expression by *Helicobacter pylori* is regulated by transcription factor nuclear factor-kappa B in gastric epithelial cells. J Immunol 160(5):2401–2407

Shibata W, Hirata Y, Yoshida H, Otsuka M, Hoshida Y, Ogura K, Maeda S, Ohmae T, Yanai A, Mitsuno Y, Seki N, Kawabe T, Omata M (2005) NF-kappaB and ERK-signaling pathways contribute to the gene expression induced by *cag* PAI-positive-*Helicobacter pylori* infection. World J Gastroenterol 11(39):6134–6143. https://doi.org/10.3748/WJG.V11.I39.6134

Shin S, Case CL, Archer KA, Nogueira CV, Kobayashi KS, Flavell RA, Roy CR, Zamboni DS (2008) Type IV secretion-dependent activation of host MAP kinases induces an increased proinflammatory cytokine response to *Legionella pneumophila*. PLoS Pathog 4(11):e1000220. https://doi.org/10.1371/journal.ppat.1000220

Sokolova O, Maubach G, Naumann M (2014) MEKK3 and TAK1 synergize to activate IKK complex in *Helicobacter pylori* infection. Biochim Biophys Acta—Mol Cell Res 1843(4):715–724. https://doi.org/10.1016/j.bbamcr.2014.01.006

Sorbara MT, Philpott DJ (2011) Peptidoglycan: a critical activator of the mammalian immune system during infection and homeostasis. Immunol Rev 243(1):40–60. https://doi.org/10.1111/j.1600-065X.2011.01047.x

Stein M, Rappuoli R, Covacci A (2000) Tyrosine phosphorylation of the *Helicobacter pylori* CagA antigen after cag-driven host cell translocation. Proc Natl Acad Sci U S A 97(3):1263–1268. https://doi.org/10.1073/pnas.97.3.1263

Stein SC, Faber E, Bats SH, Murillo T, Speidel Y, Coombs N, Josenhans C (2017) *Helicobacter pylori* modulates host cell responses by CagT4SS-dependent translocation of an intermediate metabolite of LPS inner core heptose biosynthesis. PLoS Pathog 13(7):e1006514. https://doi.org/10.1371/journal.ppat.1006514

Strober W, Murray PJ, Kitani A, Watanabe T (2006) Signalling pathways and molecular interactions of NOD1 and NOD2. Nat Rev Immunol 6(1):9–20. https://doi.org/10.1038/nri1747

Suarez G, Romero-Gallo J, Piazuelo MB, Wang G, Maier RJ, Forsberg LS, Azadi P, Gomez MA, Correa P, Peek RM (2015) Modification of *Helicobacter pylori* peptidoglycan enhances NOD1 activation and promotes cancer of the stomach. Cancer Res 75(8):1749–1759. https://doi.org/10.1158/0008-5472.CAN-14-2291

Tan X, Wei LJ, Fan GJ, Jiang YN, Yu XP (2015) Effector responses of bovine blood neutrophils against *Escherichia coli*: Role of NOD1/NF-κB signalling pathway. Vet Immunol Immunopathol 168(1–2):68–76. https://doi.org/10.1016/j.vetimm.2015.08.010

Tanahashi T, Kita M, Kodama T, Yamaoka Y, Sawai N, Ohno T, Mitsufuji S, Wei YP, Kashima K, Imanishi J (2000) Cytokine expression and production by purified *Helicobacter pylori* urease in human gastric epithelial cells. Infect Immun 68(2):664–671. https://doi.org/10.1128/IAI.68.2.664-671.2000

Tegtmeyer N, Neddermann M, Asche CI, Backert S (2017) Subversion of host kinases: a key network in cellular signaling hijacked by *Helicobacter pylori* CagA. Mol Microbiol 105:358–372. https://doi.org/10.1111/mmi.13707

Tran LS, Tran D, De Paoli A, D'Costa K, Creed SJ, Ng GZ, Le L, Sutton P, Silke J, Nachbur U, Ferrero RL (2018) NOD1 is required for *Helicobacter pylori* induction of IL-33 responses in gastric epithelial cells. Cell Microbiol 20(5):e12826. https://doi.org/10.1111/cmi.12826

Tummuru MK, Cover TL, Blaser MJ (1993) Cloning and expression of a high-molecular-mass major antigen of *Helicobacter pylori*: evidence of linkage to cytotoxin production. Infect Immun 61(5):1799–1809

van Den Brink GR, ten Kate FJ, Ponsioen CY, Rive MM, Tytgat GN, van Deventer SJ, Peppelenbosch MP (2000) Expression and activation of NF-kappa B in the antrum of the human stomach. J Immunol 164(6):3353–3359. https://doi.org/10.4049/jimmunol.164.6.3353

Viala J, Chaput C, Boneca IG, Cardona A, Girardin SE, Moran AP, Athman R, Mémet S, Huerre MR, Coyle AJ, DiStefano PS, Sansonetti PJ, Labigne A, Bertin J, Philpott DJ, Ferrero RL (2004) Nod1 responds to peptidoglycan delivered by the *Helicobacter pylori* cag pathogenicity island. Nat Immunol 5(11):1166–1174. https://doi.org/10.1038/ni1131

Wang P, Zhang L, Jiang J-M, Ma D, Tao H-X, Yuan S-L, Wang Y-C, Wang L-C, Liang H, Zhang Z-S, Liu C-J (2012) Association of NOD1 and NOD2 genes polymorphisms with *Helicobacter pylori* related gastric cancer in a Chinese population. World J Gastroenterol 18(17):2112. https://doi.org/10.3748/wjg.v18.i17.2112

Watanabe T, Asano N, Fichtner-Feigl S, Gorelick PL, Tsuji Y, Matsumoto Y, Chiba T, Fuss IJ, Kitani A, Strober W (2010a) NOD1 contributes to mouse host defense against *Helicobacter pylori* via induction of type I IFN and activation of the ISGF3 signaling pathway. J Clin Invest 120(5):1645–1662. https://doi.org/10.1172/JCI39481

Watanabe T, Asano N, Kitani A, Fuss IJ, Chiba T, Strober W (2010b) NOD1-mediated mucosal host defense against *Helicobacter pylori*. Int J Inflam 2010:1–6. https://doi.org/10.4061/2010/476482

Yokota S, Ohnishi T, Muroi M, Tanamoto K, Fujii N, Amano K (2007) Highly-purified *Helicobacter pylori* LPS preparations induce weak inflammatory reactions and utilize Toll-like receptor 2 complex but not Toll-like receptor 4 complex. FEMS Immunol Med Microbiol 51(1):140–148. https://doi.org/10.1111/j.1574-695X.2007.00288.x

Zhan Y, Seregin SS, Chen J, Chen GY (2016) Nod1 limits colitis-associated tumorigenesis by regulating IFN-γ production. J Immunol 196(12):5121–5129. https://doi.org/10.4049/jimmunol.1501822

Zhang XS, Tegtmeyer N, Traube L, Jindal S, Perez-Perez G, Sticht H, Backert S, Blaser MJ (2015) A specific A/T polymorphism in Western tyrosine phosphorylation B-motifs regulates *Helicobacter pylori* CagA epithelial cell interaction. PloS Pathog 11:e1004621. https://doi.org/10.1371/journal.ppat.1004621

Zhao Q, Busch B, Jiménez-Soto LF, Ishikawa-Ankerhold H, Massberg S, Terradot L, Fischer W, Haas R, Blanke SR (2018) Integrin but not CEACAM receptors are dispensable for Helicobacter pylori CagA translocation. PLOS Pathogens 14(10):e1007359. https://doi.org/10.1371/journal.ppat.1007359

Zhou P, She Y, Dong N, Li P, He H, Borio A, Wu Q, Lu S, Ding X, Cao Y, Xu Y, Gao W, Dong M, Ding J, Wang D-C, Zamyatina A, Shao F (2018) Alpha-kinase 1 is a cytosolic innate immune receptor for bacterial ADP-heptose. Nature 561(7721):122–126. https://doi.org/10.1038/s41586-018-0433-3

Zimmermann S, Pfannkuch L, Al-Zeer MA, Bartfeld S, Koch M, Liu J, Rechner C, Soerensen M, Sokolova O, Zamyatina A, Kosma P, Mäurer AP, Glowinski F, Pleissner K-P, Schmid M, Brinkmann V, Karlas A, Naumann M, Rother M, Machuy N, Meyer TF (2017) ALPK1- and TIFA-dependent innate immune response triggered by the *Helicobacter pylori* type IV secretion system. Cell Rep 20(10):2384–2395. https://doi.org/10.1016/j.celrep.2017.08.039

Carbohydrate-Dependent and Antimicrobial Peptide Defence Mechanisms Against *Helicobacter pylori* Infections

Médea Padra, John Benktander, Karen Robinson and Sara K. Lindén

Contents

Abstract The human stomach is a harsh and fluctuating environment for bacteria with hazards such as gastric acid and flow through of gastric contents into the intestine. *H. pylori* gains admission to a stable niche with nutrient access from exudates when attached to the epithelial cells under the mucus layer, whereof adherence to

M. Padra · J. Benktander · S. K. Lindén (✉)
Department of Medical Biochemistry and Cell Biology, Institute of Biomedicine, Sahlgrenska Academy, University of Gothenburg, Box 435, 405 30 Gothenburg, Sweden
e-mail: sara.Linden@gu.se

K. Robinson (✉)
NIHR Nottingham Biomedical Research Centre, Nottingham University Hospitals NHS Trust and Centre for Biomolecular Sciences, University of Nottingham, University Park, Nottingham NG7 2RD, UK
e-mail: karen.robinson@nottingham.ac.uk

S. Backert (ed.), *Molecular Mechanisms of Inflammation: Induction, Resolution and Escape by Helicobacter pylori*, Current Topics in Microbiology and Immunology 421, https://doi.org/10.1007/978-3-030-15138-6_8

glycolipids and other factors provides stable and intimate attachment. To reach this niche, *H. pylori* must overcome mucosal defence mechanisms including the continuously secreted mucus layer, which provides several layers of defence: (1) mucins in the mucus layer can bind *H. pylori* and transport it away from the gastric niche with the gastric emptying, (2) mucins can inhibit *H. pylori* growth, both via glycans that can have antibiotic like function and via an aggregation-dependent mechanism, (3) antimicrobial peptides (AMPs) have antimicrobial activity and are retained in a strategic position in the mucus layer and (4) underneath the mucus layer, the membrane-bound mucins provide a second barrier, and can function as releasable decoys. Many of these functions are dependent on *H. pylori* interactions with host glycan structures, and both the host glycosylation and concentration of antimicrobial peptides change with infection and inflammation, making these interactions dynamic. Here, we review our current understanding of mucin glycan and antimicrobial peptide-dependent host defence mechanisms against *H. pylori* infection.

1 Introduction

A highly hydrated mucus layer covers the gastric mucosal surface, which is the first barrier *H. pylori* encounters when colonizing the stomach (Allen et al. 1986). The mucus layer protects the epithelial cells against chemical or mechanical damage and microbes entering the stomach (Allen et al. 1986; Walker 1985; Andrews et al. 2009; Lindén et al. 2008a). In addition, mucus contributes to surface neutralization of luminal acid by mucosal bicarbonate secretion, resulting in a pH range from acidic in the lumen to neutral at the cell surface (Flemstrom et al. 1986). The main components of the mucus layer are the highly glycosylated mucin glycoproteins, produced by mucous cells. The high level of glycosylation allows mucins to lubricate the epithelium (Crouzier et al. 2015) and prevent the degradation of the protein backbone by proteases (Garner et al. 2001). Mucin secretion can occur via a constitutive pathway continuously maintaining the mucus layer or via regulated pathways as a response to environmental stimuli (Rogers 1994). In the healthy human stomach, MUC5AC and MUC6 are the major secreted mucins, produced by the specialized cells of surface epithelium and gland, respectively (De Bolos et al. 1995). Underneath the mucus layer, the membrane-bound mucin MUC1 is a prominent feature of the gastric glycocalyx (Lindén et al. 2009). Mucus provides a matrix for a rich array of antimicrobial molecules, including antimicrobial peptides (AMPs). If mucin production is aberrant or the mucus layer eroded, the antimicrobial molecules may have impaired efficacy as they lose their strategic position.

The gastric environment that *H. pylori* is exposed to contains an extensive repertoire of glycans, both in the secreted mucus layer and on molecules attached to the epithelial cell surface (Jin et al. 2017; Natomi et al. 1993; Benktander et al. 2018), which dynamically change during infection (Lindén et al. 2008a). In turn, *H. pylori* produces a range of molecules that recognize these glycans with different carbohydrate specificities (Lindén et al. 2008b; Borén et al. 1993). Adhesion to mucins in the mucus layer and glycolipids on the epithelial surface can stimulate or inhibit bacterial growth (Skoog et al. 2012, 2017), provide access to nutrient sources and protection against physical and chemical stresses (Dunne 2002; Heukelekian and Heller 1940), and limit access to the epithelial cells by acting as releasable decoys for binding or by steric hindrance (Lindén et al. 2008a, 2009, 2010).

AMPs are essential components of the innate immune response to infection by pathogens and act as an important early barrier in host defence. These are largely cationic peptides, which exhibit a wide range of antimicrobial activity against Gram-positive and Gram-negative bacteria, fungi, protozoa and viruses. A large number of types of AMPs exist, with differing tissue distributions and antimicrobial potential (Brogden et al. 2013; Doss et al. 2010). Numerous AMPs are documented to have bactericidal activity against *H. pylori*, including members of the defensin family and the cathelicidin LL-37 (George et al. 2003; Patel et al. 2013; Isomoto et al. 2004, 2005; Nuding et al. 2013; Otte et al. 2009). Some AMPs have been reported to also act as chemo-attractants for immune cells or modulators of the host response (Brogden et al. 2013), and may play other roles in defence against *H. pylori* infection.

This review summarizes what is known about the roles of mucus, mucins and AMPs in *H. pylori* infection, their ability to protect the host and kill or affect growth of the bacteria and how these defences change during infection and inflammation.

2 Mucus, Mucins and Glycolipids

The mucus layer is the first barrier *H. pylori* encounters in the stomach. Over 258 O-glycan structures with chain lengths ranging from 1 to 14 residues have been identified on human gastric mucins, with each patient carrying 34–103 structures (Jin et al. 2017). The extensive glycosylation gives the mucins an extended conformation and a 'bottle-brush-like' appearance (Ed 1982). The structure of mucin glycans depends on glycosyltransferases expressed in the cells (Rini et al. 2009) and it varies between species (Leonard et al. 2016), individuals (Jin et al. 2017), tissue locations and health conditions (Medzihradszky et al. 2015; West et al. 2010). Mucin-type *O*-glycosylation is initiated by the addition of α-N-acetylgalactosamine (GalNAc) to the hydroxyl group of serine or threonine side chains of the folded protein (Hanisch 2001; Bennett et al. 2012) which is elongated by three regions: core region (core 1–core 8), backbone region (type-1 and type-2) and peripheral region. This latter region can be terminated by fucose, galactose, GalNAc or sialic acid residues, forming histo-blood group antigens such as A, B, H, Lewis a (Le^a), Lewis b (Le^b), Lewis x (Le^x), Lewis y (Le^y), as well as sialyl-Le^a and sialyl-Le^x structures (Green 1989). The terminal structures of mucin glycans are highly heterogeneous. The Le^a and Le^b blood group antigens mainly appear on the surface epithelium co-localized with MUC5AC, whereas the Le^x and Le^y antigens are expressed by glandular cells co-localized with MUC6 (De Bolos et al. 1995; Murata et al. 1992). Underneath the mucus layer, the membrane-bound mucin MUC1 is a prominent feature of the epithelial cell surface, extending further out from the epithelium than many other components of the glycocalyx (Lindén et al. 2009).

Glycosphingolipids are abundant amphipathic molecules in the epithelial apical cell membrane, with one carbohydrate chain linked to a ceramide. The ceramide part is composed of two non-polar lipid chains: a sphingosine and a fatty acid. The ceramide parts of the glycosphingolipids are anchored in the cell membrane with the carbohydrate chain extending out from the cell (Schnaar et al. 2009). The major non-acid glycosphingolipids in the human stomach are galactosylceramide, lactosylceramide, globotriaosylceramide and globotertraosylceramide. Furthermore, some larger fucosylated glycosphingolipids of the lacto-series are indicated. The major acid glycosphingolipids are sulphatide, GM3, GM1, GD3 and GD1a (Natomi et al. 1993). In addition, the relatively minor gangliosides GD1b, NeuAcα3neolactoteraosylceramide, NeuAcα3neolactohexaosylceramide and NeuAcα3neolactooctaosylceramide have been found in the human stomach (Benktander et al. 2018). *H. pylori* adhesion to glycosphingolipids has been studied extensively for two reasons (Tables 1 and 2). Firstly, they are a component of the cell membrane, which suggests that pathogen adhesion to glycosphingolipids provides an intimate connection to the host under the mucus layer. Secondly, glycosphingolipids carry only one glycan, which enables studies of structural adhesion specificities using these biological glycoconjugates.

3 *Helicobacter pylori* Components Involved in Adhesion

H. pylori adhesion to epithelial cells allow the bacteria to gain easy access to nutrients from host tissues (Kirschner and Blaser 1995; van Amsterdam and van der Ende 2004; Tan et al. 2011), protects the bacteria from the extreme acidity of the gastric lumen and plays a role in bacterial replication (Tan et al. 2009). *H. pylori* adhesion to host cells also triggers host inflammatory responses following infection (Alkout et al. 2000) and activates the type IV secretion system (Posselt et al. 2013; Zhang et al. 2015a, b; Backert et al. 2017; see also Chapters. "Impact of *Helicobacter pylori* Virulence Factors on the Host Immune Response and Gastric Pathology" and "Role of NOD1 and ALPK1/TIFA Signalling in Innate Immunity Against *Helicobacter pylori* Infection" of this book). The glycan environment that *H. pylori* is exposed to varies between host individuals, differs between the oral and gastric niches and changes in response to *H. pylori* infection and disease (Ho et al. 1995; Byrd et al. 1997; Sakamoto et al. 1989; Lindén et al. 2002, 2008a, b).

H. pylori produces an array of adhesive molecules with regulated expression to adapt to the dynamic microenvironment in the stomach, which are involved in colonization and initiate host-cell responses (Ishijima et al. 2011; Lindén et al. 2009; Yamaoka 2008; Rieder et al. 1997; Aberg et al. 2014; Acio-Pizzarello et al. 2017). The receptors for *H. pylori* include glycolipids and mucins that carry glycan structures mediating the binding (Lindén et al. 2002; Benktander et al. 2012), but also a range of other proteins where the moiety conferring adhesion is less certain, such

Table 1 Glycosphingolipid structures binding to BabA

Common name	Structure	References
H5 type 1	Fucα1-2Galβ1-3GlcNAcβ1-3Galβ1-4Glcβ1Cer	Benktander et al. (2012), Borén et al. (1993), Aspholm-Hurtig et al. (2004)
H6 type 4 or Globo H	Fucα1-2Galβ1-3GalNAcβ1-3Galα1-4Galβ1-4Glcβ1Cer	Benktander et al. (2012)
Le^b-6	Fucα1-2Galβ1-3(Fucα1-4)GlcNAcβ1-3Galβ1-4Glcβ1Cer	Benktander et al. (2012), Borén et al. (1993), Aspholm-Hurtig et al. (2004)
Le^b-6 (Ganglio)	Fucα1-2Galβ1-3(Fucα1-4)GalNAcβ1-4Galβ1-4Glcβ1Cer	Fagerberg et al. (2009)
B6 type 1	Galα1-3(Fucα1-2)Galβ1-3GlcNAcβ1-3Galβ1-4Glcβ1Cer	Benktander et al. (2012), Aspholm-Hurtig et al. (2004)
A6 type 1	GalNAcα1-3(Fucα1-2)Galβ1-3GlcNAcβ1-3Galβ1-4Glcβ1Cer	Benktander et al. (2012), Aspholm-Hurtig et al. (2004)
B7 type 1	Galα1-3(Fucα1-2)Galβ1-3(Fucα1-4)GlcNAcβ1-3Galβ1-4Glcβ1Cer	Benktander et al. (2012), Aspholm-Hurtig et al. (2004)
A7 type 1	GalNAcα1-3(Fucα1-2)Galβ1-3(Fucα1-4)GlcNAcβ1-3Galβ1-4Glcβ1Cer	Benktander et al. (2012), Aspholm-Hurtig et al. (2004)
A7 type 4 or Globo A	GalNAcα1-3(Fucα1-2)Galβ1-3GalNAcβ1-3Galα1-4Galβ1-4Glcβ1Cer	Benktander et al. (2012)
A8 type 1	GalNAcα1-3(Fucα1-2)Galβ1-3GlcNAcβ1-3Galβ1-3GlcNAcβ1-3Galβ1-4Glcβ1Cer	Benktander et al. (2012)
A9 type 1	GalNAcα1-3(Fucα1-2)Galβ1-3GlcNAcβ1-3(Fucα1-2)Galβ1-3GlcNAcβ1-3Galβ1-4Glcβ1Cer	Benktander et al. (2012)
Branched H10 type 1	Fucα1-2Galβ1-3GlcNAcβ1-6(Fucα1-2Galβ1-3GlcNAcβ1-3)Galβ1-3GlcNAcβ1-3Galβ1-4Glcβ1Cer	Benktander et al. (2012)
Branched H10 type 1/2	Fucα1-2Galβ1-4GlcNAcβ1-6(Fucα1-2Galβ1-3GlcNAcβ1-3)Galβ1-3GlcNAcβ1-3Galβ1-4Glcβ1Cer	Benktander et al. (2012)
Branched Le^b/H11 type 1	Fucα1-2Galβ1-4(Fucα1-3)GlcNAcβ1-6(Fucα1-2Galβ1-3GlcNAcβ1-3)Galβ1-3GlcNAcβ1-3Galβ1-4Glcβ1Cer	Benktander et al. (2012)

Table 2 Glycosphingolipid structures binding to SabA

Common name	Structure	References
Sialyl-neolactotetra	NeuAcα2-3Galβ1-4GlcNAcβ1-3Galβ1-4Glcβ1Cer	Benktander et al. (2018), Aspholm et al. (2006), Roche et al. (2004)
Sialyl-Le[a]	NeuAcα2-3Galβ1-3(Fucα1-4)GlcNAcβ1-3Galβ1-4Glcβ1Cer	Roche et al. (2004)
Sialyl-Le[x]	NeuAcα2-3Galβ1-4(Fucα1-3)GlcNAcβ1-3Galβ1-4Glcβ1Cer	Roche et al. (2004)
Sialyl-neolactohexa	NeuAcα2-3Galβ1-4GlcNAcβ1-3Galβ1-4GlcNAcβ1-3Galβ1-4Glcβ1Cer	Benktander et al. (2018), Aspholm et al. (2006), Roche et al. (2004)
VIM-2	NeuAcα2-3Galβ1-4GlcNAcβ1-3Galβ1-4(Fucα1-3)GlcNAcβ1-3Galβ1-4Glcβ1Cer	Roche et al. (2004)
	NeuAcα2-3Galβ1-4(Fucα1-3)GlcNAcβ1-3Galβ1-4(Fucα1-3)GlcNAcβ1-3Galβ1-4Glcβ1Cer	Roche et al. (2004)
Sialyl-neolactoocta	NeuAcα2-3Galβ1-4GlcNAcβ1-3Galβ1-4GlcNAcβ1-3Galβ1-4GlcNAcβ1-3Galβ1-4Glcβ1Cer	Benktander et al. (2018), Roche et al. (2004)
	Galα1-3(Fucα1-2)Galβ1-4GlcNAcβ1-6(NeuAcα2-3Galβ1-4GlcNAcβ1-3)Galβ1-4GlcNAcβ1-3Galβ1-4Glcβ1Cer	Aspholm et al. (2006), Roche et al. (2004)
	NeuAcα2-3Galβ1-4GlcNAcβ1-6(NeuAcα2-3Galβ1-4GlcNAcβ1-3)Galβ1-4GlcNAcβ1-3Galβ1-4GlcNAcβ1-3Galβ1-4Glcβ1Cer	Aspholm et al. (2006), Roche et al. (2004)

as integrins (Kwok et al. 2007; Conradi et al. 2012) and carcinoembryonic antigen-related cell adhesion molecules (CEACAMs) (Moonens et al. 2018; Tegtmeyer et al. 2019). *H. pylori* has been reported to produce a range of molecules that recognize different glycan structures (Fig. 1), including fucosylated and sialylated structures (Borén et al. 1993; Evans et al. 1988; Hirmo et al. 1996; Miller-Podraza et al. 1996), sulphated structures (Ascencio et al. 1993; Lopez-Bolanos et al. 2009), lactosylceramide (Angstrom et al. 1998) and lactotetraosylceramide (Teneberg et al. 2002). Several of the *H. pylori* molecules responsible for the adhesion belong to the outer membrane protein family, although other types of molecules also bind to host glycans.

3.1 The Blood Group-Binding Adhesin BabA

The BabA adhesin binds to Lewis B (Le^b) and other fucosylated antigens (see Table 1). The binding properties of BabA differ between strains and geographi-

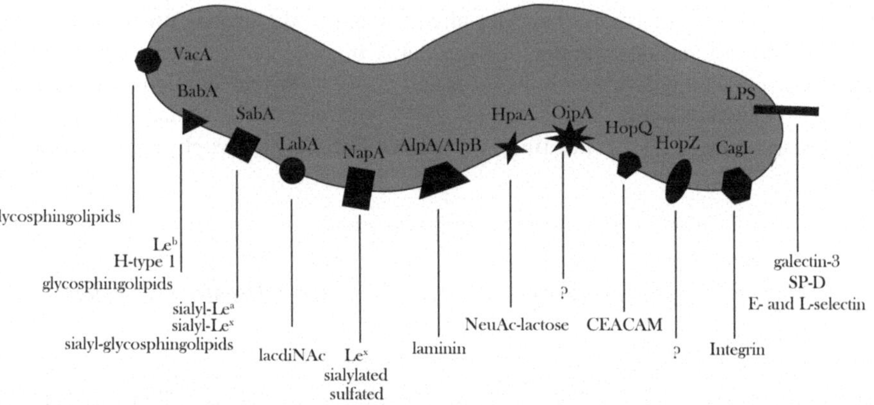

Fig. 1 ***H. pylori*** **molecules interacting with host glycans and/or participating in adhesion to the host**. Some of these are outer membrane proteins that confer *H. pylori* binding directly with host glycans (i.e. BabA and SabA), whereas LPS may participate in binding glycans via other proteins [such as TFF1 (Dunne et al. 2018)] linking the LPS with structures on host cells or mucins. The cytotoxin VacA and the neutrophil-activating protein NapA are presumably secreted before interacting with host glycans. Many other interactions also occur between the bacterial adhesins and host cells, for example HopQ interacts with CEACAM1 (Javaheri et al. 2016)

cal regions: in most of the world where blood groups A, B and O are prevalent, such as in Europe and North America, the *H. pylori* strains, designated as 'generalists', bind to A, B, H and Le^b type 1 structures. However, in the native South American population, where only blood group O is prevalent, the BabA protein has lost the ability to bind A and B structures and only binds H and Le^b type 1 structures. These types of strains are denoted as 'specialists' (Aspholm-Hurtig et al. 2004). Besides the type 1 structures, it has also been found that fucosylated structures of A and H type 4 mediate BabA binding (Benktander et al. 2012). Colonization with *H. pylori* strains expressing BabA has been associated with development of severe gastric diseases (Gerhard et al. 1999; Prinz et al. 2001; Ilver et al. 1998; Yamaoka et al. 2006).

3.2 The Sialic Acid-Binding Adhesin SabA

The sialic acid-binding adhesin (SabA) mediates adhesion to α2,3-sialylated structures, such as sialyl-Lewis a (sialyl-Le^a) and sialyl-Lewis x (sialyl-Le^x, see Table 2) (Mahdavi et al. 2002). Most of the strains carrying SabA also carrying BabA and the prevalence of SabA is higher in CagA-positive than in CagA-negative strains (Mahdavi et al. 2002). The expression of SabA has been shown to be highly variable due to regulation in the promoter and coding regions via phase regulation as well as repression by acid-responsive ArsRS two-component signal transduction system (Goodwin et al. 2008; Harvey et al. 2014; Aberg et al. 2014). This promotes a very adaptive response to changing environment in the gastric mucosa, enabling both attachment and de-attachment from SabA-binding carbohydrates. In addition, the *sabA* gene can recombine with *sabB* and *hopQ* to further regulate adhesion and expression level of SabA (Talarico et al. 2012). SabA expression has been associated with gastric cancer (GC), intestinal metaplasia and corpus atrophy (Yamaoka et al. 2006).

3.3 *Other* H. pylori *Molecules Involved in Adhesion to Host Structures*

The LabA protein from the HOP family has been suggested to specifically recognize the lacdiNAc motif (GalNAcβ1-4GlcNAc) (Rossez et al. 2014). Binding to lactotetraosylceramide (Galβ3GlcNAcβ3Galβ4Glcβ1Cer) from human and pig gastric mucosa has been identified both in *H. pylori* and *H. suis,* a close relative of zoonotic importance (Teneberg et al. 2002; Padra et al. 2018). Studies using isogenic mutants of *H. pylori* discovered roles in adhesion of other proteins, such as adherence-associated lipoprotein A and B (AlpA and AlpB) that facilitates *H. pylori* binding to the host extracellular molecule laminin and can influence the development of gastric inflammation in mongolian gerbils (Odenbreit et al. 1999; Senkovich et al. 2011). HpaA lipoprotein has been characterized as an N-acetylneuraminyllactose-binding hemagglutinin, which has an essential role in *H. pylori* colonization of the mouse stomach (Carlsohn et al. 2006). The outer inflammatory protein A (OipA) encoded by the *hopH* gene has been suggested to mediate *H. pylori* binding to gastric epithelial cells (GECs) based on the observation that *hopH* knockout mutant strains bound less to GECs than wild-type strains (Dossumbekova et al. 2006). The HopZ membrane protein of *H. pylori* has also been associated with adhesion to GECs (Peck et al. 1999), although, in another study, the lack of this protein did not affect the ability of the bacterium to colonize the stomach of guinea pigs (de Jonge et al. 2004). The host receptors for the OipA and HopZ adhesion proteins have not yet been identified.

3.4 *Lipopolysaccharide (LPS)*

The LPS on the surface of most *H. pylori* strains express Lewis blood group antigens with structures similar to the ones on host cells (Moran 2008, 2009). The LPS O-antigen side chain with Le^x structure can mediate *H. pylori* adhesion to host cells (Gerhardt et al. 2001; Edwards et al. 2000) and studies using isogenic mutant strains with modified LPS O-antigen backbone revealed the importance of Le^x-mediated adhesion in bacterial colonization (Moran et al. 2000; Logan et al. 2000). Several host structures have been suggested as binding targets for LPS: the leucocyte–endothelium adhesion molecules E- and L-Selectin (Galustian et al. 2003), the dendritic cell-specific C-type lectin (DC-SIGN) (van Die et al. 2003), galectin-3 (Fowler et al. 2006) and the trefoil factor family (TFF) protein TFF1 (Dunne et al. 2018). *H. pylori* LPS also binds to surfactant-binding protein D (SP-D) which is a C-type lectin involved in antibody-independent pathogen recognition and clearance (Murray et al. 2002). Increased level of SP-D expression has been associated with *H. pylori* infection and the binding of *H. pylori* LPS to this protein results in bacterial immobilization and aggregation (Murray et al. 2002; Khamri et al. 2005).

3.5 *Secreted* H. pylori *Proteins that Bind to Glycans*

The vacoulating cytotoxin VacA applies its cytotoxic activity after internalization into epithelial cells and binds to the short glycosphingolipids glucosylceramide, galactosylceramide, lactosylceramide, galabiosylceramide and globotriaosylceramide (Roche et al. 2007). The glycosphingolipid-related molecule sphingomyelin, which contains a ceramide, but no glycan chain, has also been shown to bind VacA and is important in cellular entry of the toxin (Gupta et al. 2010).

The *H. pylori* neutrophil-activating protein (NapA) binds to sulphated carbohydrate structures on glycolipids and salivary mucins (Namavar et al. 1998; Teneberg et al. 1997). This protein also binds to the Le^x blood-group antigen (Namavar et al. 1998) and to sialylated glycans (Teneberg et al. 1997), and is involved in transendothelial migration of neutrophils (Brisslert et al. 2005).

4 Regulation and Consequences of *H. pylori* Binding to Mucins

In the oro-gastric tract, *H. pylori* binding to mucins occurs via at least four distinct modes: blood group-binding adhesin (BabA)-mediated binding to fucosylated structures, sialic acid-binding adhesin (SabA)-dependent binding to sialylated structures, a charge/low pH-dependent-binding mechanism and via an adhesion mode that recognizes structures on salivary mucins, but not on gastric ones (Lindén et al. 2002, 2004, 2008b). Since the pH in the oral cavity niche is predominantly neutral, charge-dependent *H. pylori* binding in this environment is unlikely to occur, whereas the other three binding modes are functional (Lindén et al. 2008b). In the gastric lumen, pH varies depending on acid secretion, food or drug intake, as well as with the development of certain diseases. Hence in the stomach, binding that has a neutral pH optimum (via SabA, BabA or saliva-specific adhesion) as well as charge-dependent adhesion at acidic pH to glycans present on mucins secreted from the stomach as well as salivary mucins swallowed into the gastric juice can occur (Lindén et al. 2008b). In addition, binding via the other *H. pylori* components mentioned above may very well occur, since many epitopes recognized by these components are present among the gastric mucin glycans. The GalNAcx1-4GlcNAc binding protein LabA has potential for binding mucins (Rossez et al. 2014). However, in our laboratory, we have not detected differences in binding to mucins with or without this structure with strains with or without LabA in their genome, suggesting that the other four binding modes dominate, while glycosphingolipids with the GalNAcβ1-4GlcNAc motif has to the best of our knowledge not been found in the human stomach.

4.1 Infection Induced Changes in Mucin Production, Localization and Glycosylation

In human adult tissues from infected patients, decreased numbers of MUC5AC-positive cells and aberrantly localized MUC6 have been identified in the surface epithelium (Byrd et al. 1997; Kocer et al. 2004). However, changes in MUC5AC or MUC6 localization have not been identified in paediatric biopsies or in experimental infection in rhesus monkeys up to one year post infection (Lindén et al. 2008a, 2010), suggesting that such changes take time to develop. In contrast, changes in mucin production rate and glycosylation are rapid events with consequences for host–pathogen interactions: *H. pylori* impairs mucin production and secretion in the murine gastric mucosa both during early and chronic stages of the infection (Navabi et al. 2013). Furthermore, experimental infection in rhesus monkeys induces decreased levels of Le^b and increased levels of SLe^x in the gastric mucosa, which are particularly pronounced during the first months of infection and that in turn affects *H. pylori* adhesion to its host (Lindén et al. 2008a). These infection-related changes lead to a decrease in binding via BabA and an increase in binding via SabA (Lindén et al. 2008a; Mahdavi et al. 2002). In line with these results, human infection is associated with decreased fucosylation and increased sialylation in the gastric mucosa (Ota et al. 1998). In addition, increased expression of sialylated Lewis antigens upon *H. pylori* infection has been observed in mice, mongolian gerbils and humans (Ohno et al. 2011; Magalhaes et al. 2015; Gomes et al. 2012). Thus, in the mucus layer of a healthy human stomach, predominantly ligands for BabA are present while infection and inflammation increase the ligands for SabA and the charge-dependent binding modes (Lindén et al. 2008a, b).

4.2 Competition Between Binding to Mucins and Glycolipids

Many of the carbohydrate structures that *H. pylori* binds to are present both on glycolipids and mucins (Padra et al. 2018; Jin et al. 2017; Benktander et al. 2018; Natomi et al. 1993). Adhesion to glycolipids provides a stable and intimate attachment, whereas adhesion to mucins, regardless of whether they are membrane-bound type or secreted into the mucus layer, is an unstable niche as these are continuously shed into the acidic lumen and removed from the stomach with the gastric emptying (Fig. 2). Bacteria that penetrate the mucus layer and reach the glycocalyx encounter the large membrane-bound mucins: MUC1 can carry ligands for the BabA and SabA adhesins (depending on host genotype and inflammatory status), and for bacteria that carry these adhesins, MUC1 can act as a releasable decoy being shed from the epithelial surface into the gastric juice (Lindén et al. 2009). MUC1 also inhibits adhesion to gastric epithelial cells from MUC1 non-binding bacteria by steric hindrance (Lindén et al. 2009). The mucus layer and mucins have a mixed role during infection with *H. pylori*: on one hand, the mucus layer serves as a protective niche, protecting both *H. pylori* in the mucus layer and attached to the epithelial surface from the acidic gastric juice and presence of *babA* in the infecting strain has been associated with a more severe clinical outcome (Gerhard et al. 1999; Prinz et al. 2001; Ilver et al. 1998; Yamaoka et al. 2006). On the other hand, *H. pylori*-infected rhesus monkeys and human children secreting mucins with high *H. pylori* binding capacity develop infections with lower *H. pylori* density and gastritis grade than individuals with low binding ability (Lindén et al. 2008a, 2010), supporting the notion that the ability of secreted mucins to bind to *H. pylori* protects the gastric epithelium. Furthermore, mice lacking the Muc1 mucin are more susceptible to infection by *H. pylori* (McGuckin et al. 2007). These findings may be explained by a model where mucin-bound bacteria are removed and disseminated with the shedding of the mucus, whereas BabA can provide intimate adherence to glycolipids on the epithelial cell surface. Indeed, low producers of BabA are associated with a more severe clinical outcome compared to high producers or BabA-negative strains (Fujimoto et al. 2007), suggesting that these aspects of virulence and host defence equipoise.

4.3 Regulation of H. pylori *Adhesin Expression*

H. pylori can inhibit mucus production, thereby creating a more stable growth niche (Navabi et al. 2013); however, it also has other ways of regulating adhesion. This potentially ensures adhesion to stable targets, such as glycolipids, but avoids removal from the protected niche under the mucus when mucins are shed. An excessive binding to mucins would allow the bacteria to be washed away along with shedding mucus. Decreasing the amount of adhesin expressed would be one way to enable long-term colonization. In vitro culture of *H. pylori* together with synthetic Le^b-glycoconjugates decrease the expression of the *babA* adhesin and the correlation between binding to human gastric mucins with varying glycosylation and *babA* expression tend to be negative (Skoog et al. 2012, 2017). In rhesus monkeys, mice and gerbils, expression of babA and Le^b binding is lost early during experimental infection, either by babA phase variation or by gene conversion (Styer et al. 2010), and decreased binding or loss of binding to Le^b has also been shown to occur among human patients during chronic infection (Nell et al. 2014). SabA expression can also be affected by certain environmental factors, e.g. an acidic environment has been

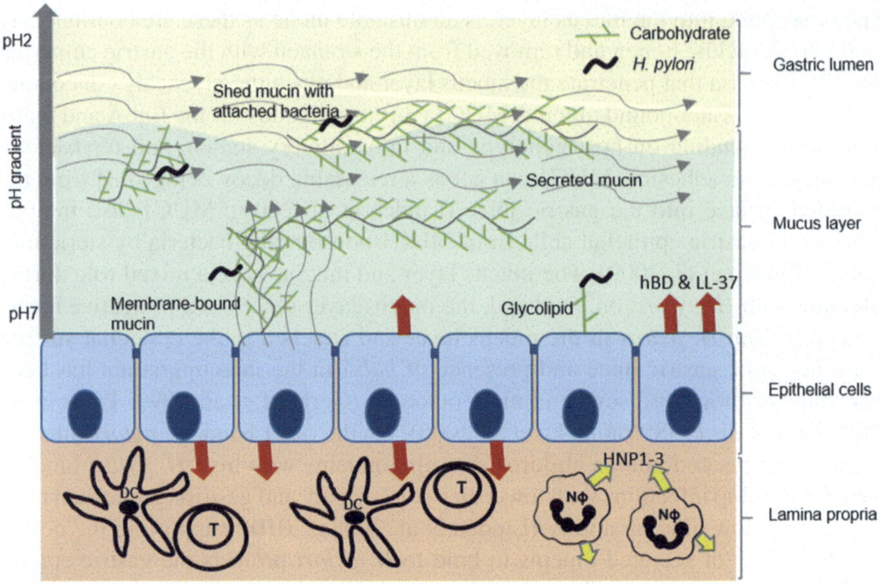

Fig. 2 Sources and mucosal location of glycoconjugates and AMPs interacting with *H. pylori*. *H. pylori* and molecules secreted from *H. pylori* (i.e. VacA and NapA) can either bind to mucins or to glycosphingolipids. In addition to acting as decoys to the intimate adherence provided by glycolipids, mucin glycans can also inhibit *H. pylori* growth. Depending on the genotype of the infecting strain as well as the glycans expressed by the host (depends on genotype and inflammatory status), *H. pylori* can bind to glycans on mucins in the mucus layer or lumen at acidic pH and to membrane-bound mucins, secreted mucins and glycosphingolipids at neutral pH close to the epithelial cells. The mucins constantly flow away from the epithelial surface, are shed into the lumen and removed from the stomach with the gastric emptying. The beta-defensins hBD1, hBD-2 and hBD3 and the LL-37 cathelicidin are largely expressed by gastric epithelial cells (red arrows), whereas hBD4 and the alpha-defensins HBD-1, HBD-2 and HBD-3 are largely expressed by neutrophils (Nφ) (yellow arrows). hBD1, hBD2 and hBD-3 attract the migration of neutrophils, immature dendritic cells (DC) and T cells (T), which exacerbates inflammation and enhances immune-mediated control of *H. pylori* colonisation. Please note that the figure is not drawn to scale: i.e. in reality, the mucus layer is much thicker and *H. pylori* smaller compared to other components in the figure

shown to have a repressing effect on SabA expression (Yamaoka et al. 2006; Merrell et al. 2003). These alterations suggest that SabA expression has a quick response to the changing gastric environment. The acid responsiveness of SabA expression has been shown to be controlled by the ArsRS two-component signal transduction system (Goodwin et al. 2008; Pflock et al. 2004). In vitro studies demonstrated that *sabA* transcription is repressed by the acid-responsive ArsS and the *H. pylori* J99 isogenic mutant lacking ArsS histidine kinase locus (J99Δ*arsS*) had a ~10-fold SabA-dependent binding to human GECs compared to the wild-type strain (Goodwin et al. 2008). The presence of ArsS also affected BabA-dependent *H. pylori* binding, with an ~80% decrease in binding to Le^b of the J99Δ*arsS* strain compared to the wild-type strain, accompanied by a slight decrease in babA gene expression (Skoog et al.

2017), suggesting that part of the effect relates to the presentation at the microbial cell surface. An additional level of binding regulation is conferred by that BabA-dependent binding to fucosylated antigens is acid sensitive but fully reversible upon pH neutralization (Bugaytsova et al. 2017). Such a mechanism would allow the release from fucosylated antigens when the mucin is shed into the acidic gastric juice, allowing the bacteria a second chance to swim back to the protected niche under the mucus layer to colonize the mucosa. However, during inflammation, when sialylated or other charged structures are present in the environment, such mobility might be hampered by binding via the charge/acidic pH-dependent binding mechanism, thus retaining the pathogen in the lumen (Lindén et al. 2004).

4.4 *Effects of Glycans on* H. pylori *Growth*

α1,4GlcNAc-capped *O*-glycans inhibit *H. pylori* growth in a dose-dependent manner (Kawakubo et al. 2004). This effect can be explained by suppressed biosynthesis of cholesteryl-alpha-D-glucopyranoside (Kawakubo et al. 2004), which is a major component of the cell wall and crucial for the survival of the bacterium (Hoshino et al. 2011). The amount of α1,4GlcNAc-capped *O*-glycans on mucins differ between individuals and between tissue locations, with the relative abundance among mucins isolated from the surface epithelium ranging from 0 to 3% and among mucins isolated from the glands from 0 to 7% (Jin et al. 2017).

The growth of *H. pylori* has been demonstrated to be stimulated or inhibited by mucins derived from different individuals and disease states (Skoog et al. 2012, 2017). Glycan analysis of these mucin samples revealed that not all the mucins, that had a growth inhibitory effect, contained terminal α1,4GlcNAc, which suggests that other factors contribute to the inhibition of bacterial growth. *H. pylori* strains with different mucin-binding abilities vary in their growth response to mucins, suggesting that bacterial adhesion to mucins is also of importance for growth (Skoog et al. 2011, 2012). In vitro growth studies showed that *H. pylori* binding to mucins carrying Le^b antigens and to Le^b-glycoconjugates has a growth inhibitory effect, which is accompanied by aggregate formation caused by binding (Skoog et al. 2017). The growth-inhibiting effect of BabA-dependent binding was confirmed by using *H. pylori* isogenic mutant lacking the *babA* adhesin, which reversed the decrease in proliferation and the formation of aggregates in the presence of Le^b antigen carrying glycans (Skoog et al. 2017).

5 Antimicrobial Peptides in the Gastric Mucosa

The defensin family of AMPs are small, arginine-rich peptides containing conserved cysteine residues that form 3-4 disulphide bonds. The family is subdivided into the α-, β- and θ-defensins based on the distribution of these cysteine residues (Ganz

2003). The x-defensins are structurally different from the α- and β-defensins, and are not expressed in humans (Ganz 2003). The only cathelicidin to be described in humans is the amphipathic α-helical LL-37 (Doss et al. 2010). Similar to these molecules, there are a number of other small peptides with cationic charge and antibacterial activity, such as elafin (Nuding et al. 2013).

The main source of the α-defensins human neutrophil peptide-1 (HNP-1), HNP-2 and HNP-3 is neutrophils, whereas human defensin-5 (HD-5) and HD-6 is strongly expressed by intestinal Paneth cells (Cunliffe 2003). Release of HNP-1, -2 and -3 from neutrophils can be induced by bacterial factors and *H. pylori* (Chalifour et al. 2004; Kocsis et al. 2009b). The concentrations of HNP-1, -2 and -3 are significantly higher in gastric juice from *H. pylori*-positive patients (see Table 3) and correlate with neutrophil density in gastric mucosal biopsies (Isomoto et al. 2004). A negative correlation has also been demonstrated with serum pepsinogen levels, indicating a link with the severity of gastric inflammation and atrophy (Nishi et al. 2005). Expression of the α-defensin HD-5 has been observed in inflamed human gastric tissue, but only when intestinal metaplasia is present (Shen et al. 2005).

Human β-defensin 1 (hBD1), hBD2 and hBD3 are abundantly expressed by numerous epithelial cell types including GECs, both in vitro and in vivo (Pero et al. 2017). There are discrepancies in the literature with regard to hBD1 expression; however, since some publications reported increased levels during infection, some showed constitutively high levels in the gastric mucosa regardless of *H. pylori* status, whilst others found decreased expression in gastric mucosal biopsies from *H. pylori*-infected patients (Bajaj-Elliott et al. 2002; Otte et al. 2009; Patel et al. 2013; Taha et al. 2005; Vordenbaumen et al. 2010; Kocsis et al. 2009a). It is possible that some of these discrepancies could be due to polymorphisms in the *DEF1* gene, or in the methodologies that are used to quantify expression at the mRNA and protein level.

Elevated hBD2 expression has been widely and consistently reported in gastric mucosal biopsies and gastric juice from *H. pylori*-infected patients, and this is correlated with inflammation and interleukin-8 (IL-8) expression (George et al. 2003; Isomoto et al. 2005; Nishi et al. 2005; Nuding et al. 2013; Patel et al. 2013; Otte et al. 2009; Bajaj-Elliott et al. 2002; Bauer et al. 2013).

Similarly, hBD3 has also been found to be expressed at significantly higher levels in gastric mucosal biopsies from *H. pylori*-infected patients (Kawauchi et al. 2006; Otte et al. 2009). More recently, a study by Bauer and co-workers (2013) reported the opposite. hBD3 protein levels were decreased in biopsies from infected patients compared to uninfected patients; however, the mRNA expression levels stayed the same (Bauer et al. 2013). This again highlights the importance of the chosen methodology for quantifying defensin expression.

hBD4 is mainly expressed by neutrophils rather than epithelial cells, and is also present at similar or slightly elevated levels in the *H. pylori*-infected inflamed gastric mucosa (Otte et al. 2009; Nuding et al. 2013). hBD5 and hBD6 have not been found in human gastric tissue (Yamaguchi et al. 2002).

LL-37 was originally identified in the granules of myeloid cells, but is expressed in epithelial cells, including those of the gastric mucosa. *LL-37* expression was found to

Table 3 AMPs found in the human gastric mucosa

Antimicrobial peptides		Association of expression in gastric biopsies with *H. pylori* status and inflammation	Measurement at mRNA and/or protein level[b]	Cellular distribution in gastric mucosal tissue	References
α-defensins	HNP-1, HNP-2 and HNP-3	Increased in gastric juice and gastric mucosal biopsies of infected patients	Protein (IHC and RIA)	Neutrophils	Isomoto et al. (2004), Nishi et al. (2005)
	HD5	Increased in inflamed gastric tissue with intestinal metaplasia	mRNA (RT-qPCR) and protein (IHC)	Epithelial cells	Shen et al. (2005)
β-defensins	hBD1	Decreased (or increased) in infected and inflamed gastric tissue[a]	mRNA (RT-qPCR) and protein (ELISA)	Epithelial cells	Patel et al. (2013), Vordenbaumen et al. (2010), Taha et al. (2005), Bajaj-Elliott et al. (2002)
	hBD2	Increased in infected and inflamed gastric tissue	mRNA (RT-qPCR) and protein (ELISA, IHC)	Epithelial cells	Bajaj-Elliott et al. (2002), Patel et al. 2013, Otte et al. (2009), Bauer et al. (2013)
	hBD3	Increased (or decreased) in infected and inflamed gastric tissue[a]	mRNA (RT-qPCR) and protein (ELISA, IHC)	Epithelial cells	Bauer et al. (2013), Kawauchi et al. (2006), Otte et al. (2009), Nuding et al. (2013)
	hBD4	Increased (or no difference) in infected and inflamed gastric tissue[a]	mRNA (RT-qPCR, ISH)	Epithelial cells	Otte et al. (2009), Nuding et al. (2013)
Cathelicidin	LL37	Increased (or no difference) in infected and inflamed gastric tissue[a]	mRNA (RT-qPCR) and protein (IHC, immunoblot)	Epithelial cells	Nuding et al. (2013), Otte et al. (2009), Hase et al. (2003)

[a]Indicates some discrepancies between studies. [b]Text in brackets indicates technique used: *RT-qPCR* reverse transcriptase quantitative *PCR ISH* in situ hybridization; *IHC* immunohistochemistry; *RIA* radioimmune assay, *ELISA* enzyme-linked immunosorbent assay

be significantly increased in the mucosa and gastric juice of *H. pylori*-positive patients (Hase et al. 2003). Otte et al. (2009) also found that *LL-37* mRNA was approximately 20-fold higher in *H. pylori*-positive gastric tissue compared to uninfected controls, whereas Nuding et al. (2013) reported that *LL-37* mRNA expression was unchanged.

6 Bactericidal Activity of AMPs Against *H. pylori*

The key function of AMPs is their antimicrobial activity. The cationic characteristics and amphipathic nature of AMPs allow them to target the membranes of microbes without causing damage to the hosts of these organisms. They have a specificity for anionic phospholipids in the outer-most membrane of bacterial cells (Ganz 2003). hBD1, hBD2 and hBD3 vary in their net charge, and the more positively charged hBD3 has the greatest antimicrobial activity (Kluver et al. 2005). LL-37 is expressed as a pro-protein and, after activation by a serine protease, the cleaved h-CAP18 fragment may also exhibit antimicrobial activity (Zaiou et al. 2003).

AMPs are thought to kill target microbes by disrupting their membrane integrity via a mechanism known as the 'Shai-Matsuzaki-Huang model'. This model involves association of the peptides with the outer membrane of the bacterium, followed by displacement of the lipid bilayer resulting in membrane disruption and pore formation, leading to cell leakage and death (Bocchinfuso et al. 2009). Other bactericidal effects have also been described for some AMPs, for example, inhibition of cell wall synthesis (de Leeuw et al. 2010) and the formation of nanonets to entrap bacteria (Chu et al. 2012).

Inverse associations between the expression of AMPs and *H. pylori* colonization densities in the gastric mucosa of patients indicate that they may control levels of bacteria in the stomach (Patel et al. 2013). A number of studies have directly assessed the bactericidal effect of AMPs against *H. pylori;* however, this is mostly with supraphysiological concentrations in the μM range as compared to the nM range in gastric juice (Uehara et al. 2003; Nuding et al. 2013; Nishi et al. 2005). George et al. (2003) reported ~50% killing of the *H. pylori* strain 26695 after a 20 min exposure to 10 μM concentrations of hBD1 and hBD2, but 80% killing with the same concentration of hBD3. There was almost 100% killing with a 1 μM hBD3. Others have also demonstrated that hBD2 has bactericidal activity against *H. pylori* (Hamanaka et al. 2001; Wehkamp et al. 2003). In further work by Nuding et al. (2013), who tested the sensitivity of *H. pylori* to 90 min incubation with 20 μg/ml (approx. 5 μM) hBD1, 2, 3, 4 and LL-37, 35–75% killing was observed with hBD3 and over 80% killing with LL-37. No significant effects were noted from exposure to hBD1, hBD2 or hBD4. Exposure to 1 μM LL-37 for 3 h was reported to induce 50% killing of *H. pylori* and this was potentiated to 100% killing when added in combination with hBD1 (Hase et al. 2003). Bactericidal activity of the α-defensins HD-5 and HD-6 has been observed with 50 μg/ml in vitro (Tanabe et al. 2008). The fact that some AMPs, which are generally not present in the infected gastric mucosa (such as hBD3 and HD-6), have strong bactericidal activities (Kawauchi et al. 2006), probably indicates that *H. pylori* has evolved resistance mechanisms to aid its long-term survival in the gastric mucosa.

There are high degrees of structural similarity between human β-defensins and their murine counterparts. Whilst similar signalling mechanisms and functions have been noted (Boughan et al. 2006; Rohrl et al. 2010), no published studies have directly assessed the bactericidal activity of murine defensins against *H. pylori.*

7 *H. pylori*-Mediated Regulation of Defensin Expression

The pathways involved in regulation of defensin expression seem to be best understood for hBD2, which is not present in the uninfected gastric mucosa, but induced in response to *cag*PAI-positive *H. pylori* in a NOD1 and NF-κB-dependent manner (Boughan et al. 2006). The role of NF-κB signalling has also been demonstrated with gastric epithelial tissue (Taha et al. 2005) and several human gastric epithelial cell lines (Bajaj-Elliott et al. 2002; O'Neil et al. 2000; Wada et al. 2001; Patel et al. 2013). Other potential promoter-binding domains have been recognized including transcription factor AP-1 and nuclear factor for IL-6 expression (NF-IL6), indicating that other pathways may be involved in the regulation of hBD2 expression (Wehkamp et al. 2004). hBD4 is also known to be induced during *H. pylori* infection via a *cag*PAI-dependent, but NF-κB-independent mechanism involving the p38 MAPK pathway (Otte et al. 2009). This is consistent with the presence of several AP-1 binding sites and lack of NF-κB binding sites in the promoter sequence of the gene (Garcia et al. 2001; Kawahara et al. 2001). The regulation of hBD3 expression is also independent of NF-κB. hBD3 is induced in the early stages of infection, via epidermal growth factor receptor (EGFR)-activated MAP kinase and JAK/STAT signalling (Bauer et al. 2012). Involvement of toll-like receptor 4 (TLR4) has also been suggested, based on the downregulation of hBD3 when this receptor is blocked (Kawauchi et al. 2006). During chronic infection, however, hBD3 is downregulated in a CagA-dependent manner. Upon translocation of CagA in host cells and its interaction with SHP-2, EGFR activation is blocked by receptor dephosphorylation, thus preventing the expression of hBD3 (Bauer et al. 2012).

hBD1 expression is known to be constitutive in many cell types including GECs. Decreased expression of hBD1 has been reported in gastric tissue from *H. Pylori*-infected patients (Patel et al. 2013; Taha et al. 2005; Vordenbaumen et al. 2010). Experiments with human GECs in vitro showed that this downregulation was mediated by *cag*PAI$^+$ *H. pylori* strains, and involves NF-κB signalling (Patel et al. 2013).

The promoter sequence of *LL37* includes binding sites for NF-κB as well as NF-IL6 and IFNγ response elements (Mookherjee et al. 2006). In agreement with this, a number of external stimuli have been reported to influence *LL37* expression, including bacterial components (Kusaka et al. 2018). In vitro studies found that increased expression of LL-37 in GECs was dependent on infection with a *cag*PAI$^+$ strain of *H. pylori* (Hase et al. 2003).

8 AMP Regulation of Immune Cell Functions

In addition to their bactericidal functions, AMPs have a broad range of biological activities on the host response. Some AMPs have been shown to exert immune cell chemotaxis, directing immune effector cells to sites of infection to enhance the clearance of infections. α-defensins, hBD3 and hBD4 are important in recruiting neutrophils. hBD1, hBD2 and hBD3 have been demonstrated to attract immature dendritic cells and memory T cells via interactions with chemokine receptors such as CC-chemokine receptor-2 (CCR2) and CCR6 (Rohrl et al. 2008, 2010; Jin et al. 2010). Very high concentrations of AMPs are needed for chemo-attraction of immune cells from the bloodstream; however, this limits their activity to the local environment (Zasloff 2007).

A number of studies have also shown a role for AMPs in immune suppression (Brogden et al. 2013). LL-37 has been observed to bind and neutralise LPS, protecting mice from LPS-induced inflammation (Rosenfeld et al. 2006). LL-37 also exhibits suppression of LPS-induced translocation of NF-κB into the nucleus of human monocytes and epithelial cells, and reduced expression levels of pro-inflammatory cytokines TNFα and IL-6 (Mookherjee et al. 2006). Immunosuppressive properties have also been ascribed to defensins. For example, hBD3 has been shown to stimulate expression of interleukin-37, a potent inhibitor of innate immunity and inflammation (Smithrithee et al. 2015). Murine BD2 mediates dendritic cell death through activation of TLR4, therefore, terminating the immune response (Biragyn et al. 2008).

AMPs are therefore important modulators of the immune response, protecting the host from potentially harmful microbes directly through their antimicrobial activity and recruitment of immune effector cells whilst protecting the host from the harmful effects of an active inflammatory response. Although there is little data on their role in GC, evidence suggests that expression of AMPs can influence cancer development and tumour progression (Al-Rayahi and Sanyi 2015).

9 AMPs and Glycans as Therapeutics for Eradication of *H. pylori*

There is now a great deal of interest in exploiting the antibacterial properties of AMPs for eradication of infections, particularly as antibiotic resistance is becoming a serious issue. Recombinant HNP1 has been used to successfully eradicate antibiotic resistant *H. pylori* in mice (Zhang et al. 2018). Others have used synthetic analogues of AMPs to eradicate *H. pylori* in vivo (Zhang et al. 2015a, b), either alone or in synergy with conventional antibiotics (Narayana et al. 2015).

Several strategies have been used to investigate glycotherapeutics against *H. pylori*. Oral administration of free oligosaccharide NeuAcα2,3Galβ1,4Glc (3′-sialyllactose) to *H. pylori*-infected rhesus monkeys resulted in clearance of the pathogen in 50% of the animals (Mysore et al. 1999), and *H. pylori* colonized mice fed with pig milk containing Le^b- and sialyl-Le^x-expressing glycoproteins had a lower pathogen density compared to animals fed with Le^b- and sialyl-Le^x-negative milk or water only (Gustafsson et al. 2006). Furthermore, metabolic labelling of bacterial glycans with an azide-containing sugar allowed selective delivery of immune stimulants to azide-covered *H. pylori* (Kaewsapsak et al. 2013).

10 Concluding Remarks

AMPs play a major role, not just in influencing the colonisation density of *H. pylori* in the gastric mucosa, but also control the immune response and are likely to have an impact on gastric carcinogenesis. It may be possible to base new antibacterial and cancer therapies on these important molecules. The effects of carbohydrates on the *H. pylori* density in the stomach is likely to be the combined effect of the competition between *H. pylori* binding to epithelial glycans conferring intimate adherence versus decoy glycans on mucins and the amount of glycans with antimicrobial activity such as α1,4GlcNAc-capped *O*-glycans or aggregation causing glycans. There may be further glycan-based mechanisms that have not been revealed yet. In addition to the glycotherapies investigated so far (administration of *H. pylori* binding glycans/glycoconjugates and glycolabelling of bacteria for immune targeting), there is also potential for other avenues, such as restoring mucin production and affecting the mucin glycosylation that may contribute to *H. pylori* eradication, possibly in combination with other therapies.

Acknowledgements This work was supported by the Swedish research council Formas (221-2013-590), the Swedish Cancer Society, the RR Julin and WM Lundgren Foundations. KR is supported by the National Institute for Health Research (NIHR), through the NIHR Nottingham Biomedical Research Centre, University of Nottingham and the University Hospitals NHS Trust. The views expressed are those of the authors and not necessarily those of the NHS, the NIHR or the Department of Health. The funders had no role in study design, data collection and analysis, decision to publish or preparation of the manuscript.

References

Aberg A, Gideonsson P, Vallstrom A, Olofsson A, Ohman C, Rakhimova L, Boren T, Engstrand L, Brannstrom K, Arnqvist A (2014) A repetitive DNA element regulates expression of the *Helicobacter pylori* sialic acid binding adhesin by a rheostat-like mechanism. PLoS Pathog 10:e1004234

Acio-Pizzarello CR, Acio AA, Choi EJ, Bond K, Kim J, Kenan AC, Chen J, Forsyth MH (2017) Determinants of the regulation of *Helicobacter pylori* adhesins include repeat sequences in both promoter and coding regions as well as the two-component system ArsRS. J Med Microbiol 66:798–807

Alkout AM, Blackwell CC, Weir DM (2000) Increased inflammatory responses of persons of blood group O to *Helicobacter pylori*. J Infect Dis 181:1364–1369
Allen A, Hutton DA, Leonard AJ, Pearson JP, Sellers LA (1986) The role of mucus in the protection of the gastroduodenal mucosa. Scand J Gastroenterol Suppl 125:71–78
Al-Rayahi IA, Sanyi RH (2015) The overlapping roles of antimicrobial peptides and complement in recruitment and activation of tumor-associated inflammatory cells. Front Immunol 6:2
Andrews GP, Laverty TP, Jones DS (2009) Mucoadhesive polymeric platforms for controlled drug delivery. Eur J Pharm Biopharm 71:505–518
Angstrom J, Teneberg S, Milh MA, Larsson T, Leonardsson I, Olsson BM, Halvarsson MO, Danielsson D, Naslund I, Ljungh A, Wadstrom T, Karlsson KA (1998) The lactosylceramide binding specificity of *Helicobacter pylori*. Glycobiology 8:297–309
Ascencio F, Fransson LA, Wadstrom T (1993) Affinity of the gastric pathogen *Helicobacter pylori* for the N-sulphated glycosaminoglycan heparan sulphate. J Med Microbiol 38:240–244
Aspholm M, Olfat FO, Nordén J, Sondén B, Lundberg C, Sjöström R, Altraja S, Odenbreit S, Haas R, Wadström T, Engstrand l, Semino-mora C, Liu H, Dubois A, Teneberg S, Arnqvist A, Borén T (2006) SabA is the *H. pylori* Hemagglutinin and is polymorphic in binding to sialylated glycans. PLoS Pathog 2:e110
Aspholm-Hurtig M, Dailide G, Lahmann M, Kalia A, Ilver D, Roche N, Vikström S, Sjöström R, Lindén S, Bäckström A, Lundberg C, Arnqvist A, Mahdavi J, Nilsson UJ, Velapatiño B, Gilman RH, Gerhard M, Alarcon T, López-Brea M, Nakazawa T, Fox JG, Correa P, Dominguez-Bello MG, Perez-Perez GI, Blaser MJ, Normark S, Carlstedt I, Oscarson S, Teneberg S, Berg de, Borén T (2004) Functional adaptation of BabA, the H. *pylori*; ABO blood group antigen binding adhesin. Science 305:519
Backert S, Haas R, Gerhard M, Naumann M (2017) The *Helicobacter pylori* type IV secretion system encoded by the cag pathogenicity island: architecture, function, and signaling. Curr Top Microbiol Immunol 413:187–220. https://doi.org/10.1007/978-3-319-75241-9_8
Bajaj-Elliott M, Fedeli P, Smith GV, Domizio P, Maher L, Ali RS, Quinn AG, Farthing MJ (2002) Modulation of host antimicrobial peptide (beta-defensins 1 and 2) expression during gastritis. Gut 51:356–361
Bauer B, Pang E, Holland C, Kessler M, Bartfeld S, Meyer TF (2012) The *Helicobacter pylori* virulence effector CagA abrogates human beta-defensin 3 expression via inactivation of EGFR signaling. Cell Host Microbe 11:576–586
Bauer B, Wex T, Kuester D, Meyer T, Malfertheiner P (2013) Differential expression of human beta defensin 2 and 3 in gastric mucosa of *Helicobacter pylori*-infected individuals. Helicobacter 18:6–12
Benktander J, Angstrom J, Breimer ME, Teneberg S (2012) Redefinition of the carbohydrate binding specificity of *Helicobacter pylori* BabA adhesin. J Biol Chem 287:31712–31724
Benktander J, Barone A, Johansson MM, Teneberg S (2018) *Helicobacter pylori* SabA binding gangliosides of human stomach. Virulence 9:738–751
Bennett EP, Mandel U, Clausen H, Gerken TA, Fritz TA, Tabak LA (2012) Control of mucin-type O-glycosylation: a classification of the polypeptide GalNAc-transferase gene family. Glycobiology 22:736–756
Biragyn A, Coscia M, Nagashima K, Sanford M, Young HA, Olkhanud P (2008) Murine beta-defensin 2 promotes TLR-4/MyD88-mediated and NF-kappaB-dependent atypical death of APCs via activation of TNFR2. J Leukoc Biol 83(4):998–1008
Bocchinfuso G, Palleschi A, Orioni B, Grande G, Formaggio F, Toniolo C, Park Y, Hahm KS, Stella L (2009) Different mechanisms of action of antimicrobial peptides: insights from fluorescence spectroscopy experiments and molecular dynamics simulations. J Pept Sci 15:550–558
Borén T, Falk P, Roth KA, Larson G, Normark S (1993) Attachment of *Helicobacter pylori* to human gastric epithelium mediated by blood group antigens. Science 262:1892–1895
Boughan PK, Argent RH, Body-Malapel M, Park JH, Ewings KE, Bowie AG, Ong SJ, Cook SJ, Sorensen OE, Manzo BA, Inohara N, Klein NJ, Nunez G, Atherton JC, Bajaj-Elliott M (2006)

Nucleotide-binding oligomerization domain-1 and epidermal growth factor receptor: critical regulators of beta-defensins during *Helicobacter pylori* infection. J Biol Chem 281:11637–11648

Brisslert M, Enarsson K, Lundin S, Karlsson A, Kusters JG, Svennerholm AM, Backert S, Quiding-Järbrink M (2005) *Helicobacter pylori* induce neutrophil transendothelial migration: role of the bacterial HP-NAP. FEMS Microbiol Lett 249(1):95–103

Brogden KA, Johnson GK, Vincent SD, Abbasi T, Vali S (2013) Oral inflammation, a role for antimicrobial peptide modulation of cytokine and chemokine responses. Expert Rev Anti Infect Ther 11:1097–1113

Bugaytsova JA, Bjornham O, Chernov YA, Gideonsson P, Henriksson S, Mendez M, Sjostrom R, Mahdavi J, Shevtsova A, Ilver D, Moonens K, Quintana-Hayashi MP, Moskalenko R, Aisenbrey C, Bylund G, Schmidt A, Aberg A, Brannstrom K, Koniger V, Vikstrom S, Rakhimova L, Hofer A, Ogren J, Liu H, Goldman MD, Whitmire JM, Aden J, Younson J, Kelly CG, Gilman RH, Chowdhury A, Mukhopadhyay AK, Nair GB, Papadakos KS, Martinez-Gonzalez B, Sgouras DN, Engstrand L, Unemo M, Danielsson D, Suerbaum S, Oscarson S, Morozova-Roche LA, Olofsson A, Grobner G, Holgersson J, Esberg A, Stromberg N, Landstrom M, Eldridge AM, Chromy BA, Hansen LM, Solnick JV, Lindén SK, Haas R, Dubois A, Merrell DS, Schedin S, Remaut H, Arnqvist A, Berg DE, Boren T (2017) *Helicobacter pylori* adapts to chronic infection and gastric disease via pH-responsive BabA-mediated adherence. Cell Host Microbe 21:376–389

Byrd JC, Yan P, Sternberg L, Yunker CK, Scheiman JM, Bresalier RS (1997) Aberrant expression of gland-type gastric mucin in the surface epithelium of *Helicobacter pylori*-infected patients. Gastroenterology 113:455–464

Carlsohn E, Nystrom J, Bolin I, Nilsson CL, Svennerholm AM (2006) HpaA is essential for *Helicobacter pylori* colonization in mice. Infect Immun 74:920–926

Chalifour A, Jeannin P, Gauchat JF, Blaecke A, Malissard M, N'Guyen T, Thieblemont N, Delneste Y (2004) Direct bacterial protein PAMP recognition by human NK cells involves TLRs and triggers alpha-defensin production. Blood 104:1778–1783

Chu H, Pazgier M, Jung G, Nuccio SP, Castillo PA, de Jong MF, Winter MG, Winter SE, Wehkamp J, Shen B, Salzman NH, Underwood MA, Tsolis RM, Young GM, Lu W, Lehrer RI, Baumler AJ, Bevins CL (2012) Human alpha-defensin 6 promotes mucosal innate immunity through self-assembled peptide nanonets. Science 337:477–481

Conradi J, Tegtmeyer N, Woźna M, Wissbrock M, Michalek C et al (2012) An RGD helper sequence in CagL of *Helicobacter pylori* assists in interactions with integrins and injection of CagA. Front Cell Infect Microbiol 2:70

Crouzier T, Boettcher K, Geonnotti AR, Kavanaugh NL, Hirsch J B, Ribbeck K, Lieleg O (2015) Modulating Mucin hydration and lubrication by deglycosylation and polyethylene glycol binding. Adv Mater Interfaces 2:1500308-n/a

Cunliffe RN (2003) Alpha-defensins in the gastrointestinal tract. Mol Immunol 40:463–467

de Bolos C, Garrido M, Real FX (1995) MUC6 apomucin shows a distinct normal tissue distribution that correlates with Lewis antigen expression in the human stomach. Gastroenterology 109:723–734

de Jonge R, Durrani Z, Rijpkema SG, Kuipers EJ, van Vliet AH, Kusters JG (2004) Role of the *Helicobacter pylori* outer-membrane proteins AlpA and AlpB in colonization of the guinea pig stomach. J Med Microbiol 53:375–379

De Leeuw E, Li C, Zeng P, Li C, Diepeveen-De Buin M, Lu WY, Breukink E, Lu W (2010) Functional interaction of human neutrophil peptide-1 with the cell wall precursor lipid II. FEBS Lett 584:1543–1548

Doss M, White MR, Tecle T, Hartshorn KL (2010) Human defensins and LL-37 in mucosal immunity. J Leukoc Biol 87:79–92

Dossumbekova A, Prinz C, Mages J, Lang R, Kusters JG, van Vliet AH, Reindl W, Backert S, Saur D, Schmid RM, Rad R (2006) *Helicobacter pylori* HopH (OipA) and bacterial pathogenicity: genetic and functional genomic analysis of hopH gene polymorphisms. J Infect Dis 194:1346–1355

Dunne WM Jr (2002) Bacterial adhesion: seen any good biofilms lately? Clin Microbiol Rev 15:155–166

Dunne C, Naughton J, Duggan G, Loughrey C, Kilcoyne M, Joshi L, Carrington S, Earley H, Backert S, Robbe Masselot C, May FEB, Clyne M (2018) Binding of *Helicobacter pylori* to human gastric mucins correlates with binding of TFF1. Microorganisms 6

Ed GJM (1982) Microscopic aspects of adhesion and lubrication. In: Proceedings of the 34th international meeting of the Societe' de Chimie Physique, Paris France, 14–18 Sept 1981

Edwards NJ, Monteiro MA, Faller G, Walsh EJ, Moran AP, Roberts IS, High NJ (2000) Lewis X structures in the O antigen side-chain promote adhesion of *Helicobacter pylori* to the gastric epithelium. Mol Microbiol 35:1530–1539

Evans DG, Evans DJ Jr, Moulds JJ, Graham DY (1988) N-acetylneuraminyllactose-binding fibrillar hemagglutinin of *Campylobacter pylori*: a putative colonization factor antigen. Infect Immun 56:2896–2906

Fagerberg D, Ångström J, Halim A, Hultberg A, Rakhimova L, Hammarström L, Borén T, Teneberg S (2009) Novel Leb-like *Helicobacter pylori*-binding glycosphingolipid created by the expression of human α-1,3/4-fucosyltransferase in FVB/N mouse stomach. Glycobiology 19:182–191

Flemstrom G, Knutson L, Kivilaakso E (1986) Gastroduodenal mucosal secretion of bicarbonate and mucus: physiological control and role in protection. Klin Wochenschr 64(Suppl 7):107–111

Fowler M, Thomas RJ, Atherton J, Roberts IS, High NJ (2006) Galectin-3 binds to *Helicobacter pylori* O-antigen: it is upregulated and rapidly secreted by gastric epithelial cells in response to *H. pylori* adhesion. Cell Microbiol 8:44–54

Fujimoto S, Olaniyi Ojo O, Arnqvist A, Wu JY, Odenbreit S, Haas R, Graham DY, Yamaoka Y (2007) *Helicobacter pylori* BabA expression, gastric mucosal injury, and clinical outcome. Clin Gastroenterol Hepatol 5:49–58

Galustian C, Elviss N, Chart H, Owen R, Feizi T (2003) Interactions of the gastrotropic bacterium *Helicobacter pylori* with the leukocyte-endothelium adhesion molecules, the selectins–a preliminary report. FEMS Immunol Med Microbiol 36:127–134

Ganz T (2003) Defensins: antimicrobial peptides of innate immunity. Nat Rev Immunol 3:710–720

Garcia JR, Krause A, Schulz S, Rodriguez-Jimenez FJ, Kluver E, Adermann K, Forssmann U, Frimpong-Boateng A, Bals R, Forssmann WG (2001) Human beta-defensin 4: a novel inducible peptide with a specific salt-sensitive spectrum of antimicrobial activity. FASEB J 15:1819–1821

Garner B, Merry AH, Royle L, Harvey DJ, Rudd PM, Thillet J (2001) Structural elucidation of the N- and O-glycans of human apolipoprotein(a): role of o-glycans in conferring protease resistance. J Biol Chem 276:22200–22208

George JT, Boughan PK, Karageorgiou H, Bajaj-Elliott M (2003) Host anti-microbial response to *Helicobacter pylori* infection. Mol Immunol 40:451–456

Gerhard M, Lehn N, Neumayer N, Boren T, Rad R, Schepp W, Miehlke S, Classen M, Prinz C (1999) Clinical relevance of the *Helicobacter pylori* gene for blood-group antigen-binding adhesin. Proc Natl Acad Sci U S A 96:12778–12783

Gerhardt M, Hirmo S, Wadström T, Miller-Podraza H, Teneberg S, Karlsson K-A, Appelmelk B, Odenbreit S, Haas R, Arnqvist A, Borén T (2001) *Helicobacter pylori*: an adherent pain in the stomach. In: Achtman MAS, S (ed) *Helicobacter pylori*: molecular and cellular biology. Horizon Scientific Press, Wymondham

Gomes J, Magalhaes A, Carvalho AS, Hernandez GE, Papp SL, Head SR, Michel V, David L, Gartner F, Touati E, Reis CA (2012) Glycophenotypic alterations induced by Pteridium aquilinum in mice gastric mucosa: synergistic effect with *Helicobacter pylori* infection. PLoS ONE 7:e38353

Goodwin AC, Weinberger DM, Ford CB, Nelson JC, Snider JD, Hall JD, Paules CI, Peek RM Jr, Forsyth MH (2008) Expression of the *Helicobacter pylori* adhesin SabA is controlled via phase variation and the ArsRS signal transduction system. Microbiology 154:2231–2240

Green C (1989) The ABO, Lewis and related blood group antigens; a review of structure and biosynthesis. FEMS Microbiol Immunol 1:321–330

Gupta VR, Wilson BA, Blanke SR (2010) Sphingomyelin is important for the cellular entry and intracellular localization of *Helicobacter pylori* VacA. Cell Microbiol 12:1517–1533

Gustafsson A, Hultberg A, Sjostrom R, Kacskovics I, Breimer ME, Boren T, Hammarstrom L, Holgersson J (2006) Carbohydrate-dependent inhibition of *Helicobacter pylori* colonization using porcine milk. Glycobiology 16:1–10

Hamanaka Y, Nakashima M, Wada A, Ito M, Kurazono H, Hojo H, Nakahara Y, Kohno S, Hirayama T, Sekine I (2001) Expression of human beta-defensin 2 (hBD-2) in *Helicobacter pylori* induced gastritis: antibacterial effect of hBD-2 against Helicobacter pylori. Gut 49:481–487

Hanisch FG (2001) O-glycosylation of the mucin type. Biol Chem 382:143–149

Harvey VC, Acio CR, Bredehoft AK, Zhu L, Hallinger DR, Quinlivan-Repasi V, Harvey SE, Forsyth MH (2014) Repetitive sequence variations in the promoter region of the adhesin-encoding gene sabA of *Helicobacter pylori* affect transcription. J Bacteriol 196:3421–3429

Hase K, Murakami M, Iimura M, Cole SP, Horibe Y, Ohtake T, Obonyo M, Gallo RL, Eckmann L, Kagnoff MF (2003) Expression of LL-37 by human gastric epithelial cells as a potential host defense mechanism against *Helicobacter pylori*. Gastroenterology 125:1613–1625

Heukelekian H, Heller A (1940) Relation between food concentration and surface for bacterial growth. J Bacteriol 40:547–558

Hirmo S, Kelm S, Schauer R, Nilsson B, Wadstrom T (1996) Adhesion of *Helicobacter pylori* strains to alpha-2,3-linked sialic acids. Glycoconj J 13:1005–1011

Ho SB, Shekels LL, Toribara NW, Kim YS, Lyftogt C, Cherwitz DL, Niehans GA (1995) Mucin gene expression in normal, preneoplastic, and neoplastic human gastric epithelium. Cancer Res 55:2681–2690

Hoshino H, Tsuchida A, Kametani K, Mori M, Nishizawa T, Suzuki T, Nakamura H, Lee H, Ito Y, Kobayashi M, Masumoto J, Fujita M, Fukuda M, Nakayama J (2011) Membrane-associated activation of cholesterol alpha-glucosyltransferase, an enzyme responsible for biosynthesis of cholesteryl-alpha-D-glucopyranoside in *Helicobacter pylori* critical for its survival. J Histochem Cytochem 59:98–105

Ilver D, Arnqvist A, Ogren J, Frick IM, Kersulyte D, Incecik ET, Berg DE, Covacci A, Engstrand L, Borén T (1998) *Helicobacter pylori* adhesin binding fucosylated histo-blood group antigens revealed by retagging. Science 279:373–377

Ishijima N, Suzuki M, Ashida H, Ichikawa Y, Kanegae Y, Saito I, Boren T, Haas R, Sasakawa C, Mimuro H (2011) BabA-mediated adherence is a potentiator of the *Helicobacter pylori* type IV secretion system activity. J Biol Chem 286:25256–25264

Isomoto H, Mukae H, Ishimoto H, Date Y, Nishi Y, Inoue K, Wada A, Hirayama T, Nakazato M, Kohno S (2004) Elevated concentrations of alpha-defensins in gastric juice of patients with *Helicobacter pylori* infection. Am J Gastroenterol 99:1916–1923

Isomoto H, Mukae H, Ishimoto H, Nishi Y, Wen CY, Wada A, Ohnita K, Hirayama T, Nakazato M, Kohno S (2005) High concentrations of human beta-defensin 2 in gastric juice of patients with *Helicobacter pylori* infection. World J Gastroenterol 11:4782–4787

Javaheri A, Kruse T, Moonens K, Mejias-Luque R, Debraekeleer A, Asche Ci, Tegtmeyer N, Kalali B, Bach Nc, Sieber Sa, Hill Dj, Koniger V, Hauck Cr, Moskalenko R, Haas R, Busch Dh, Klaile E, Slevogt H, Schmidt A, Backert S, Remaut H, Singer Bb, Gerhard M (2016) *Helicobacter pylori* adhesin HopQ engages in a virulence-enhancing interaction with human CEACAMs. Nat Microbiol 2:16189. https://doi.org/10.1038/nmicrobiol.2016.189

Jin G, Kawsar HI, Hirsch SA, Zeng C, Jia X, Feng Z, Ghosh SK, Zheng QY, Zhou A, McIntyre TM, Weinberg A (2010) An antimicrobial peptide regulates tumor-associated macrophage trafficking via the chemokine receptor CCR2, a model for tumorigenesis. PLoS ONE 5:e10993

Jin C, Kenny DT, Skoog EC, Padra M, Adamczyk B, Vitizeva V, Thorell A, Venkatakrishnan V, Lindén SK, Karlsson NG (2017) Structural diversity of human gastric Mucin Glycans. Mol Cell Proteomics 16:743–758

Kaewsapsak P, Esonu O, Dube DH (2013) Recruiting the host's immune system to target *Helicobacter pylori*'s surface glycans. ChemBioChem 14:721–726

Kawahara T, Teshima S, Oka A, Sugiyama T, Kishi K, Rokutan K (2001) Type I *Helicobacter pylori* lipopolysaccharide stimulates toll-like receptor 4 and activates mitogen oxidase 1 in gastric pit cells. Infect Immun 69:4382–4389

Kawakubo M, Ito Y, Okimura Y, Kobayashi M, Sakura K, Kasama S, Fukuda MN, Fukuda M, Katsuyama T, Nakayama J (2004) Natural antibiotic function of a human gastric mucin against *Helicobacter pylori* infection. Science 305:1003–1006

Kawauchi K, Yagihashi A, Tsuji N, Uehara N, Furuya D, Kobayashi D, Watanabe N (2006) Human beta-defensin-3 induction in *H. pylori*-infected gastric mucosal tissues. World J Gastroenterol 12:5793–5797

Khamri W, Moran AP, Worku ML, Karim QN, Walker MM, Annuk H, Ferris JA, Appelmelk BJ, Eggleton P, Reid KB, Thursz MR (2005) Variations in *Helicobacter pylori* lipopolysaccharide to evade the innate immune component surfactant protein D. Infect Immun 73:7677–7686

Kirschner DE, Blaser MJ (1995) The dynamics of *Helicobacter pylori* infection of the human stomach. J Theor Biol 176:281–290

Kluver E, Schulz-Maronde S, Scheid S, Meyer B, Forssmann WG, Adermann K (2005) Structure-activity relation of human beta-defensin 3: influence of disulfide bonds and cysteine substitution on antimicrobial activity and cytotoxicity. Biochemistry 44:9804–9816

Kocer B, Ulas M, Ustundag Y, Erdogan S, Karabeyoglu M, Yldrm O, Unal B, Cengiz O, Soran A (2004) A confirmatory report for the close interaction of *Helicobacter pylori* with gastric epithelial MUC5AC expression. J Clin Gastroenterol 38:496–502

Kocsis AK, Kiss ZF, Tiszlavicz L, Tiszlavicz Z, Mandi Y (2009a) Potential role of human beta-defensin 1 in *Helicobacter pylori*-induced gastritis. Scand J Gastroenterol 44:289–295

Kocsis AK, Ocsovszky I, Tiszlavicz L, Tiszlavicz Z, Mandi Y (2009b) *Helicobacter pylori* induces the release of alpha-defensin by human granulocytes. Inflamm Res 58:241–247

Kusaka S, Nishida A, Takahashi K, Bamba S, Yasui H, Kawahara M, Inatomi O, Sugimoto M, Andoh A (2018) Expression of human cathelicidin peptide LL-37 in inflammatory bowel disease. Clin Exp Immunol 191:96–106

Kwok T, Zabler D, Urman S, Rohde M, Hartig R, Wessler S, Misselwitz R, Berger J, Sewald N, Konig W, Backert S (2007) Helicobacter exploits integrin for type IV secretion and kinase activation. Nature 449:862–866

Leonard BC, Yanez-Soto B, Raghunathan VK, Abbott NL, Murphy CJ (2016) Species variation and spatial differences in mucin expression from corneal epithelial cells. Exp Eye Res 152:43–48

Lindén S, Nordman H, Hedenbro J, Hurtig M, Borén T, Carlstedt I (2002) Strain- and blood group-dependent binding of *Helicobacter pylori* to human gastric MUC5AC glycoforms. Gastroenterology 123:1923–1930

Lindén S, Mahdavi J, Hedenbro J, Boren T, Carlstedt I (2004) Effects of pH on *Helicobacter pylori* binding to human gastric mucins: identification of binding to non-MUC5AC mucins. Biochem J 384:263–270

Lindén S, Mahdavi J, Semino-Mora C, Olsen C, Carlstedt I, Boren T, Dubois A (2008a) Role of ABO secretor status in mucosal innate immunity and *H. pylori* infection. PLoS Pathog 4:e2

Lindén SK, Wickstrom C, Lindell G, Gilshenan K, Carlstedt I (2008b) Four modes of adhesion are used during *Helicobacter pylori* binding to human mucins in the oral and gastric niches. Helicobacter 13:81–93

Lindén SK, Sheng YH, Every AL, Miles KM, Skoog EC, Florin TH, Sutton P, McGuckin MA (2009) MUC1 limits *Helicobacter pylori* infection both by steric hindrance and by acting as a releasable decoy. PLoS Pathog 5:e1000617

Lindén S, Semino-Mora C, Liu H, Rick J, Dubois A (2010) Role of mucin Lewis status in resistance to *Helicobacter pylori* infection in pediatric patients. Helicobacter 15:251–258

Logan SM, Conlan JW, Monteiro MA, Wakarchuk WW, Altman E (2000) Functional genomics of *Helicobacter pylori*: identification of a beta-1,4 galactosyltransferase and generation of mutants with altered lipopolysaccharide. Mol Microbiol 35:1156–1167

Lopez-Bolanos CC, Guzman-Murillo MA, Ruiz-Bustos E, Ascencio F (2009) The role of heparan sulfate on adhesion of 47 and 51 kDa outer membrane proteins of *Helicobacter pylori* to gastric cancer cells. Can J Microbiol 55:450–456

Magalhaes A, Marcos-Pinto R, Nairn AV, Dela Rosa M, Ferreira RM, Junqueira-Neto S, Freitas D, Gomes J, Oliveira P, Santos MR, Marcos NT, Xiaogang W, Figueiredo C, Oliveira C, Dinis-Ribeiro M, Carneiro F, Moremen KW, David L, Reis CA (2015) *Helicobacter pylori* chronic infection and mucosal inflammation switches the human gastric glycosylation pathways. Biochim Biophys Acta 1852:1928–1939

Mahdavi J, Sonden B, Hurtig M, Olfat FO, Forsberg L, Roche N, Angstrom J, Larsson T, Teneberg S, Karlsson KA, Altraja S, Wadström T, Kersulyte D, Berg DE, Dubois A, Petersson C, Magnusson KE, Norberg T, Lindh F, Lundskog BB, Arnqvist A, Hammarstrom L, Borén T (2002) *Helicobacter pylori* SabA adhesin in persistent infection and chronic inflammation. Science 297:573–578

Mcguckin MA, Every A, Skene C, Lindén S, Mcauley J, Harbour S, Kaparakis M, Sutton P (2007) Muc1 mucin limits both *Helicobacter pylori* colonization of the murine gastric mucosa and the associated gastritis. Gastroenterology (in press)

Medzihradszky KF, Kaasik K, Chalkley RJ (2015) Tissue-specific glycosylation at the glycopeptide level. Mol Cell Proteomics 14:2103–2110

Merrell DS, Goodrich ML, Otto G, Tompkins LS, Falkow S (2003) pH-regulated gene expression of the gastric pathogen *Helicobacter pylori*. Infect Immun 71:3529–3539

Miller-Podraza H, Milh MA, Bergstrom J, Karlsson KA (1996) Recognition of glycoconjugates by *Helicobacter pylori*: an apparently high-affinity binding of human polyglycosylceramides, a second sialic acid-based specificity. Glycoconj J 13:453–460

Mookherjee N, Brown KL, Bowdish DM, Doria S, Falsafi R, Hokamp K, Roche FM, Mu R, Doho GH, Pistolic J, Powers JP, Bryan J, Brinkman FS, Hancock RE (2006) Modulation of the TLR-mediated inflammatory response by the endogenous human host defense peptide LL-37. J Immunol 176:2455–2464

Moonens K, Hamway Y, Neddermann M, Reschke M, Tegtmeyer N, Kruse T, Kammerer R, Mejias-Luque R, Singer BB, Backert S, Gerhard M, Remaut H (2018) *Helicobacter pylori* adhesin HopQ disrupts trans dimerization in human CEACAMs. EMBO J 37

Moran AP (2008) Relevance of fucosylation and Lewis antigen expression in the bacterial gastroduodenal pathogen *Helicobacter pylori*. Carbohydr Res 343:1952–1965

Moran AP (2009) Molecular mimicry of host glycosylated structures by bacteria. In: Moran AP, Holst O, Brennan PJ, And Von Itzstein M (eds) Microbial glycobiology—structures, relevance and applications. Academic Press, London

Moran AP, Sturegard E, Sjunnesson H, Wadstrom T, Hynes SO (2000) The relationship between O-chain expression and colonisation ability of *Helicobacter pylori* in a mouse model. FEMS Immunol Med Microbiol 29:263–270

Murata K, Egami H, Shibata Y, Sakamoto K, Misumi A, Ogawa M (1992) Expression of blood group-related antigens, ABH, Lewis(a), Lewis(b), Lewis(x), Lewis(y), CA19-9, and CSLEX1 in early cancer, intestinal metaplasia, and uninvolved mucosa of the stomach. Am J Clin Pathol 98:67–75

Murray E, Khamri W, Walker MM, Eggleton P, Moran AP, Ferris JA, Knapp S, Karim QN, Worku M, Strong P, Reid KB, Thursz MR (2002) Expression of surfactant protein D in the human gastric mucosa and during *Helicobacter pylori* infection. Infect Immun 70:1481–1487

Mysore JV, Wigginton T, Simon PM, Zopf D, Heman-Ackah LM, Dubois A (1999) Treatment of *Helicobacter pylori* infection in rhesus monkeys using a novel antiadhesion compound. Gastroenterology 117:1316–1325

Namavar F, Sparrius M, Veerman EC, Appelmelk BJ, Vandenbroucke-Grauls CM (1998) Neutrophil-activating protein mediates adhesion of *Helicobacter pylori* to sulfated carbohydrates on high-molecular-weight salivary mucin. Infect Immun 66:444–447

Narayana JL, Huang HN, Wu CJ, Chen JY (2015) Efficacy of the antimicrobial peptide TP4 against *Helicobacter pylori* infection: in vitro membrane perturbation via micellization and in vivo suppression of host immune responses in a mouse model. Oncotarget 6:12936–12954

Natomi H, Saitoh T, Sugano K, Iwamori M, Fukayama M, Nagai Y (1993) Systematic analysis of glycosphingolipids in the human gastrointestinal tract: enrichment of sulfatides with hydroxylated longer-chain fatty acids in the gastric and duodenal mucosa. Lipids 28:737–742

Navabi N, Johansson ME, Raghavan S, Lindén SK (2013) *Helicobacter pylori* infection impairs the mucin production rate and turnover in the murine gastric mucosa. Infect Immun 81:829–837

Nell S, Kennemann L, Schwarz S, Josenhans C, Suerbaum S (2014) Dynamics of Lewis b binding and sequence variation of the babA adhesin gene during chronic *Helicobacter pylori* infection in humans. MBio 5

Nishi Y, Isomoto H, Mukae H, Ishimoto H, Wen CY, Wada A, Ohnita K, Mizuta Y, Murata I, Hirayama T, Nakazato M, Kohno S (2005) Concentrations of alpha- and beta-defensins in gastric juice of patients with various gastroduodenal diseases. World J Gastroenterol 11:99–103

Nuding S, Gersemann M, Hosaka Y, Konietzny S, Schaefer C, Beisner J, Schroeder BO, Ostaff MJ, Saigenji K, Ott G, Schaller M, Stange EF, Wehkamp J (2013) Gastric antimicrobial peptides fail to eradicate *Helicobacter pylori* infection due to selective induction and resistance. PLoS ONE 8:e73867

Odenbreit S, Till M, Hofreuter D, Faller G, Haas R (1999) Genetic and functional characterization of the alpAB gene locus essential for the adhesion of *Helicobacter pylori* to human gastric tissue. Mol Microbiol 31:1537–1548

Ohno T, Vallstrom A, Rugge M, Ota H, Graham DY, Arnqvist A, Yamaoka Y (2011) Effects of blood group antigen-binding adhesin expression during *Helicobacter pylori* infection of *Mongolian gerbils*. J Infect Dis 203:726–735

O'Neil DA, Cole SP, Martin-Porter E, Housley MP, Liu L, Ganz T, Kagnoff MF (2000) Regulation of human beta-defensins by gastric epithelial cells in response to infection with *Helicobacter pylori* or stimulation with interleukin-1. Infect Immun 68:5412–5415

Ota H, Nakayama J, Momose M, Hayama M, Akamatsu T, Katsuyama T, Graham DY, Genta RM (1998) *Helicobacter pylori* infection produces reversible glycosylation changes to gastric mucins. Virchows Arch 433:419–426

Otte JM, Neumann HM, Brand S, Schrader H, Schmidt WE, Schmitz F (2009) Expression of beta-defensin 4 is increased in human gastritis. Eur J Clin Invest 39:126–138

Padra M, Adamczyk B, Benktander J, Flahou B, Skoog EC, Padra JT, Smet A, Jin C, Ducatelle R, Samuelsson T, Haesebrouck F, Karlsson NG, Teneberg S, Lindén SK (2018) *Helicobacter suis* binding to carbohydrates on human and porcine gastric mucins and glycolipids occurs via two modes. Virulence, pp 1–48

Patel SR, Smith K, Letley DP, Cook KW, Memon AA, Ingram RJ, Staples E, Backert S, Zaitoun AM, Atherton JC, Robinson K (2013) *Helicobacter pylori* downregulates expression of human beta-defensin 1 in the gastric mucosa in a type IV secretion-dependent fashion. Cell Microbiol 15:2080–2092

Peck B, Ortkamp M, Diehl KD, Hundt E, Knapp B (1999) Conservation, localization and expression of HopZ, a protein involved in adhesion of *Helicobacter pylori*. Nucleic Acids Res 27:3325–3333

Pero R, Coretti L, Nigro E, Lembo F, Laneri S, Lombardo B, Daniele A, Scudiero O (2017) Beta-defensins in the Fight against *Helicobacter pylori*. Molecules 22

Pflock M, Dietz P, Schar J, Beier D (2004) Genetic evidence for histidine kinase HP165 being an acid sensor of *Helicobacter pylori*. FEMS Microbiol Lett 234:51–61

Posselt G, Backert S, Wessler S (2013) The functional interplay of *Helicobacter pylori* factors with gastric epithelial cells induces a multi-step process in pathogenesis. Cell Commun Signal 11:77. https://doi.org/10.1186/1478-811x-11-77

Prinz C, Schoniger M, Rad R, Becker I, Keiditsch E, Wagenpfeil S, Classen M, Rosch T, Schepp W, Gerhard M (2001) Key importance of the *Helicobacter pylori* adherence factor blood group antigen binding adhesin during chronic gastric inflammation. Cancer Res 61:1903–1909

Rieder G, Hatz RA, Moran AP, Walz A, Stolte M, Enders G (1997) Role of adherence in interleukin-8 induction in *Helicobacter pylori*-associated gastritis. Infect Immun 65:3622–3630

Rini J, Esko J, Varki A (2009) Glycosyltransferases and Glycan-processing Enzymes. In: Varki A, Cummings RD, Esko JD, Freeze HH, Stanley P, Bertozzi CR, Hart GW, Etzler ME (eds) Essentials of glycobiology, 2nd edn. Cold Spring Harbor (NY)

Roche N, Ångström J, Hurtig M, Larsson T, Borén T, Teneberg S (2004) *Helicobacter pylori* and complex gangliosides. Infect Immun 72:1519–1529

Roche N, Ilver D, Angstrom J, Barone S, Telford JL, Teneberg S (2007) Human gastric glycosphingolipids recognized by *Helicobacter pylori* vacuolating cytotoxin VacA. Microbes Infect 9:605–614

Rogers DF (1994) Airway goblet cells: responsive and adaptable front-line defenders. Eur Respir J 7:1690–1706

Rohrl J, Yang D, Oppenheim JJ, Hehlgans T (2008) Identification and biological characterization of mouse beta-defensin 14, the orthologue of human beta-defensin 3. J Biol Chem 283:5414–5419

Rohrl J, Yang D, Oppenheim JJ, Hehlgans T (2010) Human beta-defensin 2 and 3 and their mouse orthologs induce chemotaxis through interaction with CCR2. J Immunol 184:6688–6694

Rosenfeld Y, Papo N, Shai Y (2006) Endotoxin (lipopolysaccharide) neutralization by innate immunity host-defense peptides. Peptide properties and plausible modes of action. J Biol Chem 281:1636–1643

Rossez Y, Gosset P, Boneca IG, Magalhaes A, Ecobichon C, Reis CA, Cieniewski-Bernard C, Joncquel Chevalier Curt M, Leonard R, Maes E, Sperandio B, Slomianny C, Sansonetti PJ, Michalski JC, Robbe-Masselot C (2014) The LacdiNAc-Specific adhesin LabA mediates adhesion of *Helicobacter pylori* to human gastric mucosa. J Infect Dis 210:1286–1295

Sakamoto S, Watanabe T, Tokumaru T, Takagi H, Nakazato H, Lloyd KO (1989) Expression of Lewisa, Lewisb, Lewisx, Lewisy, siayl-Lewisa, and sialyl-Lewisx blood group antigens in human gastric carcinoma and in normal gastric tissue. Cancer Res 49:745–752

Schnaar RL, Suzuki A, Stanley P (2009) Glycosphingolipids. In: Nd, Varki A, Cummings RD, Esko JD, Freeze HH, Stanley P, Bertozzi CR, Hart GW, Etzler ME (eds) Essentials of glycobiology. Cold Spring Harbor Laboratory Press, Cold Spring Harbor (NY)

Senkovich OA, Yin J, Ekshyyan V, Conant C, Traylor J, Adegboyega P, McGee DJ, Rhoads RE, Slepenkov S, Testerman TL (2011) *Helicobacter pylori* AlpA and AlpB bind host laminin and influence gastric inflammation in gerbils. Infect Immun 79:3106–3116

Shen B, Porter EM, Reynoso E, Shen C, Ghosh D, Connor JT, Drazba J, Rho HK, Gramlich TL, Li R, Ormsby AH, Sy MS, Ganz T, Bevins CL (2005) Human defensin 5 expression in intestinal metaplasia of the upper gastrointestinal tract. J Clin Pathol 58:687–694

Skoog EC, Lindberg M, Lindén SK (2011) Strain-dependent proliferation in response to human gastric mucin and adhesion properties of *Helicobacter pylori* are not affected by co-isolated Lactobacillus sp. Helicobacter 16:9–19

Skoog EC, Sjoling A, Navabi N, Holgersson J, Lundin SB, Lindén SK (2012) Human gastric mucins differently regulate *Helicobacter pylori* proliferation, gene expression and interactions with host cells. PLoS ONE 7:e36378

Skoog EC, Padra M, Aberg A, Gideonsson P, Obi I, Quintana-Hayashi MP, Arnqvist A, Lindén SK (2017) BabA dependent binding of *Helicobacter pylori* to human gastric mucins cause aggregation that inhibits proliferation and is regulated via ArsS. Sci Rep 7:40656

Smithrithee R, Niyonsaba F, Kiatsurayanon C, Ushio H, Ikeda S, Okumura K, Ogawa H (2015) Human beta-defensin-3 increases the expression of interleukin-37 through CCR6 in human keratinocytes. J Dermatol Sci 77:46–53

Styer CM, Hansen LM, Cooke CL, Gundersen AM, Choi SS, Berg DE, Benghezal M, Marshall BJ, Peek RM Jr, Boren T, Solnick JV (2010) Expression of the BabA adhesin during experimental infection with *Helicobacter pylori*. Infect Immun 78:1593–1600

Taha AS, Faccenda E, Angerson WJ, Balsitis M, Kelly RW (2005) Gastric epithelial anti-microbial peptides–histological correlation and influence of anatomical site and peptic ulcer disease. Dig Liver Dis 37:51–56

Talarico S, Whitefield SE, Fero J, Haas R, Salama NR (2012) Regulation of *Helicobacter pylori* adherence by gene conversion. Mol Microbiol 84:1050–1061

Tan S, Tompkins LS, Amieva MR (2009) *Helicobacter pylori* usurps cell polarity to turn the cell surface into a replicative niche. PLoS Pathog 5:e1000407

Tan S, Noto JM, Romero-Gallo J, Peek RM Jr, Amieva MR (2011) *Helicobacter pylori* perturbs iron trafficking in the epithelium to grow on the cell surface. PLoS Pathog 7:e1002050

Tanabe H, Sato T, Watari J, Maemoto A, Fujiya M, Kono T, Ashida T, Ayabe T, Kohgo Y (2008) Functional role of metaplastic paneth cell defensins in *Helicobacter pylori*-infected stomach. Helicobacter 13:370–379

Tegtmeyer N, Harrer A, Schmitt V, Singer BB, Backert S (2019) Expression of CEACAM1 or CEACAM5 in AZ-521 cells restores the type IV secretion deficiency for translocation of CagA by *Helicobacter pylori*. Cell Microbiol 21:e12965. https://doi.org/10.1111/cmi.12965

Teneberg S, Miller-Podraza H, Lampert HC, Evans DJ Jr Evans DG, Danielsson D, Karlsson KA (1997) Carbohydrate binding specificity of the neutrophil-activating protein of *Helicobacter pylori*. J Biol Chem 272:19067–19071

Teneberg S, Leonardsson I, Karlsson H, Jovall PA, Angstrom J, Danielsson D, Naslund I, Ljungh A, Wadstrom T, Karlsson KA (2002) Lactotetraosylceramide, a novel glycosphingolipid receptor for *Helicobacter pylori*, present in human gastric epithelium. J Biol Chem 277:19709–19719

Uehara N, Yagihashi A, Kondoh K, Tsuji N, Fujita T, Hamada H, Watanabe N (2003) Human beta-defensin-2 induction in *Helicobacter pylori*-infected gastric mucosal tissues: antimicrobial effect of overexpression. J Med Microbiol 52:41–45

van Amsterdam K, van der Ende A (2004) Nutrients released by gastric epithelial cells enhance *Helicobacter pylori* growth. Helicobacter 9:614–621

van Die I, van Vliet SJ, Nyame AK, Cummings RD, Bank CM, Appelmelk B, Geijtenbeek TB, van Kooyk Y (2003) The dendritic cell-specific C-type lectin DC-SIGN is a receptor for *Schistosoma mansoni* egg antigens and recognizes the glycan antigen Lewis x. Glycobiology 13:471–478

Vordenbaumen S, Pilic D, Otte JM, Schmitz F, Schmidt-Choudhury A (2010) Defensin-mRNA expression in the upper gastrointestinal tract is modulated in children with celiac disease and *Helicobacter pylori*-positive gastritis. J Pediatr Gastroenterol Nutr 50:596–600

Wada A, Ogushi K, Kimura T, Hojo H, Mori N, Suzuki S, Kumatori A, Se M, Nakahara Y, Nakamura M, Moss J, Hirayama T (2001) *Helicobacter pylori*-mediated transcriptional regulation of the human beta-defensin 2 gene requires NF-kappaB. Cell Microbiol 3:115–123

Walker WA (1985) Role of the mucosal barrier in toxin/microbial attachment to the gastrointestinal tract. Ciba Found Symp 112:34–56

Wehkamp J, Schmidt K, Herrlinger KR, Baxmann S, Behling S, Wohlschlager C, Feller AC, Stange EF, Fellermann K (2003) Defensin pattern in chronic gastritis: HBD-2 is differentially expressed with respect to *Helicobacter pylori* status. J Clin Pathol 56:352–357

Wehkamp J, Harder J, Wehkamp K, Wehkamp-Von Meissner B, Schlee M, Enders C, Sonnenborn U, Nuding S, Bengmark S, Fellermann K, Schroder JM, Stange EF (2004) NF-kappaB- and AP-1-mediated induction of human beta defensin-2 in intestinal epithelial cells by *Escherichia coli* Nissle 1917: a novel effect of a probiotic bacterium. Infect Immun 72:5750–5758

West MB, Segu ZM, Feasley CL, Kang P, Klouckova I, Li C, Novotny MV, West CM, Mechref Y, Hanigan MH (2010) Analysis of site-specific glycosylation of renal and hepatic gamma-glutamyl transpeptidase from normal human tissue. J Biol Chem 285:29511–29524

Yamaoka Y (2008) Roles of *Helicobacter pylori* BabA in gastroduodenal pathogenesis. World J Gastroenterol 14:4265–4272

Yamaguchi Y, Nagase T, Makita R, Fukuhara S, Tomita T, Tominaga T, Kurihara H, Ouchi Y (2002) Identification of multiple novel epididymis-specific β-defensin isoforms in humans and mice. J Immunol 169(5):2516–2523, September 1, 2002. https://doi.org/10.4049/jimmunol.169.5.2516

Yamaoka Y, Ojo O, Fujimoto S, Odenbreit S, Haas R, Gutierrez O, El-Zimaity HM, Reddy R, Arnqvist A, Graham DY (2006) *Helicobacter pylori* outer membrane proteins and gastroduodenal disease. Gut 55:775–781

Zaiou M, Nizet V, Gallo RL (2003) Antimicrobial and Protease Inhibitory Functions of the Human Cathelicidin (hCAP18/LL-37) Prosequence. J Invest Dermatol 120(5):810–816, ISSN 0022-202X, https://doi.org/10.1046/j.1523-1747.2003.12132.x

Zasloff M (2007) Antimicrobial peptides, innate immunity, and the normally sterile urinary tract. J Am Soc Nephrol 18:2810–2816
Zhang XL, Jiang AM, Ma ZY, Li XB, Xiong YY, Dou JF, Wang JF (2015a) The synthetic antimicrobial peptide pexiganan and its nanoparticles (PNPs) exhibit the anti-*Helicobacter pylori* activity in vitro and in vivo. Molecules 20:3972–3985
Zhang XS, Tegtmeyer N, Traube L, Jindal S, Perez-Perez G, Sticht H, Backert S, Blaser MJ (2015) A specific A/T polymorphism in Western tyrosine phosphorylation B-motifs regulates *Helicobacter pylori* CagA epithelial cell interaction. PloS Pathog 11:e1004621. https://doi.org/10.1371/journal.ppat.1004621
Zhang X, Jiang A, Qi B, Yu H, Xiong Y, Zhou G, Qin M, Dou J, Wang J (2018) Secretion expression of human neutrophil peptide 1 (HNP1) in *Pichia pastoris* and its functional analysis against antibiotic-resistant *Helicobacter pylori*. Appl Microbiol Biotechnol 102:4817–4827

The Sweeping Role of Cholesterol Depletion in the Persistence of *Helicobacter pylori* Infections

Pau Morey and Thomas F. Meyer

Contents

Abstract The ability of *Helicobacter pylori* to persist lifelong in the human gastric mucosa is a striking phenomenon. It is even more surprising since infection is typically associated with a vivid inflammatory response. Recent studies revealed the mechanism by which this pathogen inhibits the epithelial responses to IFN-γ and other central inflammatory cytokines in order to abolish an effective antimicrobial defense. The mechanism is based on the modification and depletion of cholesterol by the pathogen's cholesterol-α-glucosyltransferase. It abrogates the assembly of numerous cytokine receptors due to the reduction of lipid rafts. Particularly, the receptors for IFN-γ, IL-22, and IL-6 then fail to assemble properly and to activate JAK/STAT signaling. Consequently, cholesterol depletion prevents the release of antimicrobial peptides, including the highly effective β-defensin-3. Intriguingly, the inhibition is spatially restricted to heavily infected cells, while the surrounding epithelium continues to respond normally to cytokine stimulation, thus providing a platform of the intense inflammation typically observed in *H. pylori* infections. It appears that pathogen and host establish a homeostatic balance between tightly colonized and rather inflamed sites. This homeostasis is influenced by the levels of available cholesterol, which potentially exacerbate *H. pylori*-induced inflammation.

P. Morey (✉)
Instituto Universitario de Investigación en Ciencias de la Salud (IUNICS), Universidad de las Islas Baleares, Palma de Mallorca, Spain
e-mail: paumorey@gmail.com

T. F. Meyer (✉)
Department of Molecular Biology, Max Planck Institute for Infection Biology, Berlin, Germany
e-mail: meyer@mpiib-berlin.mpg.de

S. Backert (ed.), *Molecular Mechanisms of Inflammation: Induction, Resolution and Escape by Helicobacter pylori*, Current Topics in Microbiology and Immunology 421, https://doi.org/10.1007/978-3-030-15138-6_9

The observed blockage of epithelial effector mechanisms by *H. pylori* constitutes a convincing explanation for the previous failures of T-cell-based vaccination against *H. pylori*, since infected epithelial cells remain inert upon stimulation by effector cytokines. Moreover, the mechanism provides a rationale for the carcinogenic action of this pathogen in that persistent infection and chronic inflammation represent a pro-carcinogenic environment. Thus, cholesterol-α-glucosyltransferase has been revealed as a central pathogenesis determinant of *H. pylori*.

1 Introduction

Approximately half of the world's population is chronically infected with *Helicobacter pylori*. This Gram-negative bacterium has co-evolved with its human host (Linz et al. 2007) and became a highly specialized inhabitant of the gastric mucosa, a unique microbial ecosystem. Infections with *H. pylori* are usually acquired in childhood and are often not recognized. However, they may lead to severe pathologies, such as active gastritis, peptic ulcers, MALT lymphoma, or gastric cancer (GC) (Bauer and Meyer 2011), particularly becoming prevalent after decades of bacterial colonization and chronic inflammation. Therefore, the pathogen's ability of long-term persistence and immune evasion can be considered as hallmarks of pathogenesis of this infection.

When present, *H. pylori* is the most abundant microbe in the stomach (Bik et al. 2006), owing to its superb adaptive ability in this hostile environment. Amongst the prime survival mechanisms of *H. pylori* is the secretion of urease, which allows the bacteria to raise the pH in their immediate surroundings through the production of ammonium ions (Salama et al. 2013). Remarkably, *H. pylori* only survives in the highly acidic stomach lumen, which has a pH between 1 and 5, for about half an hour (Schreiber et al. 2005). However, the gastric epithelium is protected from acid by a mucus layer, where the pH is close to neutral. Using a potent flagellar system and chemotactic receptors, *H. pylori* is capable of penetrating the mucus and rapidly colonizes the vicinity of gastric epithelial cells (GECs) at the gastric lumen as well as deeply in the glands (Howitt et al. 2011; Johnson and Ottemann 2018). The subsequent interplay between bacteria and host epithelium triggers a robust and sustained immune response; however, this nurtured a long-standing enigma how *H. pylori* circumvents this immune defense. Recent studies have now revealed an unexpected and intriguing feature that effectively conquers both innate and adaptive immune effector mechanisms. Here, we discuss these recent insights.

2 Inflammation Induced by *H. pylori*

H. pylori shelters from the harsh environment of the stomach in the mucus layer close to the epithelial lining. The bacteria display multiple adhesion molecules on their surface, a small proportion of which adhere directly to the epithelial cells. The infected

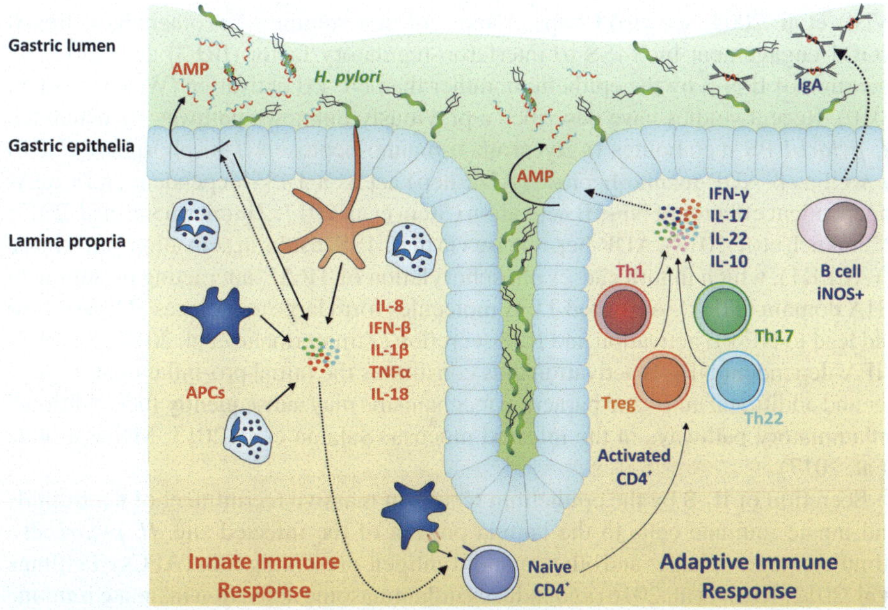

Fig. 1 Model of the immune response and defensive mechanisms elicited by *H. pylori*. Gastric epithelial cells secrete IL-8 in response to *H. pylori* infection, leading to the recruitment of innate immune cells and antigen-presenting cells (APCs) in the lamina propria. These cells further contribute to a pro-inflammatory cytokine millieu and activate naïve $CD4^+$ T cells. The adaptive immune response is dominated by the Th1 lineage, but Th17, Th22, and tolerogenic Treg subsets are also present. Together, these cells secrete specific cytokines to engage proper effectors to fight the infection, including antimicrobial peptides (AMP), which are retained in the mucus layer

cells initiate a robust innate immune response (summarized in Fig. 1), dominated by pro-inflammatory transcription factor NF-κB activation and IL-8 secretion, which is strongly augmented in the presence of a type IV secretion system (T4SS), encoded by the *cag* pathogenicity island (*cag*PAI) (Backert et al. 2015). The best-known function of the *cag*PAI T4SS is the translocation of the CagA effector protein into the cytosol of infected host cells (Posselt et al. 2013). Upon translocation, CagA is phosphorylated by host cell kinases, which enables it to exert pathogenic and likely also malignant functions (Tegtmeyer et al. 2017a). The $\alpha_5\beta_1$ integrins appear to act as host cell receptors for the T4SS via an interaction with CagL (Kwok et al. 2007; Barden et al. 2013) as well as CagA itself and CagY (Jimenez-Soto et al. 2009). In addition, the bacterial outer membrane protein HopQ has been identified as a T4SS-associated adhesin (Königer et al. 2016; Belogolova et al. 2013; Javaheri et al. 2016), which facilitates CagA translocation via binding to epithelial CEACAM receptors (Königer et al. 2016; Javaheri et al. 2016; Moonens et al. 2018; Tegtmeyer et al. 2019). For many years, it was unclear by what mechanism the T4SS causes NF-κB activation (Backert and Naumann 2010). Initial studies linked the T4SS-dependent delivery of bacterial peptidoglycan to NOD1 with NF-κB activation and IL-8 production

(Viala et al. 2004; see also Chaps. 2 and 7 of this volume). Yet others have linked NOD1 engagement by T4SS to interferon-regulatory factor (IRF3) activation and secretion of IFN-β by the epithelium, rather than NF-κB activation (Watanabe et al. 2010). Recent studies have described a previously unknown pathway by which the *H. pylori* LPS metabolites D-glycero-β-D-manno-heptose 1,7-bisphosphate (HBP) as well as β-ADP-manno-heptose (ADP-hep) act as a T4SS-dependent and CagA-independent effector of NF-κB activation (Stein et al. 2017; Zimmermann et al. 2017; Pfannkuch et al. 2018). ADP-hep release via the T4SS results in activation of α-kinase 1 (ALPK1), which in turn causes phosphorylation of TRAF-interacting protein with FHA domain (TIFA). Activated TIFA molecules form large complexes (TIFAsomes) and lead to NF-κB activation and IL-8 secretion (Zimmermann et al. 2017). ALPK1-TIFA-dependent NF-κB activation thus constitutes the initial pro-inflammatory trigger and additional auto- and paracrine mechanisms may subsequently fuel additional inflammatory pathways in the infected mucosa (Salama et al. 2013; Mejias-Luque et al. 2017).

Secretion of IL-8 by the epithelium results in massive recruitment of neutrophils and innate immune cells to the lamina propria of the infected site. *H. pylori* also stimulates macrophages and other myeloid antigen-presenting cells (APCs) (Fehlings et al. 2012; Koch et al. 2016) and induces inflammasome activation in innate immune cells (Pachathundikandi et al. 2016; Tran et al. 2017), resulting in abundant IL-1β, TNF-α, and IL-18 production in the gastric mucosa, amongst other cytokines. Stimulated APCs induce an adaptive immune response upon activation of T helper cells. Naive $CD4^+$ T cells are differentiated predominantly to the Th1 lineage, as many studies describe the central role of this cellular subset in the immune response to *H. pylori* (Itoh et al. 1999; Smythies et al. 2000; Bagheri et al. 2018). In animal models, IFN-γ produced by these cells contributes to a decreased *Helicobacter* burden (Sawai et al. 1999; Akhiani et al. 2002; Sayi et al. 2009; Morey et al. 2018). Likewise, IFN-β-defective mice exhibit a higher *H. pylori* load, associated with a decrease in epithelial CXCL10 production, suggesting a protective role for the recruitment of Th1 cells to the infection site (Watanabe et al. 2010). Th17 and Th22 cells have also been implicated in the immune response against *H. pylori*. In infected IL-23 knockout mice, Horvath and co-workers (2012) observed a lower inflammation score associated with decreased IL-17 levels and a higher bacterial load compared to wild-type mice. IL-23 also induces Th22 differentiation in *H. pylori*-infected mice as well as patients, which secrete IL-22 and contribute to the development of gastritis (Dixon et al. 2016; Zhuang et al. 2015). Finally, inducible nitric oxide synthase (iNOS)-producing plasma cells have recently been detected in considerable numbers in the *H. pylori*-infected mucosa partly producing pathogen-specific antibodies (Neumann et al. 2016).

In conclusion, the mucosa colonized by *H. pylori* represents a strongly pro-inflammatory cytokine environment, containing abundantly recruited immune cells. In spite of this, the immune response fails to clear *H. pylori*, rather supporting a chronic inflammatory condition. How *H. pylori* is capable of evading such a strong and sustained response is a question that puzzled researchers for decades.

It is well known that *H. pylori* can modulate T-cell responses by inducing tolerance (Müller et al. 2011; Pachathundikandi et al. 2013). The virulence factors VacA and GGT both prevent the activation and proliferation of $CD4^+$ T cells and drive them to differentiate toward a Treg lineage. This tolerogenic phenotype partially reduces the differentiation to Th1 and Th17 lineages and suppresses their effector functions (Gebert et al. 2003; Oertli et al. 2013; Salama et al. 2013). However, such impairment of T cell function could perhaps be overcome through protective immunization (Velin et al. 2016). Experimental vaccination studies performed in animal models highlighted the importance of Th1 and Th17 compartments (Akhiani et al. 2002; Velin et al. 2009; Hitzler et al. 2011), while B cells and antibody response did not confer protection against *H. pylori* (Ermak et al. 1998; Hitzler et al. 2011). Nonetheless, all vaccination strategies have so far failed to achieve complete eradication, despite a robust enhancement of cell-mediated immunity, suggesting that the link between $CD4^+$ T cells and relevant effector mechanisms against *H. pylori* is somehow blocked (Aebischer et al. 2010).

3 Epithelial Defense Molecules: Antimicrobial Potential Against *H. pylori*

Epithelial cells constitute the first line of defense against invading microorganisms. In the gastrointestinal tract, maintenance of the barrier function is crucial to protect the host from pathogenic microorganisms and excessive inflammatory stimuli. Epithelia contribute to immunity beyond the function of a passive physical shield, since they can adopt an active defensive role by secreting mucins and antimicrobial molecules (Ostaff et al. 2013).

In terms of protection against *H. pylori*, several observations point toward epithelial cells as important immune effectors. While immune cells accumulate in the lamina propria, *H. pylori* mostly resides in the luminal side of the mucosa (Keilberg et al. 2016), although it is also capable of penetrating sub-epithelial niches (Tegtmeyer et al. 2017b). On the luminal side of the epithelium, the most relevant antibacterial immune effectors are antimicrobial factors and secreted IgA; however, they do not seem to provide effective protection against *H. pylori* (Ermak et al. 1998). Rather, *H. pylori* persists embedded in the mucus layer in the vicinity of epithelial cells (Johnson and Ottemann 2018). Commensal bacteria in the gut, by contrast, are not able to reach the epithelia due to the accumulation of antimicrobial molecules of epithelial origin in the mucus barrier, such as the lectin RegIII-γ, or cationic antimicrobial peptides like human beta-defensins (hBDs) 1, 2, and 3 (Meyer-Hoffert et al. 2008; Vaishnava et al. 2011; Antoni et al. 2013). The presence of these molecules keeps microorganisms at bay, thereby preventing excessive inflammation (Vaishnava et al. 2011). Using novel primary gastric epithelial mucosoid cultures, it was recently demonstrated that *H. pylori* infection elicits increased antibacterial activity in the mucus layer (Boccellato et al. 2018).

Although some antimicrobial molecules of the epithelium are constitutively expressed, most are induced in response to inflammatory stimuli, such as bacterial ligands or cytokines produced by innate and adaptive immune cells (Ostaff et al. 2013). Nuding et al. (2013) studied the occurrence of antimicrobial peptides in *H. pylori*-infected gastric biopsies and observed several that were selectively induced; however, none exhibited strong activity against *H. pylori*, such as hBD3 or LL37. Those peptides that were robustly expressed in infected biopsies, on the other hand, such as hBD1, hBD2, and elafin, showed poor antimicrobial activity against *H. pylori*—possibly due to specific modifications in the lipid A part of *H. pylori* LPS (Cullen et al. 2011). The case of hBD3 is intriguing. In vitro, epithelial cells rapidly express hBD3 upon *H. pylori* infection via EGFR activation (Boughan et al. 2006; Kawauchi et al. 2006; Bauer et al. 2012a; Muhammad et al. 2016). However, this pathway becomes effectively blocked after prolonged contact with *H. pylori,* which uses its T4SS effector protein CagA to activate SHP-2 phosphatase (Bauer et al. 2012a). In an ex vivo study, hBD3 expression was even lower in infected than non-infected biopsies; however, the CagA status of infecting *H. pylori* strains appeared to have no influence on hBD3 levels, suggesting that yet another CagA-independent mechanism suppresses hBD3 (Bauer et al. 2012b). Epithelial hBD3 expression is also induced in vitro by cytokines such as IFN-γ or IL-22 and increased when cells are simultaneously co-treated with TNF-α or IL-17A, respectively (Joly et al. 2005; Albanesi et al. 2007; Wolk et al. 2004; Dixon et al. 2016). Moreover, hBD3 is upregulated in vivo under pro-inflammatory conditions such as ulcerative colitis, Crohn's disease, or bronchiolitis (Fahlgren et al. 2004; Nuding et al. 2013), which resemble the inflammatory environment found in chronic gastritis.

4 Cholesterol Glucosides in *H. pylori*: Implications for Virulence

H. pylori is one of the few bacteria presenting cholesteryl glucosides (CGs) in the membrane, accounting for 25% of total lipids (Hirai et al. 1995). Since it lacks the genes for de novo sterol biosynthesis, it acquires cholesterol from the host and exhibits positive chemotaxis toward cholesterol (Wunder et al. 2006). In the gastric mucosa, cholesterol can be found either bound to mucins or in the epithelial cell membrane, where it forms cholesterol-rich microdomains, also known as lipid rafts (Gong et al. 1990; Lai et al. 2013). Which bacterial factors directly mediate cholesterol uptake is still not fully understood. *H. pylori* phosphatidylethanolamine binds to free cholesterol present in growth medium or in host cell lipid rafts (Shimomura et al. 2012). Once attached, cholesterol is glucosylated to cholesteryl-α-D-glucopyranoside (αCG) by the membrane-bound enzyme cholesterol-α-glucosyl transferase or CGT, encoded by the *cgt* gene, also annotated as *capJ* or HP0421 (Lebrun et al. 2006). αCG is further modified by either an acyl or a phosphatidyl moiety at C6 of the glucose to yield cholesteryl-6′-O-tetradecanoyl-α-D-glucopyranoside (αCAG) and cholesteryl-6′-O-phosphatidyl-α-D-glucopyranoside (αCPG) (Lebrun et al. 2006). The three forms coexist in the bacterial membrane and form cholesterol microdomains, which require the flotillin-like protein HP0248 for correct organization (Hutton et al. 2017).

Eukaryotic lipid rafts serve as a platform for multiple cellular receptors (Simons and Sampaio 2011). Therefore, bacterial manipulation of host cholesterol microdomains can lead to important changes in cellular homeostasis. Likewise, chemical alteration of host lipid rafts might have important effects on *H. pylori* infections. Disruption of lipid rafts by methyl-β-cyclodextrin (mβCD), which depletes cholesterol from eukaryotic membranes, impairs CagA delivery, IL-8 induction, and bacterial invasion (Lai et al. 2008). mβCD treatment of epithelial cell lines also impairs the interaction between the *H. pylori* T4SS and host $\alpha_5\beta_1$ integrins, resulting in diminished NF-κB activation (Hutton et al. 2010). Similarly, VacA delivery into host cells depends on lipid raft integrity and GPI-anchored proteins (Lai et al. 2013; Ricci et al. 2000).

Disruption of the *cgt* gene results in impaired cholesterol uptake and absence of any kind of bacterial CGs (Lebrun et al. 2006; Wunder et al. 2006). Mutants in this gene have been used in a variety of in vitro models to examine the role of host cholesterol uptake in *H. pylori* pathogenesis (Lai et al. 2013). In epithelial cell lines infected with different *H. pylori* strains, CGT activity led to clustering of lipid rafts at the infection site within 2–6 h (Fig. 2) (Lai et al. 2008, 2013; Wang et al. 2012; Du et al. 2014). Under these conditions, Wang and co-workers (2012) showed that CGT activity was crucial for CagA translocation and IL-8 induction in AGS cells. Moreover, they observed that translocation of bacterial CGs into the host cell membrane influences lipid raft clustering. In contrast, upon extended in vitro infection (e.g., 24 h) CGT activity resulted in complete disruption of lipid rafts (Wunder et al. 2006; Morey et al. 2018) (Fig. 2). At this time point, a transcriptomic analysis revealed no differences in NF-κB activation between infections of primary GECs with wild-type *H. pylori* and an isogenic *cgt* mutant (Morey et al. 2018).

The presence of CGs in the bacterial outer membrane was also found to decrease phagocytosis by macrophages (Wunder et al. 2006). In a different study, CGT activity delayed *H. pylori* phagocytosis (Du et al. 2014). Compared to Δ*cgt* mutants, internalized wild-type *H. pylori* end up in vacuoles with slower phagosomal maturation, facilitating extended intracellular survival. These differential phenotypes are presumably linked to coalescence of lipid rafts, leading to a different membrane composition at the site of phagocytosis (Du et al. 2014). In T cells, *cgt* mutants have drastically diminished anti-proliferative effects following infection (Beigier-Bompadre et al. 2011). Invariant natural killer T cells are known to react to glycolipids. They recognize *H. pylori* CGs and become activated and mount an inflammatory response when challenged with either purified CGs or wild-type *H. pylori*, but not with a *cgt* mutant (Ito et al. 2013). In the same study, the authors also describe an intriguing positive correlation between the CGT activity of clinical *H. pylori* isolates and the gastric atrophy score of the respective patients (Ito et al. 2013).

Unfortunately, *H. pylori* strains' defective in CGT are not suitable for in vivo experiments. Although they do not present notable growth defects in vitro, they

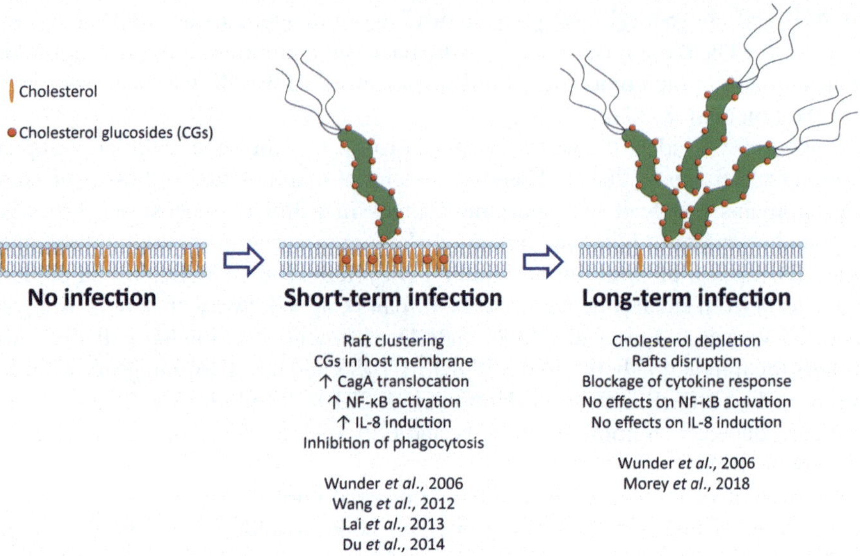

Fig. 2 Effects of CGT activity on host–pathogen interaction. *H. pylori* cells co-opt cholesterol from host cell membranes. Early interaction leads to the coalescence of cholesterol-rich microdomains, or lipid rafts, at the infection site, enhancing T4SS activity and the inflammatory response. In phagocytic cells, this process interferes with the engulfment of *H. pylori* and the phagosome maturation. Longer infection times result in depletion of cholesterol from host membranes, disrupting lipid rafts. At these time points, CGT activity does not significantly affect NF-κB activation by *H. pylori*. Instead, it avoids downstream signaling by cytokine receptors located in lipid rafts (Du et al. 2014; Lai et al. 2013; Morey et al. 2018; Wang et al. 2012; Wunder et al. 2006)

are unable to establish colonization in mice (Lai et al. 2013; Wunder et al. 2006; Morey et al. 2018). This could be due to the role CGT activity plays in preventing the bactericidal activity of metabolites such as 7-dehydrocholesterol (Shimomura et al. 2013) or docosahexaenoic acid (Correia et al. 2014), as well as antimicrobial peptides (McGee et al. 2011), which might have a higher impact in vivo. In order to study the role of CGT activity in *H. pylori* virulence, Wunder and co-workers (2006) counteracted bacterial cholesterol uptake from host cells by feeding mice a cholesterol-rich diet. These mice showed a lower *H. pylori* load compared to mice fed a normal diet. This increased protection was correlated to an increased Th1 transcriptional response. These results provided early strong evidence for a role of cholesterol extraction in *H. pylori* persistence and inflammation.

5 *H. pylori* Targets Host Cholesterol to Block the Epithelial Response to IFN-γ and Related Cytokines

CD4⁺ T cells of the mucosa orchestrate adaptive immunity via the activation of effector cells. In the stomach, GECs, which express the corresponding cytokine receptors, assume a key role as effector cells against *H. pylori*. The receptor for IFN-γ, IFNGR, consists of two subunits (IFNGR1 and IFNGR2), which require assembly within cholesterol-rich microdomains of the plasma membrane. Ligand binding triggers stimulation of the JAK/STAT signaling cascade (Blouin and Lamaze 2013), whereby phosphorylated STAT1 forms dimers that are translocated into the nucleus to induce the expression of interferon-response genes (IRGs) (Fig. 3a). Most of these genes are involved in defense and inflammation.

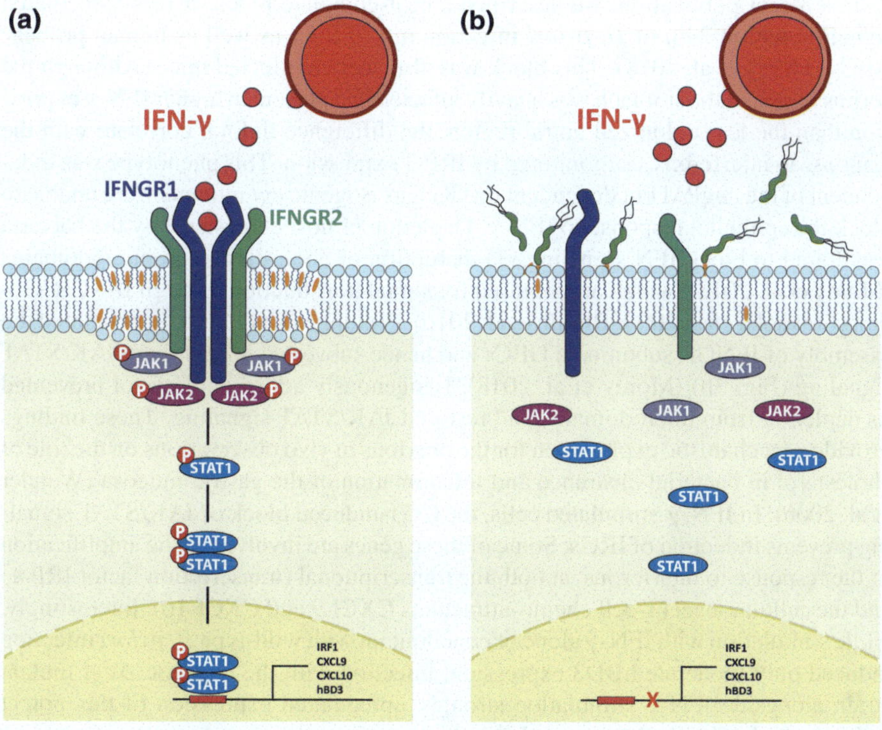

Fig. 3 CGT-mediated disruption of JAK/STAT signaling. Gastric epithelial cells display receptors for IFN-γ, whose subunits (IFNGR1/IFNGR2) are assembled in lipid rafts. **a** In normal conditions, stimulation with IFN-γ triggers JAK/STAT signaling via JAK1, JAK2, and STAT1 phosphorylation and nuclear translocation to promote expression of genes involved in inflammation and defense. **b** Prolonged *H. pylori* infection results in CGT-mediated disruption of lipid rafts and disassembly of IFNGR subunits, which prevents JAK/STAT signaling

Transcriptional analysis of micro-dissected gastric glands has shown that mucus-producing cells are engaged by the adaptive Th1 response in *H. pylori*-infected mice, activating multiple IRGs (Müller et al. 2004). In this context, an in vitro study reported an enhanced response to IFN-γ in epithelial cells upon short-term infection with *H. pylori*, via synergistic NOD1 activation via the T4SS (Allison et al. 2013). However, other reports suggest that direct interaction between *H. pylori* and epithelial cells elicits an inhibitory effect on the IFN-γ response: SHP-2 activation by CagA results in partial suppression of JAK/STAT signaling in response to IFN-γ—a phenotype that was more accentuated in East Asian compared to reference strains (Yu-Chih Wang et al. 2014). Kraft et al. (2001) also described inhibition of IFN-γ and TNF-α-induced expression of CXCL9 and CXCL10 in GECs treated with the membranous fraction of *H. pylori*. In agreement with this, cells infected with both *cag*PAI-positive and *cag*PAI-negative *H. pylori* strains fail to activate STAT1 upon IFN-γ treatment, thereby preventing induction of IRGs (Mitchell et al. 2004).

In a recent publication, we described a near-complete block of IFN-γ-mediated signaling within 24 h of *H. pylori* infection in cell lines as well as human primary GECs (Morey et al. 2018). This block was also seen in infected mice: Although the corpus region of the stomach was heavily infected and presented higher IFN-γ expression than the less colonized antral region, the difference did not correlate with the response to interferons, as monitored by IRF-1 expression. This phenotype was independent of the *cag*PAI but dependent on CGT, as isogenic *cgt* mutants were unable to block the epithelial response to IFN-γ. Depletion of host cholesterol by the bacteria was found to block IFN signaling via disruption of cholesterol-rich microdomains (Morey et al. 2018), which are required for signal transduction through IFNGR subunits (Takaoka et al. 2000; Blouin et al. 2016). Prolonged CGT activity prevented the assembly of IFNGR subunits in GECs and hence subverted downstream JAK/STAT signaling (Fig. 3b) (Morey et al. 2018). Exogenously added cholesterol prevented its depletion from microdomains and restored JAK/STAT signaling. These findings provide a mechanistic explanation for the previous in vivo observations on the role of cholesterol in bacterial clearance and inflammation of the gastric mucosa (Wunder et al. 2006). In IFN-γ-stimulated cells, the CGT-induced block of JAK/STAT signaling prevents induction of IRGs. Some of these genes are involved in the amplification of the response to interferons, at both the transcriptional (transcription factor IRF-1) and the cellular level (T-cell chemo-attractants CXCL9 and CXCL10). Interestingly, while stimulation with IFN-γ alone or concomitant with wild-type *H. pylori* infection induced only moderate hBD3 expression, infection with the isogenic Δ*cgt* mutant strain alongside IFN-γ stimulation strongly upregulated expression of this potent antimicrobial effector (Morey et al. 2018).

Although CGT-deficient *H. pylori* strains are unable to colonize mice successfully, their failure to block the response to IFN-γ leads to higher expression of IRGs following infection compared to wild-type *H. pylori* (Morey et al. 2018). Using an advanced model of human primary GECs cultured in air–liquid interface, infection with *H. pylori* could be extended for several days (Boccellato et al. 2018). Under these conditions, bacteria form localized infection foci and these infected areas fail to activate JAK/STAT signaling when treated with IFN-γ, while non-infected areas from

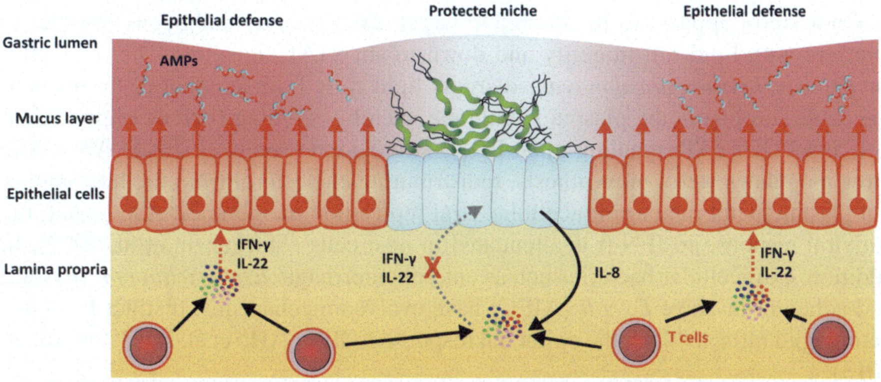

Fig. 4 Model for *H. pylori* persistence amidst chronic gastric inflammation. Bacterial colonization by *H. pylori* results in sustained inflammation. Gastric epithelial cells in direct contact with *H. pylori* contribute to the maintenance of the pro-inflammatory cytokine milieu via IL-8 secretion. However, infected cells are not responsive to the adaptive immune response due to cholesterol depletion mediated by *H. pylori*'s CGT, while neighboring cells are stimulated by the surrounding inflammation to produce epithelial defense factors such as antimicrobial peptides (AMPs). In this way, *H. pylori* generates micro-niches of diminished inflammation and epithelial response despite a global pro-inflammatory environment

the same culture exhibit a robust response (Morey et al. 2018) (Fig. 4). Nonetheless, infected cells do remain responsive to other stimuli and mount an NF-κB-induced inflammatory response (Morey et al. 2018). In animal models, *H. pylori* persists by forming micro-colonies deep in the gastric glands (Howitt et al. 2011). Once colonization is established, the resulting inflammation appears to prevent secondary *H. pylori* infections, except in already occupied glands (Keilberg et al. 2016), suggesting that the micro-niches are protected from the immune response. These observations are in striking agreement with the notion of immune-protected micro-niches at *H. pylori* colonization sites, established via cholesterol extraction (Morey et al. 2018).

The impact of cholesterol depletion on receptor function in the plasma membrane is not limited to the IFNGR but also affects receptors for other cytokines, e.g., IFN-β, IL-6, and IL-22 (Takaoka et al. 2000; Sehgal et al. 2002; Flanagan et al. 2008). Their inactivation seems to play similarly important roles in facilitating persistent *H. pylori* colonization. The block of epithelial IL-22 stimulation may be particularly important, as this cytokine almost exclusively instructs epithelial cells to activate STAT3 signaling as a key factor of defense against extracellular bacteria (Wolk et al. 2004; Zheng et al. 2008). GECs effectively respond to IL-22 by secreting antimicrobial peptides (including hBD3) and other biocidal activity against *H. pylori* (Dixon et al. 2016). However, IL-22$^{-/-}$ knockout mice infected with *H. pylori* do not show altered levels of antimicrobial peptide expression or bacterial colonization as compared to wild-type mice (Dixon et al. 2016), suggesting infection leads to a defect in epithelial response to IL-22.

Cholesterol appears to be a tangible target also for other pathogens in order to interfere with lipid raft integrity and downstream STAT signaling. The intracellular parasite *Leishmania donovani* disrupts lipid rafts in macrophages by quenching cholesterol from the plasma membrane, thereby avoiding IFNGR oligomerization and JAK/STAT signaling upon IFN-γ stimulation (Sen et al. 2011). West Nile virus modulates sterol biosynthesis, redistributing cholesterol from the host epithelial plasma membrane to intracellular viral replication membranes. As a result, the antiviral response to IFN-β is attenuated in host cells (Mackenzie et al. 2007). In addition, extracellular bacteria such as enterohemorrhagic *Escherichia coli* (EHEC) and adherent-invasive *E. coli* (AIEC) have evolved mechanisms, distinct from targeting lipid rafts, to block the epithelial response to IFN-γ (Ho et al. 2012; Ossa et al. 2013).

6 Concluding Remarks

Current antibiotic therapies against *H. pylori* register increasing failure rates and new therapies that are able to increase the eradication rate and to interfere with the malignancies initiated by this pathogen are currently a high priority (Tacconelli et al. 2018). The persistence of *H. pylori* in the gastric mucosa is associated with a chronic inflammatory response that is the basis of the morbidity associated with this pathogen. The robust and sustained innate and adaptive immune response appears to diminish the bacterial load, but fails to clear the infection. *H. pylori* has developed strategies to evade immunity by inducing tolerance and by interrupting the communication between different arms of the immune system. Although vaccination attempts achieved rescuing the immunosuppressive features of *H. pylori*, affecting T-cell maturation via the action of VacA and GGT, and raised a full T cell response (Malfertheiner et al. 2018), this response failed to engage the proper immune effectors required for *H. pylori* eradication. This is explained by the failure of infected epithelial cells to respond to critical cytokines released by the recruited immune cells. This provides a scenario in gastric glands where *H. pylori* readily persists in bacterial micro-colonies on the epithelial cell surfaces, while other regions in the stomach mucosa, including the sub-epithelial tissue, become chronically inflamed.

It now seems plausible that alterations in cholesterol homeostasis and metabolism could be linked to various processes in immunity or cancer development as well as anecdotal observations in the context of *H. pylori*-associated pathogenesis. In agreement with the findings illuminated in this review, higher levels of low-density lipoprotein cholesterol (LDL-C) in the blood correlated with increased severity of *H. pylori*-related inflammation (Kucukazman et al. 2009), while long-term treatment with statins contributed to the amelioration of gastritis (Nseir et al. 2010) and a decrease in the incidence of GC (Singh and Singh 2013). Understanding the basic features connecting *H. pylori* infection, cholesterol function, the responses of the immune system and chronic inflammation is therefore crucial for developing effective approaches for prevention and treatment of *H. pylori*-induced pathology.

References

Aebischer T, Meyer TF, Andersen LP (2010) Inflammation, immunity, and vaccines for Helicobacter. Helicobacter 15(Suppl 1):21–28. https://doi.org/10.1111/j.1523-5378.2010.00777.x

Akhiani AA, Pappo J, Kabok Z, Schon K, Gao W, Franzen LE et al (2002) Protection against *Helicobacter pylori* infection following immunization is IL-12-dependent and mediated by Th1 Cells. J Immunol 169:6977–6984. https://doi.org/10.4049/jimmunol.169.12.6977

Albanesi C, Fairchild HR, Scarponi C, Pità OD, Donald Y, Leung M et al (2007) IL-4 and IL-13 negatively regulate TNF-alpha- and IFN-gamma -Induced beta-defensin expression through STAT-6, suppressor of cytokine signaling (SOCS)-1, and SOCS-3. J Immunol 179:984–992

Allison CC, Ferrand J, McLeod L, Hassan M, Kaparakis-Liaskos M, Grubman A et al (2013) Nucleotide oligomerization domain 1 enhances IFN-γ signaling in gastric epithelial cells during *Helicobacter pylori* infection and exacerbates disease severity. J Immunol (Baltimore, Md.: 1950). https://doi.org/10.4049/jimmunol.1200591

Antoni L, Nuding S, Weller D, Gersemann M, Ott G, Wehkamp J et al (2013) Human colonic mucus is a reservoir for antimicrobial peptides. J Crohn's & colitis. https://doi.org/10.1016/j.crohns.2013.05.006

Backert S, Tegtmeyer N, Fischer W (2015) Composition, structure and function of the *Helicobacter pylori cag* pathogenicity island encoded type IV secretion system. Future Microbiol. 10(6):955–965. https://doi.org/10.2217/fmb.15.32

Backert S, Naumann M (2010) What a disorder: proinflammatory signaling pathways induced by *Helicobacter pylori*. Trends Microbiol 18(11):479–486. https://doi.org/10.1016/j.tim.2010.08.003

Bagheri N, Salimzadeh L, Shirzad H (2018) The role of T helper 1-cell response in *Helicobacter pylori*-infection. Microb Pathog 123:1–8. https://doi.org/10.1016/J.MICPATH.2018.06.033

Barden S, Lange S, Tegtmeyer N, Conradi J, Sewald N, Backert S, Niemann HH (2013) A helical RGD motif promoting cell adhesion: crystal structures of the *Helicobacter pylori* type IV secretion system pilus protein CagL. Structure. 21(11):1931–1941. https://doi.org/10.1016/j.str.2013.08.018

Bauer B, Meyer TF (2011) The human gastric pathogen *Helicobacter pylori* and its association with gastric cancer and ulcer disease. Ulcers 2011:1–23. https://doi.org/10.1155/2011/340157

Bauer B, Pang E, Holland C, Kessler M, Bartfeld S, Meyer TF (2012a) The *Helicobacter pylori* virulence effector CagA abrogates human b-defensin 3 expression via inactivation of EGFR signaling. Cell Host Microbe 11:576–586. https://doi.org/10.1016/j.chom.2012.04.013

Bauer B, Wex T, Kuester D, Meyer T, Malfertheiner P (2012b) Differential expression of human beta defensin 2 and 3 in gastric mucosa of *Helicobacter pylori*-infected individuals. Helicobacter 18:6–12. https://doi.org/10.1111/hel.12000

Beigier-Bompadre M, Moos V, Belogolova E, Allers K, Schneider T, Churin Y et al (2011) Modulation of the CD4 + T-cell response by *Helicobacter pylori* depends on known virulence factors and bacterial cholesterol and cholesterol alpha-glucoside content. J Infect Dis 204:1339–1348. https://doi.org/10.1093/infdis/jir547

Belogolova E, Bauer B, Pompaiah M, Asakura H, Brinkman V, Ertl C et al (2013) *Helicobacter pylori* outer membrane protein HopQ identified as a novel T4SS-associated virulence factor. Cell Microbiol, 15, n/a-n/a. https://doi.org/10.1111/cmi.12158

Bik EM, Eckburg PB, Gill SR, Nelson KE, Purdom EA, Francois F et al (2006) Molecular analysis of the bacterial microbiota in the human stomach. Proc Natl Acad Sci USA 103:732–737. https://doi.org/10.1073/pnas.0506655103

Blouin CM, Hamon Y, Gonnord P, Boularan C, Kagan J, Viaris de Lesegno C et al (2016) Glycosylation-dependent IFN-γR partitioning in lipid and actin nanodomains is critical for JAK activation. Cell 166:920–934. https://doi.org/10.1016/j.cell.2016.07.003

Blouin CM, Lamaze C (2013) Interferon gamma receptor: the beginning of the journey. Front Immunol 4:267. https://doi.org/10.3389/fimmu.2013.00267

Boccellato F, Woelffling, S, Imai-Matsushima A, Sanchez G, Goosmann C, Schmid M et al (2018) Polarised epithelial monolayers of the gastric mucosa reveal insights into mucosal homeostasis and defence against infection. Gut gutjnl-2017–314540. https://doi.org/10.1136/gutjnl-2017-314540

Boughan PK, Argent RH, Body-malapel M, Park J-H, Ewings KE, Bowie AG et al (2006) Nucleotide-binding oligomerization domain-1 and epidermal growth factor receptor. J Biol Chem 281:11637–11648. https://doi.org/10.1074/jbc.M510275200

Correia M, Casal S, Vinagre J, Seruca R, Figueiredo C, Touati E et al (2014) *Helicobacter pylori*'s cholesterol uptake impacts resistance to docosahexaenoic acid. Int J Med Microbiol 304:314–320. https://doi.org/10.1016/J.IJMM.2013.11.018

Cullen TW, Giles DK, Wolf LN, Ecobichon C, Boneca IG, Trent MS (2011) *Helicobacter pylori* versus the host: remodeling of the bacterial outer membrane is required for survival in the gastric mucosa. PLoS Pathog 7:e1002454. https://doi.org/10.1371/journal.ppat.1002454

Dixon BREA, Radin JN, Piazuelo MB, Contreras DC, Algood HMS, Noto J et al (2016) IL-17a and IL-22 induce expression of antimicrobials in gastrointestinal epithelial cells and may contribute to epithelial cell defense against *Helicobacter pylori*. PLoS ONE 11:e0148514. https://doi.org/10.1371/journal.pone.0148514

Du S-Y, Wang H-J, Cheng H-H, Chen S-D, Wang LH-C, Wang W-C (2014) Cholesterol glucosylation by *Helicobacter pylori* delays internalization and arrests phagosome maturation in macrophages. J Microbiol Immunol Infect = Wei mian yu gan ran za zhi https://doi.org/10.1016/j.jmii.2014.05.011

Ermak TH, Giannasca PJ, Nichols R, Myers GA, Nedrud J, Weltzin R et al (1998) Immunization of mice with urease vaccine affords protection against Helicobacter pylori infection in the absence of antibodies and is mediated by MHC class II-restricted responses. J Exp Med 188:2277–2288. https://doi.org/10.1084/JEM.188.12.2277

Fahlgren A, Hammarstrom S, Danielsson A, Hammarstrom M-L (2004) Beta-Defensin-3 and -4 in intestinal epithelial cells display increased mRNA expression in ulcerative colitis. Clin Exp Immunol 137:379–385. https://doi.org/10.1111/j.1365-2249.2004.02543.x

Fehlings M, Drobbe L, Moos V, Renner Viveros P, Hagen J, Beigier-Bompadre M et al (2012) Comparative analysis of the interaction of *Helicobacter pylori* with human dendritic cells, macrophages, and monocytes. Infect Immun 80:2724–2734. https://doi.org/10.1128/IAI.00381-12

Flanagan K, Modrusan Z, Cornelius J, Chavali A, Kasman I, Komuves L et al (2008) Intestinal epithelial cell up-regulation of LY6 molecules during colitis results in enhanced chemokine secretion. J Immunol (Baltimore, Md.: 1950), 180: 3874–3881. https://doi.org/10.4049/jimmunol.180.6.3874

Gebert B, Fischer W, Weiss E, Hoffmann R, Haas R (2003) *Helicobacter pylori* vacuolating cytotoxin inhibits T lymphocyte activation. Science (New York, N.Y.), 301: 1099–1102. https://doi.org/10.1126/science.1086871

Gong DH, Turner B, Bhaskar KR, Lamont JT (1990) Lipid binding to gastric mucin: protective effect against oxygen radicals. Am J physiol 259:G681–G686

Hirai Y, Haque M, Yoshida T, Yokota K, Yasuda T, Oguma K (1995) Unique cholesteryl glucosides in *Helicobacter pylori*: composition and structural analysis. J Bacteriol 177:5327–5333

Hitzler I, Oertli M, Becher B, Agger EM, Müller A (2011) Dendritic cells prevent rather than promote immunity conferred by a *Helicobacter* vaccine using a mycobacterial adjuvant. Gastroenterology 141:186–196.e1. https://doi.org/10.1053/J.GASTRO.2011.04.009

Ho NK, Ossa JC, Silphaduang U, Johnson R, Johnson-Henry KC, Sherman PM (2012) Enterohemorrhagic escherichia coli O157:H7 shiga toxins inhibit gamma interferon-mediated cellular activation. Infect Immun 80:2307–2315. https://doi.org/10.1128/IAI.00255-12

Horvath DJ, Washington MK, Cope VA, Algood HMS (2012) IL-23 contributes to control of chronic *Helicobacter pylori* infection and the development of T helper responses in a mouse model. Front Immunol 3:56. https://doi.org/10.3389/fimmu.2012.00056

Howitt MR, Lee JY, Lertsethtakarn P, Vogelmann R, Joubert L-M, Ottemann KM et al (2011) ChePep controls *Helicobacter pylori* infection of the gastric glands and chemotaxis in the epsilonproteobacteria. MBio 2:e00098–11. https://doi.org/10.1128/mbio.00098-11

Hutton ML, D'Costa K, Rossiter AE, Wang L, Turner L, Steer DL et al (2017) A *Helicobacter pylori* homolog of eukaryotic flotillin is involved in cholesterol accumulation, epithelial cell responses and host colonization. Front cell Infec Microbiol 7:219. https://doi.org/10.3389/fcimb.2017.00219

Hutton ML, Kaparakis-Liaskos M, Turner L, Cardona A, Kwok T, Ferrero RL (2010) *Helicobacter pylori* exploits cholesterol-rich microdomains for induction of NF-kappaB-dependent responses and peptidoglycan delivery in epithelial cells. Infect Immun 78:4523–4531. https://doi.org/10.1128/IAI.00439-10

Ito Y, Vela JL, Matsumura F, Hoshino H, Tyznik A, Lee H et al (2013) *Helicobacter pylori* cholesteryl α-glucosides contribute to its pathogenicity and immune response by natural killer T cells. PLoS ONE 8:e78191. https://doi.org/10.1371/journal.pone.0078191

Itoh T, Wakatsuki Y, Yoshida M, Usui T, Matsunaga Y, Kaneko S et al (1999) The vast majority of gastric T cells are polarized to produce T helper 1 type cytokines upon antigenic stimulation despite the absence of *Helicobacter pylori* infection. J Gastroenterol 34:560–570

Javaheri A, Kruse T, Moonens K, Mejías-Luque R, Debraekeleer A, Asche CI, Tegtmeyer N, Kalali B, Bach NC, Sieber SA, Hill DJ, Königer V, Hauck CR, Moskalenko R, Haas R, Busch DH, Klaile E, Slevogt H, Schmidt A, Backert S, Remaut H, Singer BB, Gerhard M (2016) *Helicobacter pylori* adhesin HopQ engages in a virulence-enhancing interaction with human CEACAMs. Nat Microbiol 2:16189. https://doi.org/10.1038/nmicrobiol.2016.189

Jimenez-Soto LF, Kutter S, Sewald X, Ertl C, Weiss E, Kapp U et al (2009) *Helicobacter pylori* type IV secretion apparatus exploits beta1 integrin in a novel RGD-independent manner. PLoS Pathog 5(12):e1000684. https://doi.org/10.1371/journal.ppat.1000684

Johnson KS, Ottemann KM (2018) Colonization, localization, and inflammation: the roles of *H. pylori* chemotaxis in vivo. Curr Opin Microbiol 41:51–57. https://doi.org/10.1016/j.mib.2017.11.019

Joly S, Organ CC, Johnson GK, McCray PB, Guthmiller JM (2005) Correlation between beta-defensin expression and induction profiles in gingival keratinocytes. Mol Immunol 42:1073–1084. https://doi.org/10.1016/j.molimm.2004.11.001

Kawauchi K, Yagihashi A, Tsuji N, Uehara N, Furuya D, Kobayashi D et al (2006) Human beta-defensin-3 induction in *H. pylori*-infected gastric mucosal tissues. World J Gastroenterol: WJG 12:5793–5797

Keilberg D, Zavros Y, Shepherd B, Salama NR, Ottemann KM (2016) Spatial and temporal shifts in bacterial biogeography and gland occupation during the development of a chronic infection. MBio 7:e01705–16. https://doi.org/10.1128/mbio.01705-16

Koch M, Mollenkopf H-J, Meyer TF (2016) Macrophages recognize the *Helicobacter pylori* type IV secretion system in the absence of toll-like receptor signalling. Cell Microbiol 18:137–147. https://doi.org/10.1111/cmi.12492

Königer V, Holsten L, Harrison U, Busch B, Loell E, Zhao Q et al (2016) *Helicobacter pylori* exploits human CEACAMs via HopQ for adherence and translocation of CagA. Nat Microbiol 2:16188. https://doi.org/10.1038/nmicrobiol.2016.188

Kraft M, Riedel S, Maaser C, Kucharzik T, Steinbuechel A, Domschke W et al (2001) IFN-gamma synergizes with TNF-alpha but not with viable *H. pylori* in up-regulating CXC chemokine secretion in gastric epithelial cells. Clin Exp Immunol 126:474–481

Kucukazman M, Yavuz B, Sacikara M, Asilturk Z, Ata N, Ertugrul DT et al (2009) The relationship between updated sydney system score and LDL cholesterol levels in patients infected with *Helicobacter pylori*. Dig Dis Sci 54:604–607. https://doi.org/10.1007/s10620-008-0391-y

Kwok T, Zabler D, Urman S, Rohde M, Hartig R, Wessler S, Misselwitz R, Berger J, Sewald N, König W, Backert S (2007) *Helicobacter exploits* integrin for type IV secretion and kinase activation. Nature 449:862–866. https://doi.org/10.1038/nature06187

Lai C-H, Chang Y-C, Du S-Y, Wang H-J, Kuo C-H, Fang S-H et al (2008) Cholesterol depletion reduces *Helicobacter pylori* CagA translocation and CagA-induced responses in AGS cells. Infect Immun 76:3293–3303. https://doi.org/10.1128/IAI.00365-08

Lai C-H, Hsu Y-M, Wang H-J, Wang W-C (2013) Manipulation of host cholesterol by *Helicobacter pylori* for their beneficial ecological niche. BioMedicine 3:27–33. https://doi.org/10.1016/j.biomed.2012.12.002

Lebrun A-H, Wunder C, Hildebrand J, Churin Y, Zähringer U, Lindner B et al (2006) Cloning of a cholesterol-alpha-glucosyltransferase from *Helicobacter pylori*. J Biol Chem 281:27765–27772. https://doi.org/10.1074/jbc.M603345200

Linz B, Balloux F, Moodley Y, Manica A, Liu H, Roumagnac P et al (2007) An African origin for the intimate association between humans and *Helicobacter pylori*. Nature 445:915–918. https://doi.org/10.1038/nature05562

Mackenzie JM, Khromykh AA, Parton RG (2007) Cholesterol manipulation by West Nile virus perturbs the cellular immune response. Cell Host Microbe 2:229–239. https://doi.org/10.1016/j.chom.2007.09.003

Malfertheiner P, Selgrad M, Wex T, Romi B, Borgogni E, Spensieri F, et al (2018) Efficacy, immunogenicity, and safety of a parenteral vaccine against *Helicobacter pylori* in healthy volunteers challenged with a Cag-positive strain: a randomised, placebo-controlled phase 1/2 study. Lancet Gastroenterol Hepatol. https://doi.org/10.1016/s2468-1253(18)30125-0

McGee DJ, George AE, Trainor EA, Horton KE, Hildebrandt E, Testerman TL (2011) Cholesterol enhances *Helicobacter pylori* resistance to antibiotics and LL-37. Antimicrob Agents Chemother 55:2897–2904. https://doi.org/10.1128/AAC.00016-11

Mejias-Luque R, Zoller J, Anderl F, Loew-Gil E, Vieth M, Adler T et al (2017) Lymphotoxin beta receptor signalling executes *Helicobacter pylori*-driven gastric inflammation in a T4SS-dependent manner. Gut 66(8):1369–1381. https://doi.org/10.1136/gutjnl-2015-310783

Meyer-Hoffert U, Hornef MW, Henriques-Normark B, Axelsson L-G, Midtvedt T, Pütsep K et al (2008) Secreted enteric antimicrobial activity localises to the mucus surface layer. Gut 57:764–771. https://doi.org/10.1136/gut.2007.141481

Mitchell DJ, Huynh HQ, Ceponis PJM, Jones NL, Sherman PM, Al MET et al (2004) *Helicobacter pylori* disrupts STAT1-mediated gamma interferon-induced signal transduction in epithelial cells. Infect Immun 72:537–545. https://doi.org/10.1128/IAI.72.1.537

Moonens K, Hamway Y, Neddermann M, Reschke M, Tegtmeyer N, Kruse T, Kammerer R, Mejías-Luque R, Singer BB, Backert S, Gerhard M, Remaut H (2018) *Helicobacter pylori* adhesin HopQ disrupts *trans* dimerization in human CEACAMs. EMBO J 37(13):e98665. https://doi.org/10.15252/embj.201798665

Morey P, Pfannkuch L, Pang E, Boccellato F, Sigal M, Imai-Matsushima A et al (2018) *Helicobacter pylori* depletes cholesterol in gastric glands to prevent interferon gamma signaling and escape the inflammatory response. Gastroenterology 154:1391–1404.e9. https://doi.org/10.1053/j.gastro.2017.12.008

Muhammad JS, Zaidi SF, Zhou Y, Sakurai H, Sugiyama T (2016) Novel epidermal growth factor receptor pathway mediates release of human β-defensin 3 from *Helicobacter pylori* -infected gastric epithelial cells. Pathogens Dis 74, ftv128. https://doi.org/10.1093/femspd/ftv128

Müller A, Merrell DS, Grimm J, Falkow S (2004) Profiling of microdissected gastric epithelial cells reveals a cell type-specific response to *Helicobacter pylori* infection. Gastroenterology 127(5):1446–1462. https://doi.org/10.1053/j.gastro.2004.08.054

Müller A, Oertli M, Arnold IC (2011) *H. pylori* exploits and manipulates innate and adaptive immune cell signaling pathways to establish persistent infection. Cell Commun Signal 9(1):25. https://doi.org/10.1186/1478-811x-9-25

Neumann L, Mueller M, Moos V, Heller F, Meyer TF, Loddenkemper C et al (2016) Mucosal inducible NO synthase-producing IgA + plasma cells in *Helicobacter pylori*-infected patients. J Immunol 197(5):1801–1808. https://doi.org/10.4049/jimmunol.1501330

Nseir W, Khateeb J, Tatour I, Haiek S, Samara M, Assy N (2010) Long-term statin therapy affects the severity of chronic gastritis. Helicobacter 15:510–515. https://doi.org/10.1111/j.1523-5378.2010.00803.x

Nuding S, Gersemann M, Hosaka Y, Konietzny S, Schaefer C, Beisner J et al (2013) Gastric antimicrobial peptides fail to eradicate *Helicobacter pylori* infection due to selective induction and resistance. PLoS ONE 8:e73867. https://doi.org/10.1371/journal.pone.0073867

Oertli M, Noben M, Engler DB, Semper RP, Reuter S, Maxeiner J et al (2013) *Helicobacter pylori* γ-glutamyl transpeptidase and vacuolating cytotoxin promote gastric persistence and immune tolerance. Proc Natl Acad Sci USA 110:3047–3052. https://doi.org/10.1073/pnas.1211248110

Ossa JC, Ho NK, Wine E, Leung N, Gray-Owen SD, Sherman PM (2013) Adherent-invasive *Escherichia coli* blocks interferon-γ-induced signal transducer and activator of transcription (STAT)-1 in human intestinal epithelial cells. Cell Microbiol 15:446–457. https://doi.org/10.1111/cmi.12048

Ostaff MJ, Stange EF, Wehkamp J (2013) Antimicrobial peptides and gut microbiota in homeostasis and pathology. EMBO Mol Med 5:1465–1483. https://doi.org/10.1002/emmm.201201773

Pachathundikandi SK, Müller A, Backert S (2016) Inflammasome activation by *Helicobacter pylori* and its implications for persistence and immunity. Curr Top Microbiol Immunol 397:117–131. https://doi.org/10.1007/978-3-319-41171-2_6

Pachathundikandi SK, Tegtmeyer N, Backert S (2013) Signal transduction of helicobacter pylori during interaction with host cell protein receptors of epithelial and immune cells. Gut Microbes 4(6):454–474. https://doi.org/10.4161/gmic.27001

Pfannkuch L, Hurwitz R, Traulsen J, Kosma P, Schmid M, Meyer T (2018) ADP heptose, a novel pathogen associated molecular pattern associated with *Helicobacter pylori* type 4 secretion. BioRxiv. https://doi.org/10.1101/405951

Posselt G, Backert S, Wessler S (2013) The functional interplay of *Helicobacter pylori* factors with gastric epithelial cells induces a multi-step process in pathogenesis. Cell Commun Signal 11:77. https://doi.org/10.1186/1478-811X-11-77

Ricci V, Galmiche A, Doye A, Necchi V, Solcia E, Boquet P (2000) High cell sensitivity to *Helicobacter pylori* VacA toxin depends on a GPI-anchored protein and is not blocked by inhibition of the clathrin-mediated pathway of endocytosis. Mol Biol Cell 11:3897–3909. https://doi.org/10.1091/mbc.11.11.3897

Salama NR, Hartung ML, Müller A (2013) Life in the human stomach: persistence strategies of the bacterial pathogen *Helicobacter pylori*. Nat Rev Microbiol, advance on. https://doi.org/10.1038/nrmicro3016

Sawai N, Kita M, Kodama T, Tanahashi T, Yamaoka Y, Tagawa Y et al (1999) Role of gamma interferon in *Helicobacter pylori*-induced gastric inflammatory responses in a mouse model. Infect Immun 67:279–285

Sayi A, Kohler E, Hitzler I, Arnold I, Schwendener R, Rehrauer H et al (2009) The CD4 + T cell-mediated IFN-gamma response to *Helicobacter* infection is essential for clearance and determines gastric cancer risk. J Immunol (Baltimore, Md.: 1950), 182:7085–101. https://doi.org/10.4049/jimmunol.0803293

Schreiber S, Bücker R, Groll C, Azevedo-Vethacke M, Garten D, Scheid P et al (2005) Rapid loss of motility of *Helicobacter pylori* in the gastric lumen in vivo. Infect Immun 73:1584–1589. https://doi.org/10.1128/IAI.73.3.1584-1589.2005

Sehgal PB, Guo GG, Shah M, Kumar V, Patel K (2002) Cytokine signaling: STATS in plasma membrane rafts. J Biol Chem 277:12067–12074. https://doi.org/10.1074/jbc.M200018200

Sen S, Roy K, Mukherjee S, Mukhopadhyay R, Roy S (2011) Restoration of IFNγR subunit assembly, IFNγ signaling and parasite clearance in *Leishmania donovani* infected macrophages: role of membrane cholesterol. PLoS Pathog 7:e1002229. https://doi.org/10.1371/journal.ppat.1002229

Shimomura H, Hosoda K, Hayashi S, Yokota K, Hirai Y (2012) Phosphatidylethanolamine of *Helicobacter pylori* functions as a steroid-binding lipid in the assimilation of free cholesterol and 3β-hydroxl steroids into the bacterial cell membrane. J Bacteriol 194:2658–2667. https://doi.org/10.1128/JB.00105-12

Shimomura H, Hosoda K, McGee DJ, Hayashi S, Yokota K, Hirai Y (2013) Detoxification of 7-dehydrocholesterol fatal to *Helicobacter pylori* is a novel role of cholesterol glucosylation. J Bacteriol 195:359–367. https://doi.org/10.1128/JB.01495-12

Simons K, Sampaio JL (2011) Membrane organization and lipid rafts. Cold Spring Harb Perspect Biol 3:a004697. https://doi.org/10.1101/cshperspect.a004697

Singh PP, Singh S (2013) Statins are associated with reduced risk of gastric cancer: a systematic review and meta-analysis. Ann Oncol 24(7):1721–1730. https://doi.org/10.1093/annonc/mdt150

Smythies LE, Waites KB, Lindsey JR, Harris PR, Ghiara P, Smith PD (2000) *Helicobacter pylori*-induced mucosal inflammation is Th1 mediated and exacerbated in IL-4, but not IFN-gamma, gene-deficient mice. J Immunol (Baltimore, Md.: 1950), 165:1022–1029

Stein SC, Faber E, Bats SH, Murillo T, Speidel Y, Coombs N et al (2017) *Helicobacter pylori* modulates host cell responses by CagT4SS-dependent translocation of an intermediate metabolite of LPS inner core heptose biosynthesis. PLoS Pathog 13:e1006514. https://doi.org/10.1371/journal.ppat.1006514

Tacconelli E, Carrara E, Savoldi A, Harbarth S, Mendelson M, Monnet DL et al (2018) Discovery, research, and development of new antibiotics: the WHO priority list of antibiotic-resistant bacteria and tuberculosis. Lancet Infect Dis 18:318–327. https://doi.org/10.1016/S1473-3099(17)30753-3

Takaoka A, Mitani Y, Suemori H, Sato M, Yokochi T, Noguchi S et al (2000) Cross talk between interferon-gamma and -alpha/beta signaling components in caveolar membrane domains. Science 288:2357–2360. https://doi.org/10.1126/science.288.5475.2357

Tegtmeyer N, Neddermann M, Asche CI, Backert S (2017a) Subversion of host kinases: a key network in cellular signaling hijacked by *Helicobacter pylori* CagA. Mol Microbiol 105(3):358–372. https://doi.org/10.1111/mmi.13707

Tegtmeyer N, Wessler S, Necchi V, Rohde M, Harrer A, Rau TT, Asche CI, Boehm M, Loessner H, Figueiredo C, Naumann M, Palmisano R, Solcia E, Ricci V, Backert S (2017b) *Helicobacter pylori* employs a unique basolateral type IV secretion mechanism for CagA delivery. Cell Host Microbe 22(4):552–560.e5. https://doi.org/10.1016/j.chom.2017.09.005

Tegtmeyer N, Harrer A, Schmitt V, Singer BB, Backert S (2019) Expression of CEACAM1 or CEACAM5 in AZ-521 cells restores the type IV secretion deficiency for translocation of CagA by *Helicobacter pylori*. Cell Microbiol 21:e12965. https://doi.org/10.1111/cmi.12965

Tran LS, Chonwerawong M, Ferrero RL (2017) Regulation and functions of inflammasome-mediated cytokines in *Helicobacter pylori* infection. Microbes Infect 19(9–10):449–458. https://doi.org/10.1016/j.micinf.2017.06.005

Vaishnava S, Yamamoto M, Severson KM, Ruhn KA, Yu X, Koren O et al (2011) The antibacterial lectin RegIIIgamma promotes the spatial segregation of microbiota and host in the intestine. Science (New York, N.Y.), 334: 255–8. https://doi.org/10.1126/science.1209791

Velin D, Favre L, Bernasconi E, Bachmann D, Pythoud C, Saiji E et al (2009) Interleukin-17 is a critical mediator of vaccine-induced reduction of *Helicobacter* infection in the mouse model. Gastroenterology 136:2237–2246.e1. https://doi.org/10.1053/J.GASTRO.2009.02.077

Velin D, Straubinger K, Gerhard M (2016) Inflammation, immunity, and vaccines for *Helicobacter pylori* infection. Helicobacter 21(Suppl 1):26–29. https://doi.org/10.1111/hel.12336

Viala J, Chaput C, Boneca IG, Cardona A, Girardin SE, Moran AP et al (2004) Nod1 responds to peptidoglycan delivered by the *Helicobacter pylori cag* pathogenicity island. Nat Immunol 5:1166–1174. https://doi.org/10.1038/ni1131

Wang HJ, Cheng WC, Cheng HH, Lai CH, Wang WC (2012) *Helicobacter pylori* cholesteryl glucosides interfere with host membrane phase and affect type IV secretion system function during infection in AGS cells. Mol Microbiol 83:67–84

Wang Y-C, Chen C-L, Sheu B-S, Yang Y-J, Tseng P-C, Hsieh C-Y et al (2014) *Helicobacter pylori* infection activates src homology-2 domain-containing phosphatase 2 to suppress IFN-γ signaling. J Immunol (Baltimore, Md.: 1950) 193:4149–58. https://doi.org/10.4049/jimmunol.1400594

Watanabe T, Asano N, Fichtner-Feigl S, Gorelick PL, Tsuji Y, Matsumoto Y et al (2010) NOD1 contributes to mouse host defense against *Helicobacter pylori* via induction of type I IFN and

activation of the ISGF3 signaling pathway. J Clin Investig 120:1645–1662. https://doi.org/10.1172/JCI39481
Wolk K, Kunz S, Witte E, Friedrich M, Asadullah K, Sabat R (2004) IL-22 increases the innate immunity of tissues. Immunity 21:241–254
Wunder C, Churin Y, Winau F, Warnecke D, Vieth M, Lindner B et al (2006) Cholesterol glucosylation promotes immune evasion by *Helicobacter pylori*. Nat Med 12:1030–1038. https://doi.org/10.1038/nm1480
Zheng Y, Valdez PA, Danilenko DM, Hu Y, Sa SM, Gong Q et al (2008) Interleukin-22 mediates early host defense against attaching and effacing bacterial pathogens. Nat Med 14(3):282–289. https://doi.org/10.1038/nm1720
Zhuang Y, Cheng P, Liu XF, Peng LS, Li BS, Wang TT et al (2015) A pro-inflammatory role for Th22 cells in *Helicobacter pylori*-associated gastritis. Gut 64(9):1368–1378. https://doi.org/10.1136/gutjnl-2014-307020
Zimmermann S, Pfannkuch L, Al-Zeer MA, Bartfeld S, Koch M, Liu J et al (2017) ALPK1- and TIFA-dependent innate immune response triggered by the *Helicobacter pylori* type IV secretion system. Cell Rep 20(10):2384–2395. https://doi.org/10.1016/j.celrep.2017.08.039

Helicobacter pylori Deregulates T and B Cell Signaling to Trigger Immune Evasion

Victor E. Reyes and Alex G. Peniche

Contents

Abstract *Helicobacter pylori* is a prevalent human pathogen that successfully establishes chronic infection, which leads to clinically significant gastric diseases including chronic gastritis, peptic ulcer disease (PUD), and gastric cancer (GC). *H. pylori* is able to produce a persistent infection due in large part to its ability to hijack the host immune response. The host adaptive immune response is activated to strategically and specifically attack pathogens and normally clears them from the infected host. Since B and T lymphocytes are central mediators of adaptive immunity, in this chapter we review their development and the fundamental mechanisms regulating their activation in order to understand how some of the normal processes are subverted by *H. pylori*. In this review, we place particular emphasis on the $CD4^{+}$ T cell responses, their subtypes, and regulatory mechanisms because of the expanding literature in this area related to *H. pylori*. T lymphocyte differentiation and function are finely orchestrated through a series of cell–cell interactions, which include immune

V. E. Reyes (✉) · A. G. Peniche
Department of Pediatrics, The University of Texas Medical Branch at Galveston, Galveston, TX, USA
e-mail: vreyes@utmb.edu

A. G. Peniche
e-mail: alpenich@utmb.edu

S. Backert (ed.), *Molecular Mechanisms of Inflammation: Induction, Resolution and Escape by Helicobacter pylori*, Current Topics in Microbiology and Immunology 421, https://doi.org/10.1007/978-3-030-15138-6_10

checkpoint receptors. Among the immune checkpoint receptor family, there are some with inhibitory properties that are exploited by tumor cells to facilitate their immune evasion. Gastric epithelial cells (GECs), which act as antigen-presenting cells (APCs) in the gastric mucosa, are induced by *H. pylori* to express immune checkpoint receptors known to sway T lymphocyte function and thus circumvent effective T effector lymphocyte responses. This chapter reviews these and other mechanisms used by *H. pylori* to interfere with host immunity in order to persist.

1 Introduction

Helicobacter pylori is a Gram-negative bacterium within the class of ε-proteobacteria, Campylobacterales order, and is a primary constituent of the human gastric microbiome. *H. pylori* is an important human pathogen that frequently infects during childhood and successfully establishes chronic infection in >66% of the world's population (www.CDC.gov). *H. pylori* is involved in significant clinical gastroduodenal disorders that include chronic gastritis, peptic ulcer disease (PUD), and two malignancies: gastric adenocarcinoma (GC) and mucosa-associated lymphoid tissue (MALT) lymphoma. GC remains as the third deadliest cancer worldwide with a five-year survival rate of 14% and accounts for approximately one million deaths (www.who.int; 2017 Fact Sheet).

Important to *H. pylori*'s capacity to establish chronic infection is its ability to evade or subvert innate and adaptive immune responses via multiple mechanisms. One of the earliest clues that *H. pylori* subverts the adaptive host response was the observation that $CD4^+$ T cell responses in the infected gastric mucosa were polarized to T helper (Th) 1 cells (Bamford et al. 1998b; Amedei et al. 2006), which are not optimal for extracellular bacteria as *H. pylori*. As we have studied in detail the mucosal immunity to *H. pylori*, we have gained insights that helped us to better understand how *H. pylori* induces a diverse T cell response that includes Th1, Th17, and T regulatory (Treg) cell responses. In this chapter, we will examine the following:

- A comprehensive background on the adaptive immune response. To better appreciate how those responses are altered during *H. pylori* infection, we will start by discussing the normal development of B and T lymphocytes and their activation processes and provide a brief description of the various $CD4^+$ T cell subsets.
- Extracellular receptor–ligand interactions and intracellular signal involvement.
- Finally, we will examine how these cells are affected by *H. pylori* infection, either directly or indirectly, by other cells affected by the infection—including the gastric epithelium. Most of the discussion will be on T cell activation, as another chapter in this book (Chapter "MALT Lymphoma as a Model of Chronic Inflammation-Induced Gastric Tumor Development") will provide a rich discussion of B cells, as they are the target in mucosal-associated lymphoid tissue (MALT) lymphoma.

2 B and T Lymphocyte Development

Lymphocytes are central players in the adaptive immune response, and, as are all other blood cells, they emerge during hematopoiesis from pluripotent hematopoietic stem cells (HSCs) that reside in bone marrow (Fig. 1). Hematopoiesis is a unidirectional process in which all immune cell types are generated from multipotent HSCs. Immune cells must be continuously replaced because of their limited life span, but also in response to infectious and inflammatory stimuli, by using receptors for cytokines and chemokines, as well as pathogen-associated molecular pattern (PAMP) recognition receptors (Chiba et al. 2018; Pachathundikandi et al. 2013). HSCs reside in the bone marrow microenvironment composed by osteoblasts, perivascular cells, endothelial cells, and immune cells, all of which promote HSCs proliferation through an array of cytokines like CXCL12 and stem cell factor (SCF). The differentiation of lymphocytes follows a tightly regulated process that initially transits through common lymphoid progenitor (CLP) cells (Kondo et al. 1997) that are $CD34^+$, $CD10^+$, $CD45RA^+$, and $CD24^-$ and are devoid of surface markers characteristic of T-, B-, or NK cells. CLP cells also contribute to the development of NK cells and subsets of dendritic cells (DCs). As B and T lymphocytes develop in the bone marrow and thymus, respectively, under the influence of local interactions and cytokines, they start to express distinctive surface markers, as detailed below for each lymphocyte population.

2.1 *B Lymphocyte*

B cells differentiate from CLPs in the bone marrow through a series of closely controlled stages, initially as progenitor B (pro-B) cells ($CD34^+$ $CD19^+$ $CD10^+$ TdT^+ $CD38^{++}$ $CD20^-$) when heavy chain immunoglobulin (Ig) V(D)J DNA rearrangement begins with a process that involves the recombination-activating genes (RAG) 1 and 2, as well as terminal deoxynucleotidyl transferase (TdT) enzymes. Once successfully rearranged, the Ig heavy chain forms a complex with surrogate light chains to give rise to a pre-B cell receptor in precursor B (or pre-B) cells ($CD34^-$ $CD19^+$ $CD10^\pm$ $CD38^+$ $cytIgM^+$ $CD20^\pm$). Signaling through this pre-B cell receptor induces light chain DNA rearrangement which induces membrane-bound Ig. Once B cells express surface IgM, they are known as immature B cells ($CD34^-$ $CD19^+$ $CD20^+$ $CD38^+$ $CD40^+$ $sIgM^+$) and subsequently express IgD, and are regarded as naïve mature B cells in the periphery. The final differentiation of B cells into Ig-secreting plasma cells occurs in the lymph nodes and other secondary lymphoid organs after activation by engagement of the surface Ig also known as (aka) B cell receptor (BCR) with antigen and the interaction of CD40 on their surface with CD154 on Th cells (Lou et al. 2015).

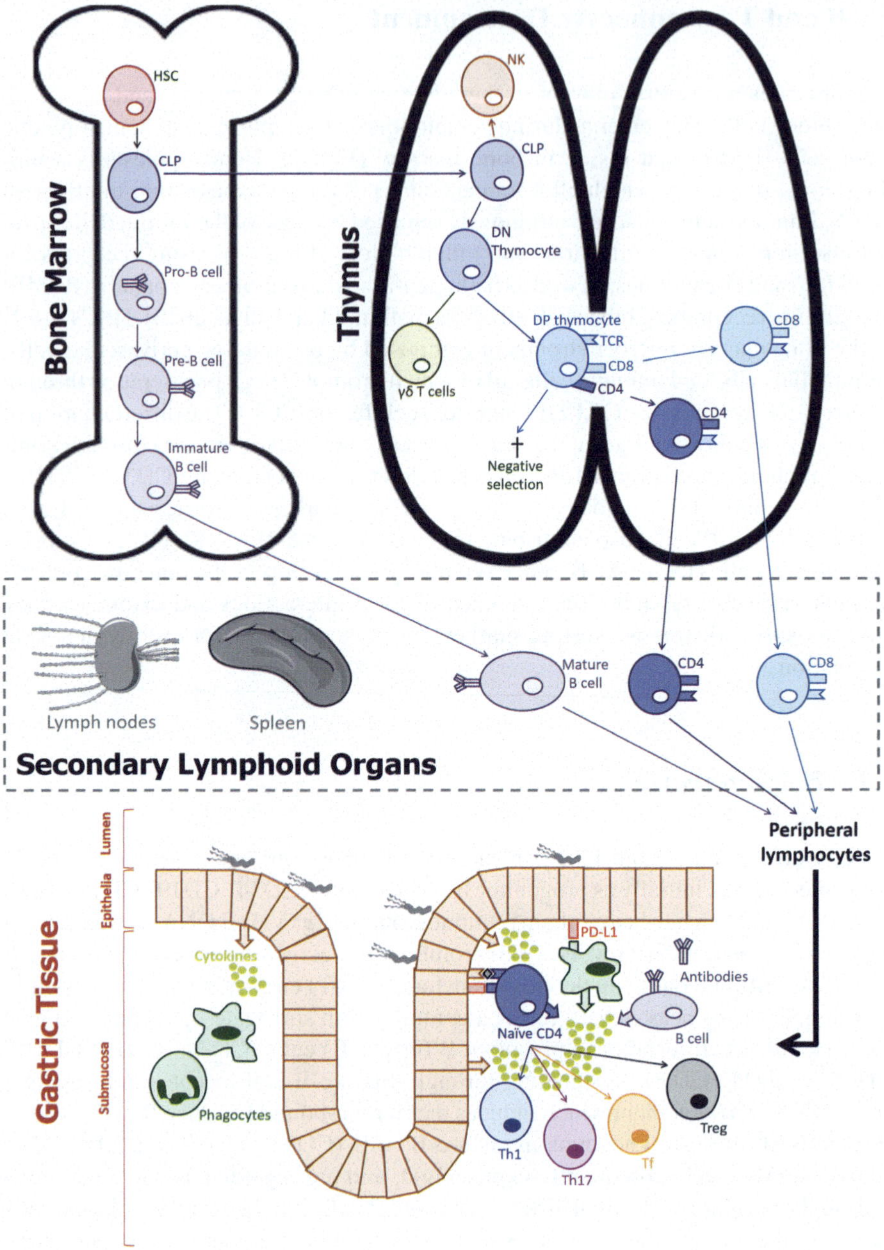
Bone Marrow
HSC
CLP
Pro-B cell
Pre-B cell
Immature B cell
Thymus
NK
CLP
DN Thymocyte
γδ T cells
DP thymocyte
TCR
CD8
CD4
Negative selection
CD8
CD4
Lymph nodes
Spleen
Mature B cell
CD4
CD8
Secondary Lymphoid Organs
Peripheral lymphocytes
Gastric Tissue
Lumen
Epithelia
Submucosa
Cytokines
Phagocytes
PD-L1
Antibodies
Naïve CD4
B cell
Th1
Th17
Tf
Treg

◂**Fig. 1** Schematic representation of lymphocyte differentiation and migration to gastric tissue. Bone marrow host hematopoietic stem cells (HSC) that progressively differentiate to rise to common lymphoid progenitors (CLP). CLPs differentiate into progenitor B cells (Pro-B cells) and double-negative (DN) thymocyte progenitors. Pro-B cells remain in bone marrow and differentiate into immature B cells that turn into mature B cells once they migrate to secondary lymphoid organs (i.e., lymph nodes and spleen). The CLP that migrate to the thymus commit to either natural killer (NK) cells or T lymphocyte lineage becoming γδ T lymphocytes or double-negative DN thymocytes. DN thymocytes undergo negative selection and only immature single positive cells survive to become $CD4^+$ or $CD8^+$ T lymphocytes capable of migrating to secondary lymphoid organs. Lymphoid cells are eventually recruited to gastric infected tissue where they become antibody-producing cells (B cells, plasmocytes) and $CD4^+$ T lymphocytes differentiate into subsets depending on environmental cues

2.2 T Lymphocyte

As with B lymphocytes, T cells have their origin in the bone marrow and share the CLP precursor, but their development occurs within the thymus following migration of CLP cells to this organ. Thymocyte precursors ($CD4^-$ $CD8^-CD7^+$ $CD45^+$) interact with stromal cells in the thymic cortex, where most thymocytes begin to rearrange their T cell receptor (TCR) β chain loci. After the β chain locus is productively rearranged and the corresponding protein expressed, this protein forms a complex with a surrogate α chain (pre-T α) and creates a complex with CD3 (von Boehmer 2005). When this complex is formed, the cells differentiate into double positive (DP, $CD4^+$ $CD8^+$) thymocytes and rearrange their α chain loci to eventually express TCR αβ on their surface. In addition to the α and β chain loci, there are γ and δ TCR loci, but only about 3–10% of thymocytes rearrange their γδ TCR loci (Weiss et al. 1986). Once thymocytes express their TCR, they undergo positive selection in the thymic cortex where the cells that recognize antigen with the corresponding class I or II human leukocyte antigen (HLA aka major histocompatibility complex, MHC) molecules with appropriate affinity survive, while those that fail to recognize antigen die by apoptosis. The surviving cells migrate to the thymic medulla where they experience another selection step. In the medulla, thymocytes interact with antigen-presenting cells (APCs: DCs and macrophages) which present self-antigens bound by (HLA) molecules and those thymocytes with a very strong affinity die by apoptosis, while those that survive downregulate either their CD4 or CD8 co-receptor to become single positive T cells (for a review see Takaba and Takayanagi 2017). The cells that emerge from the thymus into the periphery are naïve T cells that will differentiate further into distinct subsets following activation, as described below.

Although the role of the thymus in T cell differentiation, maturation, and expansion has long been recognized, extrathymic T cell differentiation and maturation have been reported in mice and humans (Lefrancois and Puddington 1995; Bandeira et al. 1991; Howie et al. 1998). Sites that have been shown to support extrathymic T cell differentiation include the gut and tonsils (Howie et al. 1998; McClory et al. 2012). It is important to bear in mind that the gut harbors the largest number of T cells in the body, where they are exposed to the largest possible antigenic challenge that includes dietary antigens and the gut microbiome. Interestingly, the gastrointestinal tract also holds unconventional populations of T cells such as intraepithelial lymphocytes (IEL), which represent an effector T cell population reported to develop extrathymically (Fichtelius 1967), as supported by their presence in athymic mice (Bandeira et al. 1991). The gastric epithelium has 5–8 IEL cells per 100 epithelial cells, and these numbers increase up to sixfold during disease states, such as gastritis (Feeley et al. 1998; Hayat et al. 1999). These cells express a CD8αα homodimer, rather than the conventional CD8αβ heterodimer expressed by peripheral T cells selected in the thymus (Ruscher et al. 2017).

After their selection in the thymus, T lymphocytes enter the circulation and travel to secondary lymphoid organs. Migration of lymphocytes to those secondary lymphoid organs hinge on their surface expression of L-selectin (CD62L), the integrin leukocyte function antigen-1 (LFA-1, αLβ2), and the CC chemokine receptor (CCR)7 (von Andrian and Mackay 2000), which permit rolling, adhesion, and extravasation of T cells through high endothelial venules in secondary lymphoid organs (lymph nodes and mucosal lymphoid organs). In those secondary lymphoid organs, they may become activated by APCs. Activated T cells expand and become either effector or memory T cells. Newly activated T cells may migrate to other tissues and specific adhesion molecules, and chemokine receptors enable them to home and bind the corresponding ligands in those tissues. For instance, T cells that migrate to the gastrointestinal mucosa require the integrin α4β7, LFA-1, and CCR9 (Michetti et al. 2000; Quiding-Jarbrink et al. 2001; Berlin et al. 1993; Zabel et al. 1999; Johansson-Lindbom et al. 2003). An important factor that determines what adhesion molecules are expressed by activated T cells is the site where they encounter antigen (Stagg et al. 2002). For instance, α4β7 expression by gastric and intestinal T cells allows them to home and bind to mucosal addressin cell adhesion molecule 1 (MAdCAM-1) expressed by high endothelial venules in the Peyer's patches and gut lamina propria (Michetti et al. 2000; Williams and Butcher 1997; Hatanaka et al. 2002).

3 B and T Lymphocyte Activation

B and T lymphocytes perform a daunting mission of recognizing from a vast universe of antigens those that are foreign to us and respond to them rapidly and specifically in spite of a very noisy background of self-antigens. The events that lead to their activation are carefully orchestrated and involve a series of extracellular signals provided via cell–cell interactions and cytokines that in turn activate intracellular signals leading to activated B and T lymphocytes. Because the events that lead to fully functional B and T lymphocytes are critical in adaptive immune responses, we will review those events below with a particular emphasis on T lymphocytes, since B lymphocytes are discussed in more detail in Chapter "MALT Lymphoma as a Model of Chronic Inflammation-Induced Gastric Tumor Development" of this book.

3.1 B Lymphocyte

Naive B lymphocytes have approximately 1.5×10^5 membrane-bound antibodies (IgM and/or IgD) that serve as B cell receptors (BCRs) to bind soluble antigens (Maddaly et al. 2010). Activation requires cross-linking of multiple monomer membrane-bound antibodies (Harwood and Batista 2010). The activation of B cells varies depending on the type of antigen and interaction with T cells. Some antigens do not require contact with T helper cells and are thus referred to as T-independent antigens. An example of these antigens is bacterial lipopolysaccharides (LPS), which at high concentrations may activate mature and immature B cells. However, the characteristic response to these antigens is "weak" in terms of antibody production and memory response and frequently results only in IgM secretion. The lack of co-stimulation (CD40L) is thought to be the reason why these antigens fail to induce class switching and increased antibody affinity (Maddaly et al. 2010). Other antigens require interaction of co-stimulatory receptors and cytokines from Th cells with B cells (acting as APCs). The co-stimulation between these cells typically occurs in secondary lymphoid organs. The binding of antigen by B cells leads to clustering of membrane-bound antibodies, and their subsequent dimerization and internalization into endosomal vesicles. Then, those B cells present peptide-laden HLA class II complexes to T cell receptors (TCRs) on antigen-specific T cells. This interaction promotes expression by B cells of the co-stimulatory molecules B7-1 (CD80) and B7-2 (CD86) which facilitate differentiation of Th cells. Activation of T cells leads to their expression CD40L which interact with CD40 on the B cells to promote their entry into the S phase. In addition, cytokines such as IL-2 and IFN-γ (Th1), and IL-4, IL-5, IL-6, IL-10, IL-13 (Th2) promote clonal expansion, antibody production, and isotype switching (from IgM to IgG) followed by differentiation into plasma cells and memory B cells (Harwood and Batista 2010).

3.2 T Lymphocyte

3.2.1 Antigen Presentation

Presentation of foreign antigens refers to the display of antigens to T cells by antigen-presenting molecules [human leukocyte antigen (HLA) class I, HLA class II or CD1] after those antigens have been appropriately processed by APCs. Antigen processing and presentation provide the host with a mechanism to constantly survey the cellular internal and external environments for the presence of potential pathogens. There are four possible pathways involved in the processing of protein antigens for presentation by either class I or class II HLA molecules. Classical antigen processing pathways include the exogenous and endogenous pathways, but autophagy and cross-presentation have expanded the possible pathways whereby antigens are processed. The location of the antigens or, in the case of replicating pathogens, the life cycle of a given pathogen determines which pathway is needed for appropriate presentation to the appropriate T cell type ($CD4^+$ versus $CD8^+$). In the case of pathogens that replicate within the cell, and whose antigens are thus synthesized endogenously, they are degraded in the cytosol into small peptides, 8–10 amino acids long, by the proteasome complex and are delivered to the lumen of the endoplasmic reticulum (ER), where nascent HLA class I molecules bind them for eventual presentation to cytotoxic $CD8^+$ T cells. In contrast, pathogens such as *H. pylori,* that replicate in the extracellular milieu, or are exogenous to the APCs, have to be endocytosed and their protein antigens processed by thiol proteases in endocytic compartments to generate peptides that will bind to HLA class II molecules for presentation to $CD4^+$ T cells (for a review, see Blum et al. 2013). More recently, autophagy and cross-presentation have been described as alternative pathways that break away from the classical pathways since autophagy captures endogenously produced antigens and delivers them to endocytic compartments where exogenous antigens are processed. Recent studies have reported that highly virulent strains of *H. pylori* noticeably affect autophagy in host GECs and macrophages (Castano-Rodriguez et al. 2015). On the other hand, cross-presentation results from the delivery of exogenously acquired antigens into the cytosol where they are processed by the proteasome and the resulting peptides are delivered to the ER lumen where they bind newly formed HLA class I (Van Kaer et al. 2017; Joffre et al. 2012).

CD1 molecules represent another group of relatively non-polymorphic antigen-presenting proteins whose genes are not present within the MHC region. In fact, CD1 are encoded in an entirely different chromosome. While human HLA genes are encoded in chromosome 6, human CD1 genes are encoded in chromosome 1. There are four human CD1 proteins (CD1a to CD1d) that also associate with β2-microglobulin. These molecules are expressed by classical APCs, and CD1d is strongly expressed by GECs. Although CD1 molecules also present antigens and their crystal structure resembles that of class I HLA molecules (Blumberg et al. 1995), they differ from class I and II HLA molecules in that they do not bind peptide antigens. Instead, CD1 molecules bind and present lipids, because their antigen binding pocket has a narrow opening, is deep, and is lined by hydrophobic residues (Ly and Moody 2014). CD1 molecules may present lipid antigens to a diverse group of T cells that include γδ TCR or αβ TCR expressing T cells, as well as invariant NK T cells (iNKT) (Adams 2014). A study by Ito et al. (2013) showed that *H. pylori* cholesteryl α-glucosides are recognized by iNKT in the stomach, which contributes to the inflammatory response that limits *H. pylori* infection (see also Chapter "The Sweeping Role of Cholesterol Depletion in the Persistence of *Helicobacter Pylori* Infections" of this book).

3.2.2 Antigen-Presenting Cells

T cells are activated by APCs able to internalize foreign antigens and process them for presentation to the T cells. Because of their role in T cell activation, APCs are crucial in orchestrating the adaptive immune response. While most nucleated cells express

HLA class I molecules, the cells that are classically referred to as professional APCs are those that express HLA class II and include DCs, macrophages, and B cells. In addition to expressing class II HLA, another important feature of these cells is their expression of the co-stimulatory molecules CD80 and CD86, whose engagement of CD28 on T cells is vital for activation of naïve T cells. Interestingly, in the gastric environment, GECs represent a non-classical APC-type, as they constitutively express class II HLA, CD80, CD86, CD74, the antigen processing cathepsins, and newer members of the B7 family, as described below (Ye et al. 1997; Fan et al. 1998, 2000; Barrera et al. 2001, 2002, 2005; Beswick et al. 2004, 2007a; Das et al. 2006). The expression by GECs of class II HLA, CD80, CD86, and CD74 increases during infection with *H. pylori* (Ye et al. 1997; Fan et al. 1998, 2000; Beswick et al. 2004, 2005). Furthermore, a recent study showed that GECs express retinoic acid, which is responsible in the induction of $\alpha_4\beta_7$ integrin and the CCR9 chemokine receptor on both $CD4^+$ and $CD8^+$ T cells, which in turn facilitates their homing to the gastrointestinal mucosa (Bimczok et al. 2015). It is worth noting that retinoic acid also influences the homing to the gastrointestinal mucosa of IgA-secreting B cells (Mora and von Andrian 2009).

3.2.3 T Cell Receptor Signaling

The recognition by the TCR of antigen-laden MHC molecules on the surface of APCs leads to the formation of an immunological synapse between both cell types (Huppa et al. 2003), but this interaction alone is insufficient to lead to T cell activation since the short cytoplasmic tail of TCRs does not allow them to deliver intracellular signals. TCR interacts closely with a complex of other membrane proteins on T cells, that are collectively referred to as CD3 (including γ-, δ-, ε-, and ζ-subunits). After TCR engagement of peptide-laden MHC molecules, the cytoplasmic domains of CD3 subunits are responsible for delivering intracellular signals. Further, CD4 and CD8 bind to conserved membrane proximal domains on the β2-domain of MHC class II (Cammarota et al. 1992) and α3 of class I MHC molecules (Devine et al. 1999), respectively. The cytoplasmic domains of CD4 and CD8 bind the Src family kinase LCK (lymphocyte-specific protein tyrosine kinase), which in turn phosphorylates the immunoreceptor tyrosine-based activation motifs (ITAMs) within the cytoplasmic domains of CD3 subunits (Love and Hayes 2010). Phosphorylation of CD3 subunits directs the recruitment of zeta-chain-associated protein kinase of 70 kDa (ZAP70). After ZAP70 is activated, it phosphorylates the linker for activation of T cells (LAT) and Src homology 2 domain-containing 76 kDa leukocyte protein (SLP76). A series of signaling proteins are recruited, leading to calcium mobilization, actin cytoskeleton reorganization, and activation of Ras guanosine triphosphate hydrolases (GTPases). As a consequence of these signaling processes, various transcription factors are activated, including nuclear factor-κB (NF-κB), activator protein 1 (AP-1), and nuclear factor of activated T cells (NFAT), which aid in directing T cell responses.

3.2.4 Co-stimulation/Co-inhibition

In addition to the signals delivered by the CD3 complex after TCR recognition of antigen, T cells must receive co-stimulation via engagement of CD28 on their surface with CD80 or CD86, also, respectively, known as B7-1 and B7-2, on the surface of APCs. Engagement of CD28 on T cells is essential for T cell activation since in the absence of the signals delivered via CD28 after binding its ligand on APCs T cells become anergic, as shown in experiments with anti-CD28 blocking antibodies (Harding et al. 1992). The intracellular signals delivered by CD28 prevent this anergic state, and they include the Tec family kinases ITK/EMT, Rlk, and Itk, as well as phosphatidylinositol 3-kinase (PI3K) (August et al. 1994; Schaeffer et al. 1999; Pages et al. 1994). The signals delivered via CD28 affect crucial events in T cells, such as transcriptional signaling, post-translational protein modifications, cytokine synthesis, and epigenetic changes that ultimately affect their phenotype and function. The ligands for CD28, CD80, and CD86 vary in their expression pattern. CD86 is constitutively expressed on APCs and is upregulated quickly during immune responses, whereas CD80 is slower in its upregulation (Lenschow et al. 1994). Both of these receptors are expressed by GECs and are upregulated during *H. pylori* infection (Ye et al. 1997). The studies by Ye et al. showed that CD86 expression was higher on GECs from *H. pylori*-infected gastric biopsy tissues compared with those from uninfected subjects (Ye et al. 1997). Another member of this family of receptors and ligands is the cytotoxic T-lymphocyte-associated antigen 4 (CTLA-4) which is expressed on activated T cells and acts as an immune checkpoint inhibitor. CTLA-4 competes for the same receptors, CD80 and CD86, and binds with higher affinity to antagonize CD28, and thus acts to provide inhibitory signals (Walunas et al. 1994). Furthermore, CTLA-4 is a key mediator of Treg function (Friedline et al. 2009). The next members of this family of immunoregulatory receptors identified, that are not constitutively expressed on resting T cells, but are induced following activation, include inducible co-stimulator (ICOS, CD278) and programmed death-1 receptor (PD-1, CD279), which provide co-stimulatory or co-inhibitory signals, respectively. The corresponding co-receptor for ICOS is ICOS-L (aka B7-H2, CD275), while PD-1 may bind two separate co-receptors: programmed death ligand-1 (PD-L1) (aka B7-H1, CD274) and PD-L2 (aka B7-DC, CD273) (Fig. 2). Interestingly, PD-L1 may also bind CD80 to deliver inhibitory signals (Park et al. 2010). PD-1 binding to any of its co-receptors results in dephosphorylation and inactivation of ZAP70 and the recruitment of Src homology 2 domain-containing tyrosine phosphatase 2 (SHP2) (Yokosuka et al. 2012), which in turn causes dephosphorylation of PI3K leading to activation of Akt (Boussiotis et al. 2014). Ligation of PD-1 may also prevent extracellular-signal-regulated kinase (ERK) activation, which may be rescued via signaling activated by exogenous IL-2, IL-7, and IL-15 (Bennett et al. 2003). The engagement of PD-1 on T cells also inhibits their cell-cycle progression and proliferation via suppression of cell-cycle regulatory genes. Additional data collected on the functions of the PD-1/PD-L1 and PD-L2 axis suggest that the role of these receptors on T cell biology extends beyond suppression of effector T cells. Studies by Allison and colleagues highlighted that not only the expression of PD-L1 and

PD-L2 on APCs is differentially upregulated, but also PD-L1 and PD-L2 may have different roles affecting Th1 and Th2 responses (Loke and Allison 2003). The interaction of PD-1 with PD-L1 may also reprogram human Th1 cells into Treg cells (Amarnath et al. 2011), and this interaction may also affect diverse CD4$^+$ T cell subsets differently (McAlees et al. 2015).

The B7 family of proteins with either co-stimulatory or co-inhibitory properties has expanded in recent years and is now collectively referred to as "immune checkpoint regulators" (Ceeraz et al. 2013), which now include ten reported members (Xu et al. 2016). This family of receptors currently includes B7-1, B7-2, B7-DC, PD-L1, ICOS-L, B7-H3, B7-H4, B7-H5, B7-H6, and B7-H7 (Fig. 2). Various members of this family of receptors are overexpressed by various forms of cancer, including GC, possibly as a mechanism of evasion of tumor immune surveillance (Cimino-Mathews et al. 2016; Chen et al. 2015; Hou et al. 2014). These observations together with their known T cell regulatory activity made these proteins attractive as targets for oncologic immunotherapy with some successes (La-Beck et al. 2015).

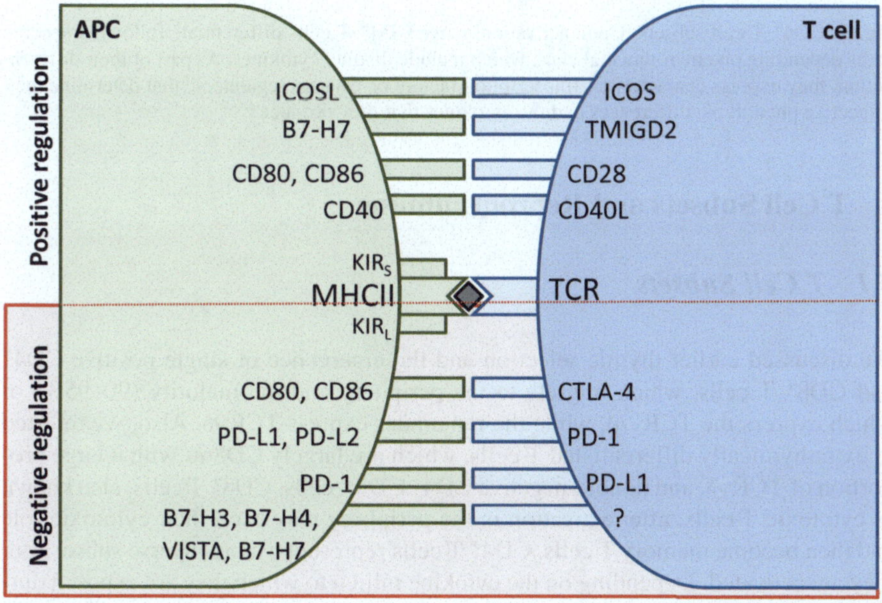

Fig. 2 Co-stimulatory and co-inhibitory receptors and their ligands. These molecules are also known as members of the B7-CD28 superfamily or immune checkpoint regulators because they affect T cell activity

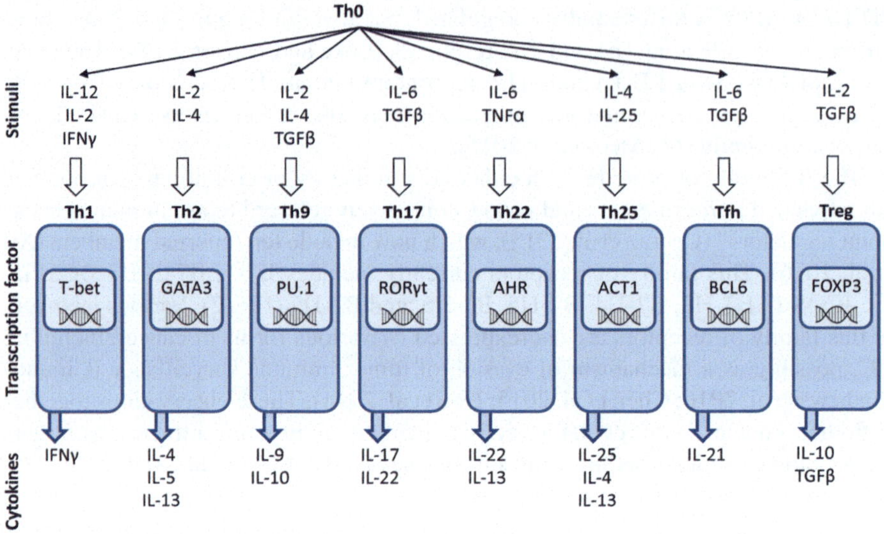

Fig. 3 CD4⁺ T cell subsets. Upon activation, naïve CD4⁺ T cells differentiate following specific paths depending on environmental cues, which include distinct cytokines. As part of their differentiation, they express characteristic transcription factors or "master regulators" that determine their respective phenotypic differences and the cytokines that they produce

4 T Cell Subsets and Reprogramming

4.1 T Cell Subsets

We discussed earlier thymic selection and the emergence of single positive $CD4^+$ and $CD8^+$ T cells, which migrate to the periphery and the majority (90–95%) of which express the TCR αβ, while the remainder express TCRγδ. Also, we referred to extrathymically differentiated T cells, which are largely CD8αα with a large proportion of TCRγδ, and double-negative $CD4^-CD8^-$ cells. $CD8^+$ T cells, also known as cytotoxic T cells, after activation in the periphery may exert their cytotoxic role and then become memory T cells. $CD4^+$ T cells represent a more diverse subset after they are activated. Depending on the cytokine milieu to which they are exposed during their activation by APCs, $CD4^+$ T cells are programmed into distinct subsets with the expression of characteristic transcription factors and cytokine profiles, which in turn allow them to exert distinct functions. Currently, the $CD4^+$ T cell subsets that have been defined include Th1, Th2, Th3, Th9, Th17, Th22, Th25, follicle helper T cells (Tfh), and Treg (Fig. 3). Interestingly, the literature on the immune response to *H. pylori* has been inclusive of most of these subsets.

4.1.1 Th1 Cells

The first subsets of T lymphocytes studied in the context of the host response to *H. pylori* were Th1 and Th2 (Karttunen et al. 1990; Bamford et al. 1998b). Each CD4$^+$ T cell subset is characterized by the expression of a specialized cytokine gene under regulation by subset-defining transcription factors. Th1 is induced to differentiate by IL-12 from APCs (Hsieh et al. 1993), which induce signal transducers and the activator of transcription 4 (STAT4) or STAT1. These STATs lead to the expression of the transcription factor T-bet, regarded as the master regulator of Th1 cells (Szabo et al. 2000), and synthesis of IFN-γ, although neither of them is unique to Th1 cells. T-bet activates the *ifn-γ* gene by binding directly to its promoter (Jenner et al. 2009) and silences *il4* gene expression (Djuretic et al. 2007). It is widely accepted that the role of Th1 cells is to foster cell-mediated immunity against intracellular pathogens.

4.1.2 Th2 and Th25 Cells

Th2 cells are induced to differentiate in the presence of IL-4, which induces STAT6 phosphorylation. Phospho-STAT6 promotes the expression of the transcription factor GATA3, which in turn leads Th2 cells to produce IL-4, IL-5, and IL-13 (Scheinman and Avni 2009). GATA3 directly represses the *ifn-γ* gene (Chang and Aune 2007; Djuretic et al. 2007). Further proof of the importance of GATA3 in Th2 cell development was obtained in studies in which GATA3 was deleted from T cells and those cells failed to differentiate into the Th2 lineage, while its overexpression in Th1 cells caused them to reprogram into Th2 cells (Pai et al. 2004; Zhang et al. 1997). Th2 cells are central in humoral immunity and host responses to helminth infections; however, they are chief contributors to the pathogenesis of allergic inflammatory diseases (Nakayama et al. 2017). The literature suggests the existence of a novel IL-25-producing T cell subset designated as Th25 cells, which seem to be closely related to the Th2 cell lineage (Swaidani et al. 2011), as both cell types need IL-4 for cytokine production and IL-25 (also known as IL-17E) enhances cytokine production (Fort et al. 2001). These cells are regulated by the transcription factor Act1 and were shown to induce non-lymphoid cells to synthesize Th2 cytokines during infection with helminths (Swaidani et al. 2011; Fallon et al. 2006), and possibly to extracellular pathogens, in general, as suggested by a recent study (de Sousa and Quaresma 2018). Fallon et al. (2006) reported that *il25*$^{-/-}$ mice were inefficient at eliminating the gastrointestinal nematode *Nippostrongylus brasiliensis*. To date, there are no studies demonstrating the involvement of Th25 cells in *H. pylori* gastric inflammation, as might be expected given their linkage to Th2 cell lineage and recent emergence of Th25 cells.

4.1.3 Th17

Th17 was initially described as a distinct Th subset in the last decade (Harrington et al. 2005; Park et al. 2005), and this lineage of Th cells has the retinoic acid receptor-related orphan receptor-γt (RORγt) as their master regulator (Ivanov et al. 2006). Their differentiation involves either IL-1β (Sutton et al. 2006; Pachathundikandi et al. 2016), IL-23 (Harrington et al. 2005), or the combination of IL-6 and TGF-β (Mangan et al. 2006). Th17 cells acquired their designation because of their ability to synthesize IL-17, both IL-17A and IL-17F (Harrington et al. 2005). IL-17 is a pro-inflammatory cytokine which acts both on hematopoietic and non-hematopoietic cells and induces antibacterial peptides, pro-inflammatory cytokines, chemokines, and prostaglandins. Among the chemokines induced by IL-17 are CXCL1, CXCL2, CXCL5, and IL-8, which promote neutrophil recruitment (Laan et al. 1999; Delyria et al. 2009), as well as CCL20, a chemokine important in cell recruitment to mucosal surfaces (Acosta-Rodriguez et al. 2007). These cells are linked to inflammation and autoimmunity (Langrish et al. 2005), as well as immunity to extracellular microbes, such as *H. pylori*, and their importance in immunity to mucosal pathogens has been highlighted in published studies (Khader et al. 2009). The differentiation of Th17 cells is inhibited by IL-27 (Hirahara et al. 2012), which also promotes Th1 cell differentiation (Yoshida et al. 2001).

4.1.4 Th22 and Th9 Cells

Th22 and Th9 cells are recently characterized $CD4^+$ Th subsets. Akin to Th17, which was so designated because of their production of IL-17, Th22 produces IL-22, whose expression was previously linked to Th17 cells, but now it is accepted that Th17 (Liang et al. 2006; Kreymborg et al. 2007) and NK cells (Cupedo et al. 2009; Crellin et al. 2010), in addition to Th22, produce IL-22, although the latter secrete the highest levels. In contrast to Th17 cells, which produce both cytokines, Th22 cells do not secrete IL-17A (Eyerich et al. 2009). Though they have been found within infected tissues and multiple inflammatory states, their role in immunity has not been well characterized due to the difficulty in culturing them in vitro, but that may change soon after a recent report described their generation in vitro in the absence of Th17 cells (Plank et al. 2017). IL-22 aids in the control of mucosal infections through the induction of inflammatory mediators and antimicrobial peptides (Rutz et al. 2013). A recent report correlated IL-22-induced antimicrobial peptides with vaccine-induced protection against *H. pylori* in mice (Moyat et al. 2017). As noted above, Th9 cells also represent a recently described subset of effector T cells whose differentiation from naïve T cells depends on transforming TGF-β and IL-4 (Dardalhon et al. 2008). This subset of effector T cells has a complex requirement of different transcription factors that include STAT6, PU.1, IRF4, and GATA3 (Chang et al. 2010; Staudt et al. 2010; Goswami et al. 2012). While their function in vivo is not clearly outlined, the available data suggest their involvement in atopy, as IL-9 promotes mast cell growth and induces changes in mast cell gene expression (Brough et al. 2014; Kearley et al.

2011). Though a proteomic study showed that IL-9 was elevated in the mucosa of *H. pylori*-positive GC samples (Ellmark et al. 2006), the role of Th9 in protection against *H. pylori* or associated pathogenesis is not clear. However, it has been suggested that IL-9 could be limiting the pro-inflammatory activity of Th17 cells since IL-9-deficient Th17 cells induce more severe autoimmune gastritis (Stephens et al. 2011). Interestingly, Th9 cells have recently been implicated in inflammatory bowel disease—more specifically in ulcerative colitis (Gerlach et al. 2015).

4.1.5 Tfh Cells

Tfh cells are a $CD4^+$ subset specialized in providing B cell help while sustaining enduring antibody responses in germinal centers of secondary lymphoid organs. Tfh cells are distinct from other $CD4^+$ T cell subsets by the expression of their hallmark CXCR5, and the transcription factor essential for their differentiation, B cell lymphoma-6 (BCL-6). Newly activated $CD4^+$ T cells when exposed to IL-6 are induced to differentiate into Tfh by signaling through the IL-6 receptor (IL-6R/gp130), which elicits Bcl6 expression (Nurieva et al. 2009). In turn, Bcl6 elicits early CXCR5 expression and the Tfh migrates to the B cell follicle border (Choi et al. 2011). Initially, these Tfh cells are induced by DCs and macrophages, but eventually the main APCs that they encounter are antigen-specific B cells in the follicle, interfollicular zone, and the T-B border. Their interaction with B cells is significant since B cells express ICOS-L, which provides co-stimulatory signals via ICOS on Tfh cells, which are essential for their complete differentiation (Choi et al. 2011). These T cells are important in immunity against infectious agents as antibody responses are critical in immune responses to most pathogens.

4.1.6 Treg Cells

Treg cells are $CD4^+$ T cells characterized by a high surface expression of CD25 (IL-2 receptor α chain), in addition to the expression of the transcription factor forkhead box P3 (FoxP3) (Hori et al. 2003). They represent about 5–15% of all $CD4^+$ T cells in the body, and there are two populations of Treg cells, which develop in different sites. Natural Treg (nTreg) cells undergo thymic maturation while induced Treg (iTreg) cells mature post-thymically (Rodriguez-Perea et al. 2016). The latter population, iTreg, is represented by two subsets that include Tr1, which lack FoxP3 and secrete IL-10 (Vieira et al. 2004), and Th3 that are $FoxP3^+$ and secrete TGF-β (Weiner 2001). Their foremost function is to suppress immunity by limiting extent and intensity of an immune response, and to maintain peripheral self-tolerance. This became evident by an experiment of nature in which humans with dysfunctional FoxP3 develop a condition known as immunodysregulation polyendocrinopathy enteropathy X-linked (IPEX) syndrome. This syndrome is characterized by a series of autoimmune disorders in various parts of the body such as the intestines, skin, and endocrine glands. Treg cells inhibit effector T cells through cell–cell contact and through the cytokines that they produce, which include IL-10, TGF-β and IL-35 (Rodriguez-Perea et al. 2016; Jafarzadeh et al. 2015). The gastrointestinal mucosa is a body site where Treg cells are found at a high frequency as they aid in maintaining immune tolerance and are important in preventing intestinal inflammation (Izcue et al. 2009).

5 Reprogramming and Plasticity

Since the original definition of Th1 or Th2 effector cells was based on their cytokine production profiles, Th effector subsets were regarded as being terminally differentiated following a linear and unalterable process—each subset with a distinctive cytokine profile. However, studies in vitro initially suggested that those Th1 and Th2 cells could be induced to produce cytokines characteristic of the other subset when cultured under conditions that would promote the opposite subset. For instance, Th1 cells secreted IL-4 when they were cultured under Th2 culture conditions (Zhu et al. 2004). Similar observations were made with Th17 and Treg cells. Treg were reported to produce IL-17 after culture with IL-6, and they also upregulated RORγt expression (Yang et al. 2008). Furthermore, Treg cells may self-induce into IL-17-producing cells in the presence of IL-6 if TGF-β is absent (Xu et al. 2007) and Th17 may revert in vivo and in vitro into Th1 cells, as demonstrated by various independent groups (Martin-Orozco et al. 2009; Bending et al. 2009). These and similar observations uncovered the ability of "differentiated" T cells to convert to another phenotype leading to the hypothesis that T cells have phenotypic plasticity that is influenced by environmental cues. Thus, this ability of $CD4^+$ T cells to become reprogrammed and acquire features of other T cell subsets is now referred to as T cell plasticity. $CD4^+$ T cell plasticity can be modulated by a combination of extracellular and intracellular signals (DuPage and Bluestone 2016). The extracellular cues that may influence plasticity of $CD4^+$ T cells include the available cytokine milieu with the signaling that is activated, intensity of the TCR signaling, and signals activated by co-stimulator/co-inhibitor receptors. Plasticity may also be influenced intracellularly by signaling cascades, cell metabolism, and transcription factor (i.e., master regulators: FoxP3, RORγ, etc.) regulation. During infection with *H. pylori*, most of these extracellular and intracellular regulators are altered or used by the bacterium. For instance, our studies showed that infection with *H. pylori* results in the induction of the co-stimulatory molecules CD80, CD86, as well as the co-inhibitory receptor PD-L1 on GECs (Beswick et al. 2007b; Das et al. 2006; Ye et al. 1997), which not only influence the activation of T cells, but also promote their reprogramming (Beswick et al. 2007a). TGF-β, which stimulates both Th17 and Treg cells, is also produced by *H. pylori*-infected GECs, in a response that is dependent on the virulence genes *vacA* and *cagA* (Beswick et al. 2011).

6 *H. pylori* Induction and Evasion of the Host Immune Response

Pathogens that establish infections for life possess characteristics in their interactions with the human host that permit prolonged colonization periods, even in the presence of immune responses. In the case of *H. pylori,* the bacteria are adapted to colonize a distinctive niche that is hostile to most other microorganisms. Its ability to establish persistent infection with associated chronic inflammation predisposes the host to develop clinically significant gastric diseases, such as PUD and GC. The inflammatory response reflects the induction of host immunity, but *H. pylori* has an arsenal of mechanisms that enable successful evasion of innate and adaptive immunity in order to persist within the human gastric mucosa. Because the adaptive immune response is highly specific and is responsible for lasting immunity, we will focus the discussion on the adaptive immune response with an emphasis on how *H. pylori* subverts lymphocytes. Since T cells are activated by their interactions with APCs, to better understand how T cell responses are affected, we will also discuss the influence of *H. pylori* on classical APCs and the epithelium in their interactions with T cells.

Macrophages, DCs, B cells, and GECs are influenced during infection by *H. pylori,* and, in turn, they contribute to the mucosal response that takes place. Although *H. pylori* is not invasive, the bacterium and its products come in contact with cells in the lamina propria. Thus, the infected gastric mucosa has a significant influx of immune cells that include macrophages, DCs, and lymphocytes. Macrophages are recruited to the *H. pylori*-infected gastric mucosa and aid in the production of pro-inflammatory cytokines and chemokines (Dzierzanowska-Fangrat et al. 2008). Depending on how macrophages are activated they are functionally polarized as either M1 (classically activated by IFNγ and bacterial products and are pro-inflammatory), or M2 (alternatively activated by IL-4, IL-10, or IL-13 and are associated with wound healing and tissue repair) (Murray 2017). Analysis of gastric macrophages in *H. pylori*-infected mice showed that they were polarized to M1, and in humans, they showed a mixed M1/M2 phenotype, but in atrophic gastritis macrophages, they were also M1 (Quiding-Jarbrink et al. 2010). However, the work by Wilson's group has shown that macrophages in *H. pylori*-infected mice show activation of the arginase/ornithine decarboxylase pathway (Lewis et al. 2011), which is a feature associated with M2 macrophages. Further, another group reported the presence of $CD68^{+}CD163^{+}Stabilin\text{-}1^{+}$ (M2) macrophages in the lamina propria of *H. pylori*-infected patients (Fehlings et al. 2012). Wilson's group also showed that *H. pylori* induces the heme oxygenase-1 (HO-1) gene in macrophages. HO-1 is an anti-inflammatory and antioxidant enzyme. This response was elicited by phosphorylated CagA and signaling that involves the activation of p38 and NF (erythroid-derived 2)-like 2 (NRF2) (Gobert et al. 2014). The activation of HO-1 in *H. pylori*-infected macrophages fosters a switch to regulatory macrophages able to dampen immune responses. Macrophages are also considered as key promoters in the differentiation of Th17 cells within the *H. pylori*-infected mucosa. Macrophages exposed to *H. pylori* or urease secrete pro-Th17 cytokines (Zhuang et al. 2011). Furthermore, two independent studies using two different mouse models of autoimmune disease identified B cell activating factor (BAFF aka B-lymphocyte stimulator, BLyS, and TNF-superfamily member 13B) of the TNF-α family as a promoter for Th17 responses (Zhou et al. 2011; Lai Kwan et al. 2008). One report suggested that BAFF was directly involved in these responses, while the other report suggested that BAFF acts as a modulator of the cytokine milieu that would, in turn, affect the induction and function of Th17 cells. Munari et al. (2014) showed that IL-17 and BAFF levels are elevated in the mucosa of *H. pylori*-infected patients, and the increase of these two cytokines hinges on the presence of *H. pylori*. Macrophages in the gastric mucosa of patients are a major source of BAFF, which causes pro-Th17 cytokine production in a reactive oxygen species (ROS)-dependent manner. Taken together, all these reported observations suggest that *H. pylori* may affect macrophage polarity in multiple ways, and these macrophages may in turn contribute to the diverse Th cell responses reported during the infection. However, an important property for macrophages to affect T cells directly is by being able to phagocytose and process *H. pylori* antigens for presentation of the antigens to $CD4^{+}$ T cells.

Macrophages readily internalize *H. pylori*, but the bacteria avoid phagocytic killing. Virulent type 1 strains of *H. pylori* were found to disturb phagosome maturation and induce formation of anomalous vacuoles referred to as megasomes (Allen et al. 2000; Zheng and Jones 2003). Normal maturation of phagosomes occurs in stepwise fashion in which phagosomes fuse with early endosomes, late endosomes, and lysosomes. The intravacuolar pH decreases with each stage in order to allow for activation of the lysosomal proteases needed for antigen processing (Desjardins et al. 1994). However, *H. pylori* stops phagosome maturation, preventing it from attaining its full degradative capacity, which in turn allows for extended *H. pylori* intracellular survival (Allen 1999). Experiments with isogenic *vacA* and urease mutant strains to infect murine macrophages and macrophage cell lines pointed to their role in extending the survival of *H. pylori* (Zheng and Jones 2003; Schwartz and Allen 2006). The ability of *H. pylori* VacA to perturb the endocytic traffic at a late stage was initially described by Rappuoli's and Montecucco's groups using elegant cell biology methods (Papini et al. 1994).

DCs are robust APCs and a major immune cell type connecting both innate and adaptive immune responses. DCs are also among the cell types affected by *H. pylori* during infection and thus represent an important tool in the arsenal used by *H. pylori* to subvert host immunity since DCs may also function as different subsets that differentially regulate T cell functions. Among the various effects that *H. pylori* has on DCs include the induction of cytokines, such as IL-12, IL-23, and TNF-α, which are associated with Th1 responses (Amedei et al. 2006), in addition to a panel of other pro-inflammatory cytokines and chemokines (Kranzer et al. 2004). *H. pylori* has also been reported to promote monocyte maturation into DCs with increased HLA class II expression. An important virulence factor in promoting these responses is the neutrophil-activating protein NapA (Pachathundikandi et al. 2015). As the name implies, it also affects neutrophils and was initially labeled as such because it was reported to induce a high production of oxygen radicals from neutrophils (Evans et al. 1995a, b). In vitro studies showed that NapA alone, added to in vitro cultures, could significantly limit development of Th2 clones to antigens such as tetanus toxoid (TT) and mite allergen. Interestingly, in those studies, most (89%) of the allergen-specific Th clones were Th2 clones in the absence of NapA, but in its presence their frequency decreased to only a small fraction (29%), while Th1 clones increased considerably (D'Elios et al. 2007). Because of NapA's potential to reprogram antigen-specific Th2 cell responses to polarized Th1, its possible use as an immunomodulator in Th2 diseases, such as atopy, has been suggested (Reyes and Beswick 2007). As discussed below, among the subsets of $CD4^+$ T cells that have been found to infiltrate the *H. pylori*-infected gastric mucosa are Treg and Th17 cells, but their balance is skewed toward a regulatory response. The effect that *H. pylori* has on the Treg/Th17 balance appears to be exerted via DCs (Kao et al. 2010). Studies conducted by Anne Muller's group using bone marrow-derived DCs exposed to *H. pylori* and co-cultured with $CD4^+$ T cells and a cocktail of anti-CD3, IL-2, and TGF-β showed that they induced more $CD25^+FoxP3^+CD4^+$ T cells than naïve DCs (not exposed to *H. pylori*) as determined by flow cytometry (Oertli et al. 2012). Interestingly, mesenteric lymph node (MLN)-DCs that were immunomagnetically isolated from *H. pylori*-infected mice also promoted the development of a large percentage of $CD25^+FoxP3^+CD4^+$ T cells in co-cultures with naïve $CD4^+$ T cells (Oertli et al. 2012). These and other similar observations suggest that *H. pylori* induces tolerogenic properties in DCs.

Presentation of *H. pylori* antigens by DCs not only activates T cells, but also indirectly promotes B cell activation through CD40-CD40L interactions between lymphocytes (Guindi 2000). The exact role of B lymphocytes in the development of anti-*H. pylori* immunity remains ill-defined, although *H. pylori*-carriers are known to develop strong local and systemic *H. pylori*-specific IgA and IgG antibody production (Futagami et al. 1998; Nurgalieva et al. 2005; Portal-Celhay and Perez-Perez 2006). Since infected individuals have elevated serum Ig titers to *H. pylori*, this response has been used to detect *H. pylori* infection, although IgG antibodies are not considered reliable indicators of current infection. The elicited antibodies fail to control *H. pylori* (Ermak et al. 1998). Early studies with a murine model of *H. pylori* infection examined the protective role of B cells by intragastric administration of *H. pylori*-specific IgA antibodies simultaneously with *Helicobacter felis* bacteria into germ-free mice. After infection with *H. felis*, the investigators observed a reduction of 70% of the number of colonized mice at 4 weeks post-infection (Czinn et al. 1993). In addition, experiments using mice deficient in IgA or immunoglobulin (μMT) that were immunized with urease and lysates of *H. pylori* or *H. felis*, later challenged with *H. pylori*, showed no differences in gastric colonization by *H. pylori* during the acute phase of infection (Ermak et al. 1998; Blanchard et al. 1999; Pappo et al. 1999; Akhiani et al. 2004, 2005). However, analysis of the chronic phase of infection (>8 wk p.i.) showed that μMT mice were able to clear the *H. pylori* infection with signs of severe gastritis, whereas the wild-type mice presented extensive *H. pylori* colonization with mild gastric inflammation (Blanchard et al. 1999). Overall, these reports showed that vaccine-induced immunity is elicited in comparable levels in wild-type and antibody- or B-lymphocyte-deficient mice. Interestingly, T cells from wild-type, IgA- and μMT-deficient mice produced comparably high levels of IFN-γ, whereas the levels of IL-10 produced were significantly higher in wild-type mice than in the deficient mice (Akhiani et al. 2004, 2005). The use of IL-10/IgA double knockout mice helped to further examine the role of inflammation in controlling *H. pylori* colonization. These double knockout mice were 1.2-log significantly less colonized by *H. pylori* than mice deficient only in IL-10, which in turn were less colonized than wild-type mice. These observations led to the view that B cells and/or antibodies may have a pathological effect by promoting chronic inflammation.

IL-10 is among the immune signaling molecules made by B cells and has been linked with downregulation of protective T cell responses. IL-10 is significantly elevated in the gastric mucosa of patients and mice infected with *H. pylori* (Bodger et al. 2001). This cytokine is used by regulatory T and B cells to limit the inflammatory response. Mice deficient in IL-10 had a 100-fold reduction of *H. pylori* colonization in comparison with wild-type mice (Chen et al. 2001; Ismail et al. 2003). B cells can be activated directly by other mechanisms, including TLR, BCR, and cytokines receptors. BCR and TLR7/9 activation by nucleic acid–protein complexes, originating from chronic infection, and associated inflammation, initiates B cell activation via MyD88/NF-κB (Farinha and Gascoyne 2005; Fukata et al. 2008). Interestingly, experiments with murine B cells exposed to *H. pylori* extracts upregulated CD80 and IL-10 production via TLR2/MyD88 activation and promoted differentiation of naïve $CD4^+$ T cells into IL-10-producing $CD4^+CD25^+$ Treg cells, with suppressive activity in vitro through CD40/CD40L (Sayi et al. 2011; Smith 2014) (Fig. 4). Therefore, B cells can be activated pro-regulatory (IL-10 production) cooperating with T cells in the suppression of immunopathological inflammation associated with *H. pylori* infection. IFN-α is another cytokine made by plasmacytoid DC antigen-1 $(PDCA-1)^+$ B cells and found to suppress *H. pylori*-induced gastritis, and downregulate Th1-type cytokines (Otani et al. 2012). Interestingly, IFN-α administration to *H. pylori*-infected mice reduced neutrophil infiltration and levels of TNF-α and IFN-γ (Otani et al. 2012). Gastric samples from *H. pylori*-infected patients showed significantly increased IFN-α and IgM in their sera, as well as $PDCA-1^+$ B cells compared to controls (Ma et al. 2016). In addition, $PDCA-1^+$ B cells were more frequent in *H. pylori*- infected patients suffering from atrophic gastritis or peptic ulcers in comparison with non-atrophic gastritis patients (Ma et al. 2016).

The cellular infiltrate within the *H. pylori*-infected gastric mucosa includes both $CD4^+$ and $CD8^+$ T cells, which are significantly increased in the neck, pit, and gland regions, as noted in gastric biopsy sections (Nurgalieva et al. 2005; Bamford et al. 1998a). In early studies, we and others reported that the response is polarized to Th1 cells (Haeberle et al. 1997; Bamford et al. 1998a; Karttunen et al. 1990), which was an early indication that the immune response to *H. pylori* is misguided since Th1 cells influence cell-mediated immunity, which is inadequate against extracellular pathogens, such as *H. pylori*. In fact, Th1 cells seem to aid in pathogenesis, as supported by observations in human carriers, suggesting Th1 participation in *H. pylori*-associated lesions (Robinson et al. 2008). The presence of Th17 and Treg cells in the infected gastric mucosa has been reported by various independent groups (Jang 2010; Shi et al. 2010; Zhang et al. 2008; Lundgren et al. 2003, 2005). Further, in the *H. pylori*-infected gastric mucosa there is a marked infiltration of $CD4^+$ T cells with abnormal Th17/Treg cell ratios (Gil et al. 2014; Lundgren et al. 2003, 2005).

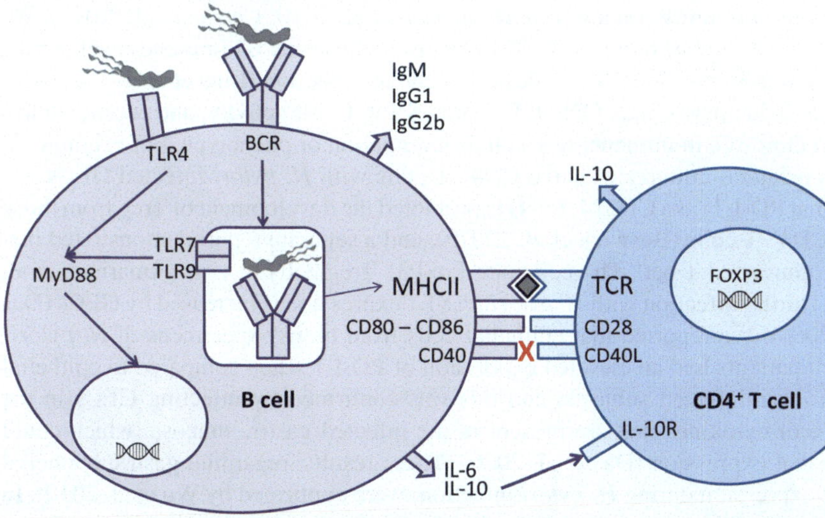

Fig. 4 *H. pylori* upregulate the expression of CD80 and IL-10 production via TLRs on B cells. B cells exposed to *H. pylori* upregulate receptors and cytokines that then promote Treg cell differentiation

$CD25^+/CD4^+$ versus $FoxP3^+/CD4^+$ cells frequencies vary significantly depending on the type of disease and severity (Cheng et al. 2012). FoxP3 is a master regulator of Treg cells whose frequency is significantly higher in GC patients than in patients with other *H. pylori*-related gastric diseases (Cheng et al. 2012). An increase in Treg cells leads to higher bacterial density and contributes to the development of atrophic gastritis and GC progression by suppressing anti-tumor effector T cells. Treg cells in the gastric mucosa helped explain earlier reports of T cell hyporesponsiveness of T cells from *H. pylori*-infected subjects when restimulated with *H. pylori* antigens, as compared to T cells from uninfected individuals (Fan et al. 1994; Karttunen et al. 1990). As Th17 cells are important in immune-mediated clearance of extracellular bacteria, their presence in the *H. pylori*-infected mucosa is expected. In fact, mouse immunization studies reported the contribution of Th17 cells, and a robust IL-17 secretion in protection against *H. pylori* (Delyria et al. 2009), but in those studies, the vaccinated mice were challenged with the *H. pylori* SS1 strain, which is defective in the type 4 secretion system. As explained below in some detail, our studies showed that a functional type 4 secretion system is important in *H. pylori*'s ability to evade Th17 cell responses (Lina et al. 2013). In infected mice fully virulent *H. pylori* inhibits Th17 and tips the balance to Treg cells (Kao et al. 2010). The Treg/Th17 balance is essential to immune homeostasis.

T cell activity is also shaped by immune checkpoint receptors expressed on immune cells that deliver inhibitory signals (Ceeraz et al. 2013). As mentioned earlier, the B7 family of co-stimulatory/co-inhibitory receptors has emerged as central in immune regulation, keeping a subtle balance between immune potency and suppression of autoimmunity (reviewed in (Francisco et al. 2010; Ceeraz et al. 2013)). We showed that *H. pylori* regulates GEC expression of various B7 immune checkpoints, which in turn impact local T cell development and function (Lina et al. 2013, 2015). These proteins perform as ON/OFF switches for T cell activity, and recent studies suggest their role in influencing T cell differentiation or phenotype. For example, in studies using co-cultures of naïve $CD4^+$ T cells with *H. pylori*-infected GECs, we noted that PD-L1 (aka CD274, B7-H1) promoted the development of Treg from those naïve $CD4^+$ T cells (Beswick et al. 2007b), and a separate report demonstrated that PD-L1 converted $T\text{-bet}^+$ Th1 cells into $FoxP3^+$ Treg cells in vivo (Amarnath et al. 2011). During infection with *H. pylori*, PD-L1 expression is increased by GECs (Das et al. 2006). We reported that epithelial cells from biopsy specimens of *H. pylori*-infected patients had an elevated expression of PD-L1 when compared to epithelial cells from uninfected subjects, and this was confirmed by infecting GECs in the absence of cytokines that are present in the infected gastric mucosa, which could induce that expression (Das et al. 2006). These results, regarding gastric epithelial PD-L1 expression during *H. pylori* infection, were confirmed by Wu et al. 2010). In subsequent studies, we observed that *H. pylori* infection, besides eliciting increased expression of PD-L1, also leads to a reduced expression of ICOS-L, which is the only positive co-stimulator known to act on activated or memory T cells (Lina et al. 2013). These findings suggested that *H. pylori* uses the epithelium to create a prime inhibitory scenario for Th effector cells by altering the expression of these proteins with profound immunomodulatory effects. These responses are partially dependent on *H. pylori* CagA and peptidoglycan translocated by the type IV secretion system (Posselt et al. 2013; Backert et al. 2015; Zhang et al. 2015) (Fig. 5). CagA was found to reduce ICOS-L expression by activating the p70 S6 kinase pathway. CagA contributes to the *H. pylori*-mediated activation of the mTOR/p70 S6 kinase pathway. The serine/threonine protein kinase mTOR acts downstream from PI3K/Akt and controls activation of p70 S6 kinase. The role of p70 S6 in downregulation of ICOS-L by the *cagA*$^+$ *H. pylori* strains was confirmed by adding to the cultures rapamycin, a specific inhibitor of p70 S6 kinase/mTOR. Because the ICOS-L–ICOS interaction is critical for Th17 cell development, maintenance, and function (Paulos et al. 2010), *H. pylori* is able to evade Th17 cell-mediated clearance by modifying ICOS-L expression as demonstrated in in vivo studies (Lina et al. 2013). The B7 family of "checkpoint regulators" (Ceeraz et al. 2013) affect adaptive immunity beyond T cell activation, as described above. They impact T cell differentiation, cytokine production, and reprogramming (Kuang et al. 2014; Lee et al. 2013; Ishiwata et al. 2010). As T lymphocytes play a key role in adaptive immunity, *H. pylori*'s influence on the expression of immune checkpoints may be pivotal in persistent infection and pathogenicity. It is worth noting that tumor cell expression of checkpoint inhibitors promotes tumor immune evasion and growth by inducing "exhaustion" of effector T cells (Wherry 2011), and the ability of *H. pylori* to alter the expression of these molecules may allow *H. pylori* to aid developing neoplastic cells to escape immune surveillance mechanisms.

Besides altering the local mucosal environment by modulating the expression of key immunoregulatory molecules or production of cytokines by the gastric epithelium

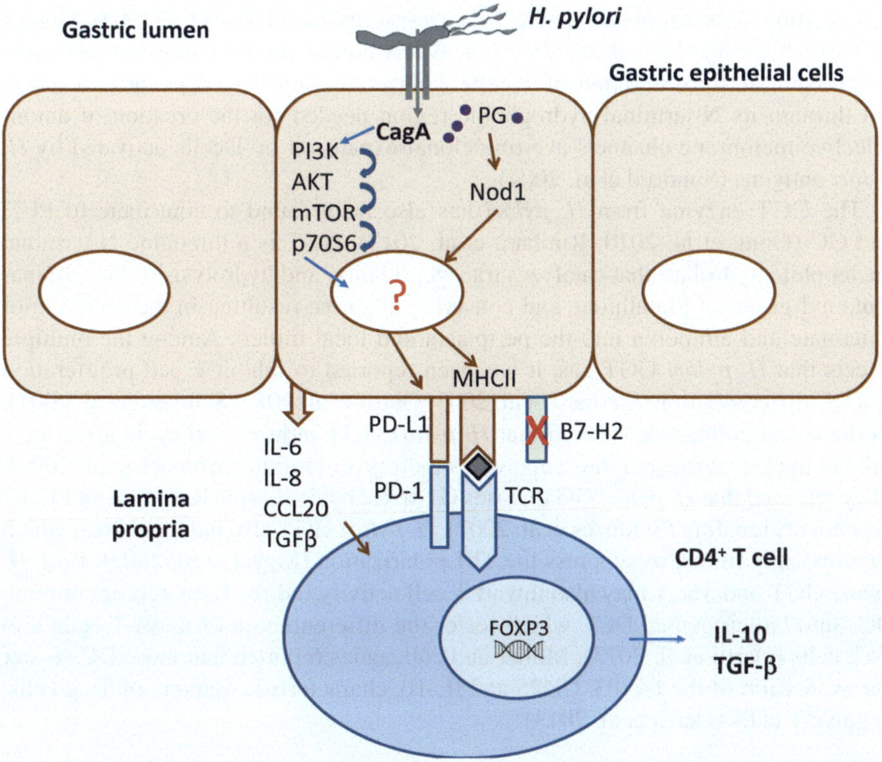

Fig. 5 *H. pylori* CagA and peptidoglycan translocated by the type IV secretion system into GECs promote a suppressive environment. *H. pylori* CagA and peptidoglycan injected into GECs lead to a reduction of B7-H2 expression by activating the p70 S6 kinase pathway. *H. pylori* CagA also promotes PD-L1 (B7-H1) expression by GECs. Both of these responses combined promote a suppressive environment because the ICOS-L–ICOS interaction is critical for Th17 cell development, maintenance, and function and in the absence of the interaction Th17, cells fail to develop. PD-L1 provides inhibitory signals for effector Th cells and promotes differentiation of Treg cells (Lina et al. 2013)

and immune cells, *H. pylori* is also able to directly inhibit CD4$^+$ T cells. *H. pylori*'s VacA toxin and γ-glutamyl-transpeptidase (GGT) have been reported to hinder T cell activation (Sundrud et al. 2004; Boncristiano et al. 2003). Both of these toxins are secreted products of *H. pylori*. VacA uses CD18 (β2-integrin) as a receptor on T cells (Sewald et al. 2008). VacA is internalized after the cytoplasmic domain of CD18 is phosphorylated by protein kinase C (Sewald et al. 2011). *H. pylori* VacA impedes T cell signaling and proliferation by promoting the arrest of the cells cycle at G1/S. *H. pylori* VacA mediates this effect by interfering with the TCR and IL-2 signaling pathways at the level of the $Ca_2{}^+$/calmodulin-dependent phosphatase calcineurin. By this mechanism *H. pylori* VacA prevents translocation of the important T cell transcription factor NFAT (nuclear factor of activated T cells) into the nucleus of T cells leading to the suppression of *il-2* gene transcription (Gebert et al. 2003). Studies by Cover's group showed that *H. pylori* VacA constrains IL-2-induced cell-cycle progression and proliferation of T cells without altering IL-2-dependent survival, but through its N-terminal hydrophobic region needed for the creation of anion-selective membrane channels averting clonal expansion of T cells activated by *H. pylori* antigens (Sundrud et al. 2004).

The GGT enzyme from *H. pylori* has also been found to contribute to PUD and GC (Gong et al. 2010; Rimbara et al. 2013). GGT is a threonine N-terminal nucleophile hydrolase that catalyzes transpeptidation and hydrolysis of the gamma-glutamyl group of glutathione and converts glutamine resulting in the secretion of glutamate and ammonia into the periplasm and local milieu. Among the multiple effects that *H. pylori* GGT has, it has been reported to inhibit T cell proliferation and DC differentiation (Gerhard et al. 2005; Oertli et al. 2013; Schmees et al. 2007). Gerhard and colleagues showed that *H. pylori* GGT induces cell-cycle arrest in T cells at the G1 phase and thus suppresses their proliferation (Schmees et al. 2007). They reported that *H. pylori* GGT causes G1 arrest by disrupting Ras- and not PI3 K-dependent signaling (Schmees et al. 2007). *H. pylori* GGT also induces Cox2, which paradoxically may also suppress the Th1 polarization (Meyer et al. 2003). Both *H. pylori* GGT and VacA may also thwart T cell activity indirectly by reprogramming DCs into "tolerogenic" DCs, which foster the differentiation of naïve T cells into Treg cells (Oertli et al. 2013). Muller and colleagues reported that those DCs foster the expression of the FoxP3, CD25 and IL-10, characteristic markers of Treg cells, in naïve T cells (Oertli et al. 2013).

7 Concluding Remarks

Although the incidence of *H. pylori* infection has been decreasing due to enhancements in living conditions, the global prevalence of *H. pylori* remains high. In North America, approximately one-third of all adults are infected, while in developing regions, almost half of the population carries *H. pylori* (Eusebi et al. 2014). Thus, *H. pylori* remains an important human pathogen associated with significant clinical disease. Over the last few years, we have learned substantially regarding its diverse mechanisms to surreptitiously maneuver the host immune response in order to maintain persistent infection that may last a lifetime. Because the diseases associated with its infection remain a significant public health concern, due to their associated morbidity and mortality, and because of the increasing incidence of antibiotic resistance, there is a clear need for an effective vaccine that allows the host to surmount the multiple strategies used by *H. pylori* to thwart the host adaptive responses reviewed in this chapter. Since T lymphocytes are arguably the most essential cells in adaptive immunity, *H. pylori*'s impact on the expression of crucial receptors that control T lymphocyte function or tolerance is decisive in bacterial persistence and pathogenesis. Thus, in order to reduce the incidence of this important human pathogen through vaccination, we clearly need to better understand how it manipulates the host immune armamentarium in order to effectively and appropriately steer it in directions that favor the host over the pathogen.

References

Acosta-Rodriguez EV, Rivino L, Geginat J, Jarrossay D, Gattorno M, Lanzavecchia A, Sallusto F, Napolitani G (2007) Surface phenotype and antigenic specificity of human interleukin 17-producing T helper memory cells. Nat Immunol 8(6):639–646. https://doi.org/10.1038/ni1467

Adams EJ (2014) Lipid presentation by human CD1 molecules and the diverse T cell populations that respond to them. Curr Opin Immunol 26:1–6. https://doi.org/10.1016/j.coi.2013.09.005

Akhiani AA, Schon K, Franzen LE, Pappo J, Lycke N (2004) *Helicobacter pylori*-specific antibodies impair the development of gastritis, facilitate bacterial colonization, and counteract resistance against infection. J Immunol 172(8):5024–5033

Akhiani AA, Stensson A, Schon K, Lycke NY (2005) IgA antibodies impair resistance against *Helicobacter pylori* infection: studies on immune evasion in IL-10-deficient mice. J Immunol 174(12):8144–8153

Allen LA (1999) Intracellular niches for extracellular bacteria: lessons from *Helicobacter pylori*. J Leukoc Biol 66(5):753–756

Allen LA, Schlesinger LS, Kang B (2000) Virulent strains of *Helicobacter pylori* demonstrate delayed phagocytosis and stimulate homotypic phagosome fusion in macrophages. J Exp Med 191(1):115–128

Amarnath S, Mangus CW, Wang JC, Wei F, He A, Kapoor V, Foley JE, Massey PR, Felizardo TC, Riley JL, Levine BL, June CH, Medin JA, Fowler DH (2011) The PDL1-PD1 axis converts human Th1 cells into regulatory T cells. Sci Transl Med 3(111):111–120. https://doi.org/10.1126/scitranslmed.3003130

Amedei A, Cappon A, Codolo G, Cabrelle A, Polenghi A, Benagiano M, Tasca E, Azzurri A, D'Elios MM, Del PG, de Bernard M (2006) The neutrophil-activating protein of *Helicobacter pylori* promotes Th1 immune responses. J Clin Invest 116(4):1092–1101. https://doi.org/10.1172/JCI27177

August A, Gibson S, Kawakami Y, Kawakami T, Mills GB, Dupont B (1994) CD28 is associated with and induces the immediate tyrosine phosphorylation and activation of the Tec family kinase ITK/EMT in the human Jurkat leukemic T-cell line. Proc Natl Acad Sci U S A 91(20):9347–9351

Backert S, Tegtmeyer N, Fischer W (2015) Composition, structure and function of the *Helicobacter pylori cag* pathogenicity island encoded type IV secretion system. Future Microbiol. 10(6):955–965. https://doi.org/10.2217/fmb.15.32

Bamford KB, Fan X, Crowe SE, Leary JF, Gourley WK, Luthra GK, Brooks EG, Graham DY, Reyes VE, Ernst PB (1998a) Lymphocytes in the human gastric mucosa during *Helicobacter pylori* have a T helper cell 1 phenotype. Gastro 114(3):482–492

Bamford KB, Fan XJ, Crowe SE, Leary JF, Gourley WK, Luthra G, Brooks EG, Graham DY, Reyes VE, Ernst PB (1998b) Lymphocytes during infection with *Helicobacter pylori* have a helper 1 (Th1) phenotype. Gastroenterology 114:1–12

Bandeira A, Itohara S, Bonneville M, Burlen-Defranoux O, Mota-Santos T, Coutinho A, Tonegawa S (1991) Extrathymic origin of intestinal intraepithelial lymphocytes bearing T-cell antigen receptor gamma delta. Proc Natl Acad Sci U S A 88(1):43–47

Barrera C, Espejo R, Reyes VE (2002) Differential glycosylation of MHC class II molecules on gastric epithelial cells: Implications in local immune responses. Hum Immunol 63:384–393

Barrera C, Ye G, Espejo R, Gunasena S, Almanza R, Leary J, Crowe S, Ernst P, Reyes VE (2001) Expression of cathepsins B, L, S, and D by gastric epithelial cells implicates them as antigen presenting cells in local immune responses. Hum Immunol 62(10):1081–1091

Barrera CA, Beswick EJ, Bland DA, Espejo R, Mifflin RC, Adegboyega P, Crowe SE, Ernst PB, Reyes VE (2005) Polarized expression of CD74 by gastric epithelial cells. J Histochem Cytochem 53(12):1481–1489. https://doi.org/10.1369/jhc.4A6552.2005

Bending D, De la Pena H, Veldhoen M, Phillips JM, Uyttenhove C, Stockinger B, Cooke A (2009) Highly purified Th17 cells from BDC2.5NOD mice convert into Th1-like cells in NOD/SCID recipient mice. J Clin Invest 119(3):565–572. https://doi.org/10.1172/jci37865

Bennett F, Luxenberg D, Ling V, Wang IM, Marquette K, Lowe D, Khan N, Veldman G, Jacobs KA, Valge-Archer VE, Collins M, Carreno BM (2003) Program death-1 engagement upon TCR activation has distinct effects on costimulation and cytokine-driven proliferation: attenuation of ICOS, IL-4, and IL-21, but not CD28, IL-7, and IL-15 responses. J Immunol 170(2):711–718

Berlin C, Berg EL, Briskin MJ, Andrew DP, Kilshaw PJ, Holzmann B, Weissman IL, Hamann A, Butcher EC (1993) Alpha 4 beta 7 integrin mediates lymphocyte binding to the mucosal vascular addressin MAdCAM-1. Cell 74(1):185–195

Beswick EJ, Bland D, Das S, Suarez G, Sierra J, Reyes VE (2004) *Helicobacter pylori* urease binds to CD74 and stimulates gastric epithelial cell responses associated with pathogenesis. Gastro 126(4):A401

Beswick EJ, Bland DA, Suarez G, Barrera CA, Fan XJ, Reyes VE (2005) *Helicobacter pylori* binds to CD74 on gastric epithelial cells and stimulates interleukin-8 production. Infect Immun 73(5):2736–2743. https://doi.org/10.1128/IAI.73.5.2736-2743.2005

Beswick EJ, Pinchuk IV, Das S, Powell DW, Reyes VE (2007a) B7-H1 expression on gastric epithelial cells after *Helicobacter pylori* exposure promotes the development of $CD4^+$ $CD25^+$ $FoxP3^+$ regulatory T cells. Infect Immun 75(9):4334–4341

Beswick EJ, Pinchuk IV, Das S, Powell DW, Reyes VE (2007b) Expression of the programmed death ligand 1, B7-H1, on gastric epithelial cells after *Helicobacter pylori* exposure promotes development of $CD4^+$ $CD25^+$ $FoxP3^+$ regulatory T cells. Infect Immun 75(9):4334–4341. https://doi.org/10.1128/IAI.00553-07

Beswick EJ, Pinchuk IV, Earley RB, Schmitt DA, Reyes VE (2011) The role of gastric epithelial cell-derived TGF-{beta} in reduced $CD4^+$ T cell proliferation and development of regulatory T cells during *Helicobacter pylori* infection. Infect Immun 79(7):2737–2745. https://doi.org/10.1128/IAI.01146-10

Bimczok D, Kao JY, Zhang M, Cochrun S, Mannon P, Peter S, Wilcox CM, Monkemuller KE, Harris PR, Grams JM, Stahl RD, Smith PD, Smythies LE (2015) Human gastric epithelial cells contribute to gastric immune regulation by providing retinoic acid to dendritic cells. Mucosal Immunol 8(3):533–544. https://doi.org/10.1038/mi.2014.86

Blanchard TG, Czinn SJ, Redline RW, Sigmund N, Harriman G, Nedrud JG (1999) Antibody-independent protective mucosal immunity to gastric *helicobacter* infection in mice. Cell Immunol 191(1):74–80. https://doi.org/10.1006/cimm.1998.1421

Blum JS, Wearsch PA, Cresswell P (2013) Pathways of antigen processing. Annu Rev Immunol 31:443–473. https://doi.org/10.1146/annurev-immunol-032712-095910

Blumberg RS, Gerdes D, Chott A, Porcelli SA, Balk SP (1995) Structure and function of the CD1 family of MHC-like cell surface proteins. Immunol Rev 147:5–29

Bodger K, Bromelow K, Wyatt JI, Heatley RV (2001) Interleukin 10 in Helicobacter pylori associated gastritis: immunohistochemical localisation and in vitro effects on cytokine secretion. J Clin Path 54(4):285–92

Boncristiano M, Paccani SR, Barone S, Ulivieri C, Patrussi L, Ilver D, Amedei A, D'Elios MM, Telford JL, Baldari CT (2003) The *Helicobacter pylori* vacuolating toxin inhibits T cell activation by two independent mechanisms. J Exp Med 198(12):1887–1897. https://doi.org/10.1084/jem.20030621

Boussiotis VA, Chatterjee P, Li L (2014) Biochemical signaling of PD-1 on T cells and its functional implications. Cancer J 20(4):265–271. https://doi.org/10.1097/PPO.0000000000000059

Brough HA, Cousins DJ, Munteanu A, Wong YF, Sudra A, Makinson K, Stephens AC, Arno M, Ciortuz L, Lack G, Turcanu V (2014) IL-9 is a key component of memory Th cell peanut-specific responses from children with peanut allergy. J Allergy Clin Immunol 134(6):1329–1338. https://doi.org/10.1016/j.jaci.2014.06.032

Cammarota G, Scheirle A, Takacs B, Doran DM, Knorr R, Bannwarth W, Guardiola J, Sinigaglia F (1992) Identification of a CD4 binding site on the beta 2 domain of HLA- DR molecules. Nature 356:799–801. https://doi.org/10.1038/356799a0

Castano-Rodriguez N, Kaakoush NO, Goh KL, Fock KM, Mitchell HM (2015) Autophagy in *Helicobacter pylori* infection and related gastric cancer. Helicobacter 20(5):353–369. https://doi.org/10.1111/hel.12211

Ceeraz S, Nowak EC, Noelle RJ (2013) B7 family checkpoint regulators in immune regulation and disease. Trends Immunol 34(11):556–563. https://doi.org/10.1016/j.it.2013.07.003

Chang HC, Sehra S, Goswami R, Yao W, Yu Q, Stritesky GL, Jabeen R, McKinley C, Ahyi AN, Han L, Nguyen ET, Robertson MJ, Perumal NB, Tepper RS, Nutt SL, Kaplan MH (2010) The transcription factor PU.1 is required for the development of IL-9-producing T cells and allergic inflammation. Nat Immunol 11(6):527–534. https://doi.org/10.1038/ni.1867

Chang S, Aune TM (2007) Dynamic changes in histone-methylation 'marks' across the locus encoding interferon-gamma during the differentiation of T helper type 2 cells. Nat Immunol 8(7):723–731. https://doi.org/10.1038/ni1473

Chen L, Chen J, Xu B, Wang Q, Zhou W, Zhang G, Sun J, Shi L, Pei H, Wu C, Jiang J (2015) B7-H3 expression associates with tumor invasion and patient's poor survival in human esophageal cancer. Am J Transl Res 7(12):2646–2660

Chen W, Shu D, Chadwick VS (2001) *Helicobacter pylori* infection: mechanism of colonization and functional dyspepsia Reduced colonization of gastric mucosa by *Helicobacter pylori* in mice deficient in interleukin-10. J Gastroenterol Hepatol 16(4):377–383

Cheng HH, Tseng GY, Yang HB, Wang HJ, Lin HJ, Wang WC (2012) Increased numbers of Foxp3-positive regulatory T cells in gastritis, peptic ulcer and gastric adenocarcinoma. World J Gastroenterol 18(1):34–43. https://doi.org/10.3748/wjg.v18.i1.34

Chiba Y, Mizoguchi I, Hasegawa H, Ohashi M, Orii N, Nagai T, Sugahara M, Miyamoto Y, Xu M, Owaki T, Yoshimoto T (2018) Regulation of myelopoiesis by proinflammatory cytokines in infectious diseases. Cell Mol Life Sci 75(8):1363–1376. https://doi.org/10.1007/s00018-017-2724-5

Choi YS, Kageyama R, Eto D, Escobar TC, Johnston RJ, Monticelli L, Lao C, Crotty S (2011) ICOS receptor instructs T follicular helper cell versus effector cell differentiation via induction of the transcriptional repressor Bcl6. Immunity 34(6):932–946. https://doi.org/10.1016/j.immuni.2011.03.023

Cimino-Mathews A, Thompson E, Taube JM, Ye X, Lu Y, Meeker A, Xu H, Sharma R, Lecksell K, Cornish TC, Cuka N, Argani P, Emens LA (2016) PD-L1 (B7-H1) expression and the immune tumor microenvironment in primary and metastatic breast carcinomas. Hum Pathol 47(1):52–63. https://doi.org/10.1016/j.humpath.2015.09.003

Crellin NK, Trifari S, Kaplan CD, Cupedo T, Spits H (2010) Human NKp44$^+$ IL-22$^+$ cells and LTi-like cells constitute a stable RORC$^+$ lineage distinct from conventional natural killer cells. J Exp Med 207(2):281–290. https://doi.org/10.1084/jem.20091509

Cupedo T, Crellin NK, Papazian N, Rombouts EJ, Weijer K, Grogan JL, Fibbe WE, Cornelissen JJ, Spits H (2009) Human fetal lymphoid tissue-inducer cells are interleukin 17-producing precursors to RORC$^+$ CD127$^+$ natural killer-like cells. Nat Immunol 10(1):66–74. https://doi.org/10.1038/ni.1668

Czinn SJ, Cai A, Nedrud JG (1993) Protection of germ-free mice from infection by *Helicobacter felis* after active oral or passive IgA immunization. Vaccine 11(6):637–642

D'Elios MM, Amedei A, Cappon A, Del PG, de Bernard M (2007) The neutrophil-activating protein of *Helicobacter pylori* (HP-NAP) as an immune modulating agent. FEMS Immunol Med Microbiol 50(2):157–164. https://doi.org/10.1111/j.1574-695X.2007.00258.x

Dardalhon V, Awasthi A, Kwon H, Galileos G, Gao W, Sobel RA, Mitsdoerffer M, Strom TB, Elyaman W, Ho IC, Khoury S, Oukka M, Kuchroo VK (2008) IL-4 inhibits TGF-beta-induced

Foxp3$^+$ T cells and together with TGF-beta, generates IL-9$^+$ IL-10$^+$ Foxp3$^{(-)}$ effector T cells. Nat Immunol 9(12):1347–1355. https://doi.org/10.1038/ni.1677

Das S, Suarez G, Beswick EJ, Sierra JC, Graham DY, Reyes VE (2006) Expression of B7-H1 on gastric epithelial cells: its potential role in regulating T cells during *Helicobacter pylori* infection. J Immunol 176(5):3000–3009

de Sousa JR, Quaresma JAS (2018) The role of T helper 25 cells in the immune response to *Mycobacterium leprae*. J Am Acad Dermatol 78(5):1009–1011. https://doi.org/10.1016/j.jaad.2017.11.025

Delyria ES, Redline RW, Blanchard TG (2009) Vaccination of mice against *H pylori* induces a strong Th-17 response and immunity that is neutrophil dependent. Gastro 136(1):247–256. https://doi.org/10.1053/j.gastro.2008.09.017

Desjardins M, Huber LA, Parton RG, Griffiths G (1994) Biogenesis of phagolysosomes proceeds through a sequential series of interactions with the endocytic apparatus. J Cell Biol 124(5):677–688

Devine L, Sun J, Barr MR, Kavathas PB (1999) Orientation of the Ig domains of CD8 alpha beta relative to MHC class I. J Immunol 162(2):846–851

Djuretic IM, Levanon D, Negreanu V, Groner Y, Rao A, Ansel KM (2007) Transcription factors T-bet and Runx3 cooperate to activate Ifng and silence Il4 in T helper type 1 cells. Nat Immunol 8(2):145–153. https://doi.org/10.1038/ni1424

DuPage M, Bluestone JA (2016) Harnessing the plasticity of CD4$^{(+)}$ T cells to treat immune-mediated disease. Nat Rev Immunol 16(3):149–163. https://doi.org/10.1038/nri.2015.18

Dzierzanowska-Fangrat K, Michalkiewicz J, Cielecka-Kuszyk J, Nowak M, Celinska-Cedro D, Rozynek E, Dzierzanowska D, Crabtree JE (2008) Enhanced gastric IL-18 mRNA expression in *Helicobacter pylori*-infected children is associated with macrophage infiltration, IL-8, and IL-1 beta mRNA expression. Eur J Gastroenterol Hepatol 20(4):314–319. https://doi.org/10.1097/MEG.0b013e3282f340da

Ellmark P, Ingvarsson J, Carlsson A, Lundin BS, Wingren C, Borrebaeck CA (2006) Identification of protein expression signatures associated with *Helicobacter pylori* infection and gastric adenocarcinoma using recombinant antibody microarrays. Mol Cell Proteomics 5(9):1638–1646. https://doi.org/10.1074/mcp.M600170-MCP200

Ermak TH, Giannasca PJ, Nichols R, Myers GA, Nedrud J, Weltzin R, Lee CK, Kleanthous H, Monath TP (1998) Immunization of mice with urease vaccine affords protection against *Helicobacter pylori* infection in the absence of antibodies and is mediated by MHC class II-restricted responses. J Exp Med 188(12):2277–2288

Eusebi LH, Zagari RM, Bazzoli F (2014) Epidemiology of *Helicobacter pylori* infection. Helicobacter 19(Suppl 1):1–5. https://doi.org/10.1111/hel.12165

Evans DJ Jr, Evans DG, Lampert HC, Nakano H (1995a) Identification of four new prokaryotic bacterioferritins, from *Helicobacter pylori*, Anabaena variabilis, Bacillus subtilis and Treponema pallidum, by analysis of gene sequences. Gene 153(1):123–127

Evans DJ Jr, Evans DG, Takemura T, Nakano H, Lampert HC, Graham DY, Granger DN, Kvietys PR (1995b) Characterization of a *Helicobacter pylori* neutrophil-activating protein. Infect Immun 63(6):2213–2220

Eyerich S, Eyerich K, Pennino D, Carbone T, Nasorri F, Pallotta S, Cianfarani F, Odorisio T, Traidl-Hoffmann C, Behrendt H, Durham SR, Schmidt-Weber CB, Cavani A (2009) Th22 cells represent a distinct human T cell subset involved in epidermal immunity and remodeling. J Clin Invest 119(12):3573–3585. https://doi.org/10.1172/JCI40202

Fallon PG, Ballantyne SJ, Mangan NE, Barlow JL, Dasvarma A, Hewett DR, McIlgorm A, Jolin HE, McKenzie AN (2006) Identification of an interleukin (IL)-25-dependent cell population that provides IL-4, IL-5, and IL-13 at the onset of helminth expulsion. J Exp Med 203(4):1105–1116. https://doi.org/10.1084/jem.20051615

Fan X, Gunasena H, Cheng Z, Espejo R, Crowe SE, Ernst PB, Reyes VE (2000) *Helicobacter pylori* urease binds to class II MHC on gastric epithelial cells and induces their apoptosis. J Immunol 165(4):1918–1924

Fan XJ, Chua A, Shahi CN, Mcdevitt J, Keeling PWN, Kelleher D (1994) Gastric T lymphocyte responses to *Helicobacter pylori* in patients with *H-pylori* colonisation. Gut 35:1379–1384

Fan XJ, Crowe SE, Behar S, Gunasena H, Ye G, Haeberle H, Van Houten N, Gourley WK, Ernst PB, Reyes VE (1998) The effect of class II major histocompatibility complex expression on adherence of *Helicobacter pylori* and induction of apoptosis in gastric epithelial cells: A mechanism for T helper cell type 1-mediated damage. J Exp Med 187(10):1659–1669

Farinha P, Gascoyne RD (2005) Molecular pathogenesis of mucosa-associated lymphoid tissue lymphoma. J Clin Oncol 23(26):6370–6378. https://doi.org/10.1200/JCO.2005.05.011

Feeley KM, Heneghan MA, Stevens FM, McCarthy CF (1998) Lymphocytic gastritis and coeliac disease: evidence of a positive association. J Clin Pathol 51(3):207–210

Fehlings M, Drobbe L, Moos V, Renner VP, Hagen J, Beigier-Bompadre M, Pang E, Belogolova E, Churin Y, Schneider T, Meyer TF, Aebischer T, Ignatius R (2012) Comparative analysis of the interaction of *Helicobacter pylori* with human dendritic cells, macrophages, and monocytes. Infect Immun 80(8):2724–2734. https://doi.org/10.1128/IAI.00381-12

Fichtelius KE (1967) The mammalian equivalent to bursa Fabricii of birds. Exp Cell Res 46(1):231–234

Fort MM, Cheung J, Yen D, Li J, Zurawski SM, Lo S, Menon S, Clifford T, Hunte B, Lesley R, Muchamuel T, Hurst SD, Zurawski G, Leach MW, Gorman DM, Rennick DM (2001) IL-25 induces IL-4, IL-5, and IL-13 and Th2-associated pathologies *in vivo*. Immunity 15(6):985–995

Francisco LM, Sage PT, Sharpe AH (2010) The PD-1 pathway in tolerance and autoimmunity. Immunol Rev 236:219–242. https://doi.org/10.1111/j.1600-065X.2010.00923.x

Friedline RH, Brown DS, Nguyen H, Kornfeld H, Lee J, Zhang Y, Appleby M, Der SD, Kang J, Chambers CA (2009) $CD4^+$ regulatory T cells require CTLA-4 for the maintenance of systemic tolerance. J Exp Med 206(2):421–434. https://doi.org/10.1084/jem.20081811

Fukata M, Breglio K, Chen A, Vamadevan AS, Goo T, Hsu D, Conduah D, Xu R, Abreu MT (2008) The myeloid differentiation factor 88 (MyD88) is required for $CD4^+$ T cell effector function in a murine model of inflammatory bowel disease. J Immunol 180(3):1886–1894

Futagami S, Takahashi H, Norose Y, Kobayashi M (1998) Systemic and local immune responses against *Helicobacter pylori* urease in patients with chronic gastritis: distinct IgA and IgG productive sites. Gut 43(2):168–175

Gebert B, Fischer W, Weiss E, Hoffmann R, Haas R (2003) *Helicobacter pylori* vacuolating cytotoxin inhibits T lymphocyte activation. Science 301(5636):1099–1102. https://doi.org/10.1126/science.1086871

Gerhard M, Schmees C, Voland P, Endres N, Sander M, Reindl W, Rad R, Oelsner M, Decker T, Mempel M, Hengst L, Prinz C (2005) A secreted low-molecular-weight protein from *Helicobacter pylori* induces cell-cycle arrest of T cells. Gastro 128(5):1327–1339

Gerlach K, McKenzie AN, Neurath MF, Weigmann B (2015) IL-9 regulates intestinal barrier function in experimental T cell-mediated colitis. Tissue Barriers 3(1–2):e983777. https://doi.org/10.4161/21688370.2014.983777

Gil JH, Seo JW, Cho MS, Ahn JH, Sung HY (2014) Role of Treg and Th17 cells of the gastric mucosa in children with *Helicobacter pylori* gastritis. J Pediatr Gastroenterol Nutr 58(2):252–258. https://doi.org/10.1097/MPG.0000000000000194

Gobert AP, Verriere T, Asim M, Barry DP, Piazuelo MB, de Sablet T, Delgado AG, Bravo LE, Correa P, Peek RM Jr, Chaturvedi R, Wilson KT (2014) Heme oxygenase-1 dysregulates macrophage polarization and the immune response to *Helicobacter pylori*. J Immunol 193(6):3013–3022. https://doi.org/10.4049/jimmunol.1401075

Gong M, Ling SS, Lui SY, Yeoh KG, Ho B (2010) *Helicobacter pylori* gamma-glutamyl transpeptidase is a pathogenic factor in the development of peptic ulcer disease. Gastro 139(2):564–573. https://doi.org/10.1053/j.gastro.2010.03.050

Goswami R, Jabeen R, Yagi R, Pham D, Zhu J, Goenka S, Kaplan MH (2012) STAT6-dependent regulation of Th9 development. J Immunol 188(3):968–975. https://doi.org/10.4049/jimmunol.1102840

Guindi M (2000) Role of activated host T cells in the promotion of MALT lymphoma growth. Semin Cancer Biol 10(5):341–344. https://doi.org/10.1006/scbi.2000.0351

Haeberle HA, Kubin M, Bamford KB, Garofalo R, Graham DY, El Zaatari F, Karttunen R, Crowe SE, Reyes VE, Ernst PB (1997) Differential stimulation of interleukin-12 (IL-12) and IL-10 by live and killed *Helicobacter pylori in vitro* and association of IL-12 production with gamma interferon-producing T cells in the human gastric mucosa. Infect Immun 65(10):4229–4235

Harding FA, McArthur JG, Gross JA, Raulet DH, Allison JP (1992) CD28-mediated signalling co-stimulates murine T cells and prevents induction of anergy in T-cell clones. Nature 356:607–609. https://doi.org/10.1038/356607a0

Harrington LE, Hatton RD, Mangan PR, Turner H, Murphy TL, Murphy KM, Weaver CT (2005) Interleukin 17-producing $CD4^+$ effector T cells develop via a lineage distinct from the T helper type 1 and 2 lineages. Nat Immunol 6(11):1123–1132. https://doi.org/10.1038/ni1254

Harwood NE, Batista FD (2010) Early events in B cell activation. Annu Rev Immunol 28:185–210. https://doi.org/10.1146/annurev-immunol-030409-101216

Hatanaka K, Hokari R, Matsuzaki K, Kato S, Kawaguchi A, Nagao S, Suzuki H, Miyazaki K, Sekizuka E, Nagata H, Ishii H, Miura S (2002) Increased expression of mucosal address in cell adhesion molecule-1 (MAdCAM-1) and lymphocyte recruitment in murine gastritis induced by *Helicobacter pylori*. Clin Exp Immunol 130(2):183–189

Hayat M, Arora DS, Dixon MF, Clark B, O'Mahony S (1999) Effects of *Helicobacter pylori* eradication on the natural history of lymphocytic gastritis. Gut 45(4):495–498

Hirahara K, Ghoreschi K, Yang XP, Takahashi H, Laurence A, Vahedi G, Sciume G, Hall AO, Dupont CD, Francisco LM, Chen Q, Tanaka M, Kanno Y, Sun HW, Sharpe AH, Hunter CA, O'Shea JJ (2012) Interleukin-27 priming of T cells controls IL-17 production in trans via induction of the ligand PD-L1. Immunity 36(6):1017–1030. https://doi.org/10.1016/j.immuni.2012.03.024

Hori S, Nomura T, Sakaguchi S (2003) Control of regulatory T cell development by the transcription factor Foxp3. Science 299(5609):1057–1061

Hou J, Yu Z, Xiang R, Li C, Wang L, Chen S, Li Q, Chen M, Wang L (2014) Correlation between infiltration of $FOXP3^+$ regulatory T cells and expression of B7-H1 in the tumor tissues of gastric cancer. Exp Mol Pathol 96(3):284–291. https://doi.org/10.1016/j.yexmp.2014.03.005

Howie D, Spencer J, DeLord D, Pitzalis C, Wathen NC, Dogan A, Akbar A, MacDonald TT (1998) Extrathymic T cell differentiation in the human intestine early in life. J Immunol 161(11):5862–5872

Hsieh CS, Macatonia SE, Tripp CS, Wolf SF, O'garra A, Murphy KM (1993) Development of Th1 $CD4^+$ T cells through IL-12 produced by Listeria-induced macrophages. Science 260(5107):547–549

Huppa JB, Gleimer M, Sumen C, Davis MM (2003) Continuous T cell receptor signaling required for synapse maintenance and full effector potential. Nat Immunol 4(8):749–755. https://doi.org/10.1038/ni951

Ishiwata K, Watanabe N, Guo M, Tomihara K, Brumlik MJ, Yagita H, Pardoll D, Chen L, Shin T (2010) Costimulator B7-DC attenuates strong Th2 responses induced by Nippostrongylus brasiliensis. J Immunol 184(4):2086–2094. https://doi.org/10.4049/jimmunol.0804051

Ismail HF, Fick P, Zhang J, Lynch RG, Berg DJ (2003) Depletion of neutrophils in IL-10(−/−) mice delays clearance of gastric *Helicobacter* infection and decreases the Th1 immune response to *Helicobacter*. J Immunol 170(7):3782–3789

Ito Y, Vela JL, Matsumura F, Hoshino H, Tyznik A, Lee H, Girardi E, Zajonc DM, Liddington R, Kobayashi M, Bao X, Bugaytsova J, Boren T, Jin R, Zong Y, Seeberger PH, Nakayama J, Kronenberg M, Fukuda M (2013) *Helicobacter pylori* cholesteryl alpha-glucosides contribute to its pathogenicity and immune response by natural killer T cells. PLoS ONE 8(12):e78191. https://doi.org/10.1371/journal.pone.0078191

Ivanov II, McKenzie BS, Zhou L, Tadokoro CE, Lepelley A, Lafaille JJ, Cua DJ, Littman DR (2006) The orphan nuclear receptor RORgammat directs the differentiation program of proinflammatory $IL\text{-}17^+$ T helper cells. Cell 126(6):1121–1133. https://doi.org/10.1016/j.cell.2006.07.035

Izcue A, Coombes JL, Powrie F (2009) Regulatory lymphocytes and intestinal inflammation. Annu Rev Immunol 27:313–338. https://doi.org/10.1146/annurev.immunol.021908.132657

Jafarzadeh A, Jamali M, Mahdavi R, Ebrahimi HA, Hajghani H, Khosravimashizi A, Nemati M, Najafipour H, Sheikhi A, Mohammadi MM, Daneshvar H (2015) Circulating levels of interleukin-35 in patients with multiple sclerosis: evaluation of the influences of FOXP3 gene polymorphism and treatment program. J Mol Neurosci 55(4):891–897. https://doi.org/10.1007/s12031-014-0443-z

Jang TJ (2010) The number of Foxp3-positive regulatory T cells is increased in *Helicobacter pylori* gastritis and gastric cancer. Pathol Res Pract 206(1):34–38. https://doi.org/10.1016/j.prp.2009.07.019

Jenner RG, Townsend MJ, Jackson I, Sun K, Bouwman RD, Young RA, Glimcher LH, Lord GM (2009) The transcription factors T-bet and GATA-3 control alternative pathways of T-cell differentiation through a shared set of target genes. Proc Natl Acad Sci U S A 106(42):17876–17881. https://doi.org/10.1073/pnas.0909357106

Joffre OP, Segura E, Savina A, Amigorena S (2012) Cross-presentation by dendritic cells. Nat Rev Immunol 12(8):557–569. https://doi.org/10.1038/nri3254

Johansson-Lindbom B, Svensson M, Wurbel MA, Malissen B, Marquez G, Agace W (2003) Selective generation of gut tropic T cells in gut-associated lymphoid tissue (GALT): requirement for GALT dendritic cells and adjuvant. J Exp Med 198(6):963–969. https://doi.org/10.1084/jem.20031244

Kao JY, Zhang M, Miller MJ, Mills JC, Wang B, Liu M, Eaton KA, Zou W, Berndt BE, Cole TS, Takeuchi T, Owyang SY, Luther J (2010) *Helicobacter pylori* immune escape is mediated by dendritic cell-induced Treg skewing and Th17 suppression in mice. Gastro 138(3):1046–1054. https://doi.org/10.1053/j.gastro.2009.11.043

Karttunen R, Andersson G, Poikonen K, Kosunen TU, Kartunen T, Juutinen K, Niemela S (1990) *Helicobacter pylori* induces lymphocyte activation in peripheral blood cultures. Clin Exp Immunol 82:485–488

Kearley J, Erjefalt JS, Andersson C, Benjamin E, Jones CP, Robichaud A, Pegorier S, Brewah Y, Burwell TJ, Bjermer L, Kiener PA, Kolbeck R, Lloyd CM, Coyle AJ, Humbles AA (2011) IL-9 governs allergen-induced mast cell numbers in the lung and chronic remodeling of the airways. Am J Respir Crit Care Med 183(7):865–875. https://doi.org/10.1164/rccm.200909-1462OC

Khader SA, Gaffen SL, Kolls JK (2009) Th17 cells at the crossroads of innate and adaptive immunity against infectious diseases at the mucosa. Mucosal Immunol 2(5):403–411. https://doi.org/10.1038/mi.2009.100

Kondo M, Weissman IL, Akashi K (1997) Identification of clonogenic common lymphoid progenitors in mouse bone marrow. Cell 91(5):661–672

Kranzer K, Eckhardt A, Aigner M, Knoll G, Deml L, Speth C, Lehn N, Rehli M, Schneider-Brachert W (2004) Induction of maturation and cytokine release of human dendritic cells by *Helicobacter pylori*. Infect Immun 72(8):4416–4423. https://doi.org/10.1128/IAI.72.8.4416-4423.2004

Kreymborg K, Etzensperger R, Dumoutier L, Haak S, Rebollo A, Buch T, Heppner FL, Renauld JC, Becher B (2007) IL-22 is expressed by Th17 cells in an IL-23-dependent fashion, but not required for the development of autoimmune encephalomyelitis. J Immunol 179(12):8098–8104

Kuang DM, Xiao X, Zhao Q, Chen MM, Li XF, Liu RX, Wei Y, Ouyang FZ, Chen DP, Wu Y, Lao XM, Deng H, Zheng L (2014) B7-H1-expressing antigen-presenting cells mediate polarization of protumorigenic Th22 subsets. J Clin Invest 124(10):4657–4667. https://doi.org/10.1172/JCI74381

La-Beck NM, Jean GW, Huynh C, Alzghari SK, Lowe DB (2015) Immune checkpoint inhibitors: new insights and current place in cancer therapy. Pharmacotherapy 35(10):963–976. https://doi.org/10.1002/phar.1643

Laan M, Cui ZH, Hoshino H, Lotvall J, Sjostrand M, Gruenert DC, Skoogh BE, Linden A (1999) Neutrophil recruitment by human IL-17 via C-X-C chemokine release in the airways. J Immunol 162(4):2347–2352

Lai Kwan LQ, King Hung KO, Zheng BJ, Lu L (2008) Local BAFF gene silencing suppresses Th17-cell generation and ameliorates autoimmune arthritis. Proc Natl Acad Sci U S A 105(39):14993–14998. https://doi.org/10.1073/pnas.0806044105

Langrish CL, Chen Y, Blumenschein WM, Mattson J, Basham B, Sedgwick JD, McClanahan T, Kastelein RA, Cua DJ (2005) IL-23 drives a pathogenic T cell population that induces autoimmune inflammation. J Exp Med 201(2):233–240. https://doi.org/10.1084/jem.20041257

Lee IF, Wang X, Hao J, Akhoundsadegh N, Chen L, Liu L, Langermann S, Ou D, Warnock GL (2013) B7-H4.Ig inhibits the development of type 1 diabetes by regulating Th17 cells in NOD mice. Cell Immunol 282(1):1–8. https://doi.org/10.1016/j.cellimm.2013.03.005

Lefrancois L, Puddington L (1995) Extrathymic intestinal T-cell development: virtual reality? Immunol Today 16(1):16–21

Lenschow DJ, Sperling AI, Cooke MP, Freeman G, Rhee L, Decker DC, Gray G, Nadler LM, Goodnow CC, Bluestone JA (1994) Differential up-regulation of the B7-1 and B7-2 costimulatory molecules after Ig receptor engagement by antigen. J Immunol 153:1990–1997

Lewis ND, Asim M, Barry DP, de Sablet T, Singh K, Piazuelo MB, Gobert AP, Chaturvedi R, Wilson KT (2011) Immune evasion by *Helicobacter pylori* is mediated by induction of macrophage arginase II. J Immunol 186(6):3632–3641. https://doi.org/10.4049/jimmunol.1003431

Liang SC, Tan XY, Luxenberg DP, Karim R, Dunussi-Joannopoulos K, Collins M, Fouser LA (2006) Interleukin (IL)-22 and IL-17 are coexpressed by Th17 cells and cooperatively enhance expression of antimicrobial peptides. J Exp Med 203(10):2271–2279. https://doi.org/10.1084/jem.20061308

Lina TT, Alzahrani S, House J, Yamaoka Y, Sharpe AH, Rampy BA, Pinchuk IV, Reyes VE (2015) *Helicobacter pylori* cag pathogenicity island's role in B7-H1 induction and immune evasion. PLoS ONE 10(3):e0121841. https://doi.org/10.1371/journal.pone.0121841

Lina TT, Pinchuk IV, House J, Yamaoka Y, Graham DY, Beswick EJ, Reyes VE (2013) CagA-dependent downregulation of B7-H2 expression on gastric mucosa and inhibition of Th17 responses during *Helicobacter pylori* infection. J Immunol 191(7):3838–3846. https://doi.org/10.4049/jimmunol.1300524

Loke P, Allison JP (2003) PD-L1 and PD-L2 are differentially regulated by Th1 and Th2 cells. Proc Natl Acad Sci U S A 100(9):5336–5341. https://doi.org/10.1073/pnas.0931259100

Lou Z, Casali P, Xu Z (2015) Regulation of B cell differentiation by intracellular membrane-associated proteins and microRNAs: role in the antibody response. Front Immunol 6:537. https://doi.org/10.3389/fimmu.2015.00537

Love PE, Hayes SM (2010) ITAM-mediated signaling by the T-cell antigen receptor. Cold Spring Harb Perspect Biol 2(6):a002485. https://doi.org/10.1101/cshperspect.a002485

Lundgren A, Stromberg E, Sjoling A, Lindholm C, Enarsson K, Edebo A, Johnsson E, Suri-Payer E, Larsson P, Rudin A, Svennerholm AM, Lundin BS (2005) Mucosal FOXP3-expressing $CD4^+$ CD25 high regulatory T cells in *Helicobacter pylori*-infected patients. Infect Immun 73(1):523–531. https://doi.org/10.1128/IAI.73.1.523-531.2005

Lundgren A, Suri-Payer E, Enarsson K, Svennerholm AM, Lundin BS (2003) *Helicobacter pylori*-specific $CD4^+$ CD25 high regulatory T cells suppress memory T-cell responses to *H. pylori* in infected individuals. Infect Immun 71(4):1755–1762

Ly D, Moody DB (2014) The CD1 size problem: lipid antigens, ligands, and scaffolds. Cell Mol Life Sci 71(16):3069–3079. https://doi.org/10.1007/s00018-014-1603-6

Ma Z, Liu G, Zhang M, Li M, Liu Y, Yanfang J (2016) *Helicobacter pylori* infection increases frequency of PDCA-$1^{(+)}$ (CD$317^{(+)}$) B-cell subsets. Arch Med Res 47(2):96–104. https://doi.org/10.1016/j.arcmed.2016.04.003

Maddaly R, Pai G, Balaji S, Sivaramakrishnan P, Srinivasan L, Sunder SS, Paul SF (2010) Receptors and signaling mechanisms for B-lymphocyte activation, proliferation and differentiation–insights from both *in vivo* and *in vitro* approaches. FEBS Lett 584(24):4883–4894. https://doi.org/10.1016/j.febslet.2010.08.022

Mangan PR, Harrington LE, O'Quinn DB, Helms WS, Bullard DC, Elson CO, Hatton RD, Wahl SM, Schoeb TR, Weaver CT (2006) Transforming growth factor-beta induces development of the T(H)17 lineage1. Nature 441(7090):231–234. https://doi.org/10.1038/nature04754

Martin-Orozco N, Chung Y, Chang SH, Wang YH, Dong C (2009) Th17 cells promote pancreatic inflammation but only induce diabetes efficiently in lymphopenic hosts after conversion into Th1 cells. Eur J Immunol 39(1):216–224. https://doi.org/10.1002/eji.200838475

McAlees JW, Lajoie S, Dienger K, Sproles AA, Richgels PK, Yang Y, Khodoun M, Azuma M, Yagita H, Fulkerson PC, Wills-Karp M, Lewkowich IP (2015) Differential control of $CD4^{(+)}$ T-cell subsets by the PD-1/PD-L1 axis in a mouse model of allergic asthma. Eur J Immunol 45(4):1019–1029. https://doi.org/10.1002/eji.201444778

McClory S, Hughes T, Freud AG, Briercheck EL, Martin C, Trimboli AJ, Yu J, Zhang X, Leone G, Nuovo G, Caligiuri MA (2012) Evidence for a stepwise program of extrathymic T cell development within the human tonsil. J Clin Invest 122(4):1403–1415. https://doi.org/10.1172/JCI46125

Meyer F, Ramanujam KS, Gobert AP, James SP, Wilson KT (2003) Cutting edge: cyclooxygenase-2 activation suppresses Th1 polarization in response to *Helicobacter pylori*. J Immunol 171(8):3913–3917

Michetti M, Kelly CP, Kraehenbuhl JP, Bouzourene H, Michetti P (2000) Gastric mucosal alpha(4)beta(7)-integrin-positive CD4 T lymphocytes and immune protection against *helicobacter* infection in mice. Gastro 119(1):109–118

Mora JR, von Andrian UH (2009) Role of retinoic acid in the imprinting of gut-homing IgA-secreting cells. Semin Immunol 21(1):28–35. https://doi.org/10.1016/j.smim.2008.08.002

Moyat M, Bouzourene H, Ouyang W, Iovanna J, Renauld JC, Velin D (2017) IL-22-induced antimicrobial peptides are key determinants of mucosal vaccine-induced protection against *H. pylori* in mice. Mucosal Immunol 10(1):271–281. https://doi.org/10.1038/mi.2016.38

Munari F, Fassan M, Capitani N, Codolo G, Vila-Caballer M, Pizzi M, Rugge M, Della BC, Troilo A, D'Elios S, Baldari CT, D'Elios MM, de Bernard M (2014) Cytokine BAFF released by *Helicobacter pylori*-infected macrophages triggers the Th17 response in human chronic gastritis. J Immunol 193(11):5584–5594. https://doi.org/10.4049/jimmunol.1302865

Murray PJ (2017) Macrophage polarization. Annu Rev Physiol 79:541–566. https://doi.org/10.1146/annurev-physiol-022516-034339

Nakayama T, Hirahara K, Onodera A, Endo Y, Hosokawa H, Shinoda K, Tumes DJ, Okamoto Y (2017) Th2 cells in health and disease. Annu Rev Immunol 35:53–84. https://doi.org/10.1146/annurev-immunol-051116-052350

Nurgalieva ZZ, Conner ME, Opekun AR, Zheng CQ, Elliott SN, Ernst PB, Osato M, Estes MK, Graham DY (2005) B-cell and T-cell immune responses to experimental *Helicobacter pylori* infection in humans. Infect Immun 73(5):2999–3006. https://doi.org/10.1128/IAI.73.5.2999-3006.2005

Nurieva RI, Chung Y, Martinez GJ, Yang XO, Tanaka S, Matskevitch TD, Wang YH, Dong C (2009) Bcl6 mediates the development of T follicular helper cells. Science 325(5943):1001–1005. https://doi.org/10.1126/science.1176676

Oertli M, Noben M, Engler DB, Semper RP, Reuter S, Maxeiner J, Gerhard M, Taube C, Muller A (2013) *Helicobacter pylori* gamma-glutamyl transpeptidase and vacuolating cytotoxin promote gastric persistence and immune tolerance. Proc Natl Acad Sci U S A 110(8):3047–3052. https://doi.org/10.1073/pnas.1211248110

Oertli M, Sundquist M, Hitzler I, Engler DB, Arnold IC, Reuter S, Maxeiner J, Hansson M, Taube C, Quiding-Jarbrink M, Muller A (2012) DC-derived IL-18 drives Treg differentiation, murine *Helicobacter pylori*-specific immune tolerance, and asthma protection. J Clin Invest 122(3):1082–1096. https://doi.org/10.1172/JCI61029

Otani K, Tanigawa T, Watanabe T, Nadatani Y, Sogawa M, Yamagami H, Shiba M, Watanabe K, Tominaga K, Fujiwara Y, Arakawa T (2012) Toll-like receptor 9 signaling has anti-inflammatory effects on the early phase of *Helicobacter pylori*-induced gastritis. Biochem Biophys Res Commun 426(3):342–349. https://doi.org/10.1016/j.bbrc.2012.08.080

Pachathundikandi SK, Tegtmeyer N, Backert S (2013) Signal transduction of *Helicobacter pylori* during interaction with host cell protein receptors of epithelial and immune cells. Gut Microbes 4(6):454–474. https://doi.org/10.4161/gmic.27001

Pachathundikandi SK, Müller A, Backert S (2016) Inflammasome activation by *Helicobacter pylori* and its implications for persistence and immunity. Curr Top Microbiol Immunol 397:117–131. https://doi.org/10.1007/978-3-319-41171-2_6

Pachathundikandi SK, Lind J, Tegtmeyer N, El-Omar EM, Backert S (2015) Interplay of the gastric pathogen *Helicobacter pylori* with toll-like receptors. Biomed Res Int 2015:192420. https://doi.org/10.1155/2015/192420

Pages F, Ragueneau M, Rottapel R, Truneh A, Nunes J, Imbert J, Olive D (1994) Binding of phosphatidylinositol-3-OH kinase to CD28 is required for T-cell signalling. Nature 369(6478):327–329. https://doi.org/10.1038/369327a0

Pai SY, Truitt ML, Ho IC (2004) GATA-3 deficiency abrogates the development and maintenance of T helper type 2 cells. Proc Natl Acad Sci U S A 101(7):1993–1998. https://doi.org/10.1073/pnas.0308697100

Papini E, de Bernard M, Milia E, Bugnoli M, Zerial M, Rappuoli R, Montecucco C (1994) Cellular vacuoles induced by *Helicobacter pylor*i originate from late endosomal compartments. Proc Natl Acad Sci USA 91(21):9720–9724

Pappo J, Torrey D, Castriotta L, Savinainen A, Kabok Z, Ibraghimov A (1999) *Helicobacter pylori* infection in immunized mice lacking major histocompatibility complex class I and class II functions. Infect Immun 67(1):337–341

Park H, Li Z, Yang XO, Chang SH, Nurieva R, Wang YH, Wang Y, Hood L, Zhu Z, Tian Q, Dong C (2005) A distinct lineage of CD4 T cells regulates tissue inflammation by producing interleukin 17. Nat Immunol 6(11):1133–1141. https://doi.org/10.1038/ni1261

Park JJ, Omiya R, Matsumura Y, Sakoda Y, Kuramasu A, Augustine MM, Yao S, Tsushima F, Narazaki H, Anand S, Liu Y, Strome SE, Chen L, Tamada K (2010) B7-H1/CD80 interaction is required for the induction and maintenance of peripheral T-cell tolerance. Blood 116(8):1291–1298. https://doi.org/10.1182/blood-2010-01-265975

Paulos CM, Carpenito C, Plesa G, Suhoski MM, Varela-Rohena A, Golovina TN, Carroll RG, Riley JL, June CH (2010) The inducible costimulator (ICOS) is critical for the development of human T(H)17 cells. Sci Transl Med 2(55):55–78. https://doi.org/10.1126/scitranslmed.3000448

Plank MW, Kaiko GE, Maltby S, Weaver J, Tay HL, Shen W, Wilson MS, Durum SK, Foster PS (2017) Th22 cells form a distinct Th lineage from Th17 cells *in vitro* with unique transcriptional properties and Tbet-dependent Th1 plasticity. J Immunol 198(5):2182–2190. https://doi.org/10.4049/jimmunol.1601480

Portal-Celhay C, Perez-Perez GI (2006) Immune responses to *Helicobacter pylori* colonization: mechanisms and clinical outcomes. Clin Sci (Lond) 110(3):305–314. https://doi.org/10.1042/CS20050232

Posselt G, Backert S, Wessler S (2013) The functional interplay of *Helicobacter pylori* factors with gastric epithelial cells induces a multi-step process in pathogenesis. Cell Commun Signal 11:77. https://doi.org/10.1186/1478-811X-11-77

Quiding-Jarbrink M, Ahlstedt I, Lindholm C, Johansson EL, Lonroth H (2001) Homing commitment of lymphocytes activated in the human gastric and intestinal mucosa. Gut 49(4):519–525

Quiding-Jarbrink M, Raghavan S, Sundquist M (2010) Enhanced M1 macrophage polarization in human *helicobacter pylori*-associated atrophic gastritis and in vaccinated mice. PLoS ONE 5(11):e15018. https://doi.org/10.1371/journal.pone.0015018

Reyes VE, Beswick EJ (2007) *Helicobacter pylori* neutrophil activating protein's potential as tool in therapeutic immune modulation. Expert Opin Ther Pat 17(10):1315–1320. https://doi.org/10.1517/13543776.17.10.1315

Rimbara E, Mori S, Kim H, Shibayama K (2013) Role of gamma-glutamyltranspeptidase in the pathogenesis of *Helicobacter pylori* infection. Microbiol Immunol 57(10):665–673. https://doi.org/10.1111/1348-0421.12089

Robinson K, Kenefeck R, Pidgeon EL, Shakib S, Patel S, Polson RJ, Zaitoun AM, Atherton JC (2008) *Helicobacter pylori*-induced peptic ulcer disease is associated with inadequate regulatory T cell responses. Gut 57(10):1375–1385. https://doi.org/10.1136/gut.2007.137539

Rodriguez-Perea AL, Arcia ED, Rueda CM, Velilla PA (2016) Phenotypical characterization of regulatory T cells in humans and rodents. Clin Exp Immunol 185(3):281–291. https://doi.org/10.1111/cei.12804

Ruscher R, Kummer RL, Lee YJ, Jameson SC, Hogquist KA (2017) CD8alphaalpha intraepithelial lymphocytes arise from two main thymic precursors. Nat Immunol 18(7):771–779. https://doi.org/10.1038/ni.3751

Rutz S, Eidenschenk C, Ouyang W (2013) IL-22, not simply a Th17 cytokine. Immunol Rev 252(1):116–132. https://doi.org/10.1111/imr.12027

Sayi A, Kohler E, Toller IM, Flavell RA, Muller W, Roers A, Muller A (2011) TLR-2-activated B cells suppress *Helicobacter*-induced preneoplastic gastric immunopathology by inducing T regulatory-1 cells. J Immunol 186(2):878–890. https://doi.org/10.4049/jimmunol.1002269

Schaeffer EM, Debnath J, Yap G, McVicar D, Liao XC, Littman DR, Sher A, Varmus HE, Lenardo MJ, Schwartzberg PL (1999) Requirement for Tec kinases Rlk and Itk in T cell receptor signaling and immunity. Science 284(5414):638–641

Scheinman EJ, Avni O (2009) Transcriptional regulation of GATA3 in T helper cells by the integrated activities of transcription factors downstream of the interleukin-4 receptor and T cell receptor. J Biol Chem 284(5):3037–3048. https://doi.org/10.1074/jbc.M807302200

Schmees C, Prinz C, Treptau T, Rad R, Hengst L, Voland P, Bauer S, Brenner L, Schmid RM, Gerhard M (2007) Inhibition of T-cell proliferation by *Helicobacter pylori* gamma-glutamyl transpeptidase. Gastro 132(5):1820–1833. https://doi.org/10.1053/j.gastro.2007.02.031

Schwartz JT, Allen LA (2006) Role of urease in megasome formation and *Helicobacter pylori* survival in macrophages. J Leukoc Biol 79(6):1214–1225. https://doi.org/10.1189/jlb.0106030

Sewald X, Gebert-Vogl B, Prassl S, Barwig I, Weiss E, Fabbri M, Osicka R, Schiemann M, Busch DH, Semmrich M, Holzmann B, Sebo P, Haas R (2008) Integrin subunit CD18 Is the T-lymphocyte receptor for the *Helicobacter pylori* vacuolating cytotoxin. Cell Host Microbe 3(1):20–29. https://doi.org/10.1016/j.chom.2007.11.003

Sewald X, Jimenez-Soto L, Haas R (2011) PKC-dependent endocytosis of the *Helicobacter pylori* vacuolating cytotoxin in primary T lymphocytes. Cell Microbiol 13(3):482–496. https://doi.org/10.1111/j.1462-5822.2010.01551.x

Shi Y, Liu XF, Zhuang Y, Zhang JY, Liu T, Yin Z, Wu C, Mao XH, Jia KR, Wang FJ, Guo H, Flavell RA, Zhao Z, Liu KY, Xiao B, Guo Y, Zhang WJ, Zhou WY, Guo G, Zou QM (2010) *Helicobacter pylori*-induced Th17 responses modulate Th1 cell responses, benefit bacterial growth, and contribute to pathology in mice. J Immunol 184(9):5121–5129. https://doi.org/10.4049/jimmunol.0901115

Smith SM (2014) Role of toll-like receptors in *Helicobacter pylori* infection and immunity. World J Gastrointest Pathophysiol 5(3):133–146. https://doi.org/10.4291/wjgp.v5.i3.133

Stagg AJ, Kamm MA, Knight SC (2002) Intestinal dendritic cells increase T cell expression of alpha4beta7 integrin. Eur J Immunol 32(5):1445–1454. https://doi.org/10.1002/1521-4141(200205)32:5%3c1445:AID-IMMU1445%3e3.0.CO;2-E

Staudt V, Bothur E, Klein M, Lingnau K, Reuter S, Grebe N, Gerlitzki B, Hoffmann M, Ulges A, Taube C, Dehzad N, Becker M, Stassen M, Steinborn A, Lohoff M, Schild H, Schmitt E, Bopp T (2010) Interferon-regulatory factor 4 is essential for the developmental program of T helper 9 cells. Immunity 33(2):192–202. https://doi.org/10.1016/j.immuni.2010.07.014

Stephens GL, Swerdlow B, Benjamin E, Coyle AJ, Humbles A, Kolbeck R, Fung M (2011) IL-9 is a Th17-derived cytokine that limits pathogenic activity in organ-specific autoimmune disease. Eur J Immunol 41(4):952–962. https://doi.org/10.1002/eji.201040879

Sundrud MS, Torres VJ, Unutmaz D, Cover TL (2004) Inhibition of primary human T cell proliferation by *Helicobacter pylori* vacuolating toxin (VacA) is independent of VacA effects on IL-2 secretion. Proc Natl Acad Sci U S A 101(20):7727–7732. https://doi.org/10.1073/pnas.0401528101

Sutton C, Brereton C, Keogh B, Mills KH, Lavelle EC (2006) A crucial role for interleukin (IL)-1 in the induction of IL-17-producing T cells that mediate autoimmune encephalomyelitis. J Exp Med 203(7):1685–1691. https://doi.org/10.1084/jem.20060285

Swaidani S, Bulek K, Kang Z, Gulen MF, Liu C, Yin W, Abbadi A, Aronica M, Li X (2011) T cell-derived Act1 is necessary for IL-25-mediated Th2 responses and allergic airway inflammation. J Immunol 187(6):3155–3164. https://doi.org/10.4049/jimmunol.1002790

Szabo SJ, Kim ST, Costa GL, Zhang X, Fathman CG, Glimcher LH (2000) A novel transcription factor, T-bet, directs Th1 lineage commitment. Cell 100(6):655–669

Takaba H, Takayanagi H (2017) The mechanisms of T cell selection in the thymus. Trends Immunol 38(11):805–816. https://doi.org/10.1016/j.it.2017.07.010

Van Kaer L, Parekh VV, Postoak JL, Wu L (2017) Role of autophagy in MHC class I-restricted antigen presentation. Mol Immunol S0161–5890(17):330550–330553. https://doi.org/10.1016/j.molimm.2017.10.021

Vieira PL, Christensen JR, Minaee S, O'Neill EJ, Barrat FJ, Boonstra A, Barthlott T, Stockinger B, Wraith DC, O'garra A (2004) IL-10-secreting regulatory T cells do not express Foxp3 but have comparable regulatory function to naturally occurring $CD4^+$ $CD25^+$ regulatory T cells. J Immunol 172(10):5986–5993

von Andrian UH, Mackay CR (2000). T-cell function and migration. Two sides of the same coin. N Engl J Med 343(14):1020–1034. https://doi.org/10.1056/nejm200010053431407

von Boehmer H (2005) Unique features of the pre-T-cell receptor alpha-chain: not just a surrogate. Nat Rev Immunol 5(7):571–577. https://doi.org/10.1038/nri1636

Walunas TL, Lenschow DJ, Bakker CY, Linsley PS, Freeman GJ, Green JM, Thompson CB, Bluestone JA (1994) CTLA-4 can function as a negative regulator of T cell activation. Immunity 1(5):405–413

Weiner HL (2001) Induction and mechanism of action of transforming growth factor-beta-secreting Th3 regulatory cells. Immunol Rev 182:207–214

Weiss A, Newton M, Crommie D (1986) Expression of T3 in association with a molecule distinct from the T-cell antigen receptor heterodimer. Proc Natl Acad Sci U S A 83(18):6998–7002

Wherry EJ (2011) T cell exhaustion. Nat Immunol 12(6):492–499. https://doi.org/10.1038/ni.2035

Williams MB, Butcher EC (1997) Homing of naive and memory T lymphocyte subsets to Peyer's patches, lymph nodes, and spleen. J Immunol 159(4):1746–1752

Wu YY, Lin CW, Cheng KS, Lin C, Wang YM, Lin IT, Chou YH, Hsu PN (2010) Increased programmed death-ligand-1 expression in human gastric epithelial cells in *Helicobacter pylori* infection. Clin Exp Immunol 161(3):551–559. https://doi.org/10.1111/j.1365-2249.2010.04217.x

Xu L, Kitani A, Fuss I, Strober W (2007) Cutting edge: regulatory T cells induce $CD4^+$ CD25-$Foxp3^-$ T cells or are self-induced to become Th17 cells in the absence of exogenous TGF-beta. J Immunol 178(11):6725–6729

Xu Z, Shen J, Wang MH, Yi T, Yu Y, Zhu Y, Chen B, Chen J, Li L, Li M, Zuo J, Jiang H, Zhou D, Luan J, Xiao Z (2016) Comprehensive molecular profiling of the B7 family of immune-regulatory ligands in breast cancer. Oncoimmunology 5(8):e1207841. https://doi.org/10.1080/2162402X.2016.1207841

Yang XO, Nurieva R, Martinez GJ, Kang HS, Chung Y, Pappu BP, Shah B, Chang SH, Schluns KS, Watowich SS, Feng XH, Jetten AM, Dong C (2008) Molecular antagonism and plasticity of regulatory and inflammatory T cell programs. Immunity 29(1):44–56. https://doi.org/10.1016/j.immuni.2008.05.007

Ye G, Barrera C, Fan XJ, Gourley WK, Crowe SE, Ernst PB, Reyes VE (1997) Expression of B7-1 and B7-2 costimulatory molecules by human gastric epithelial cells—potential role in $CD4^{(+)}$ T cell activation during *Helicobacter pylori* infection. J Clin Invest 99(7):1628–1636. https://doi.org/10.1172/JCI119325

Yokosuka T, Takamatsu M, Kobayashi-Imanishi W, Hashimoto-Tane A, Azuma M, Saito T (2012) Programmed cell death 1 forms negative costimulatory microclusters that directly inhibit T cell receptor signaling by recruiting phosphatase SHP2. J Exp Med 209(6):1201–1217. https://doi.org/10.1084/jem.20112741

Yoshida H, Hamano S, Senaldi G, Covey T, Faggioni R, Mu S, Xia M, Wakeham AC, Nishina H, Potter J, Saris CJ, Mak TW (2001) WSX-1 is required for the initiation of Th1 responses and resistance to *L. major* infection. Immunity 15(4):569–578

Zabel BA, Agace WW, Campbell JJ, Heath HM, Parent D, Roberts AI, Ebert EC, Kassam N, Qin S, Zovko M, LaRosa GJ, Yang LL, Soler D, Butcher EC, Ponath PD, Parker CM, Andrew DP (1999) Human G protein-coupled receptor GPR-9-6/CC chemokine receptor 9 is selectively expressed on intestinal homing T lymphocytes, mucosal lymphocytes, and thymocytes and is required for thymus-expressed chemokine-mediated chemotaxis. J Exp Med 190(9):1241–1256

Zhang B, Rong G, Wei H, Zhang M, Bi J, Ma L, Xue X, Wei G, Liu X, Fang G (2008) The prevalence of Th17 cells in patients with gastric cancer. Biochem Biophys Res Commun 374(3):533–537. https://doi.org/10.1016/j.bbrc.2008.07.060

Zhang DH, Cohn L, Ray P, Bottomly K, Ray A (1997) Transcription factor GATA-3 is differentially expressed in murine Th1 and Th2 cells and controls Th2-specific expression of the interleukin-5 gene. J Biol Chem 272(34):21597–21603

Zhang XS, Tegtmeyer N, Traube L, Jindal S, Perez-Perez G, Sticht H, Backert S, Blaser MJ (2015) A specific A/T polymorphism in Western tyrosine phosphorylation B-motifs regulates *Helicobacter pylori* CagA epithelial cell interaction. PloS Pathog 11:e1004621. https://doi.org/10.1371/journal.ppat.1004621

Zheng PY, Jones NL (2003) *Helicobacter pylori* strains expressing the vacuolating cytotoxin interrupt phagosome maturation in macrophages by recruiting and retaining TACO (coronin 1) protein. Cell Microbiol 5(1):25–40. https://doi.org/10.1371/journal.pone.0023629

Zhou X, Xia Z, Lan Q, Wang J, Su W, Han YP, Fan H, Liu Z, Stohl W, Zheng SG (2011) BAFF promotes Th17 cells and aggravates experimental autoimmune encephalomyelitis. PLoS ONE 6(8):e23629

Zhu J, Min B, Hu-Li J, Watson CJ, Grinberg A, Wang Q, Killeen N, Urban JF Jr, Guo L, Paul WE (2004) Conditional deletion of Gata3 shows its essential function in T(H)1-T(H)2 responses. Nat Immunol 5(11):1157–1165. https://doi.org/10.1038/ni1128

Zhuang Y, Shi Y, Liu XF, Zhang JY, Liu T, Fan X, Luo J, Wu C, Yu S, Chen L, Luo P, Guo G, Liu Z, Tang B, Mao XH, Guo Y, Zou QM (2011) *Helicobacter pylori*-infected macrophages induce Th17 cell differentiation. Immunobiology 216(1–2):200–207. https://doi.org/10.1016/j.imbio.2010.05.005

Mechanisms of Inflammasome Signaling, microRNA Induction and Resolution of Inflammation by *Helicobacter pylori*

Suneesh Kumar Pachathundikandi, Nicole Blaser and Steffen Backert

Contents

Abstract Inflammasome-controlled transcription and subsequent cleavage-mediated activation of mature IL-1β and IL-18 cytokines exemplify a crucial innate immune mechanism to combat intruding pathogens. *Helicobacter pylori* represents a predominant persistent infection in humans, affecting approximately half of the population worldwide, and is associated with the development of chronic gastritis, peptic ulcer disease, and gastric cancer. Studies in knockout mice have demonstrated that the pro-inflammatory cytokine IL-1β plays a central role in gastric tumorigenesis. Infection by *H. pylori* was recently reported to stimulate the inflammasome both in cells of the mouse and human immune systems. Using mouse models and in vitro cultured cell systems, the bacterial pathogenicity factors and molecular mechanisms of inflammasome activation have been analyzed. On the one hand, it appears that *H. pylori*-stimulated IL-1β production is triggered by engagement of the immune receptors TLR2 and NLRP3, and caspase-1. On the other hand, microRNA hsa-miR-223-3p is induced by the bacteria, which controls the expression of NLRP3. This regulating effect by *H. pylori* on microRNA expression was also described for more than 60 additionally identified microRNAs, indicating a prominent role for inflammatory and other responses. Besides TLR2, TLR9 becomes activated by *H. pylori* DNA and further TLR10 stimulated by the bacteria induce the secretion of IL-8 and TNF, respectively. Interestingly, TLR-dependent pathways can accelerate both pro- and anti-inflammatory responses during *H. pylori* infection. Balancing

---

S. K. Pachathundikandi · N. Blaser · S. Backert (✉)
Department of Biology, Division of Microbiology, Friedrich Alexander University Erlangen-Nuremberg, Staudtstr. 5, 91058 Erlangen, Germany
e-mail: Steffen.Backert@fau.de

S. Backert (ed.), *Molecular Mechanisms of Inflammation: Induction, Resolution and Escape by Helicobacter pylori*, Current Topics in Microbiology and Immunology 421, https://doi.org/10.1007/978-3-030-15138-6_11

from a pro-inflammation to anti-inflammation phenotype results in a reduction in immune attack, allowing *H. pylori* to persistently colonize and to survive in the gastric niche. In this chapter, we will pinpoint the role of *H. pylori* in TLR- and NLRP3 inflammasome-dependent signaling together with the differential functions of pro- and anti-inflammatory cytokines. Moreover, the impact of microRNAs on *H. pylori*–host interaction will be discussed, and its role in resolution of infection versus chronic infection, as well as in gastric disease development.

1 Introduction

Mechanisms initiating inflammatory effects in response to bacterial infections are mostly associated with the activation of innate immune signaling complexes, so-called inflammasomes (Man 2018). These inflammasomes constitute multimeric protein complexes and generally consist of a pattern recognition receptor (PRR) for sensing pathogen-associated molecular patterns (PAMPs) and pro-caspase-1 (Hara et al. 2013). Based on the bacterial ligand specificity, subcellular localization of the receptor, and the involved signaling pathway, PRRs are classified into two main families—the Toll-like receptors (TLRs) and nucleotide-binding-domain leucine-rich repeat containing receptors (NLRs) (Martinon et al. 2002; Mogensen 2009; Hara et al. 2013). The PRR being part of the inflammasome is decisive for the classification of the various inflammasomes (Hara et al. 2013). Following assembly of the inflammasome, the complex becomes activated in a two-step manner. First, nuclear factor kappa-light-chain-enhancer of activated B cells (NF-κB) or other transcription factors induce the enhanced expression of proforms of several pro-inflammatory cytokines or interleukins (ILs), such as IL-1β or IL-18 (Mariathasan and Monack 2007). Subsequently, pro-caspase-1 becomes activated by the inflammasome complex itself or by a second signal, leading to the processing of IL-1β or IL-18 precursor proteins by matured caspase-1 (Martinon et al. 2002; Dinarello 2009). As a result, inflammasome activation promotes the secretion of the bioactive forms of both of these interleukins, and moreover, a specific form of cell death, called pyroptosis (Martinon et al. 2002; Dinarello 2009; Miao et al. 2011). NLRP3, NLRP1, and NLRC4 from the nucleotide-binding-domain (NOD)-like receptor family are known to play an essential role in the inflammasome activation by various Gram-negative bacteria, such as *H. pylori* or *Salmonella typhimurium*, or by recognition of different bacterial products (Mariathasan et al. 2006; Man 2018). Inflammasome activation and the associated increased IL-1β concentrations were shown to encourage the pathogenesis of chronic and acute diseases in the gastrointestinal tract, such as inflammatory bowel disease, gastritis, Crohn's disease, or gastric cancer (GC) (Guo et al. 2015). In addition, another study has demonstrated that the uncontrolled activation of TLRs leading to continuous stimulation of the immune system favors chronic inflammatory diseases (Andreakos et al. 2004).

A prominent cause for peptic ulcer disease (PUD), GC, mucosa associated lymphoid tissue (MALT) lymphoma, and other illnesses is the Gram-negative gastrointestinal bacterium *H. pylori* (Atherton and Blaser 2009; Polk and Peek 2010; Salama et al. 2013; Pachathundikandi et al. 2013). A recent review summarizing the meta-data analysis of the *H. pylori* global prevalence has indicated that 44.3–60.3% percent of the global population was infected with this pathogen in 2015 (Sjomina et al. 2018). A correlation between the levels of secreted IL-1β in the stomach with gastric inflammation and the generation of GC has been indicated (Posselt et al. 2013; Kameoka et al. 2016). *H. pylori* was shown to modulate the IL-1β secretion through activation of TLR2 by lipopolysaccharide (LPS), and other virulence factors, such as vacuolating cytotoxin A (VacA) or urease, and thus promotes the NLRP3 inflammasome assembly and activation (Pachathundikandi et al. 2016). Moreover, *H. pylori* affects the host innate immune response not only by activating TLRs, but also by direct injection of the effector protein CagA in gastric epithelial cells (GECs) using a type IV secretion system (T4SS) (Backert et al. 2008, 2015; Zhang et al. 2015). In this case, translocated CagA is associated with IL-1β mRNA production and inflammasome assembly (Kameoka et al. 2016). Thus, a series of *H. pylori* virulence factors was shown to be associated with inflammasome activation, which promotes bacterial clearance, and hence seemed to be associated with resolution of inflammation (Naumann et al. 2017).

MicroRNAs (miRNAs) are small endogenous noncoding RNA molecules, which function by targeting mRNA for translational repression or decay to regulate post-transcriptional gene expression. miRNAs were discovered in 1993 for their role in development of *Caenorhabditis elegans* (Lee et al. 1993; Wightman et al. 1993). Further studies in the last decade identified many different miRNAs and a leading database represented 38,589 entries of hairpin precursor miRNAs, expressing 48,885 mature miRNA products, in 271 species (http://www.mirbase.org/). There are several other databases, which have compiled the functions of verified miRNAs for their roles in cellular pathways. These studies also revealed the importance of miRNAs on maintaining homeostasis, development, causing diseases of different kinds, controlling infections and immune activities (Alvarez-Garcia and Miska 2005; Giza et al. 2014; Cadamuro et al. 2014; Zhou et al. 2018). In this context, we looked for the multitude of effects managed by miRNAs during *H. pylori* infection. *H. pylori* infection attracts a strong immune reaction from the host and eventually turns into the production of different types of adaptive immune responses (Dunn et al. 1997; White et al. 2015; Gobert and Wilson 2016). *H. pylori* was reported to be sensed by many different PRRs in GECs, neutrophils, macrophages, and dendritic cells (DCs). The various downstream signaling pathways activate several transcription factors including NF-κB, activator protein 1 (AP-1), and interferon regulatory factor (IRF) to produce various chemokines, cytokines, and other inflammatory gene products for mounting innate and adaptive immune responses (Gobert and Wilson 2016). The most evident adaptive immune response is a Th1-mediated cytotoxic phenotype, but in the course of infection it will also induce Th2 and Th17 responses and Treg population also fill-up the infected mucosa (Wilson and Crabtree 2007; White et al. 2015; see also Chapter "Inflammation, Immunity and Vaccine Development for the Gastric Pathogen *Helicobacter pylori*" of this book). This complex mixed adaptive immune response makes *H. pylori* infection a combination of "friend or foe" responses and plays a crucial role in determining the development of associated diseases, and also has some benefits on controlling ectopic inflammatory conditions.

H. pylori infection induces myriads of host receptor-mediated signal transduction cascades in GECs and immune cells (Backert and Naumann 2010; Pachathundikandi et al. 2015; Pachathundikandi and Backert 2018a). The primary mechanism of innate immune receptor-mediated signaling is to activate the production of various inflammatory mediators for control of the infection (Wilson and Crabtree 2007; White et al. 2015). However, successful and long-term colonizing bacteria like *H. pylori* need to circumvent these efforts by the host immune system. The central process of inflammation is carried out through the production of several soluble inflammatory mediators called cytokines and chemokines (Gobert and Wilson 2016). Pro-inflammatory mediators take the lead at the initial phases to coordinate a full attack on invading pathogens; however, uncontrolled pro-inflammatory actions are detrimental to host tissue. There are anti-inflammatory mechanisms through cytokines, other mediators, and cellular responses. *H. pylori* has been found to manipulate these mechanisms by inducing both pro- and anti-inflammatory mediators, which ultimately dampen the proper functioning of the immune system and supports long-term colonization and persistence in the gastric mucosa of infected individuals (Wilson and Crabtree 2007; White et al. 2015). In this review, we aimed to focus on the role of *H. pylori* infection and host immune responses including cytokine signaling, inflammasome activation, miRNA induction and further impact on the resolution of inflammation.

2 Inflammasome Activation Through NOD2 and TLR2 Signal Transduction

In response to bacterial infections, PRRs being mainly expressed in immune cells recognize foreign PAMPs and promote pro- and anti-inflammatory responses (Bryant et al. 2010; Franchi et al. 2012). The best-studied PRRs are the transmembrane TLRs (Bafica et al. 2005). Next to the TLRs, NOD-like receptors are the main group of bacterial sensors (Kersse et al. 2011; Franchi et al. 2012). In general, many recent reports have shown a stimulatory effect of the bacterial cell wall component peptidoglycan on NOD2, resulting in a NOD2 stimulation, which further leads to the production of pro-inflammatory cytokines and the induction of bacterial killing (Cooney et al. 2010; Travassos et al. 2010; Bansal and Balaji 2011). In addition, a study analyzing the infection by the Gram-negative, obligate intracellular pathogen *Ehrlichia chaffeensis* in murine models, examined that the course of lethal infection was associated with an increased expression of the PRRs TLR2, NOD1 or NOD2 (Chattoraj et al. 2013). In the course of lethal Ehrlichiosis, NOD2 contributed to the dysregulated immune response and immunopathology, while TLR2 seemed to be responsible for the clearance of the pathogens when NOD2 is absent (Chattoraj et al. 2013). In contrast, Hedl and Abraham (2013) have indicated a NOD2-dependent enhancement of bacterial killing in intestinal macrophages in vivo. Thereby, a crucial role of IL-1 for the increased bacterial clearance in response to chronic NOD2 stimulation was determined (Hedl and Abraham 2013). Investigating the effect of NOD2 on the immune response, several studies found that NOD2 modulates NF-κB and MAPK pathways leading to the production of cytokines, chemokines, and antimicrobial peptides (Girardin et al. 2003; Latz et al. 2013). However, recent studies have also shown that NOD2 is able to bind pro-caspase-1 leading further to inflammasome activation (summarized by Mariathasan and Monack 2007). Using human monocytes, a crucial role of NOD2 on IL-1β production and caspase1-dependent maturation was determined (Ferwerda et al. 2008). In response to muramyl dipeptide, it was shown that the association of NOD2 with the inflammasome sensor protein NALP1 resulted in the formation of an inflammasome complex activating caspase-1 (Hsu et al. 2008). Thus, the complex formation of NOD2 with NALP1 might be crucial for the NOD2-dependent effects on IL-1β processing, suggesting a more decisive role of NOD2 in caspase-1 activation than in the modulation of NF-κB and MAPK pathways (Hsu et al. 2008). Consistent with that, Hedl and Abraham (2013) have shown that NOD2-induced inflammasome activation was specifically dependent on the inflammasome sensor proteins NLRP3, NLRP1, and ASC in primary human monocyte-derived macrophages. As TLR2 binds the same bacterial molecules as NOD2, these PRRs seemed to crossregulate the functions of each other and act in a co-stimulatory manner (Lamkanfi 2011; Mercier et al. 2012). Moreover, TLRs were shown to be more effective in activating the NF-κB- and MAPK-dependent pathways (Maeda 2005). Stimulation of macrophages with LPS, known as a ligand for TLR4, leads to an enhanced IL-1β production, presumably promoted by TLR4- or TLR3-based pathways (Maelfait et al. 2008). In addition, Rapsinski and colleagues (2015) have shown that macrophages lacking TLR2 exhibit a decreased IL-1β production and further, the internalization of TLR2 ligands, such as bacterial amyloid curli, is crucial for the effective stimulation of the NLRP3 inflammasome. The TLR2-dependent NLRP3 inflammasome activation and resulting pro-inflammatory cytokine secretion in human cells are shown in Fig. 1a. However, the stimulation of NLRP3 by activated TLR2 was not examined for all TLR2 ligands, suggesting that may be ligand-specific differences in TLR2-based inflammasome activation (Rapsinski et al. 2015).

During *H. pylori* infection of mice, TLR2 signaling was reported to be the rate-limiting step in the activation of the NLRP3 inflammasome (Koch et al. 2015). A study using a transposon mutant library of *H. pylori* revealed that urease deficient bacteria were no longer able of such activation in DCs (Koch et al. 2015). Infection of DCs with specific Δ*ure*A and Δ*ure*B mutants pinpointed the role of UreB in this activation (Koch et al. 2015). However, pro-IL-1β expression was induced through *H. pylori* LPS-mediated activation of TLR4, as particular LPS mutants failed to do so (Koch et al. 2015). This shows the need for concerted action of TLRs in the activation of the inflammasome during *H. pylori* infection. The *tlr2* and *nlrp3* deficient mice showed an increase in interferon-gamma (IFNγ)-producing $CD4^+$ cells and reduced

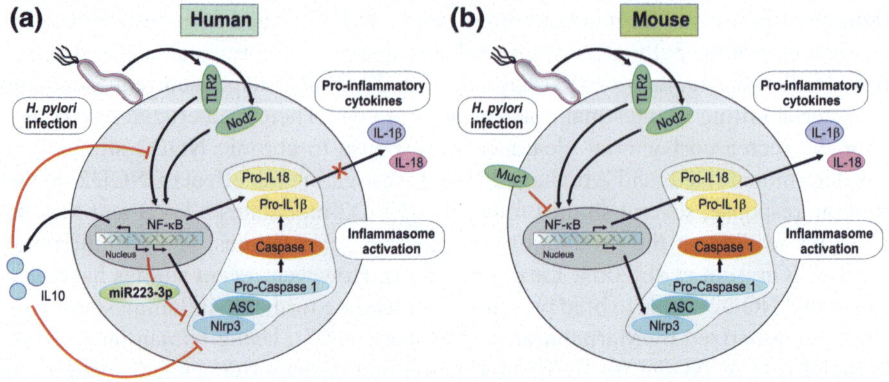

Fig. 1 Signal transduction events leading to inflammasome activation by *Helicobacter pylori.* The pathogen stimulates inflammasome formation in human (**a**) and mouse (**b**) dendritic cells, macrophages or monocytes. Host cell receptors exploited in both systems comprise TLR2 and NOD2, which are targeted by numerous bacterial factors as discussed in the text. A major downstream signaling cascade is proposed to activate the transcription factor NF-κB. This leads to the expression of NLRP3 and two interleukins, the IL-1β and IL-18 pro-forms. Engagement of NLRP3 and the adapter protein ASC by a second signal stimulates the autoproteolytical cleavage of procaspase-1, followed by processing and release of the mature IL-1β and IL-18 cytokines. However, various differences between the human and mouse infection systems have been noted. In human cells, stimulation of NF-κB leads to secretion of anti-inflammatory mediator IL-10. IL-10, in turn, was shown to block NF-κB activation and has a hampering effect on NLRP3 inflammasome activation, resulting in a reduced secretion of pro-inflammatory cytokines. Moreover, NLRP3 inflammasome component expression and thereby activation could be inhibited by human miR223-3p. In contrast, in mice an IL-10-dependent regulation on NLRP3 inflammasome components expression was not yet demonstrated. However, NF-κB could be inhibited by Muc1 in mice, resulting in a hampered NLRP3 inflammasome activity, and thus pro-inflammatory cytokine response (Ng et al. 2016). Abbreviations used: TLR: toll-like receptor; NOD: nucleotide-binding-domain, IL: interleukin, NF-κB: nuclear factor kappa-light-chain-enhancer of activated B cells, ASC: apoptosis-associated speck-like protein containing CARD

the colonization of *H. pylori* (Koch et al. 2015). Mice deficient in either *casp1, il18,* or *il18r* also revealed diminished *H. pylori* colonization rates. This indicated that TLR2-mediated NLRP3 inflammasome activation in mice benefits *H. pylori* colonization and persistence (summarized in Fig. 1b). Wild-type mice infected with Δ*ureB* mutants had more IFNγ-producing $CD4^+$ cells compared to animals infected with wild-type bacteria; likewise, *tlr2* or *nlrp3* deficient mice produced lower levels of these cells when colonized with either type of bacteria (Koch et al. 2015). Animals of an allergic asthma model revealed reduced pathologies when neonatal mice were infected with wild-type *H. pylori,* but Δ*ureB* mutant infection or IL-18 blocking abolished this effect (Koch et al. 2015). Furthermore, expression of *il1β* and *nlrp3* was found to depend on TLR2 and NOD2 in *H. pylori*-infected BMDCs. *Tlr2* or *Nod2* deficiency in mouse BMDCs significantly reduced *H. pylori*-induced activation of the inflammasome (Kim et al. 2013). We found that TLR2 overexpression in HEK293 cells differentially induced *IL-1β* expression when infected with *H. pylori*

(Pachathundikandi and Backert 2016). In conclusion, TLR2, along with NOD2-mediated activation of the NLRP3 inflammasome, depends on *H. pylori* factors such as UreB, LPS, and CagL. Together, these studies indicate a determining role of TLR2 in pro-inflammation and bacterial clearance in response to various bacterial infections. The co-stimulatory functions between TLR2 and NOD2 seemed to be critical for the outcome of an infection.

3 Role of microRNAs in *H. pylori* Interaction with Host Cells

H. pylori-infected gastric mucosa at different stages of associated diseases showed that pro-inflammatory cytokines *IL-1β, IL-6, TNFA,* and chemokine *IL-8* gene expression were correlated with progression of gastritis and atrophic gastritis to intestinal metaplasia. In some cases, this was correlated with the *H. pylori*-stimulated expression of microRNAs (Table 1), having diverse effects on many cell types (Fig. 2). In a comparative analysis of a group of 29 miRNAs from the above samples, Isomoto et al. (2012) found that let7b, miR-200c, miR-375, and miR-103 were correlated with the expression levels of *IL-1β, IL-6, IL-8,* and *TNFA,* respectively. However, miR-223 negatively correlated, except for *IL-8* (Isomoto et al. 2012). The expression of *TNFA, IL-6, IL-12A,* and *IL-2* genes were significantly increased in *H. pylori*-positive samples before eradication, while *IL-1β* and *TGFBRII* expressions did not change significantly. *TNFA* and *IL-6* levels were decreased, but *TGFBRII* expression increased in the treated group. miR-103, miR-181c, miR-370, and miR-375 expression were downregulated in both groups, but increased after eradication. *IL-6* mRNA was negatively correlated with miR-103 and miR-370, while *IL-12A* with all miRNAs studied except miR-223. However, miR-223 correlated with *IL-2* in the infected group and *IL-1β* in the non-infected cohort. The concentration of TNF and IL-6 cytokines increased in inflammatory cells, while *H. pylori* eradication decreased IL-12-p40. However, TGFβ-RII was increased in controls, but downregulated in infection and increased on eradication, which suggests compromised TGFβ action (Rossi et al. 2016). Recently, we found that miR-223-3p targets NLRP3 for regulating its concentration in *H. pylori*-infected THP1 cells, which regulates inflammasome function and responses (Pachathundikandi and Backert 2018b). Xiao et al. (2009) observed that *H. pylori* infection upregulated miR-155 by NF-κB- and AP-1 dependent mechanisms and suppressed inhibitor κB kinase ε (IKKε), Sma- and Mad-related protein 2 (SMAD2) and Fas-associated death domain protein (FADD) for IL-8 and CXCL1 (GROα) regulation. *H. pylori* infection also targets an apoptotic pathway through upregulation of miR-155 in mouse bone marrow-derived macrophages (BMMs), which was dependent on the *cag*-encoded T4SS, but independent of CagA, TLR2/4, and NOD1/2 signaling (Koch et al. 2012). On the other hand, miR-155 reduced the intracellular survival of *H. pylori* through enhancing autophagy and increased expression positively correlated with higher immunohistochemical score of inflammation. However, miR-155 levels decreased in *H. pylori*-associated metaplasia, which suggests a prognostic value in GC (Wu et al. 2016). Apart from the above study, it was shown that *H. pylori* upregulated miR-30b and miR-30d for autophagy suppression and increased intracellular survival (Tang et al. 2012; Yang et al. 2016). miR-146a and miR-155 were increased in *H. pylori* infection and that was more prominent in children when compared to adults (Cortés-Márquez et al. 2018). *H. pylori* can induce miR-146a in GECs in a NF-κB-dependent mechanism, while IL-8, TNF, and IL-1β secreted during infection might also contribute for this induction (Li et al. 2012). Moreover, PTGS2 (prostaglandin endoperoxide synthase 2) was upregulated in *H. pylori*-infected GECs (Liu et al. 2013). However, overexpression of miR-146a regulated these pro-inflammatory chemokine/cytokines such as IL-8, CXCL1 (GROα), CCL20 (MIP3α), TNF, IL-1β, and PTGS2 (Liu et al. 2010, 2013; Li et al. 2012). *H. pylori*-induced miR-146a expression might regulate the Treg/Th1 population in infected tissues of individuals and thereby the outcome of associated diseases (Lu et al. 2010). miR-146a was also found to enhance the apoptosis in *H. pylori*-infected GC cell lines, whereas high intra-tumoral expression negatively correlated with lymph node metastasis in *H. pylori*-positive GC (Wu et al. 2014). *H. pylori*-induced expression of IL-6 was suppressed by subsequent overexpression of miR-146b and miR-155 in infected gastric tissues and cell lines (Cheng et al. 2015). Furthermore, miR-196, miR-127-5p, miR-206, miR-216, miR-488, miR-195, and miR-181 were increased in *H. pylori* infection. In these studies, miR-196 and miR-488 were found to regulate the IL-6 secretion from infected cells (Chung et al. 2017). *H. pylori* infection also downregulated miR-4270 and that increased the expression of *CD300E*. This activation of CD300E by *H. pylori* or agonistic antibody treatment drastically reduced the expression of MHCII molecules and antigen presentation, which prevented activation of T cells (Pagliari et al. 2017). *H. pylori* downregulated muscle-specific miR-1 and miR-133 in gastric tissue and myoblasts for increased cell proliferation and gastric emptying (Saito et al. 2011).

The translocated T4SS effector protein CagA suppressed let7 miRNA through epigenetic modifications, leading to increased Ras oncoprotein expression and signaling in infected murine cells (Hayashi et al. 2013). Furthermore, CagA downregulated the expression of miR-320a and miR-4496 in infected gastric adenocarcinoma cell lines, both miRNAs have a predicted target in the 3'UTR region of β-catenin. Downregulation of these miRNAs in *H. pylori* infection increased β-catenin and its target genes of cancer initiation/associated markers such as CD44, CD166, epithelial cell adhesion molecule (Epi-CAM), musashi RNA binding protein-1 (Msi-1), phospholipase D1 (PLD1), and cyclin D1. CagA-induced chemo-resistance and gastric tumorigenesis were attenuated by combinatorial therapy with miR-320a and miR-4496 and 5-fluorouracil by targeting cancer initiation/associated markers and ATP-binding cassette subfamily member G2 (ABCG2) efflux transporter (Kang et al. 2017). Matsushima et al. (2011), through microarray and RT-PCR validation, showed that 31 genes were differentially expressed in the human stomach infected with *H. pylori* in comparison with non-infected controls. CagA seems also involved in the downregulation of let7 family members in *H. pylori* infection (Matsushima et al. 2011). In a microarray experiment, CagA was found to differentially express 61

Table 1 Regulation and outcome of microRNA induction by *H. pylori*

miRNA	Cytokine/chemokine signalling	References
↑ miR-146a	↓ IKKε, SAMD2, FADD, CXCL1, IL-8, IL-6	Xiao et al. (2009)
↑ miR-155	↓ IKKε, SAMD2, FADD, CXCL1, IL-8, IL-6	Liu et al. (2010), Li et al. (2012), Liu et al. (2013)
↑ miR-223-3p	↓ NLRP3, IL-1β ↑ IL-2	Pachathundikandi and Backert (2018b), Rossi et al. (2016)
↑miR-196	↓ IL-6	Chung et al. (2017)
↑miR-488	↓ IL-6	Chung et al. (2017)
↓ let7b	↑ IL-1β, TLR4	Isomoto et al. (2012), Zhu et al. (2012), Rossi et al. (2016)
↓miR-103	↑ TNF	Isomoto et al. (2012), Rossi et al. (2016)
↓miR-200c	↑ IL-6	Isomoto et al. (2012), Rossi et al. (2016)
↓miR-370	↑ IL-6, IL-12	Rossi et al. (2016)
↓ miR-375	↑ IL-8, IL-12	Isomoto et al. (2012)
↓ miR-7	↑ IL-1β, IL-8	Kong et al. (2012)
	Apoptosis/autophagy/survival/antigen presentation	
↑ miR-146a	↑ Apoptosis	Wu et al. (2014)
↑ miR-155	↓ Apoptosis ↑ Autophagy ↓ Survival	Koch et al. (2012), Wu et al. (2016)
↑ miR-30b	↓ Autophagy ↑ Survival	Tang et al. (2012)
↑ miR-30d	↓ Autophagy ↑ Survival	Yang et al. (2016)
↓ miR-4270	↓ CD300E, MHCII ↓ Antigen presentation	Pagliari et al. (2017)
	Tumorigenesis/Gastric cancer/metastasis	
↑ miR-21	↑ RECK ↑ in gastric cancer	Zhang et al. (2008), Shiotani et al. (2012)
↑ miR-27b	↑ FZD7 ↓Wnt signaling	Geng et al. (2016)
↑miR-194	↑ in gastric cancer	Shiotani et al. (2012)
↑miR-196	↑ in gastric cancer	Shiotani et al. (2012)
↑miR-17/92 cluster	↑ in gastric cancer	Shiotani et al. (2012)

(continued)

Table 1 (continued)

miRNA	Cytokine/chemokine signalling	References
↑miR-106b-93-25 cluster	↑ in gastric cancer	Shiotani et al. (2012)
↑miR-18-5p	↑ in gastric cancer	Lee et al. (2017)
↑miR-29b-1-5p	↓PHLPP1 ↑ MMP2, MMP9	Datta et al. (2018)
↑miR-135b-5p	↓KLF4 ↑ in gastric cancer↑ Cisplatin resistance	Shao et al. (2019)
↑miR-143-3p	↓AKT↑ Cell proliferation ↑ in gastric cancer	Wang et al. (2017)
↑ miR-146a	↑ lymph node metastasis in gastric cancer	Wu et al. (2014)
↑miR-1289	↓H + K + ATPase α subunit ↑ hypochlorhydria	Zhang et al. (2014)
↓ miR-223-3p	↑ in gastric cancer	Shiotani et al. (2012)
↓ miR-7	↑ in gastric cancer	Kong et al. (2012)
↓miR-106b	↑ JAK1/STAT3 signaling	Ye et al. (2015)
↓miR-375	↑ JAK1/STAT3 signaling ↑ BCL2, TWSIT1 ↑ Cell proliferation, migration	Ye et al. (2015), Miao et al. (2014)
↓miR-203	↑ CASK ↑ Gastric cancer progression	Zhou et al. (2014b)
↓miR-210	↑STMN1, DIMT1 ↑ Gastric tumorigenesis	Kiga et al. (2014)
↓miR-152	↑PDL1(B7H1) ↑ in gastric cancer	Xie et al. (2017)
↓miR-200b	↑PDL1(B7H1) ↑ in gastric cancer	Xie et al. (2017)
↓Let7	↑ Ras oncoprotein	Hayashi et al. (2013)
↓ miR-320a	↑ β-catenin	Kang et al. (2017)
↓ miR-4496	↑ β-catenin	Kang et al. (2017)
↓miR-320	↑ MCL1 ↓ Apoptosis ↑ Neoplasia	Noto et al. (2013)
↓miR-1	↑ Cell proliferation ↑ gastric emptying	Saito et al. (2011)
↓miR-133	↑ Cell proliferation ↑ gastric emptying	Saito et al. (2011)

Abbreviations used ↑ Upregulated expression, ↓ Downregulated expression

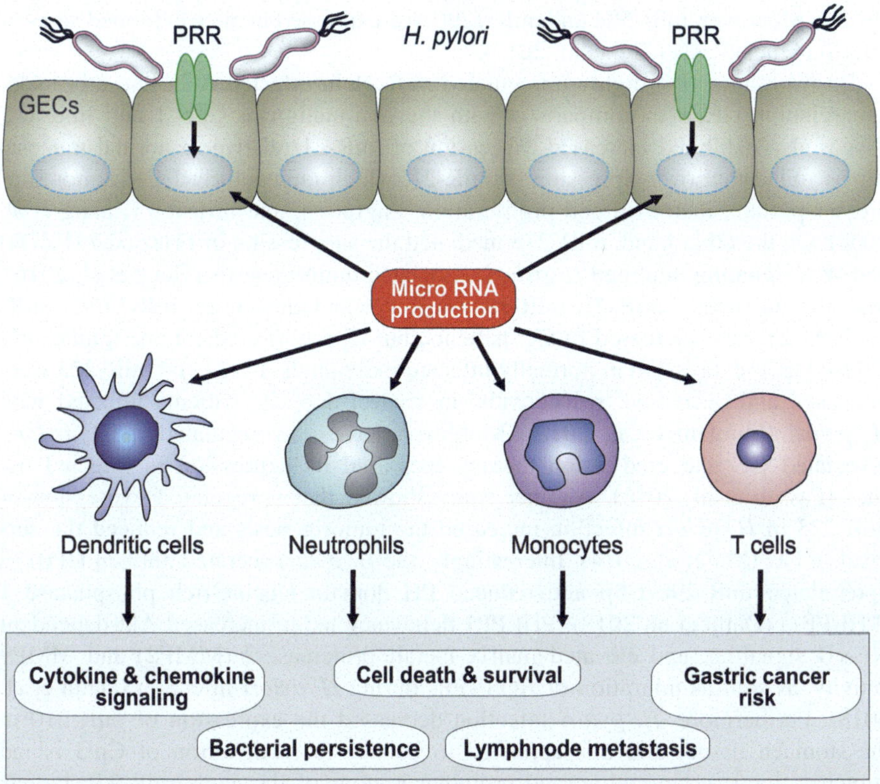

Fig. 2 Infection with *Helicobacter pylori* induces various signaling events, some of which include the production of micro RNAs. Micro RNAs can be produced by gastric epithelial cells (GECs) as well as immune cells such as dendritic cells, neutrophils, monocytes and T cells as listed in Table 1. These various micro RNAs have been reported to be associated with diverse phenotypic outcome of the infection including changed cytokine signaling, cell death or survival, bacterial persistence, and disease development such as gastric cancer and lymph node metastasis. For more details, see text

miRNAs in infected GECs and found that tumor suppressor miR-320 got downregulated and caused upregulation of the anti-apoptotic protein myeloid cell leukemia-1 (MCL-1). MCL-1 overexpression induced the neoplastic progression in CagA-positive strain-infected GECs, which paves the way for gastric carcinogenesis (Noto et al. 2013). miR-1289 expression was significantly increased in gastric biopsies of patients infected with CagA$^+$ strains compared to CagA$^-$ isolates. *H. pylori*-induced expression of miR-1289 was implicated in hypochlorhydria due to targeting of the H^+K^+ ATPase α subunit (Zhang et al. 2014). CagA-dependent let7b downregulation increased the TLR4 expression and signaling for inflammation and other immune responses in *H. pylori* infection. Transgenic expression of CagA in AGS cells upregulated miR-584 and miR-1290 and both have targets on *Foxa1* (Forkhead box A1). Downregulation of Foxa1 resulted in epithelial–mesenchymal transition

(EMT). Moreover, miR-584 and miR-1290 overexpressed in mice exhibited signs of intestinal metaplasia (Zhu et al. 2012).

H. pylori infection highly upregulated miR-21 in infected AGS and other adenocarcinoma GECs in comparison with the non-malignant GES-1 cell line. GC tissues also exhibited increased expression of miR-21 relative to normal mucosa. It was found that overexpression of miR-21, which targets the RECK tumor suppressor protein, increased cell proliferation, migration, and invasion (Zhang et al. 2008). On the other hand, miR-27b mediated the suppression of Frizzled 7 (FZD7) and Wnt signaling inhibited *H. pylori*-associated tumorigenesis (Geng et al. 2016). MicroRNAs such as miR-21, miR-194, miR-196 and clusters of miR-17/92, miR-106b-93-25 were increased in GC patients, but *H. pylori* eradication significantly reduced their expression in normally infected individuals. Let7d and miR-223 were increased and decreased, respectively, in control and GC patients infected with *H. pylori* (Shiotani et al. 2012). Moreover, let7c downregulated in *H. pylori*-associated GC and eradication therapy increased its expression in infected tissues (Fassan et al. 2016). Another study showed that upregulated expression of miR-223 in *H. pylori* infection increased the tumorigenesis and reduced the survival in GC (Ma et al. 2014). Interestingly, the *H. pylori* secreted antigen HP0175 upregulated miR-29b-1-5p and reduced PH domain leucine-rich phosphatase 1 (PHLPP1) (Datta et al. 2018). PHLPP1 deficiency led to increased Akt-dependent NF-κB signaling, and elevated matrix metalloproteinase-2 (MMP2) and MMP9 activity, as well as migration of AGS cells during *H. pylori* infection (Datta et al. 2018). Furthermore, *H. pylori* infection decreased the expression of miR-210 in the stomach tissue, and this appeared to be due to methylation of CpG island stretches that overlap coding and promoter regions of the gene. miR-210 targets like Stathmin-1 (STMN1) and DIM1 Dimethyladenosine Transferase 1 Homolog (DIMT1) upregulated expression, which increased the cellular proliferation, chances for associated gastric diseases and cancer development (Kiga et al. 2014). Similarly, tumor suppressor miR-7 downregulated during infection, which was not regulated epigenetically but dependent on inflammation-induced macrophage-derived factors. miR-7 expression inversely related to IL-1β and TNF expression in GC tissues when compared to non-tumor tissues. Moreover, overexpression of miR-7 in GC cells reduced the proliferation and colony formation (Kong et al. 2012). miR-18a-5p was differentially expressed in *H. pylori*-related GC tissue compared to non-*H. pylori* GC, whereas miR-135b-5p and miR-196a-5p expressed in both cancers (Lee et al. 2017). Furthermore, *H. pylori* LPS induced the downregulation of miR-106b and miR-375 in GECs. miR-106b downregulation led to activation of Janus kinase 1 (JAK1)/signal transducer and activator of transcription-3 (STAT3) signaling. In addition, miR-375 downregulation resulted in increased MDM2 Proto-Oncogene (MDM2) expression and Sp1 Transcription Factor (SP1) activation, which also caused p63/Dicer-dependent feedback inhibition of these miRNAs (Ye et al. 2015). In another finding, miR-375 expression downregulation induced by *H. pylori* resulted in heightened JAK1/STAT3 signaling, due to reduced targeting of JAK1 by miR-375. Dysregulated JAK1/STAT3 signaling in *H. pylori*-infected GECs increased the expression of B cell lymphoma 2 (BCL2) and twist family

BHLH transcription factor 1 (TWIST1), which caused increased cellular proliferation and migration (Miao et al. 2014). *H. pylori* infection and TNF treatment upregulated miR-135b-5p expression through a NF-κB-dependent mechanism, and in turn, downregulated target Kruppel-like factor 4 (KLF4) in GC cell lines. The human and mouse GC tissues also showed expression of upregulated miR-135b-5p. It was also found that miR-135b-5p overexpression led to resistance against cisplastin, a chemotherapeutic agent, in GC cells (Shao et al. 2019). miR-143-3p was upregulated in *H. pylori*-associated GC tissues when compared to uninfected GC. miR-143-3p has roles in controlling cell proliferation, migration, and invasion through targeting AKT serine/threonine kinase 2 (AKT2) (Wang et al. 2017). miR-203 was downregulated in *H. pylori*-infected normal gastric tissues, tumor tissues, and cancer cell lines, which caused increased expression of Ca/Calmodulin-dependent serine protein kinase (CASK) and sex-determining region Y box 4 (SOX4) for GC progression (Zhou et al. 2014a, b). *H. pylori* also increased the expression of PDL1/B7H1 (Programmed death ligand 1) by downregulated expression of miR-152 and miR-200b in GC cell lines. *H. pylori*-positive human gastric tumor tissues have also shown high PDL1(B7H1) expression by downregulating these miRNAs (Xie et al. 2017). RT-PCR validated miRNAs, among 219 differentially expressed, showed that miR-99b-3p, miR-564, and miR-638 were significantly increased in *H. pylori*-associated GC (Chang et al. 2015a). *H. pylori* infection significantly increased the gene methylation of *miR-124a1, miR-124a2, and miR-124a3* in normally infected and GC patients (Ando et al. 2009). *H. pylori* infection of individuals with miR-146 single nucleotide polymorphism (SNP) rs2910164 CC had a significantly increased risk of intestinal metaplasia and dysplasia compared to GG (Song et al. 2013). miR-212-3p and miR-361-3p were downregulated in *H. pylori*-infected Barrett's esophagus. In addition, *H. pylori* extract treated esophagal epithelial cell cultures also did the same for upregulation of the target oncoproteins caudal-type homeobox protein 2 (CDX2) and cyclooxygenase 2 (COX2) (Teng et al. 2018). The above data confirms the role of *H. pylori*-induced miRNA upregulation or downregulation in persistent infection, activation of various inflammatory responses, and in the pathogenesis of *H. pylori*-associated GC progression. More studies and better understanding will definitely help to use miRNAs in the diagnosis or treatment of associated pathologies in this bacterial infection (summarized in Fig. 2).

4 *H. pylori* Manipulates TLR Signaling and Bacterial Recognition

TLRs constitute a group of transmembrane proteins with three recognizable domains of leucine-rich repeats in the extracellular/extravacuolar region, transmembrane region, and intracellular/intravacuolar TIR domain-containing region (Beutler 2008, see also Chapter "Role of NOD1 and ALPK1/TIFA Signalling in Innate Immunity Against *Helicobacter pylori* Infection" of this book). Several ligands, also known as pathogen/microbe-associated molecular patterns (PAMPs/MAMPs), including glycolipids, lipopeptides, proteins, nucleic acids, and synthetic compounds were identified to interact with these receptors for transducing specific signals (Takeuchi and Akira 2010). Ligand binding and TLR dimerization induces the binding of cytoplasmic adaptor molecules, MyD88 or TRIF, in the TIR domain to recruit kinases and signal-transducing factors for activation of different transcription factors and pro- or anti-inflammatory gene expression (Beutler 2008; Takeuchi and Akira 2010). Research in the last two decades has identified different aspects of *H. pylori* interaction with TLRs. In addition to activation of TLR signaling, *H. pylori* mechanisms were also modulating or manipulating this important innate response system of the host (Pachathundikandi et al. 2015).

In comparison with highly active *Escherichia coli* LPS, specific LPS modification by *H. pylori* prevents detection by the host and reduces the actions of host antimicrobial peptides. Lipid-A biosynthesis in the periplasmic membrane of *H. pylori* produces hexa-acylated lipid-A. However, passage through the periplasmic space to the outer membrane exposes *H. pylori* LPS to the activity of different enzymes which forms modified LPS (Tran et al. 2006; Stead et al. 2008; Cullen et al. 2011). This modified *H. pylori* LPS was found to be 1000–10,000-fold less sensitive than *E. coli* LPS for TLR4 detection (Birkholz et al. 1993; Pérez-Pérez et al. 1995; Moran et al. 1997). Modified *H. pylori* LPS imparts resistance to the action of positively charged antimicrobial peptides (further discussed in Chapter "The Sweeping Role of Cholesterol Depletion in the Persistence of *Helicobacter pylori* Infections") due to the decreased negative charge after removal of phosphate groups (Moran et al. 1997; Tran et al. 2006; Cullen et al. 2011). Furthermore, *H. pylori* LPS was found to activate TLR2 and TLR10 during infection. *H. pylori* LPS preferred TLR2 instead of TLR4, and however, TLR2 detection was dependent on TLR2/TLR1, TRL2/TLR6, or TLR2/TLR10 heterodimers (Yokota et al. 2007; Nagashima et al. 2015). Although, there are studies showing that *H. pylori* LPS might activate TLR4, and however, intact *H. pylori* and related *Helicobacters* induce signals through TLR2 for IL-8 production (Mandell et al. 2004). Moreover, it was shown that overexpression of CD14 or MD2 recovered the responsiveness to *H. pylori* LPS in GECs, but not necessarily in macrophages (Maeda et al. 2001; Su et al. 2003; Ishihara et al. 2004). During *H. pylori* infection, TLR2 was reported to induce anti-inflammatory cytokine IL-10 production, while *tlr2* deficient mice more efficiently cleared infection than wild-type (Rad et al. 2009; Panthel et al. 2003). We found that TLR2 overexpressing human HEK293 cells differentially induced the expression of *IL-1β* (Pachathundikandi and Backert 2016) (Fig. 1a). The *H. pylori* proteins NapA, Hsp60, HpaA, and UreB were also reported to activate TLR2 in cellular infection assays or animal models (Amedei et al. 2006; Zhao et al. 2007; Lindgren et al. 2011; Koch et al. 2015).

Bacterial flagellin is probably the best-known protein PAMP and functions as the ligand for TLR5. Many of the β- and γ-proteobacteria flagellins were identified as strong ligands for TLR5. The *Salmonella* flagellin FliC is the most potent ligand identified for TLR5, so far. The binding of FliC to TLR5 occurs through the three helices in the D1 domain laterally (Smith et al. 2003a). FliC-bound TLR5 activates MyD88-dependent signaling for the activation of NF-κB and AP-1 to induce the expression of responsive genes (Smith et al. 2003a; Rhee et al. 2004). Gewirtz et al. (2004) showed that gastric adenocarcinoma cells infected with Δ*fla*A mutant *H. pylori* did not inhibit the secretion of IL-8. Furthermore, co-incubation of purified recombinant FlaA of *H. pylori* with GECs was found to be a very weak inducer of IL-8 expression or p38 activation. Moreover, flagellin from different α- and ε-proteobacteria was found to be unresponsive on activating TLR5 (Andersen-Nissen et al. 2005). As a member of the ε-proteobacteria, *H. pylori* flagellin was reported to be evading recognition of TLR5 and this evasion was attributed to the amino acid sequences 89–96 of the N-terminal D1 domain (Andersen-Nissen et al. 2005). In addition to these earlier studies, we have also identified induced expression of TLR5 in *H. pylori* infection (Smith et al. 2003b; Torok et al. 2005; Pachathundikandi et al. 2011). The infection of TLR5 overexpressing cells induced increased IL-8 and TNF secretion during *H. pylori* infection, when compared to non-infected control and infected parental HEK293 cells. The actual *H. pylori* ligand for TLR5, however, remains unknown (Pachathundikandi et al. 2011).

H. pylori DNA or RNA induced strong cytokine secretion in DCs from wild-type and *TLR2/4* deficient mice, but this response was abrogated in mice with *TLR2/4/9* deficiencies. *TLR9* single deficient cells secreted higher amounts of cytokines when compared to wild-type on exposure to *H. pylori* DNA or RNA (Rad et al. 2007). *H. pylori* RNA detection by DCs was dependent on TLR8 alone, or possibly in combination with TLR7. In addition, exposure of DCs to *H. pylori* RNA resulted in IFNγ production that was independent of TRIF or MyD88, but dependent on RIG1 activation (Rad et al. 2009). Furthermore, a recent discovery showed that *H. pylori* DNA detection by TLR9 in eukaryotic cells required T4SS delivery of chromosomal DNA (Varga et al. 2016). Remarkably, *H. pylori* uses both camouflaging and direct interaction for different level of sensing through TLRs during infection. One interesting aspect of this TLR interaction is that both pro- and anti-inflammatory responses were induced during host interaction, which suggests *H. pylori* does not want a suicide. Through manipulation of TLR interaction and further signaling, this bacterium ensures long-term survival in a difficult niche.

5 Pro- and Anti-inflammatory Cytokine Signaling by *H. pylori*

H. pylori initial contact with the epithelium activates pro-inflammatory responses through activation of several cellular signaling pathways in GECs. The subsequent activation of NF-κB and AP-1 pathways led to the production of chemokines and cytokines, especially IL-8 and IL-6 (Gionchetti et al. 1994; Basso et al. 1996; Odenbreit et al. 2006). However, their exact role in immunity against this pathogen is not well represented in the literature. *H. pylori* factors Hsp60, OipA, NapA, and *cag* T4SS were reported to be involved with the induction of IL-6 during infection (Gobert et al. 2004; Lu et al. 2005; Tsai et al. 2015; Odenbreit et al. 2006; Sugimoto et al. 2011). Recently, it was found that *H. pylori*-induced IL-6 activated STAT3 phosphorylation through its receptor in GC cell lines (Piao et al. 2016). *H. pylori* TNF inducing factor (Tipα) was also described to activate the IL6/STAT3 pathway in GC cell lines (Chen et al. 2017). Moreover, some studies suggest an IL-4/IL-13 axis in *H. pylori* infection (Lucas et al. 2001; Marotti et al. 2008).

IL-21 expression was observed to be enhanced in the *H. pylori*-infected gastric mucosa and found to activate GECs expressing IL-21R to produce MMP2 and MMP9 (Caruso et al. 2007). Moreover, IL-21 produced from major histocompatibility complex II (MHCII)-expressing gastric myofibroblasts isolated from infected tissue and GC were found to differentiate $CD4^+$ T cells to the Th17 phenotype in a co-culture, with the help of secreted cytokines IL-6 and TGFβ (Pinchuk et al. 2013). *H. pylori* infection of *il21* deficient mice exhibited increased colonization, reduced gastritis and pro-inflammatory cytokines/chemokines in comparison with wild-type mice littermates. Experimental and in silico modeling have predicted IL-21 induced STAT1 and STAT3 signaling played a significant role in the Th1 and Th17 responses, respectively, on controlling *H. pylori* colonization (Carbo et al. 2014). IL-23 in *H. pylori* infection was also found to influence Th1 and Th17 responses as *il23* deficient mice exhibited high colonization but with minimal pathology (Caruso et al. 2008; Horvath et al. 2012). IL-32 levels were increased in *H. pylori*-infected gastric tissue, GCs, and GC cell lines. AGS cells expressed higher IL-32 levels during infection, and this was dependent on *cag* status and NF-κB activation. IL-32 knockdown generated by siRNA suppressed *H. pylori*-induced chemokine production in AGS cells (Sakitani et al. 2012). However, Peng et al. (2014) reported that IL-32 expression in *H. pylori* infection was dependent on secreted IL-1β and TNF, as neutralizing antibodies against these cytokines weakened the expression of IL-32 accordingly.

The IL-17/Th17 axis plays a major role on the inflammation responses against *H. pylori* infection (see Chapter "Inflammation, Immunity and Vaccine Development for the Gastric Pathogen *Helicobacter pylori*" of this book). IL-17A contributes to *H. pylori*-associated pathophysiology as suppression of Th17 responses resulted in the control of the gastritis (Shi et al. 2010; Serrano et al. 2013; Munari et al. 2014). A Th1/Th17 mixed response is characteristic of *H. pylori*-infected gastric mucosa. IFNγ production by T cells determines the differentiation of cytotoxic Th1 responses. In most *H. pylori* infections, both of these cytotoxic immune cells work to control colonization, but with the consequence of severe immune pathologies (Itoh et al. 2003; Shi et al. 2010). Infection of *il17* deficient mice suggests that Th17 responses against *H. pylori* modulate the Th1 response against *H. pylori* (Shi et al. 2010). Allison et al. (2013) showed that NOD1 potentiated IFNγ signaling and induced increased expression of IL-8 (CXCL8) and CXCL10 (IP-10). *H. pylori* increases IFNγ secretion and IFNγ signaling was found to be regulated in GECs and immune cells through CagA-dependent Src homology 2 domain-containing phosphatase 2 (SHP2) Y-542 phosphorylation (Wang et al. 2014). Moreover, *H. pylori*-mediated cholesterol depletion from lipid rafts disrupted IFNγ signaling in primary GECs and cell lines (Morey et al. 2018). In addition to exacerbated responses through Th1/Th17 cell populations, *H. pylori*-mediated changes regulate this process to limit tissue destruction, which might jeopardize its persistence. The reduction in *H. pylori* load in immunized *ifnγ* deficient mice was mediated through IL-17A. Furthermore, neutralization of IL-17A abolished the vaccine protection and mild gastritis (Sjökvist Ottsjö et al. 2015).

Early studies on *H. pylori* pathogenesis observed that relative expression of IL-12 and IL-10 in *H. pylori* gastritis favors IFNγ producing Th1 cell differentiation (Haeberle et al. 1997; Bauditz et al. 1999). Infection of human peripheral blood-derived DCs with wild-type *H. pylori* secreted IL-12, but no IL-6 and IL-10, while a T4SS-defective Δ*cag*E mutant suppressed this effect (Guiney et al. 2003). However, in comparison with *Acinetobacter lwoffii*-infected DCs, a less infective gastric pathogen, *H. pylori* infection produced lesser amount of IL-12 and IL-10 in mouse bone marrow-derived DCs (BMDCs). Moreover, whole cell lysate of *H. pylori* was found to decrease IL-12 secretion from *A. lwoffii*-infected DCs (Kao et al. 2006). It was reported that IL-12 production was decreased at higher inoculum of *H. pylori* infection with mouse DCs, while IL-10 secretion increased, and these effects were independent of IL-1β (Obonyo et al. 2006). Neutralization of IL-12 in ex vivo cultured *H. pylori*-infected gastric biopsies reduced IFNγ production through downregulation of phosphorylated STAT4 and T-bet (Pellicanò et al. 2007). Gastric tissue resident natural killer (NK) cells were also reported to induce IFNγ production through an IL-12 synergistic effect during *H. pylori* infection (Yun et al. 2005). *H. pylori*-induced NK cell IFNγ production was dependent on membrane bound lipoprotein HpaA and TLR2 signaling (Lindgren et al. 2011). Ding et al. (2013) observed that IL-12 cytokine therapy could induce Th1 immunity for reducing the *H. pylori* colonization even in the absence of immunization or Th17 responses. NapA induced the production of IL-12 and IL-23 in primary human neutrophils and monocytes and induced a Th1 response in T cell clones prepared from healthy donors. Additionally, NapA shifted the Th2 response to Th1 in cells obtained from allergic donors (Amedei et al. 2006). IL-25 (IL-17E) signaling-induced Th2 responses significantly reduced the inflammation in mice during infection with *H. pylori,* while IL-23 signaling increased this inflammation (Horvath et al. 2012, 2013). *H. pylori* interaction with gastric DCs during infection induced the production and secretion of IL-23 and resulted in Th22 cell polarization. IL-22 and CXCL2 secretion in *H. pylori* infection attracted myeloid suppressor cells, which ultimately favored a pro-inflammatory state for development of gastritis (Zhuang et al. 2015).

Gastric mucosa infected with *H. pylori* increased the production of IL-10 in GECs and lamina propria mononuclear cells (Bodger et al. 2001). It was also shown in the *il10* deficient gastric mucosa that infection by *H. pylori* resulted in significantly reduced bacterial load compared to heterozygous littermates. The infected *il10* deficient gastric mucosa revealed increased neutrophils, mononuclear cells, as well as serum IgA and IgG concentrations, which supports a reduction in bacterial load (Chen et al. 2001). In addition, DC-SIGN signaling reverted TLR-mediated pro-inflammatory cytokine expression to favor anti-inflammatory cytokine IL-10 production in *H. pylori* infection (Gringhuis et al. 2007, 2009). Human monocyte-derived DCs infected with *H. pylori* produced IL-10 in a DC-SIGN-, TLR2-, and TLR4-dependent manner. Moreover, p38 and NF-κB-mediated signaling decreased histone modification and led to upregulated expression of IL-10 during *H. pylori* infection (Chang et al. 2012). Furthermore, *H. pylori* infection induced a Th2 response through DC-SIGN-mediated activation of NF-κB. Fucose-containing moieties from *H. pylori* and *Schistosoma mansoni* induced DC-SIGN signaling for phosphorylation of lymphocyte-specific protein 1 (LSP1) and IKKε activation, which resulted in the interaction of B cell lymphoma encoded protein 3 (Bcl3) and p50 NF-κB subunit in the nucleus. This atypical NF-κB activation downregulated proinflammatory cytokine production and upregulated anti-inflammatory cytokine IL-10 and chemokines for specific recruitment of Th2 cells (Gringhuis et al. 2014). Moreover, *H. pylori* gamma-glutamyltranspeptidase (GGT)- and VacA-dependent IL-10 production from DCs protected against experimental asthma in mice (Engler et al. 2014). CagA was found to modulate IL-10 secretion and STAT3-mediated human DC tolerization, which favors Treg development and reduced immune pathologies but dampens Th1-mediated bacterial clearance (Kaebisch et al. 2014). Oertli et al. (2013) reported a role for IL-18 in Treg cell population production. Furthermore, IL-10 secretion from infected mouse DCs suppressed maturation through STAT3 activation and reduced IL-1β secretion (Rizzuti et al. 2015). In concurrence, we found that IL-10 can suppress the NLRP3 and IL-1β expression in *H. pylori*-infected human macrophages (Pachathundikandi and Backert 2018b). The TLR2-dependent upregulation of IL-10 secretion and inhibitory effect on the NLRP3 inflammasome expression and activation in human cells is depicted in Fig. 1a.

TGFβ has also been implicated in *H. pylori* immunopathogenesis (Wu et al. 1998; Li and Li 2006; Kandulski et al. 2008). It was observed that gastric tissue TGFβ and its receptor levels were decreased in patients infected with *H. pylori,* which would affect TGFβ-mediated anti-inflammatory activities on T cells, B cells, macrophages, and NK cells (Shih 2005; Jo et al. 2010). Translocated CagA was found to interfere with TGFβ signaling through interaction with SMAD3 and SMAD4, which prevented nuclear translocation. Moreover, CagA abrogated TGFβ-mediated suppression of chemokines IL-8 and CXCL1 to CXCL3 expression in GECs (Nguyen et al. 2015). TGFβ-treated and *H. pylori*-infected cells exhibited induced expression of EMT markers (E-cadherin, Twist, Snail, Slug, and Vimentin) in GC cell lines. EMT and induced expression of TGFβ during *H. pylori* infection was dependent on CagE (Chang et al. 2015b). It was observed that *H. pylori*-infected children had increased TGFβ, IL-10, FOXP3 expression, and Treg density, which resulted in dysregulated gastric microbiota (Brawner et al. 2017). Most studies have shown increased mRNA expression of TGFβ after *H. pylori* infection, and TGFβ receptor expression was also found to be reduced in some other studies. Moreover, CagA was found to interfere with TGFβ signaling, therefore, complicating our understanding of this cytokine's induced effects in *H. pylori* infection.

H. pylori infection increased the expression of chemokines like GROα (CXCL1), IL-8 (CXCL8), MIG (CXCL9), and IP-10 (CXCL10) in the gastric mucosa of infected patients (Yamaoka et al. 1998; Eck et al. 2000; Shimoyama et al. 2002). In addition, $CCR5^+$ cells were increased in the gastric mucosa of *H. pylori*-infected children in comparison with healthy control mucosa (Krauss-Etschmann et al. 2003). However, CXCR1 and CXCR2 receptor expression was downregulated in neutrophils infected with *H. pylori* (Schmausser et al. 2004). *H. pylori* co-culture with $CD3^+$ γδ T cells produced increased secretion of CCL4 (MIP1β), and CCL5 (RANTES) but

Δ*cag*A mutant infection exhibited significantly less secretion (Romi et al. 2011). Activation of NOD1 by *H. pylori* resulted in responses that were augmented by IFNγ to produce various chemokines such as IL-8, CXCL10, CCL2, CCL3, CCL4, and CCL5 (Allison et al. 2013). CXCL1-3, CXCL8 and CCL20 were significantly increased in the *H. pylori*-infected human gastric mucosa. Moreover, CCL20 expression from GECs was dependent on T4SS-mediated NF-κB signaling (Cook et al. 2014). Peripheral blood and gastric Treg cells from infected patients also had increased expression of the receptors for the above chemokines such as CXCR1, CXCR2, and CCR6, respectively (Cook et al. 2014). Moreover, CCR2 (the receptor for CCL2) was implicated in the recruitment of monocytes, macrophages and a subset of DCs during *H. pylori* infection of mice (Arnold et al. 2017). *H. pylori*-infected primary GECs induced the expression of CXCL1-3, CXCL5, CXCL8, and CCL20, and the AGS cell line revealed a prominent expression of CXCL5 and CXCL8. This chemokine induction response was dependent on the *cag* T4SS status in the early phase, but was independent in the late phase of infection (Mustapha et al. 2014). *H. pylori* arginase Δ*rocF* mutants induced more IL-8 in GECs than wild-type bacteria, which confirms a role for immunosuppression during infection (Kim et al. 2012). More recently, it was reported that SNPs in the *cag*A gene promoter sequence resulted in differences in IL-8 secretion; the +59 motif increased the secretion of IL-8 compared to other sequences (Ferreira et al. 2016). The *cag* T4SS plays an important role in IL-8 production and was found to be mediated through the NF-κB pathway. CagY-dependent recombination events, however, lead to loss of T4SS activity and suppressed IL-8 secretion (Barrozo et al. 2016; dela Pena-Ponce et al. 2017). In addition, the above cytokine/chemokine genetic polymorphisms play an important role on deciding the inflammatory status in *H. pylori* infection. The *IL-1B* gene polymorphism at positions -511, -31 and +3954 were extensively studied and implicated in hypochlorhydria and GC development associated with *H. pylori* infection (El-Omar et al. 2000, 2003; see also Chapter "Genetic Polymorphisms in Inflammatory and Other Regulators in Gastric Cancer: Risks and Clinical Consequences" of this book). *IL-6, IL-8, IL-10,* and *TNFA* were also implicated for dysregulated production and severity of *H. pylori*-associated diseases as described elsewhere (Miftahussurur and Yamaoka 2015). As described above, *H. pylori* induces different cytokine and chemokine responses, which determine the character of the activation of innate and adaptive immune cells. Generally, the myriad of activities of these mediators favors Th1/Th17 cell-mediated immunity in the adaptive response against *H. pylori*. However, evasion strategies of *H. pylori* and induction of anti-inflammatory mechanisms in the host immune system try to balance or shift pro- to anti-inflammation, and thereby reduce the immune attack, which controls and reduces tissue damage. This limited attack and low tissue destruction allows *H. pylori* to colonize and survive there for longer periods.

6 Resolution of Inflammation by *H. pylori*

As described above, *H. pylori* infection goes through different levels of cross-talk with the host immune system. This continuous engagement of the bacterium with its host changes both systems, which have evolved to favor persistent colonization (see model in Fig. 3). The corresponding literature clearly shows the attempts of the host to clear the infection, mainly concentrated through Th1 immunity (Amedei et al. 2006; Wilson and Crabtree 2007; White et al. 2015). Th17 immunity also plays some original or supporting role in this process (Shi et al. 2010; Serrano et al. 2013; Munari et al. 2014). On the other hand, the regulatory mechanisms of the host are manipulated during *H. pylori* infection. Anti-inflammatory cytokines and cellular activities were activated in the acute or long-term infection, and this protected mice from other inflammatory conditions (Wilson and Crabtree 2007; Engler et al. 2014; Kaebisch et al. 2014; White et al. 2015). This suggests the existence of counter-acting responses of inflammation and anti-inflammation during *H. pylori* infection, but how and when this becomes activated and coordinated ultimately decides resolution of inflammation and homeostasis (Fig. 3). Epidemiological findings showed only a small percentage of people which develop associated diseases in colonized individuals. It was reported to be dependent on many factors including bacterial virulence factors, host genetic polymorphisms, environmental factors, diet and others (Amieva and El-Omar 2008; Atherton and Blaser 2009; Salama et al. 2013). Initial colonization with *H. pylori* results in the activation of PRRs. The expression of a number of pro-inflammatory mediators such as CXCL10, CXCL5, CXCL13, CXCL2, TFN and IL-8 become upregulated and manipulated the immune response by DCs and monocytes (Fig. 3). The acute inflammatory stage is marked by infiltration of neutrophils and monocytes at the inflamed site, producing classically activated M1 macrophages and resulting in the production of more pro-inflammatory mediators, such as IL-1β, IL-8, or IL-12. Subsequent development of adaptive immunity against the invader creates a memory system to prevent future attack (Fig. 3). At the onset of resolution, neutrophils start to produce microparticles, and production of lipid pro-inflammatory mediators switches to production of pro-resolution lipid mediators such as lipoxins, resolvins, maresins, and protectins (Serhan 2017). This stage of resolution is characterized by reduced secretion of pro-inflammatory cytokines and chemokines. In addition, more anti-inflammatory cytokines such as IL-10 and TGFβ are produced during this stage and thereby induces more M2 macrophages. These cells are necessary for resolution to gradually restore homeostasis with assistance of other factors. (Ortega-Gómez et al. 2013; Sugimoto et al. 2016) (Fig. 3). *H. pylori* infection is a chronic condition that in humans typically starts during childhood and lasts for a lifetime, if not eradicated by antibiotic therapy. Interestingly, not all infected individuals develop associated complications like PUD, GC, or MALT lymphoma. However, gastritis or local inflammation of the gastric mucosa appears in most infected individuals (Atherton and Blaser 2009; Polk and Peek 2010; Salama et al. 2013). The colonization of gastric mucosa by *H. pylori* and its interaction with the epithelium produces strong chemokine responses and attracts large amounts of neutrophils and other immune cells to the site of infection (Salama et al. 2013; Gobert and Wilson 2016). It is well-known that long-term infection with *H. pylori* produces a robust Th1 immune response, which can control the infection to a certain degree. The presence of the pro-resolution factors, such as Treg cells and IL-10, was demonstrated, but this was insufficient for total clearance and resolution in many cases, but sufficient to reduce the immunopathologies associated with *H. pylori* (Kao et al. 2010; Cook et al. 2014; Hussain et al. 2016). Increase of activated macrophages and T cells and a hampered apoptosis together with disturbed phagocytosis and autophagy describe chronic inflammation during infection with *H. pylori* (Fig. 3).

H. pylori infection of humans results in a mixed M1/M2 macrophage response, whereas in the mouse infection model M1 polarization occurred (Quiding-Järbrink et al. 2010; Gobert et al. 2014; Beceiro et al. 2017). This partial pro-resolution in humans may be due to mixed immune effector cell populations, which hinders bacterial clearance and resolution of the infection. *H. pylori* infection of *il10* deficient mice showed enhanced immune responses and inflammation with a highly significant reduction in bacterial colonization during the early phase, while it reduced gastritis and colonization levels at prolonged infection (Matsumoto et al. 2005). A reduction in IL-10 producing Treg cells might increase the chance of bacterial eradication and assist resolution in *H. pylori* infection (Matsumoto et al. 2005). It was also shown that infecting *irf1* deficient mice did not result in gastritis or atrophy, despite colonization with very high numbers of bacteria. These mice also failed to produce Th1 and Th2 responses, which correlated with reduced immune pathology (Sommer et al. 2001). TLRs play an important role on detection of *H. pylori* and orchestrating innate immune responses, which pave the way for adaptive immunity like Th1 responses. However, *tlr2* deficient mice efficiently cleared *H. pylori* infection compared to wild-type mice, and this may be due to lack of TLR2-mediated anti-inflammatory responses (Panthel et al. 2003). Adoptive transfer of normal splenocytes to *H. pylori*-infected severe combined immune deficiency (SCID) mice cleared the infection and inflammation resolution occurred. Wild-type mice, however, developed gastritis and reduced bacterial colonization, but bacteria were not cleared nor was resolution reached throughout the study period (Eaton and Mefford 2001). Mice immunized against *H. pylori,* cleared infection and the gastritis resolved, while the unimmunized counterparts reduced colonization, but had gastritis even after one year (Garhart et al. 2003). *H. pylori* LPS induced gastric inflammation was resolved by pharmacological activation of peroxisome proliferator-activated receptor γ (PPARγ) in an animal model (Slomiany and Slomiany 2017). *H. pylori* downregulates Caveolin 1 (Cav1) in GECs and *cav1* deficient mice infection showed severe inflammation and lower colonization when compared to wild-type littermates (Hitkova et al. 2013). There are multiple levels of involvement in *H. pylori*-induced gastric inflammation and associated pathologies, and modulation of both colonization and inflammation. The above data predicts that targeting crucial checkpoints of inflammation during infection could resolve *H. pylori* infection and opens new field of therapeutic approaches.

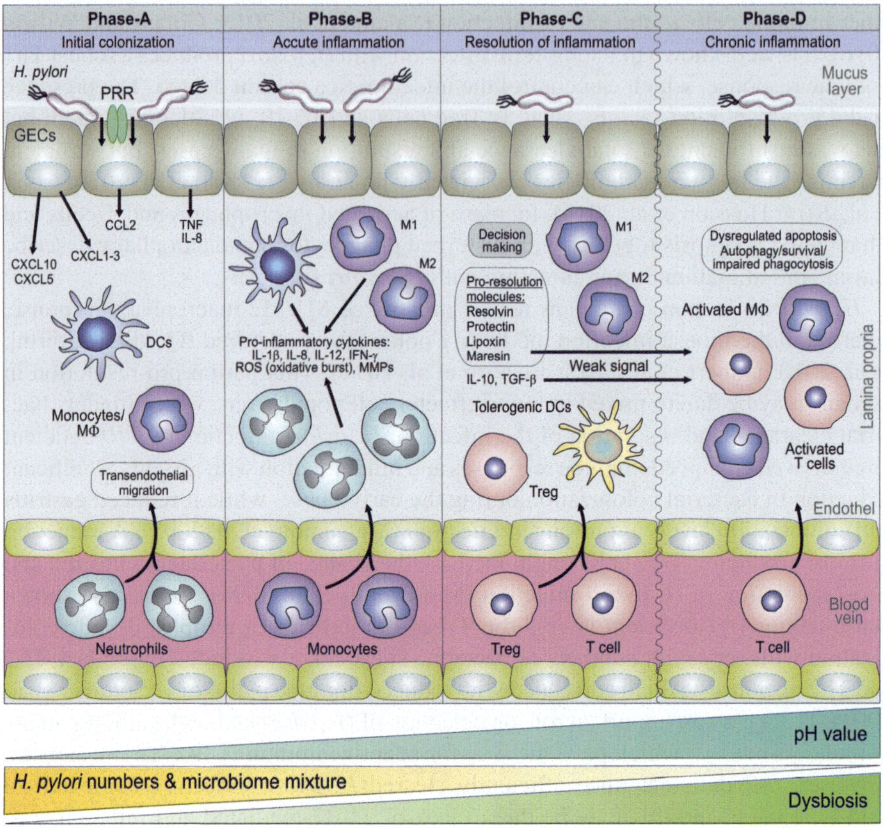

Fig. 3 Four-phase model of inflammation induced by infection of *Helicobacter pylori* in the human stomach. The initial colonization leads to sensing of the bacterium by pathogen recognition receptors (PRRs) and indicated cytokine/chemokine release (Phase A). This mediator release leads to infiltration of various indicated immune cells into the lamina propria (Brisslert et al. 2005), resulting in an acute inflammation associated with oxidative burst and release of metalloproteases (MMPs) (Phase B). Pro-resolution molecules and cells are produced during *H. pylori* infection, however, weakening signals or insufficient numbers of cells for resolution of inflammation may occur during the course of infection. Tolerogenic DCs, M2 macrophages, Tregs, IL-10, TGF-β and other resolution molecules are present at this stage of infection, whereas efferocyte recruitment and neutrophil depletion were not reported yet. This leads to a quasi-state of resolution for reduced infection associated immune pathologies, but apparently pave the way for impaired bacterial clearance (Phase C). The clearance of *H. pylori* is undermined. This scenario is accompanied by impaired processes such as phagocytosis, autophagy and cell survival, finally resulting in persistent, chronic inflammation of the gastric mucosa, which can last lifelong in patients (Phase D). The entire process proceeds over years or even decades, and is associated with a gradually decrease of gastric pH, reduction of *H. pylori* colonization and microbiome dysbiosis over time. Abbreviations used: GECs: gastric epithelial cells, CCL: C-C motif chemokine ligand, CXCL: C-X-C motif chemokine ligand, TNF: tumor necrosis factor, IL: interleukin, DCs: dendritic cells, Mϕ, M1, M2: macrophages, IFN: interferon, ROS: reactive oxygen species, TGF: transforming growth factor

7 Concluding Remarks

H. pylori infection is a chronic condition in humans, which typically starts during childhood and lasts for a lifetime. Not all infected individuals develop associated complications like PUD, GC, or MALT lymphoma. However, local inflammation of the gastric mucosa and gastritis appears in most infected individuals. *H. pylori* infection induces myriads of host receptor-mediated signaling pathways in GECs and immune cells. As a result, various cytokines and chemokines like IL-1β, IL-18, IL-6, IL-12, CXCL1-3, CXCL8 (IL-8), CXCL10, CCL2-5, and CCL20 are expressed in the infected gastric mucosa. Additionally, IL-12, IL-13, IL-17, IL-21, IL-23, IL-32, IFNγ, and TGFβ were also found to play a role at different stages of *H. pylori* infection. The complex milieu of these cytokines and chemokines influence and modulate the development of adaptive immunity against *H. pylori*. The major adaptive responses for *H. pylori* infection culminate in Th1 immunity, and however, the interplay of bacterial and host factors also induce Th2 and Treg responses in most of the individuals. *H. pylori* infection of *tlr2* and *il-10* deficient mice cleared infection and immune pathologies and produced resolution of homeostasis. However, mice deficient in *il-21*, *il-23,* and *irf1* revealed higher colonization but with reduced host immune pathology. Various miRNAs are regulated during *H. pylori* infection and involved with persistent infection, activation of various inflammatory responses and in the pathogenesis of *H. pylori*-associated GC progression. The resolution of inflammation in *H. pylori* is a complex process, but the existing data help to understand certain checkpoints for modulation of infection or inflammation to achieve therapeutic intervention.

Acknowledgements This work was supported by the German Science Foundation (project A04 in CRC-1181 to S.B.).
Conflict of Interest The authors declare no conflict of interest.

References

Allison CC, Ferrand J, McLeod L, Hassan M, Kaparakis-Liaskos M, Grubman A, Bhathal PS, Dev A, Sievert W, Jenkins BJ, Ferrero RL (2013) Nucleotide oligomerization domain 1 enhances IFN-γ signaling in gastric epithelial cells during *Helicobacter pylori* infection and exacerbates disease severity. J Immunol 190:3706–3715. https://doi.org/10.4049/jimmunol.1200591

Alvarez-Garcia I, Miska EA (2005) MicroRNA functions in animal development and human disease. Development 132:4653–4662. https://doi.org/10.1242/dev.02073

Amedei A, Cappon A, Codolo G, Cabrelle A, Polenghi A, Benagiano M, Tasca E, Azzurri A, D'Elios MM, Del Prete G, de Bernard M (2006) The neutrophil-activating protein of *Helicobacter pylori* promotes Th1 immune responses. J Clin Invest 116:1092–1101. https://doi.org/10.1172/jci27177

Amieva MR, El-Omar EM (2008) Host-bacterial interactions in *Helicobacter pylori* infection. Gastroenterology 134:306–323. https://doi.org/10.1053/j.gastro.2007.11.009

Andersen-Nissen E, Smith KD, Strobe KL, Barrett SLR, Cookson BT, Logan SM, Aderem A (2005) Evasion of Toll-like receptor 5 by flagellated bacteria. Proc Natl Acad Sci USA 102:9247–9252. https://doi.org/10.1073/pnas.0502040102

Ando T, Yoshida T, Enomoto S, Asada K, Tatematsu M, Ichinose M, Sugiyama T, Ushijima T (2009) DNA methylation of microRNA genes in gastric mucosae of gastric cancer patients: its possible involvement in the formation of epigenetic field defect. Int J Cancer 124:2367–2374. https://doi.org/10.1002/ijc.24219

Andreakos E, Foxwell B, Feldmann M (2004) Is targeting Toll-like receptors and their signaling pathway a useful therapeutic approach to modulating cytokine-driven inflammation? Immunol Rev 202:250–265. https://doi.org/10.1111/j.0105-2896.2004.00202.x

Arnold IC, Zhang X, Urban S, Artola-Borán M, Manz MG, Ottemann KM, Müller A (2017) NLRP3 controls the development of gastrointestinal $CD11b^{+}$ dendritic cells in the steady state and during chronic bacterial infection. Cell Reports 21:3860–3872. https://doi.org/10.1016/j.celrep.2017.12.015

Atherton JC, Blaser MJ (2009) Coadaptation of *Helicobacter pylori* and humans: ancient history, modern implications. J Clin Invest 119:2475–2487. https://doi.org/10.1172/jci38605

Backert S, Feller SM, Wessler S (2008) Emerging roles of Abl family tyrosine kinases in microbial pathogenesis. Trends Biochem Sci 33(2):80–90. https://doi.org/10.1016/j.tibs.2007.10.006

Backert S, Naumann M (2010) What a disorder: proinflammatory signaling pathways induced by *Helicobacter pylori*. Trends Microbiol 18:479–486. https://doi.org/10.1016/j.tim.2010.08.003

Backert S, Tegtmeyer N, Fischer W (2015) Composition, structure and function of the *Helicobacter pylori cag* pathogenicity island encoded type IV secretion system. Future Microbiol 10(6):955–965. https://doi.org/10.2217/fmb.15.32

Bafica A, Scanga CA, Feng CG, Leifer C, Cheever A, Sher A (2005) TLR9 regulates Th1 responses and cooperates with TLR2 in mediating optimal resistance to *Mycobacterium tuberculosis*. J Exp Med 202:1715–1724. https://doi.org/10.1084/jem.20051782

Bansal K, Balaji KN (2011) Intracellular pathogen sensor NOD2 programs macrophages to trigger notch1 activation. J Biol Chem 286:5823–5835. https://doi.org/10.1074/jbc.m110.192393

Barrozo RM, Hansen LM, Lam AM, Skoog EC, Martin ME, Cai LP, Lin Y, Latoscha A, Suerbaum S, Canfield DR, Solnick JV (2016) CagY is an immune-sensitive regulator of the *Helicobacter pylori* type IV secretion system. Gastroenterology 151:1164–1175.e3. https://doi.org/10.1053/j.gastro.2016.08.014

Basso D, Scrigner M, Toma A, Navaglia F, Di Mario F, Rugge M, Plebani M (1996) *Helicobacter pylori* infection enhances mucosal interleukin-1 beta, interleukin-6, and the soluble receptor of interleukin-2. Int J Clin Lab Res 26:207–210

Bauditz J, Ortner M, Bierbaum M, Niedobitek G, Lochs H, Schreiber S (1999) Production of IL-12 in gastritis relates to infection with *Helicobacter pylori*. Clin Exp Immunol 117:316–323

Beceiro S, Radin JN, Chatuvedi R, Piazuelo MB, Horvarth DJ, Cortado H, Gu Y, Dixon B, Gu C, Lange I, Koomoa D-L, Wilson KT, Algood HMS, Partida-Sánchez S (2017) TRPM2 ion channels regulate macrophage polarization and gastric inflammation during *Helicobacter pylori* infection. Mucosal Immunol 10:493–507. https://doi.org/10.1038/mi.2016.60

Beutler BA (2008) TLRs and innate immunity. Blood 113:1399–1407. https://doi.org/10.1182/blood-2008-07-019307

Birkholz S, Knipp U, Nietzki C, Adamek RJ, Opferkuch W (1993) Immunological activity of lipopolysaccharide of *Helicobacter pylori* on human peripheral mononuclear blood cells in comparison to lipopolysaccharides of other intestinal bacteria. FEMS Immunol Med Microbiol 6:317–324. https://doi.org/10.1111/j.1574-695x.1993.tb00344.x

Bodger K, Bromelow K, Wyatt JI, Heatley RV (2001) Interleukin 10 in *Helicobacter pylori* associated gastritis: immunohistochemical localisation and *in vitro* effects on cytokine secretion. J Clin Pathol 54:285–292

Brawner KM, Kumar R, Serrano CA, Ptacek T, Lefkowitz E, Morrow CD, Zhi D, Kyanam-Kabir-Baig KR, Smythies LE, Harris PR, Smith PD (2017) Helicobacter pylori infection is associated with an altered gastric microbiota in children. Mucosal Immunol 10:1169–1177. https://doi.org/10.1038/mi.2016.131

Brisslert M, Enarsson K, Lundin S, Karlsson A, Kusters JG, Svennerholm AM, Backert S, Quiding-Järbrink M (2005) *Helicobacter pylori* induce neutrophil transendothelial migration: role of the bacterial HP-NAP. FEMS Microbiol Lett 249(1):95–103

Bryant CE, Spring DR, Gangloff M, Gay NJ (2010) The molecular basis of the host response to lipopolysaccharide. Nat Rev Microbiol 8:8–14. https://doi.org/10.1038/nrmicro2266

Cadamuro ACT, Rossi AFT, Maniezzo NM, Silva AE (2014) *Helicobacter pylori* infection: host immune response, implications on gene expression and microRNAs. World J Gastroenterol 20:1424–1437. https://doi.org/10.3748/wjg.v20.i6.1424

Carbo A, Olivares-Villagómez D, Hontecillas R, Bassaganya-Riera J, Chaturvedi R, Piazuelo MB, Delgado A, Washington MK, Wilson KT, Algood HMS (2014) Systems modeling of the role of interleukin-21 in the maintenance of effector CD4$^+$ T cell responses during chronic *Helicobacter pylori* infection. MBio 5:e01243–e01214. https://doi.org/10.1128/mbio.01243-14

Caruso R, Fina D, Paoluzi OA, Del Vecchio Blanco G, Stolfi C, Rizzo A, Caprioli F, Sarra M, Andrei F, Fantini MC, MacDonald TT, Pallone F, Monteleone G (2008) IL-23-mediated regulation of IL-17 production in *Helicobacter pylori*-infected gastric mucosa. Eur J Immunol 38:470–478. https://doi.org/10.1002/eji.200737635

Caruso R, Fina D, Peluso I, Fantini MC, Tosti C, Del Vecchio Blanco G, Paoluzi OA, Caprioli F, Andrei F, Stolfi C, Romano M, Ricci V, MacDonald TT, Pallone F, Monteleone G (2007) IL-21 is highly produced in *Helicobacter pylori*-infected gastric mucosa and promotes gelatinases synthesis. J Immunol 178:5957–5965

Chang H, Kim N, Park JH, Nam RH, Choi YJ, Lee HS, Yoon H, Shin CM, Park YS, Kim JM, Lee DH (2015a) Different microRNA expression levels in gastric cancer depending on Helicobacter pylori infection. Gut Liver 9:188–196. https://doi.org/10.5009/gnl13371

Chang H, Kim N, Park JH, Nam RH, Choi YJ, Park SM, Choi YJ, Yoon H, Shin CM, Lee DH (2015b) *Helicobacter pylori* might induce TGF-β1-mediated EMT by means of cagE. Helicobacter 20:438–448. https://doi.org/10.1111/hel.12220

Chang L-L, Wang S-W, Wu I-C, Yu F-J, Su Y-C, Chen Y-P, Wu D-C, Kuo C-H, Hung C-H (2012) Impaired dendritic cell maturation and IL-10 production following *H. pylori* stimulation in gastric cancer patients. Appl Microbiol Biotechnol 96:211–220. https://doi.org/10.1007/s00253-012-4034-z

Chattoraj P, Yang Q, Khandai A, Al-Hendy O, Ismail N (2013) TLR2 and Nod2 mediate resistance or susceptibility to fatal intracellular *Ehrlichia* infection in murine models of ehrlichiosis. PLoS ONE 8:e58514. https://doi.org/10.1371/journal.pone.0058514

Chen G, Tang N, Wang C, Xiao L, Yu M, Zhao L, Cai H, Han L, Xie C, Zhang Y (2017) TNF-α-inducing protein of *Helicobacter pylori* induces epithelial-mesenchymal transition (EMT) in gastric cancer cells through activation of IL-6/STAT3 signaling pathway. Biochem Biophys Res Commun 484:311–317. https://doi.org/10.1016/j.bbrc.2017.01.110

Chen W, Shu D, Chadwick VS (2001) *Helicobacter pylori* infection: mechanism of colonization and functional dyspepsia Reduced colonization of gastric mucosa by *Helicobacter pylori* in mice deficient in interleukin-10. J Gastroenterol Hepatol 16:377–383

Cheng SF, Li L, Wang LM (2015) miR-155 and miR-146b negatively regulates IL6 in *Helicobacter pylori* (cagA+) infected gastroduodenal ulcer. Eur Rev Med Pharmacol Sci 19:607–613

Chung J-W, Jeong SH, Lee SM, Pak JH, Lee GH, Jeong J-Y, Kim J-H (2017) Expression of microRNA in host cells infected with *Helicobacter pylori*. Gut Liver 11:392–400. https://doi.org/10.5009/gnl16265

Cook KW, Letley DP, Ingram RJM, Staples E, Skjoldmose H, Atherton JC, Robinson K (2014) CCL20/CCR6-mediated migration of regulatory T cells to the *Helicobacter pylori*-infected human gastric mucosa. Gut 63:1550–1559. https://doi.org/10.1136/gutjnl-2013-306253

Cooney R, Baker J, Brain O, Danis B, Pichulik T, Allan P, Ferguson DJP, Campbell BJ, Jewell D, Simmons A (2010) NOD2 stimulation induces autophagy in dendritic cells influencing bacterial handling and antigen presentation. Nat Med 16:90–97. https://doi.org/10.1038/nm.2069

Cortés-Márquez AC, Mendoza-Elizalde S, Arenas-Huertero F, Trillo-Tinoco J, Valencia-Mayoral P, Consuelo-Sánchez A, Zarate-Franco J, Dionicio-Avendaño AR, de Jesús Herrera-Esquivel J,

Recinos-Carrera EG, Colín-Valverde C, Rivera-Gutiérrez S, Reyes-López A, Vigueras-Galindo JC, Velázquez-Guadarrama N (2018) Differential expression of miRNA-146a and miRNA-155 in gastritis induced by *Helicobacter pylori* infection in paediatric patients, adults, and an animal model. BMC Infect Dis 18:463. https://doi.org/10.1186/s12879-018-3368-2

Cullen TW, Giles DK, Wolf LN, Ecobichon C, Boneca IG, Trent MS (2011) *Helicobacter pylori* versus the host: remodeling of the bacterial outer membrane is required for survival in the gastric mucosa. PLoS Pathog 7:e1002454. https://doi.org/10.1371/journal.ppat.1002454

Datta C, Subuddhi A, Kumar M, Lepcha TT, Chakraborty S, Jana K, Ghosh Z, Mukhopadhyay AK, Basu J, Kundu M (2018) Genome-wide mRNA-miRNA profiling uncovers a role of the microRNA miR-29b-1-5p/PHLPP1 signalling pathway in *Helicobacter pylori*-driven matrix metalloproteinase production in gastric epithelial cells. Cell Microbiol 20:e12859. https://doi.org/10.1111/cmi.12859

dela Pena-Ponce MG, Jimenez MT, Hansen LM, Solnick JV, Miller LA (2017) The *Helicobacter pylori* type IV secretion system promotes IL-8 synthesis in a model of pediatric airway epithelium via p 38 MAP kinase. PLOS ONE 12:e0183324. https://doi.org/10.1371/journal.pone.0183324

Dinarello CA (2009) Immunological and inflammatory functions of the Interleukin-1 family. Annu Rev Immunol 27:519–550. https://doi.org/10.1146/annurev.immunol.021908.132612

Ding H, Nedrud JG, Blanchard TG, Zagorski BM, Li G, Shiu J, Xu J, Czinn SJ (2013) Th1-mediated immunity against *Helicobacter pylori* can compensate for lack of Th17 cells and can protect mice in the absence of immunization. PLoS ONE 8:e69384. https://doi.org/10.1371/journal.pone.0069384

Dunn BE, Cohen H, Blaser MJ (1997) *Helicobacter pylori*. Clin Microbiol Rev 10(4):720–741

Eaton KA, Mefford ME (2001) Cure of *Helicobacter pylori* infection and resolution of gastritis by adoptive transfer of splenocytes in mice. Infect Immun 69:1025–1031. https://doi.org/10.1128/iai.69.2.1025-1031.2001

Eck M, Schmausser B, Scheller K, Toksoy A, Kraus M, Menzel T, Müller-Hermelink HK, Gillitzer R (2000) CXC chemokines Gro(alpha)/IL-8 and IP-10/MIG in H*elicobacter pylori* gastritis. Clin Exp Immunol 122:192–199

El-Omar EM, Carrington M, Chow WH, McColl KE, Bream JH, Young HA, Herrera J, Lissowska J, Yuan CC, Rothman N, Lanyon G, Martin M, Fraumeni JF, Rabkin CS (2000) Interleukin-1 polymorphisms associated with increased risk of gastric cancer. Nature 404:398–402. https://doi.org/10.1038/35006081

El-Omar EM, Rabkin CS, Gammon MD, Vaughan TL, Risch HA, Schoenberg JB, Stanford JL, Mayne ST, Goedert J, Blot WJ, Fraumeni JF, Chow W-H (2003) Increased risk of noncardia gastric cancer associated with proinflammatory cytokine gene polymorphisms. Gastroenterology 124:1193–1201

Engler DB, Reuter S, van Wijck Y, Urban S, Kyburz A, Maxeiner J, Martin H, Yogev N, Waisman A, Gerhard M, Cover TL, Taube C, Müller A (2014) Effective treatment of allergic airway inflammation with Helicobacter pylori immunomodulators requires BATF3-dependent dendritic cells and IL-10. Proc Natl Acad Sci USA 111:11810–11815. https://doi.org/10.1073/pnas.1410579111

Fassan M, Saraggi D, Balsamo L, Cascione L, Castoro C, Coati I, De Bernard M, Farinati F, Guzzardo V, Valeri N, Zambon CF, Rugge M (2016) Let-7c down-regulation in *Helicobacter pylori*-related gastric carcinogenesis. Oncotarget 7:4915–4924. https://doi.org/10.18632/oncotarget.6642

Ferreira RM, Pinto-Ribeiro I, Wen X, Marcos-Pinto R, Dinis-Ribeiro M, Carneiro F, Figueiredo C (2016) *Helicobacter pylori cagA* promoter region sequences influence CagA expression and Interleukin 8 secretion. J Infect Dis 213:669–673. https://doi.org/10.1093/infdis/jiv467

Ferwerda G, Kramer M, de Jong D, Piccini A, Joosten LA, DevesaGiner I, Girardin SE, Adema GJ, van der Meer JWM, Kullberg B-J, Rubartelli A, Netea MG (2008) Engagement of NOD2 has a dual effect on proIL-1β mRNA transcription and secretion of bioactive IL-1β. Eur J Immunol 38:184–191. https://doi.org/10.1002/eji.200737103

Franchi L, Muñoz-Planillo R, Núñez G (2012) Sensing and reacting to microbes through the inflammasomes. Nat Immunol 13:325–332. https://doi.org/10.1038/ni.2231

Garhart CA, Heinzel FP, Czinn SJ, Nedrud JG (2003) Vaccine-induced reduction of *Helicobacter pylori* colonization in mice is interleukin-12 dependent but gamma interferon and inducible nitric oxide synthase independent. Infect Immun 71:910–921

Geng Y, Lu X, Wu X, Xue L, Wang X, Xu J (2016) MicroRNA-27b suppresses *Helicobacter pylori*-induced gastric tumorigenesis through negatively regulating Frizzled7. Oncol Rep 35:2441–2450. https://doi.org/10.3892/or.2016.4572

Gewirtz AT, Yu Y, Krishna US, Israel DA, Lyons SL, Peek RM (2004) *Helicobacter pylori* flagellin evades toll-like receptor 5-mediated innate immunity. J Infect Dis 189:1914–1920. https://doi.org/10.1086/386289

Gionchetti P, Vaira D, Campieri M, Holton J, Menegatti M, Belluzzi A, Bertinelli E, Ferretti M, Brignola C, Miglioli M (1994) Enhanced mucosal interleukin-6 and -8 in Helicobacter pylori-positive dyspeptic patients. Am J Gastroenterol 89:883–887

Girardin SE, Boneca IG, Viala J, Chamaillard M, Labigne A, Thomas G, Philpott DJ, Sansonetti PJ (2003) Nod2 is a general sensor of peptidoglycan through muramyl dipeptide (MDP) detection. J Biol Chem 278:8869–8872. https://doi.org/10.1074/jbc.c200651200

Giza DE, Vasilescu C, Calin GA (2014) Key principles of miRNA involvement in human diseases. Discoveries (Craiova) 2:e34. https://doi.org/10.15190/d.2014.26

Gobert AP, Bambou J-C, Werts C, Balloy V, Chignard M, Moran AP, Ferrero RL (2004) *Helicobacter pylori* heat shock protein 60 mediates Interleukin-6 production by macrophages via a Toll-like receptor (TLR)-2-, TLR-4-, and myeloid differentiation factor 88-independent mechanism. J Biol Chem 279:245–250. https://doi.org/10.1074/jbc.m307858200

Gobert AP, Verriere T, Asim M, Barry DP, Piazuelo MB, de Sablet T, Delgado AG, Bravo LE, Correa P, Peek RM Jr, Chaturvedi R, Wilson KT (2014) Heme oxygenase-1 dysregulates macrophage polarization and the immune response to *Helicobacter pylori*. J Immunol 193:3013–3022. https://doi.org/10.4049/jimmunol.1401075

Gobert AP, Wilson KT (2016) The immune battle against *Helicobacter pylori* infection: no offense. Trends Microbiol 24:366–376. https://doi.org/10.1016/j.tim.2016.02.005

Gringhuis SI, den Dunnen J, Litjens M, van der Vlist M, Geijtenbeek TBH (2009) Carbohydrate-specific signaling through the DC-SIGN signalosome tailors immunity to Mycobacterium tuberculosis, HIV-1 and *Helicobacter pylori*. Nat Immunol 10:1081–1088. https://doi.org/10.1038/ni.1778

Gringhuis SI, den Dunnen J, Litjens M, van het Hof B, van Kooyk Y, Geijtenbeek TBH (2007) C-type lectin DC-SIGN modulates Toll-like receptor signaling via Raf-1 Kinase-dependent acetylation of transcription factor NF-κB. Immunity 26:605–616. https://doi.org/10.1016/j.immuni.2007.03.012

Gringhuis SI, Kaptein TM, Wevers BA, Mesman AW, Geijtenbeek TBH (2014) Fucose-specific DC-SIGN signalling directs T helper cell type-2 responses via IKKε- and CYLD-dependent Bcl3 activation. Nature Communications 5. https://doi.org/10.1038/ncomms4898

Guiney DG, Hasegawa P, Cole SP (2003) Helicobacter pylori preferentially induces interleukin 12 (IL-12) rather than IL-6 or IL-10 in human dendritic cells. Infect Immun 71:4163–4166

Guo H, Callaway JB, Ting JP-Y (2015) Inflammasomes: mechanism of action, role in disease and therapeutics. Nat Med 21:677–687. https://doi.org/10.1038/nm.3893

Haeberle HA, Kubin M, Bamford KB, Garofalo R, Graham DY, El-Zaatari F, Karttunen R, Crowe SE, Reyes VE, Ernst PB (1997) Differential stimulation of interleukin-12 (IL-12) and IL-10 by live and killed Helicobacter pylori in vitro and association of IL-12 production with gamma interferon-producing T cells in the human gastric mucosa. Infect Immun 65:4229–4235

Hara H, Tsuchiya K, Kawamura I, Fang R, Hernandez-Cuellar E, Shen Y, Mizuguchi J, Schweighoffer E, Tybulewicz V, Mitsuyama M (2013) Phosphorylation of the adaptor ASC acts as a molecular switch that controls the formation of speck-like aggregates and inflammasome activity. Nat Immunol 14:1247–1255. https://doi.org/10.1038/ni.2749

Hayashi Y, Tsujii M, Wang J, Kondo J, Akasaka T, Jin Y, Li W, Nakamura T, Nishida T, Iijima H, Tsuji S, Kawano S, Hayashi N, Takehara T (2013) CagA mediates epigenetic regulation

to attenuate let-7 expression in *Helicobacter pylori*-related carcinogenesis. Gut 62:1536–1546. https://doi.org/10.1136/gutjnl-2011-301625
Hedl M, Abraham C (2013) NLRP1 and NLRP3 inflammasomes are essential for distinct outcomes of decreased cytokines but enhanced bacterial killing upon chronic Nod2 stimulation. Am J Physiol-Gastrointest Liver Physiol 304:G583–G596. https://doi.org/10.1152/ajpgi.00297.2012
Hitkova I, Yuan G, Anderl F, Gerhard M, Kirchner T, Reu S, Röcken C, Schäfer C, Schmid RM, Vogelmann R, Ebert MPA, Burgermeister E (2013) Caveolin-1 protects B6129 mice against *Helicobacter pylori* gastritis. PLoS Pathog 9:e1003251. https://doi.org/10.1371/journal.ppat.1003251
Horvath DJ, Radin JN, Cho SH, Washington MK, Algood HMS (2013) The Interleukin-17 Receptor B subunit is essential for the Th2 response to *Helicobacter pylori*, but not for control of bacterial burden. PLoS ONE 8:e60363. https://doi.org/10.1371/journal.pone.0060363
Horvath DJ, Washington MK, Cope VA, Algood HMS (2012) IL-23 contributes to control of chronic *Helicobacter pylori* infection and the development of T Helper responses in a mouse model. Front Immunol 3:56. https://doi.org/10.3389/fimmu.2012.00056
Hsu L-C, Ali SR, McGillivray S, Tseng P-H, Mariathasan S, Humke EW, Eckmann L, Powell JJ, Nizet V, Dixit VM, Karin M (2008) A NOD2-NALP1 complex mediates caspase-1-dependent IL-1 secretion in response to *Bacillus anthracis* infection and muramyl dipeptide. Proc Natl Acad Sci 105:7803–7808. https://doi.org/10.1073/pnas.0802726105
Hussain K, Letley DP, Greenaway AB, Kenefeck R, Winter JA, Tomlinson W, Rhead J, Staples E, Kaneko K, Atherton JC, Robinson K (2016) *Helicobacter pylori*-mediated protection from allergy is associated with IL-10-secreting peripheral blood regulatory T cells. Front Immunol 7. https://doi.org/10.3389/fimmu.2016.00071
Ishihara S, Rumi MAK, Kadowaki Y, Ortega-Cava CF, Yuki T, Yoshino N, Miyaoka Y, Kazumori H, Ishimura N, Amano Y, Kinoshita Y (2004) Essential role of MD-2 in TLR4-dependent signaling during Helicobacter pylori-associated gastritis. J Immunol 173:1406–1416
Isomoto H, Matsushima K, Inoue N, Hayashi T, Nakayama T, Kunizaki M, Hidaka S, Nakayama M, Hisatsune J, Nakashima M, Nagayasu T, Nakao K, Hirayama T (2012) Interweaving microRNAs and proinflammatory cytokines in gastric mucosa with reference to *H. pylori* infection. J Clin Immunol 32:290–299. https://doi.org/10.1007/s10875-011-9626-3
Itoh T, Yoshida M, Chiba T, Kita T, Wakatsuki Y (2003) A coordinated cytotoxic effect of IFN-gamma and cross-reactive antibodies in the pathogenesis of Helicobacter pylori gastritis. Helicobacter 8:268–278
Jo Y, Han SU, Kim YJ, Kim JH, Kim ST, Kim S-J, Hahm K-B (2010) Suppressed gastric mucosal TGF-β1 increases susceptibility to *H. pylori*-induced gastric inflammation and ulceration: A stupid host defense response. Gut and Liver 4:43–53. https://doi.org/10.5009/gnl.2010.4.1.43
Kaebisch R, Mejías-Luque R, Prinz C, Gerhard M (2014) Helicobacter pylori cytotoxin-associated gene A impairs human dendritic cell maturation and function through IL-10-mediated activation of STAT3. J Immunol 192:316–323. https://doi.org/10.4049/jimmunol.1302476
Kameoka S, Kameyama T, Hayashi T, Sato S, Ohnishi N, Hayashi T, Murata-Kamiya N, Higashi H, Hatakeyama M, Takaoka A (2016) *Helicobacter pylori* induces IL-1β protein through the inflammasome activation in differentiated macrophagic cells. Biomed Res 37:21–27. https://doi.org/10.2220/biomedres.37.21
Kandulski A, Wex T, Kuester D, Peitz U, Gebert I, Roessner A, Malfertheiner P (2008) Naturally occurring regulatory T cells ($CD4^+$, CD25 high, $FOXP3^+$) in the antrum and cardia are associated with higher *Helicobacter pylori* colonization and increased gene expression of TGF-beta1. Helicobacter 13:295–303. https://doi.org/10.1111/j.1523-5378.2008.00612.x
Kang DW, Yang ES, Noh YN, Hwang WC, Jo S-Y, Suh Y-A, Park WS, Choi K-Y, Min DS (2017) MicroRNA-320a and microRNA-4496 attenuate *Helicobacter pylori* cytotoxin-associated gene A (CagA)-induced cancer-initiating potential and chemoresistance by targeting β-catenin and ATP-binding cassette, subfamily G, member 2. J Pathol 241:614–625. https://doi.org/10.1002/path.4866
Kao JY, Rathinavelu S, Eaton KA, Bai L, Zavros Y, Takami M, Pierzchala A, Merchant JL (2006) *Helicobacter pylori*-secreted factors inhibit dendritic cell IL-12 secretion: a mechanism of inef-

fective host defense. Am J Physiol Gastrointest Liver Physiol 291:G73–G81. https://doi.org/10.1152/ajpgi.00139.2005

Kao JY, Zhang M, Miller MJ, Mills JC, Wang B, Liu M, Eaton KA, Zou W, Berndt BE, Cole TS, Takeuchi T, Owyang SY, Luther J (2010) *Helicobacter pylori* immune escape is mediated by dendritic cell-induced Treg skewing and Th17 suppression in mice. Gastroenterology 138:1046–1054. https://doi.org/10.1053/j.gastro.2009.11.043

Kersse K, Bertrand MJM, Lamkanfi M, Vandenabeele P (2011) NOD-like receptors and the innate immune system: Coping with danger, damage and death. Cytokine Growth Factor Rev 22:257–276. https://doi.org/10.1016/j.cytogfr.2011.09.003

Kiga K, Mimuro H, Suzuki M, Shinozaki-Ushiku A, Kobayashi T, Sanada T, Kim M, Ogawa M, Iwasaki YW, Kayo H, Fukuda-Yuzawa Y, Yashiro M, Fukayama M, Fukao T, Sasakawa C (2014) Epigenetic silencing of miR-210 increases the proliferation of gastric epithelium during chronic *Helicobacter pylori* infection. Nat Commun 5:4497. https://doi.org/10.1038/ncomms5497

Kim D-J, Park J-H, Franchi L, Backert S, Núñez G (2013) The Cag pathogenicity island and interaction between TLR2/NOD2 and NLRP3 regulate IL-1β production in *Helicobacter pylori* infected dendritic cells: Immunity to infection. Eur J Immunol 43:2650–2658. https://doi.org/10.1002/eji.201243281

Kim SH, Sierra RA, McGee DJ, Zabaleta J (2012) Transcriptional profiling of gastric epithelial cells infected with wild type or arginase-deficient *Helicobacter pylori*. BMC Microbiol 12:175. https://doi.org/10.1186/1471-2180-12-175

Koch KN, Hartung ML, Urban S, Kyburz A, Bahlmann AS, Lind J, Backert S, Taube C, Müller A (2015) *Helicobacter* urease-induced activation of the TLR2/NLRP3/IL-18 axis protects against asthma. J Clin Invest 125:3297–3302. https://doi.org/10.1172/jci79337

Koch M, Mollenkopf H-J, Klemm U, Meyer TF (2012) Induction of microRNA-155 is TLR- and type IV secretion system-dependent in macrophages and inhibits DNA-damage induced apoptosis. Proc Natl Acad Sci USA 109:E1153–E1162. https://doi.org/10.1073/pnas.1116125109

Kong D, Piao Y-S, Yamashita S, Oshima H, Oguma K, Fushida S, Fujimura T, Minamoto T, Seno H, Yamada Y, Satou K, Ushijima T, Ishikawa T-O, Oshima M (2012) Inflammation-induced repression of tumor suppressor miR-7 in gastric tumor cells. Oncogene 31:3949–3960. https://doi.org/10.1038/onc.2011.558

Krauss-Etschmann S, Sammler E, Koletzko S, Konstantopoulos N, Aust D, Gebert B, Luckow B, Reinhardt D, Schendel DJ (2003) Chemokine receptor 5 expression in gastric mucosa of *Helicobacter pylori*-infected and noninfected children. Clin Diagn Lab Immunol 10:22–29

Lamkanfi M (2011) Emerging inflammasome effector mechanisms. Nat Rev Immunol 11:213–220. https://doi.org/10.1038/nri2936

Latz E, Xiao TS, Stutz A (2013) Activation and regulation of the inflammasomes. Nat Rev Immunol 13:397–411. https://doi.org/10.1038/nri3452

Lee JW, Kim N, Park JH, Kim HJ, Chang H, Kim JM, Kim J-W, Lee DH (2017) Differential microRNA expression between gastric cancer tissue and non-cancerous gastric mucosa according to *Helicobacter pylori* Status. J Cancer Prev 22:33–39. https://doi.org/10.15430/jcp.2017.22.1.33

Lee RC, Feinbaum RL, Ambros V (1993) The C. elegans heterochronic gene lin-4 encodes small RNAs with antisense complementarity to lin-14. Cell 75:843–854

Li N, Xu X, Xiao B, Zhu E-D, Li B-S, Liu Z, Tang B, Zou Q-M, Liang H-P, Mao X-H (2012) *Helicobacter pylori* related proinflammatory cytokines contribute to the induction of miR-146a in human gastric epithelial cells. Mol Biol Rep 39:4655–4661. https://doi.org/10.1007/s11033-011-1257-5

Li Z, Li J (2006) Local expressions of TGF-beta1, TGF-beta1RI, CTGF, and Smad-7 in *Helicobacter pylori*-associated gastritis. Scand J Gastroenterol 41:1007–1012. https://doi.org/10.1080/00365520600554477

Lindgren A, Pavlovic V, Flach C-F, Sjöling A, Lundin S (2011) Interferon-gamma secretion is induced in IL-12 stimulated human NK cells by recognition of *Helicobacter pylori* or TLR2 ligands. Innate Immun 17:191–203. https://doi.org/10.1177/1753425909357970

Liu Z, Wang D, Hu Y, Zhou G, Zhu C, Yu Q, Chi Y, Cao Y, Jia C, Zou Q (2013) MicroRNA-146a negatively regulates PTGS2 expression induced by *Helicobacter pylori* in human gastric epithelial cells. J Gastroenterol 48:86–92. https://doi.org/10.1007/s00535-012-0609-9

Liu Z, Xiao B, Tang B, Li B, Li N, Zhu E, Guo G, Gu J, Zhuang Y, Liu X, Ding H, Zhao X, Guo H, Mao X, Zou Q (2010) Up-regulated microRNA-146a negatively modulate *Helicobacter pylori*-induced inflammatory response in human gastric epithelial cells. Microbes Infect 12:854–863. https://doi.org/10.1016/j.micinf.2010.06.002

Lu H, Wu JY, Kudo T, Ohno T, Graham DY, Yamaoka Y (2005) Regulation of interleukin-6 promoter activation in gastric epithelial cells infected with *Helicobacter pylori*. Mol Biol Cell 16:4954–4966. https://doi.org/10.1091/mbc.e05-05-0426

Lu L-F, Boldin MP, Chaudhry A, Lin L-L, Taganov KD, Hanada T, Yoshimura A, Baltimore D, Rudensky AY (2010) Function of miR-146a in controlling Treg cell-mediated regulation of Th1 responses. Cell 142:914–929. https://doi.org/10.1016/j.cell.2010.08.012

Lucas B, Bumann D, Walduck A, Koesling J, Develioglu L, Meyer TF, Aebischer T (2001) Adoptive transfer of $CD4^+$ T cells specific for subunit A of *Helicobacter pylori* urease reduces H. pylori stomach colonization in mice in the absence of interleukin-4 (IL-4)/IL-13 receptor signaling. Infect Immun 69:1714–1721. https://doi.org/10.1128/iai.69.3.1714-1721.2001

Ma L, Chen Y, Zhang B, Liu G (2014) Increased microRNA-223 in *Helicobacter pylori*-associated gastric cancer contributed to cancer cell proliferation and migration. Biosci Biotechnol Biochem 78:602–608. https://doi.org/10.1080/09168451.2014.895661

Maeda S (2005) Nod2 mutation in Crohn's disease potentiates NF- B activity and IL-1 processing. Science 307:734–738. https://doi.org/10.1126/science.1103685

Maeda S, Akanuma M, Mitsuno Y, Hirata Y, Ogura K, Yoshida H, Shiratori Y, Omata M (2001) Distinct mechanism of *Helicobacter pylori* mediated NF-kappa B activation between gastric cancer cells and monocytic cells. J Biol Chem 276:44856–44864. https://doi.org/10.1074/jbc.m105381200

Maelfait J, Vercammen E, Janssens S, Schotte P, Haegman M, Magez S, Beyaert R (2008) Stimulation of Toll-like receptor 3 and 4 induces interleukin-1β maturation by caspase-8. J Exp Med 205:1967–1973. https://doi.org/10.1084/jem.20071632

Man SM (2018) Inflammasomes in the gastrointestinal tract: infection, cancer and gut microbiota homeostasis. Nat Rev Gastroenterol Hepatol 15:721–737. https://doi.org/10.1038/s41575-018-0054-1

Mandell L, Moran AP, Cocchiarella A, Houghton J, Taylor N, Fox JG, Wang TC, Kurt-Jones EA (2004) Intact gram-negative *Helicobacter pylori*, *Helicobacter felis*, and *Helicobacter hepaticus* bacteria activate innate immunity via toll-like receptor 2 but not toll-like receptor 4. Infect Immun 72:6446–6454. https://doi.org/10.1128/iai.72.11.6446-6454.2004

Mariathasan S, Monack DM (2007) Inflammasome adaptors and sensors: intracellular regulators of infection and inflammation. Nat Rev Immunol 7:31–40. https://doi.org/10.1038/nri1997

Mariathasan S, Weiss DS, Newton K, McBride J, O'Rourke K, Roose-Girma M, Lee WP, Weinrauch Y, Monack DM, Dixit VM (2006) Cryopyrin activates the inflammasome in response to toxins and ATP. Nature 440:228–232. https://doi.org/10.1038/nature04515

Marotti B, Rocco A, De Colibus P, Compare D, de Nucci G, Staibano S, Tatangelo F, Romano M, Nardone G (2008) Interleukin-13 mucosal production in *Helicobacter pylori*-related gastric diseases. Dig Liver Dis 40:240–247. https://doi.org/10.1016/j.dld.2007.11.021

Martinon F, Burns K, Tschopp J (2002) The inflammasome: a molecular platform triggering activation of inflammatory caspases and processing of proIL-beta. Mol Cell 10:417–426

Matsumoto Y, Blanchard TG, Drakes ML, Basu M, Redline RW, Levine AD, Czinn SJ (2005) Eradication of *Helicobacter pylori* and resolution of gastritis in the gastric mucosa of IL-10-deficient mice. Helicobacter 10:407–415. https://doi.org/10.1111/j.1523-5378.2005.00349.x

Matsushima K, Isomoto H, Inoue N, Nakayama T, Hayashi T, Nakayama M, Nakao K, Hirayama T, Kohno S (2011) microRNA signatures in *Helicobacter pylori*-infected gastric mucosa. Int J Cancer 128:361–370. https://doi.org/10.1002/ijc.25348

Mercier BC, Ventre E, Fogeron M-L, Debaud A-L, Tomkowiak M, Marvel J, Bonnefoy N (2012) NOD1 cooperates with TLR2 to enhance T cell receptor-mediated activation in CD8 T cells. PLoS ONE 7:e42170. https://doi.org/10.1371/journal.pone.0042170

Miao EA, Rajan JV, Aderem A (2011) Caspase-1-induced pyroptotic cell death: Caspase-1-induced pyroptotic cell death. Immunol Rev 243:206–214. https://doi.org/10.1111/j.1600-065x.2011.01044.x

Miao L, Liu K, Xie M, Xing Y, Xi T (2014) miR-375 inhibits *Helicobacter pylori*-induced gastric carcinogenesis by blocking JAK2-STAT3 signaling. Cancer Immunol Immunother 63:699–711. https://doi.org/10.1007/s00262-014-1550-y

Miftahussurur M, Yamaoka Y (2015) *Helicobacter pylori* virulence genes and host genetic polymorphisms as risk factors for peptic ulcer disease. Expert Rev Gastroenterol Hepatol 9:1535–1547. https://doi.org/10.1586/17474124.2015.1095089

Mogensen TH (2009) Pathogen recognition and inflammatory signaling in innate immune defenses. Clin Microbiol Rev 22:240–273. https://doi.org/10.1128/cmr.00046-08

Moran AP, Lindner B, Walsh EJ (1997) Structural characterization of the lipid A component of *Helicobacter pylori* rough- and smooth-form lipopolysaccharides. J Bacteriol 179:6453–6463

Morey P, Pfannkuch L, Pang E, Boccellato F, Sigal M, Imai-Matsushima A, Dyer V, Koch M, Mollenkopf H-J, Schlaermann P, Meyer TF (2018) *Helicobacter pylori* depletes cholesterol in gastric glands to prevent interferon gamma signaling and escape the inflammatory response. Gastroenterology 154:1391–1404.e9. https://doi.org/10.1053/j.gastro.2017.12.008

Munari F, Fassan M, Capitani N, Codolo G, Vila-Caballer M, Pizzi M, Rugge M, Della Bella C, Troilo A, D'Elios S, Baldari CT, D'Elios MM, de Bernard M (2014) Cytokine BAFF released by *Helicobacter pylori*-infected macrophages triggers the Th17 response in human chronic gastritis. J Immunol 193:5584–5594. https://doi.org/10.4049/jimmunol.1302865

Mustapha P, Paris I, Garcia M, Tran CT, Cremniter J, Garnier M, Faure J-P, Barthes T, Boneca IG, Morel F, Lecron J-C, Burucoa C, Bodet C (2014) Chemokines and antimicrobial peptides have a *cag*—dependent early response to *Helicobacter pylori* infection in primary human gastric epithelial cells. Infect Immun 82:2881–2889. https://doi.org/10.1128/iai.01517-13

Nagashima H, Iwatani S, Cruz M, Jiménez Abreu JA, Uchida T, Mahachai V, Vilaichone R-K, Graham DY, Yamaoka Y (2015) Toll-like receptor 10 in *Helicobacter pylori* infection. J Infect Dis 212:1666–1676. https://doi.org/10.1093/infdis/jiv270

Naumann M, Sokolova O, Tegtmeyer N, Backert S (2017) *Helicobacter pylori*: a paradigm pathogen for subverting host cell signal transmission. Trends Microbiol 25(4):316–328. https://doi.org/10.1016/j.tim.2016.12.004

Ng GZ, Menheniott TR, Every AL, Stent A, Judd LM, Chionh YT, Dhar P, Komen JC, Giraud AS, Wang TC, McGuckin MA, Sutton P (2016) The MUC1 mucin protects against *Helicobacter pylori* pathogenesis in mice by regulation of the NLRP3 inflammasome. Gut 65:1087–1099. https://doi.org/10.1136/gutjnl-2014-307175

Nguyen TT, Kim S-J, Park JM, Hahm KB, Lee H-J (2015) Repressed TGF-β signaling through CagA-Smad3 interaction as pathogenic mechanisms of *Helicobacter pylori*-associated gastritis. J Clin Biochem Nutr 57:113–120. https://doi.org/10.3164/jcbn.15-38

Noto JM, Piazuelo MB, Chaturvedi R, Bartel CA, Thatcher EJ, Delgado A, Romero-Gallo J, Wilson KT, Correa P, Patton JG, Peek RM (2013) Strain-specific suppression of microRNA-320 by carcinogenic *Helicobacter pylori* promotes expression of the antiapoptotic protein Mcl-1. Am J Physiol Gastrointest Liver Physiol 305:G786–G796. https://doi.org/10.1152/ajpgi.00279.2013

Obonyo M, Cole SP, Datta SK, Guiney DG (2006) Evidence for interleukin-1-independent stimulation of interleukin-12 and down-regulation by interleukin-10 in *Helicobacter pylori*-infected murine dendritic cells deficient in the interleukin-1 receptor. FEMS Immunol Med Microbiol 47:414–419. https://doi.org/10.1111/j.1574-695x.2006.00105.x

Odenbreit S, Linder S, Gebert-Vogl B, Rieder G, Moran AP, Haas R (2006) Interleukin-6 induction by *Helicobacter pylori* in human macrophages is dependent on phagocytosis. Helicobacter 11:196–207

Oertli M, Noben M, Engler DB, Semper RP, Reuter S, Maxeiner J, Gerhard M, Taube C, Müller A (2013) *Helicobacter pylori* γ-glutamyl transpeptidase and vacuolating cytotoxin promote gastric persistence and immune tolerance. Proc Natl Acad Sci USA 110:3047–3052. https://doi.org/10.1073/pnas.1211248110

Ortega-Gómez A, Perretti M, Soehnlein O (2013) Resolution of inflammation: an integrated view. EMBO Mol Med 5(5):661–674. https://doi.org/10.1002/emmm.201202382

Pachathundikandi K, Backert S (2018a) Heptose 1,7-bisphosphate directed TIFA oligomerization: a novel PAMP-recognizing signaling platform in the control of bacterial infections. Gastroenterology 154:778–783. https://doi.org/10.1053/j.gastro.2018.01.009

Pachathundikandi SK, Backert S (2016) Differential expression of interleukin 1β during *Helicobacter pylori* infection of Toll-like Receptor 2 (TLR2)- and TLR10-expressing HEK293 cell lines. J Infect Dis 214:166–167. https://doi.org/10.1093/infdis/jiw154

Pachathundikandi SK, Backert S (2018b) *Helicobacter pylori* controls NLRP3 expression by regulating hsa-miR-223-3p and IL-10 in cultured and primary human immune cells. Innate Immun 24:11–23. https://doi.org/10.1177/1753425917738043

Pachathundikandi SK, Brandt S, Madassery J, Backert S (2011) Induction of TLR-2 and TLR-5 expression by *Helicobacter pylori* switches *cag*PAI-dependent signalling leading to the secretion of IL-8 and TNF-α. PLoS ONE 6:e19614. https://doi.org/10.1371/journal.pone.0019614

Pachathundikandi SK, Lind J, Tegtmeyer N, El-Omar EM, Backert S (2015) Interplay of the gastric pathogen *Helicobacter pylori* with Toll-like receptors. Biomed Res Int 2015:192420. https://doi.org/10.1155/2015/192420

Pachathundikandi SK, Müller A, Backert S (2016) Inflammasome activation by *Helicobacter pylori* and its implications for persistence and immunity. Curr Top Microbiol Immunol 397:117–31. https://doi.org/10.1007/978-3-319-41171-2_6

Pachathundikandi SK, Tegtmeyer N, Backert S (2013) Signal transduction of *Helicobacter pylori* during interaction with host cell protein receptors of epithelial and immune cells. Gut Microbes 4:454–474. https://doi.org/10.4161/gmic.27001

Pagliari M, Munari F, Toffoletto M, Lonardi S, Chemello F, Codolo G, Millino C, Della Bella C, Pacchioni B, Vermi W, Fassan M, de Bernard M, Cagnin S (2017) *Helicobacter pylori* affects the antigen presentation activity of macrophages modulating the expression of the immune receptor CD300E through miR-4270. Front Immunol 8:1288. https://doi.org/10.3389/fimmu.2017.01288

Panthel K, Faller G, Haas R (2003) Colonization of C57BL/6J and BALB/c wild-type and knockout mice with Helicobacter pylori: effect of vaccination and implications for innate and acquired immunity. Infect Immun 71:794–800

Pellicanò A, Sebkova L, Monteleone G, Guarnieri G, Imeneo M, Pallone F, Luzza F (2007) Interleukin-12 drives the Th1 signaling pathway in Helicobacter pylori-infected human gastric mucosa. Infect Immun 75:1738–1744. https://doi.org/10.1128/iai.01446-06

Peng L-S, Zhuang Y, Li W-H, Zhou Y-Y, Wang T-T, Chen N, Cheng P, Li B-S, Guo H, Yang S-M, Chen W-S, Zou Q-M (2014) Elevated interleukin-32 expression is associated with *Helicobacter pylori*-related gastritis. PLoS ONE 9:e88270. https://doi.org/10.1371/journal.pone.0088270

Pérez-Pérez GI, Shepherd VL, Morrow JD, Blaser MJ (1995) Activation of human THP-1 cells and rat bone marrow-derived macrophages by *Helicobacter pylori* lipopolysaccharide. Infect Immun 63:1183–1187

Piao J-Y, Lee HG, Kim S-J, Kim D-H, Han H-J, Ngo H-K-C, Park S-A, Woo J-H, Lee J-S, Na H-K, Cha Y-N, Surh Y-J (2016) *Helicobacter pylori* activates IL-6-STAT3 signaling in human gastric cancer cells: potential roles for reactive oxygen species. Helicobacter 21:405–416. https://doi.org/10.1111/hel.12298

Pinchuk IV, Morris KT, Nofchissey RA, Earley RB, Wu J-Y, Ma TY, Beswick EJ (2013) Stromal cells induce Th17 during *Helicobacter pylori* infection and in the gastric tumor microenvironment. PLoS ONE 8:e53798. https://doi.org/10.1371/journal.pone.0053798

Polk DB, Peek RM (2010) *Helicobacter pylori*: gastric cancer and beyond. Nat Rev Cancer 10:403–414. https://doi.org/10.1038/nrc2857

Posselt G, Backert S, Wessler S (2013) The functional interplay of *Helicobacter pylori* factors with gastric epithelial cells induces a multi-step process in pathogenesis. Cell Commun Signal 11:77. https://doi.org/10.1186/1478-811x-11-77

Quiding-Järbrink M, Raghavan S, Sundquist M (2010) Enhanced M1 macrophage polarization in human *Helicobacter pylori* associated atrophic gastritis and in vaccinated mice. PLoS ONE 5:e15018. https://doi.org/10.1371/journal.pone.0015018

Rad R, Ballhorn W, Voland P, Eisenächer K, Mages J, Rad L, Ferstl R, Lang R, Wagner H, Schmid RM, Bauer S, Prinz C, Kirschning CJ, Krug A (2009) Extracellular and intracellular pattern recognition receptors cooperate in the recognition of *Helicobacter pylori*. Gastroenterology 136:2247–2257. https://doi.org/10.1053/j.gastro.2009.02.066

Rad R, Brenner L, Krug A, Voland P, Mages J, Lang R, Schwendy S, Reindl W, Dossumbekova A, Ballhorn W, Wagner H, Schmid RM, Bauer S, Prinz C (2007) Toll-like receptor-dependent activation of antigen-presenting cells affects adaptive immunity to *Helicobacter pylori*. Gastroenterology 133:150–163.e3. https://doi.org/10.1053/j.gastro.2007.04.071

Rapsinski GJ, Wynosky-Dolfi MA, Oppong GO, Tursi SA, Wilson RP, Brodsky IE, Tükel Ç (2015) Toll-like receptor 2 and NLRP3 cooperate to recognize a functional bacterial amyloid, curli. Infect Immun 83:693–701. https://doi.org/10.1128/iai.02370-14

Rhee SH, Keates AC, Moyer MP, Pothoulakis C (2004) MEK is a key modulator for TLR5-induced interleukin-8 and MIP3alpha gene expression in non-transformed human colonic epithelial cells. J Biol Chem 279:25179–25188. https://doi.org/10.1074/jbc.m400967200

Rizzuti D, Ang M, Sokollik C, Wu T, Abdullah M, Greenfield L, Fattouh R, Reardon C, Tang M, Diao J, Schindler C, Cattral M, Jones NL (2015) *Helicobacter pylori* inhibits dendritic cell maturation via interleukin-10-mediated activation of the signal transducer and activator of transcription 3 pathway. J Innate Immun 7:199–211. https://doi.org/10.1159/000368232

Romi B, Soldaini E, Pancotto L, Castellino F, Del Giudice G, Schiavetti F (2011) *Helicobacter pylori* induces activation of human peripheral $\gamma\delta^+$ T lymphocytes. PLoS ONE 6:e19324. https://doi.org/10.1371/journal.pone.0019324

Rossi AFT, Cadamuro ACT, Biselli-Périco JM, Leite KRM, Severino FE, Reis PP, Cordeiro JA, Silva AE (2016) Interaction between inflammatory mediators and miRNAs in *Helicobacter pylori* infection. Cell Microbiol 18:1444–1458. https://doi.org/10.1111/cmi.12587

Saito Y, Suzuki H, Tsugawa H, Suzuki S, Matsuzaki J, Hirata K, Hibi T (2011) Dysfunctional gastric emptying with down-regulation of muscle-specific microRNAs in *Helicobacter pylori*-infected mice. Gastroenterology 140:189–198. https://doi.org/10.1053/j.gastro.2010.08.044

Sakitani K, Hirata Y, Hayakawa Y, Serizawa T, Nakata W, Takahashi R, Kinoshita H, Sakamoto K, Nakagawa H, Akanuma M, Yoshida H, Maeda S, Koike K (2012) Role of interleukin-32 in *Helicobacter pylori*-induced gastric inflammation. Infect Immun 80:3795–3803. https://doi.org/10.1128/iai.00637-12

Salama NR, Hartung ML, Müller A (2013) Life in the human stomach: persistence strategies of the bacterial pathogen *Helicobacter pylori*. Nat Rev Microbiol 11:385–399. https://doi.org/10.1038/nrmicro3016

Schmausser B, Josenhans C, Endrich S, Suerbaum S, Sitaru C, Andrulis M, Brändlein S, Rieckmann P, Müller-Hermelink HK, Eck M (2004) Downregulation of CXCR1 and CXCR2 expression on human neutrophils by *Helicobacter pylori*: a new pathomechanism in *H. pylori* infection? Infect Immun 72:6773–6779. https://doi.org/10.1128/iai.72.12.6773-6779.2004

Serhan CN (2017) Treating inflammation and infection in the 21st century: new hints from decoding resolution mediators and mechanisms. FASEB J 31:1273–1288. https://doi.org/10.1096/fj.201601222r

Serrano C, Wright SW, Bimczok D, Shaffer CL, Cover TL, Venegas A, Salazar MG, Smythies LE, Harris PR, Smith PD (2013) Downregulated Th17 responses are associated with reduced gastritis in *Helicobacter pylori*-infected children. Mucosal Immunol 6:950–959. https://doi.org/10.1038/mi.2012.133

Shao L, Chen Z, Soutto M, Zhu S, Lu H, Romero-Gallo J, Peek R, Zhang S, El-Rifai W (2019) Helicobacter pylori-induced miR-135b-5p promotes cisplatin resistance in gastric cancer. FASEB J 33:264–274. https://doi.org/10.1096/fj.201701456rr

Shi Y, Liu X-F, Zhuang Y, Zhang J-Y, Liu T, Yin Z, Wu C, Mao X-H, Jia K-R, Wang F-J, Guo H, Flavell RA, Zhao Z, Liu K-Y, Xiao B, Guo Y, Zhang W-J, Zhou W-Y, Guo G, Zou Q-M (2010) *Helicobacter pylori*-induced Th17 responses modulate Th1 cell responses, benefit bacterial growth, and contribute to pathology in mice. J Immunol 184:5121–5129. https://doi.org/10.4049/jimmunol.0901115

Shih S-C (2005) Expression patterns of transforming growth factor-beta and its receptors in gastric mucosa of patients with refractory gastric ulcer. World J Gastroenterol 11:136. https://doi.org/10.3748/wjg.v11.i1.136

Shimoyama T, Fukuda S, Liu Q, Nakaji S, Fukuda Y, Sugawara K (2002) Production of chemokines and reactive oxygen species by human neutrophils stimulated by *Helicobacter pylori*. Helicobacter 7:170–174

Shiotani A, Uedo N, Iishi H, Murao T, Kanzaki T, Kimura Y, Kamada T, Kusunoki H, Inoue K, Haruma K (2012) *Helicobacter pylori* eradication did not improve dysregulation of specific oncogenic miRNAs in intestinal metaplastic glands. J Gastroenterol 47:988–998. https://doi.org/10.1007/s00535-012-0562-7

Sjökvist Ottsjö L, Flach C-F, Nilsson S, Malefyt R de W, Walduck AK, Raghavan S (2015) Defining the roles of IFN-γ and IL-17A in inflammation and protection against *Helicobacter pylori* infection. PLoS ONE 10:e0131444. https://doi.org/10.1371/journal.pone.0131444

Sjomina O, Pavlova J, Niv Y, Leja M (2018) Epidemiology of *Helicobacter pylori* infection. Helicobacter 23:e12514. https://doi.org/10.1111/hel.12514

Slomiany BL, Slomiany A (2017) Role of LPS-elicited signaling in triggering gastric mucosal inflammatory responses to *Helicobacter pylori*: modulatory effect of ghrelin. Inflammopharmacology 25:415–429. https://doi.org/10.1007/s10787-017-0360-1

Smith KD, Andersen-Nissen E, Hayashi F, Strobe K, Bergman MA, Barrett SL, Cookson BT, Aderem A (2003a) Toll-like receptor 5 recognizes a conserved site on flagellin required for protofilament formation and bacterial motility. Nat Immunol 4:1247–1253. https://doi.org/10.1038/ni1011

Smith MF, Mitchell A, Li G, Ding S, Fitzmaurice AM, Ryan K, Crowe S, Goldberg JB (2003b) Toll-like receptor (TLR) 2 and TLR5, but not TLR4, are required for Helicobacter pylori-induced NF-kappa B activation and chemokine expression by epithelial cells. J Biol Chem 278:32552–32560. https://doi.org/10.1074/jbc.m305536200

Sommer F, Faller G, Röllinghoff M, Kirchner T, Mak TW, Lohoff M (2001) Lack of gastritis and of an adaptive immune response in interferon regulatory factor-1-deficient mice infected with *Helicobacter pylori*. Eur J Immunol 31:396–402. https://doi.org/10.1002/1521-4141(200102)31:2%3c396:aid-immu396%3e3.0.co;2-y

Song M, Su H, Zhang L, Ma J, Li J, Pan K, You W (2013) Genetic polymorphisms of miR-146a and miR-27a, *Helicobacter pylori* infection, and risk of gastric lesions in a Chinese population. PLoS ONE 8:e61250. https://doi.org/10.1371/journal.pone.0061250

Stead CM, Beasley A, Cotter RJ, Trent MS (2008) Deciphering the unusual acylation pattern of *Helicobacter pylori* lipid A. J Bacteriol 190:7012–7021. https://doi.org/10.1128/jb.00667-08

Su B, Ceponis PJM, Lebel S, Huynh H, Sherman PM (2003) *Helicobacter pylori* activates Toll-like receptor 4 expression in gastrointestinal epithelial cells. Infect Immun 71:3496–3502

Sugimoto M, Ohno T, Graham DY, Yamaoka Y (2011) *Helicobacter pylori* outer membrane proteins on gastric mucosal interleukin 6 and 11 expression in Mongolian gerbils: OMPs of H. pylori and gerbils. J Gastroenterol Hepatol 26:1677–1684. https://doi.org/10.1111/j.1440-1746.2011.06817.x

Sugimoto MA, Sousa LP, Pinho V, Perretti M, Teixeira MM (2016) Resolution of inflammation: what controls its onset? Front Immunol 7:160. https://doi.org/10.3389/fimmu.2016.00160

Takeuchi O, Akira S (2010) Pattern recognition receptors and inflammation. Cell 140:805–820. https://doi.org/10.1016/j.cell.2010.01.022

Tang B, Li N, Gu J, Zhuang Y, Li Q, Wang HG, Fang Y, Yu B, Zhang JY, Xie QH,Chen L, Jiang XJ, Xiao B, Zou QM, Mao XH (2012) Compromised autophagy by MIR30B benefits the intracellular survival of *Helicobacter pylori*. Autophagy 8:1045–1057.https://doi.org/10.4161/auto.20159

Teng G, Dai Y, Chu Y, Li J, Zhang H, Wu T, Shuai X, Wang W (2018) *Helicobacter pylori* induces caudal-type homeobox protein 2 and cyclooxygenase 2 expression by modulating microRNAs in esophageal epithelial cells. Cancer Sci 109:297–307. https://doi.org/10.1111/cas.13462

Torok AM, Bouton AH, Goldberg JB (2005) *Helicobacter pylori* induces interleukin-8 secretion by Toll-like receptor 2- and Toll-like receptor 5-dependent and -independent pathways. Infect Immun 73:1523–1531. https://doi.org/10.1128/iai.73.3.1523-1531.2005

Tran AX, Whittimore JD, Wyrick PB, McGrath SC, Cotter RJ, Trent MS (2006) The lipid A 1-phosphatase of *Helicobacter pylori* is required for resistance to the antimicrobial peptide polymyxin. J Bacteriol 188:4531–4541. https://doi.org/10.1128/jb.00146-06

Travassos LH, Carneiro LAM, Ramjeet M, Hussey S, Kim Y-G, Magalhães JG, Yuan L, Soares F, Chea E, Le Bourhis L, Boneca IG, Allaoui A, Jones NL, Nuñez G, Girardin SE, Philpott DJ (2010) Nod1 and Nod2 direct autophagy by recruiting ATG16L1 to the plasma membrane at the site of bacterial entry. Nat Immunol 11:55–62. https://doi.org/10.1038/ni.1823

Tsai C-C, Kuo T-Y, Hong Z-W, Yeh Y-C, Shih K-S, Du S-Y, Fu H-W (2015) *Helicobacter pylori* neutrophil-activating protein induces release of histamine and interleukin-6 through G protein-mediated MAPKs and PI3 K/Akt pathways in HMC-1 cells. Virulence 6:755–765. https://doi.org/10.1080/21505594.2015.1043505

Varga MG, Shaffer CL, Sierra JC, Suarez G, Piazuelo MB, Whitaker ME, Romero-Gallo J, Krishna US, Delgado A, Gomez MA, Good JAD, Almqvist F, Skaar EP, Correa P, Wilson KT, Hadjifrangiskou M, Peek RM (2016) Pathogenic *Helicobacter pylori* strains translocate DNA and activate TLR9 via the cancer-associated cag type IV secretion system. Oncogene 35:6262–6269. https://doi.org/10.1038/onc.2016.158

Wang F, Liu J, Zou Y, Jiao Y, Huang Y, Fan L, Li X, Yu H, He C, Wei W, Wang H, Sun G (2017) MicroRNA-143-3p, up-regulated in *Helicobacter pylori*-positive gastric cancer, suppresses tumor growth, migration and invasion by directly targeting AKT2. Oncotarget 8:28711–28724. https://doi.org/10.18632/oncotarget.15646

Wang Y-C, Chen C-L, Sheu B-S, Yang Y-J, Tseng P-C, Hsieh C-Y, Lin C-F (2014) *Helicobacter pylori* infection activates Src homology-2 domain-containing phosphatase 2 to suppress IFN-γ signaling. J Immunol 193:4149–4158. https://doi.org/10.4049/jimmunol.1400594

White JR, Winter JA, Robinson K (2015) Differential inflammatory response to *Helicobacter pylori* infection: etiology and clinical outcomes. J Inflamm Res 8:137–147. https://doi.org/10.2147/jir.s64888

Wightman B, Ha I, Ruvkun G (1993) Posttranscriptional regulation of the heterochronic gene lin-14 by lin-4 mediates temporal pattern formation in C. elegans. Cell 75:855–862

Wilson KT, Crabtree JE (2007) Immunology of *Helicobacter pylori*: insights into the failure of the immune response and perspectives on vaccine studies. Gastroenterology 133:288–308. https://doi.org/10.1053/j.gastro.2007.05.008

Wu K, Yang L, Li C, Zhu C-H, Wang X, Yao Y, Jia Y-J (2014) MicroRNA-146a enhances Helicobacter pylori induced cell apoptosis in human gastric cancer epithelial cells. Asian Pac J Cancer Prev 15:5583–5586

Wu MS, Lee CW, Shun CT, Wang HP, Lee WJ, Sheu JC, Lin JT (1998) Clinicopathological significance of altered loci of replication error and microsatellite instability-associated mutations in gastric cancer. Cancer Res 58:1494–1497

Wu K, Zhu C, Yao Y, Wang X, Song J, Zhai J (2016) MicroRNA-155-enhanced autophagy in human gastric epithelial cell in response to *Helicobacter pylori*. Saudi J Gastroenterol 22:30–36. https://doi.org/10.4103/1319-3767.173756

Xiao B, Liu Z, Li B-S, Tang B, Li W, Guo G, Shi Y, Wang F, Wu Y, Tong W-D, Guo H, Mao X-H, Zou Q-M (2009) Induction of microRNA-155 during *Helicobacter pylori* infection and its negative regulatory role in the inflammatory response. J Infect Dis 200:916–925. https://doi.org/10.1086/605443

Xie G, Li W, Li R, Wu K, Zhao E, Zhang Y, Zhang P, Shi L, Wang D, Yin Y, Deng R, Tao K (2017) Helicobacter pylori promote B7-H1 expression by suppressing miR-152 and miR-200b in gastric cancer cells. PLoS ONE 12:e0168822. https://doi.org/10.1371/journal.pone.0168822

Yamaoka Y, Kita M, Kodama T, Sawai N, Tanahashi T, Kashima K, Imanishi J (1998) Chemokines in the gastric mucosa in *Helicobacter pylori* infection. Gut 42:609–617

Yang X-J, Si R-H, Liang Y-H, Ma B-Q, Jiang Z-B, Wang B, Gao P (2016) Mir-30d increases intracellular survival of *Helicobacter pylori* through inhibition of autophagy pathway. World J Gastroenterol 22:3978–3991. https://doi.org/10.3748/wjg.v22.i15.3978

Ye F, Tang C, Shi W, Qian J, Xiao S, Gu M, Dang Y, Liu J, Chen Y, Shi R, Zhang G (2015) A MDM2-dependent positive-feedback loop is involved in inhibition of miR-375 and miR-106b induced by *Helicobacter pylori* lipopolysaccharide. Int J Cancer 136:2120–2131. https://doi.org/10.1002/ijc.29268

Yokota S, Ohnishi T, Muroi M, Tanamoto K, Fujii N, Amano K (2007) Highly-purified *Helicobacter pylori* LPS preparations induce weak inflammatory reactions and utilize Toll-like receptor 2 complex but not Toll-like receptor 4 complex. FEMS Immunol Med Microbiol 51:140–148. https://doi.org/10.1111/j.1574-695x.2007.00288.x

Yun CH, Lundgren A, Azem J, Sjöling A, Holmgren J, Svennerholm A-M, Lundin BS (2005) Natural killer cells and *Helicobacter pylori* infection: bacterial antigens and interleukin-12 act synergistically to induce gamma interferon production. Infect Immun 73:1482–1490. https://doi.org/10.1128/iai.73.3.1482-1490.2005

Zhang XS, Tegtmeyer N, Traube L, Jindal S, Perez-Perez G, Sticht H, Backert S, Blaser MJ (2015) A specific A/T polymorphism in Western tyrosine phosphorylation B-motifs regulates *Helicobacter pylori* CagA epithelial cell interaction. PloS Pathog 11:1004621. https://doi.org/10.1371/journal.ppat.1004621

Zhang YM, Noto JM, Hammond CE, Barth JL, Argraves WS, Backert S, Peek RM Jr,Smolka AJ (2014) *Helicobacter pylori*-induced posttranscriptional regulation of H-K-ATPase α-subunit gene expression by miRNA. Am J Physiol Gastrointest Liver Physiol 306:G606-613.https://doi.org/10.1152/ajpgi.00333.2013

Zhang Z, Li Z, Gao C, Chen P, Chen J, Liu W, Xiao S, Lu H (2008) miR-21 plays a pivotal role in gastric cancer pathogenesis and progression. Lab Invest 88:1358–1366. https://doi.org/10.1038/labinvest.2008.94

Zhao Y, Yokota K, Ayada K, Yamamoto Y, Okada T, Shen L, Oguma K (2007) *Helicobacter pylori* heat-shock protein 60 induces interleukin-8 via a Toll-like receptor (TLR)2 and mitogen-activated protein (MAP) kinase pathway in human monocytes. J Med Microbiol 56:154–164. https://doi.org/10.1099/jmm.0.46882-0

Zhou X, Li L, Su J, Zhang G (2014a) Decreased miR-204 in *Helicobacter pylori*-associated gastric cancer promotes cancer cell proliferation and invasion by targeting SOX4. PLoS ONE 9:e101457. https://doi.org/10.1371/journal.pone.0101457

Zhou X, Li X, Wu M (2018) miRNAs reshape immunity and inflammatory responses in bacterial infection. Signal Transduct Target Ther 3:14. https://doi.org/10.1038/s41392-018-0006-9

Zhou X, Xu G, Yin C, Jin W, Zhang G (2014b) Down-regulation of miR-203 induced by *Helicobacter pylori* infection promotes the proliferation and invasion of gastric cancer by targeting CASK. Oncotarget 5:11631–11640. https://doi.org/10.18632/oncotarget.2600

Zhu Y, Jiang Q, Lou X, Ji X, Wen Z, Wu J, Tao H, Jiang T, He W, Wang C, Du Q, Zheng S, Mao J, Huang J (2012) microRNAs up-regulated by CagA of *Helicobacter pylori* induce intestinal metaplasia of gastric epithelial cells. PLoS ONE 7:e35147. https://doi.org/10.1371/journal.pone.0035147

Zhuang Y, Cheng P, Liu X, Peng L, Li B, Wang T, Chen N, Li W, Shi Y, Chen W, Pang KC, Zeng M, Mao X, Yang S, Guo H, Guo G, Liu T, Zuo Q, Yang H, Yang L, Mao F, Lv Y, Zou Q (2015) A pro-inflammatory role for Th22 cells in *Helicobacter pylori*-associated gastritis. Gut 64:1368–1378. https://doi.org/10.1136/gutjnl-2014-307020

Impact of the Gastrointestinal Microbiome in Health and Disease: Co-evolution with the Host Immune System

Georgina L. Hold and Richard Hansen

Contents

Abstract Microbes within the gastrointestinal tract communicate with each other and with the host, which has profound effects on health and disease development. Only now, it is becoming apparent that how and when we acquire our own unique collection of "gut microbes" and also how we choose to maintain them is fundamental to our health. *Helicobacter pylori* is the most common bacterial infection worldwide, colonizing around half of the world's population, and is the major risk factor for gastric adenocarcinoma. More recently, it has also been shown to have some beneficial effects in terms of protecting against the development of other diseases. Here, we review the current knowledge on how *H. pylori* has shaped gastrointestinal microbiota colonization and the host immune system with specific focus on the impact of *H. pylori* on the various microbiome niches of the gastrointestinal tract. We discuss how the presence of *H. pylori* influences the physiology of three major regions within the gastrointestinal tract—specifically the oesophagus, stomach and colon. We pay particular attention to the role of *H. pylori* under chronic inflammatory conditions including the development of cancer. With increased incidence of diseases

G. L. Hold (✉)
Microbiome Research Centre, St. George and Sutherland Clinical School,
University of New South Wales, Sydney, Australia
e-mail: georgina.hold@unsw.edu.au

R. Hansen
Department of Paediatric Gastroenterology,
Glasgow Children's Hospital, Glasgow, Scotland, UK
e-mail: richard.hansen@nhs.net

S. Backert (ed.), *Molecular Mechanisms of Inflammation: Induction, Resolution and Escape by Helicobacter pylori*, Current Topics in Microbiology and Immunology 421, https://doi.org/10.1007/978-3-030-15138-6_12

such as eosinophilic oesophagitis, oesophageal adenocarcinoma and squamous cell carcinoma being attributed to the decline in *H. pylori*, their disease pathogenesis in light of changing *H. pylori* colonization is also discussed.

1 Introduction

Helicobacter pylori infection remains a health problem for about half of the world's population (Smolka and Schubert 2017). Chronic infection can result in both hypo- or hyperchlorhydria, depending upon the gastric site of infection. Most patients who are chronically infected develop a pangastritis, which is associated with reduced acid secretion, predisposing them to increased risk of gastric adenocarcinoma, although in reality this risk affects only ~1% of chronically infected individuals (Smith et al. 2006). Hyperacidity on the other hand is classically associated with peptic ulcer disease (PUD). Thus, historically, there was no positive outcome in terms of being infected with *H. pylori* and eradication was considered the best option. More recently, increasing evidence from population studies has highlighted the protective effect of reduced acidity in *H. pylori*-infected individuals in terms of conditions including gastroesophageal reflux disease (GORD), Barrett's oesophagus (BO), eosinophilic oesophagitis (EoE) and even oesophageal adenocarcinoma (OAC) and inflammatory bowel disease (IBD) (Corley et al. 2008; Islami and Kamangar 2008; Sonnenberg et al. 2010; Sonnenberg et al. 2017; Castaño-Rodríguez et al. 2017b). Conversely, however, other studies have also now confirmed a positive association between colorectal cancer and *H. pylori* infection (Nam et al. 2017; Lee et al. 2016; Posselt et al. 2013). This has re-opened the debate regarding *H. pylori* infection and its impact on gastric acid secretion and on host health in general. As changes in the gastrointestinal microbiome are increasingly linked with human health, understanding more about the interlink between *H. pylori* and the microbiome, and how best to exploit this relationship to promote gut and microbiome homeostasis, becomes increasingly important. Here, we discuss our current understanding of the impact of *H. pylori* infection on the gastrointestinal microbiome and how the presence of this microbe influences acquisition of our microbiome and our health status.

2 Early Life Colonization and *H. pylori* Infection in Childhood

Current thinking on microbial colonization suggests that the human gut progresses through a staged process of accruing organisms, with subsequent development and enrichment of the microbiome "ecosystem" throughout early childhood. The gut is near-sterile at birth, with a few pioneer organisms identifiable in the first stool samples, but the microbiome rapidly develops from there, first with aerobic organisms, then progressively by anaerobic colonization (Palmer et al. 2007; Rautava et al. 2012; Hansen et al. 2015). Important early life influences on this colonization include gestational age; vaginal delivery versus Caesarean section; breast versus formula feeding; timing and method of weaning to solid foods; and potentially antibiotic exposure in infancy (Fig. 1; Dominguez-Bello et al. 2010; Madan et al. 2016; Backhed et al. 2015; Yassour et al. 2016; Chernikova et al. 2018). Few longitudinal studies have examined the long-term impact of each, but the microbiome appears to reach an adult-like climax community around the age of three years (Koenig et al. 2011; Yatsunenko et al. 2012). The seminal study to date, examining the age of *H. pylori* acquisition explored seroconversion in a cohort of prospectively followed children, suggested the peak period of acquisition is also within the first five years of life (Malaty et al. 2002). It is now difficult to map the historic prevalence of *H. pylori* colonization in humans during this period of life, but it was likely very high and arguably "normal" in terms of expected colonization. Indeed, it is possible to track ancient human migrations across the globe by studying the change in *H. pylori* genomes; such is the historical importance of the organism to human biology (Atherton and Blaser 2009). In contrast, prevalence of *H. pylori* at age six in children of the Western hemisphere is now probably <10% (den Hollander et al. 2015). The potential impact of *H. pylori*, in terms of altering the gastrointestinal ecosystem for subsequent colonization by other microbes, has barely been explored. The possibility of *H. pylori* leading to reduced acid production and a diminished gastric acid barrier, and immunoregulation leading to increased immune tolerance of other microbial colonists suggests a potentially important impact on development of the microbiome in *H. pylori* colonized children (Smolka and Backert 2012). Irrespectively, one can speculate that with the decline in *H. pylori* prevalence in recent decades, Western gut colonization has also diverged from the historical "normal", whilst a number of microbiome-associated diseases have emerged and risen in both incidence and prevalence. Indeed, whilst not exploring *H. pylori* per se, De Filippo et al. (2010) described the faecal microbiome of two groups of children, those raised in Italy and those raised in Burkina Faso, and described entirely different patterns of colonization, with the Western microbiome dominated by Firmicutes and the African microbiome by Bacteroidetes. The authors did not explore *H. pylori* prevalence within this study, and indeed there are an enormous number of differences between the two groupings; however, the findings clearly demonstrate how far the Western microbiome has shifted from the historic and ancient human pattern. Targeted studies examining the impact of *H. pylori* on microbial colonization during the first three years of childhood, taking into account key milestones in early infancy and dietary changes over time, are urgently needed to help address its specific role on subsequent microbial colonization.

3 *H. pylori* and Immunoregulation

Whilst the pervading message regarding *H. pylori* around the end of last century was "the only good *H. pylori* is a dead *H. pylori*" (Graham 1997), a new concept regard-

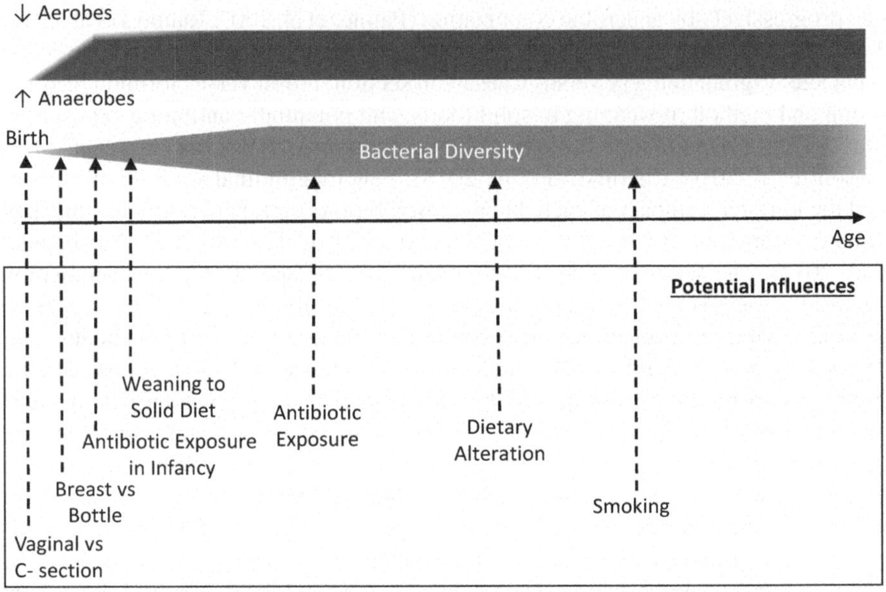

Fig. 1 Influences on microbial colonization of the human gastrointestinal tract. Following birth, the gut microbiome develops rapidly, first with aerobic organisms, then progressive anaerobic colonization. Important early life influences on this colonization include gestational age; vaginal delivery versus Caesarean section; breast versus formula feeding; timing and method of weaning to solid foods; and potentially antibiotic exposure in infancy. Subsequent antibiotic exposures and lifestyle factors including dietary alterations and smoking all impact on gut microbial diversity

ing the possible immunoregulatory role of the organism has emerged and gained prominence, largely over the last decade. Whilst there is undoubtedly a negative association between *H. pylori* seropositivity and IBD, first described in 1994, but subsequently confirmed by multiple other investigators and subsequent meta-analysis, an exact mechanism has not been described to date (El-Omar et al. 1994; Luther et al. 2010). Carriers of *H. pylori* certainly seem to confer an immunoregulatory phenotype with increased expression of Foxhead box protein 3 (FoxP3) compared with controls (Rad et al. 2006). The immune interplay is, however, more complex with both immunostimulatory and immunoregulatory components, so it is difficult to fully disentangle from the pathophysiology of another immunologically complex disease (O'Keeffe and Moran 2008). Irrespective, a compelling mouse study demonstrated that the presence of *H. pylori* in the stomach appears to modulate the immune response to infection with *Salmonella typhimurium*, increasing mesenteric IL-10 production and diminishing the colonic TH-17 response, an effect recently replicated in dextran sulphate models of mouse colitis (Higgins et al. 2011; Zhang et al. 2018). Another study has shown that different gastrointestinal microbiota has contrasting gastric pathology, microbial and host immune responses (Ge et al. 2018). Two other hypotheses worth considering, when discussing the negative association with IBD, are an immunization-type effect against other initiating organisms, possibly of the *Helicobacter* genus or a similar clade, or simply an epiphenomenon (Hansen et al. 2011). Certainly, IBD is a disease associated with cleanliness, whereas *H. pylori* carriage is more closely associated with overcrowding and reduced sanitation in a given community (Elliott et al. 2000; Mendall et al. 1992).

The other important immunoregulatory *H. pylori* aspect relates to atopic disease. Chen and Blaser (2008) used retrospective data from the National Health and Nutrition Examination Survey to explore the overlap between *H. pylori* seropositivity and atopic disease in 3327 participants under 20 years old. They identified a negative association between *H. pylori* and allergic rhinitis in both the 3–13 and 3–19 year olds, but interestingly a negative association with asthma only in the younger age grouping of 3–13 year olds. Whether this relates to a window for an immunoregulatory benefit from *H. pylori* in early childhood, or whether secondary drivers of asthma arrive in later childhood requires further exploration. Of course, the epiphenomenon of cleanliness re asthma and poor sanity re *H. pylori* is equally applicable to any asthma association and was acknowledged by the authors of this piece. Further to this, however, an asthma mouse model has also shown a benefit to *H. pylori* colonization in terms of FoxP3 expression and phenotype in an ovalbumin stimulation model, but with benefit conferred more strongly in neonatally *H. pylori*-infected mice (Arnold et al. 2011). Interestingly, antibiotic eradication of *H. pylori* prior to the challenge model seemed to abrogate the benefit. Therefore, two compelling messages emerge from this study—firstly, that the age of *H. pylori* infection appears fundamentally important to its immunoregulatory role; secondly, that the presence of live organisms and ongoing interaction with the immune system is necessary for continued benefit. Whilst extrapolating these findings from a rodent model of asthma to human immunological diseases is challenging, there is an urgent need to understand the complex interplay between *H. pylori*, the human gut microbiome and the immune system during early childhood, so we can better understand the pathophysiology and shifting epidemiology of chronic immune-mediated illnesses. Indeed, overlapping this with our current understanding of the gut microbiome and asthma pathophysiology, the work of Arrieta (2015) has demonstrated four key organisms whose abundance at 3 months (but not one year) of age is protective against the development of atopy and wheeze at one year (itself predictive of asthma)—*Veillonella*, *Lachnospira*, *Rothia* and *Faecalibacterium*. Inoculating mice with these organisms conferred a benefit even to their offspring in terms of their response to the ovalbumin challenge. Whilst not an *H. pylori*-related study as such, this important work does also neatly demonstrate the importance of the microbiome and age of exposure/colonization in terms of conferring an immunological benefit to the host. How colonization with these organisms interacts with the presence or absence of *H. pylori* would add greatly to our understanding of this important and developing area.

4 *Helicobacter pylori* and the Oesophageal Microbiome

The oesophageal microbiome has strong comparability with the oral microbiome, comprising the same six major phyla (Firmicutes, Bacteroides, Actinobacteria, Proteobacteria, Fusobacteria and TM7—a candidate phylum of ubiquitous yet mainly uncultured bacterial species) (Snider et al. 2016; Pei et al. 2004). One of the largest scale surveys of the oral microbiome, prior to the next generation sequencing era, was conducted by Yang et al. (2009). They analysed 6800 16S rRNA gene clones from 34 subjects and found that the microbiota of the normal oesophagus was dominated by Firmicutes, with *Streptococcus* being the most dominant genus. GORD affects between 25 and 40% of the population and is the result of gastric acid refluxing into the oesophagus and causing inflammation. If left untreated GORD can progress, through various pathological stages, towards OAC. GORD is not necessarily due to excess acid but more likely increased oesophageal exposure to acid due to increased reflux of stomach contents, impaired acid clearance or acid hypersensitivity.

One important aspect of descriptive microbiome study that warrants attention is the association/causation conundrum. This is of course highly prevalent in all medical scientific research, but is of particular importance in microbiome research, where a microbial cataloguing approach is generally applied to two groups: health/disease, affected/unaffected tissue, before/after treatment, etc. The alterations in microbes between the groupings are then reported, and biological significance is often attributed as a result. To truly test biological significance and look at causation over association, follow-on studies should work with isolates of key identified organisms to further delineate and describe the mechanisms by which they are linked to disease/health. This step is rarely undertaken and is a major weakness of human microbiome study to date. The other angle to approach these studies from is that the microbial changes are secondary to the disease state and not biologically linked, so, for instance, inflamed tissue offers a distinct ecological niche which in turn facilitates the growth of a distinct microbiome to take advantage of the different microenvironment and available nutrients. This is the association side of the conundrum. We have deliberately not dissected association from causation within this review and have chosen to report study findings as originally discussed by the authors. Further work is needed to disentangle association from causation within all microbiome studies and is a major unmet need within the field.

OAC develops from the premalignant lesion Barrett's oesophagus (BO), which comprises a characteristic intestinal metaplasia within the distal oesophagus. BO is associated with >40-fold increased risk for OAC development, although the progression rate of BO to OAC is very low at ~0.1–0.3% per year (Hvid-Jensen et al. 2011; Bhat et al. 2011). This offers a significant window of opportunity in terms of understanding the journey from reflux oesophagitis and BO to OAC, with the possibility of identifying biomarkers or therapeutic targets to prevent OAC development. The microbiota signature of the inflamed oesophagus (patients had either reflux oesophagitis or Barrett's oesophagus) has been shown to contain larger numbers of *Bacteroidetes, Proteobacteria, Fusobacteria, Neisseria, Campylobacter, Veillonella* and *Spirochaetes* with a reduction in the *Streptococcus* family. These trends have been reproduced in other studies, reliably demonstrating an alteration in the oesophageal microbiota in reflux disease which most likely reflects physiological changes due to excess gastric acid as well as inflammatory mediators (Yang et al. 2009; Pei et al. 2005).

When considering the microbiome associated with OAC, the evidence is limited to date and there is a clear need for further research in this area. The OAC microbiota is associated with reduced microbial diversity, but contains a higher abundance of *Bifidobacteria, Bacteroides, Fusobacteria, Veillonella, Staphylococcus* and *Lactobacilli*, and decreased abundance of *Campylobacter* when compared with BO samples. Some OAC samples have been shown to be dominated by a single bacterial species belonging to the order Lactobacillales. However, the increased abundance of the various groups was also seen in healthy control samples; therefore, further research is required before an OAC microbiome signature can be defined (Blackett et al. 2013; Elliott et al. 2017).

Oesophageal squamous cell carcinoma (OSCC) accounts for approximately 90% of oesophageal cancer cases worldwide. Although OAC accounts for most oesophageal cancer cases in the USA and Europe, OSCC is the most common type in Asia, Africa, South America and amongst African-Americans (Cook et al. 2009). A study of 325 resected oesophageal cancer specimens (92% of which were OSCC) showed an increase in *Fusobacterium nucleatum* in cancer specimens compared to normal mucosa, which was also associated with shorter survival (Yamamura et al. 2016). Other studies have suggested that squamous dysplasia is linked with reduced bacterial richness, and general alterations in the microbiota have been documented, but the findings are generally inconsistent between studies and require further robustly designed and sufficiently powered studies to address this issue effectively and to then begin to assess causation (May and Abrams 2018).

Microbiome alterations have been assessed in patients with eosinophilic oesophagitis (EoE) (Benitez et al. 2015; Harris et al. 2015). Increases in bacteria belonging to the Proteobacteria phylum including *Neisseria, Corynebacterium* and *Haemophilus* in EoE patients were seen across the two studies although there was no consensus between studies. Decreases in *Streptococcus* and *Atopobium* were also seen in comparison with control patients. There is compelling evidence, from several different geographical locations, showing an inverse association between *H. pylori* infection and EoE (Dellon and Hirano 2018). Although mechanistic detail is lacking, the decrease in *H. pylori* prevalence in the last 20 years matches the trend of increased EoE incidence and one can hypothesize that this may relate to its population-level reduction in acid secretion or its influence on colonization by other important organisms.

In summary, the oesophageal microbiome is altered in disease states including Barrett's oesophagus, GORD, oesophageal cancer and also EoE. Whether *H. pylori* infection plays a role in the development of OAC remains a matter of debate, although epidemiological evidence clearly defines a negative association between *H. pylori* infection and OAC. The evidence for *H. pylori* involvement in OSCC is even less apparent than OAC. What is clear however is that the impact on host physiology and the ensuing effect on the oesophageal microbiome requires further investigation.

5 *Helicobacter pylori* and the Gastric Microbiome

Gastric cancer (GC) remains the most commonly diagnosed gastrointestinal malignancy in the world despite the fact that we have known about the strongest risk factor, *H. pylori* infection, for over 30 years. This infection, which currently affects more than 50% of the world's population, is the major risk factor for non-cardia, non-diffuse GC and is thought to contribute to ~75% of all GC cases (Smith et al. 2006). The gastric microbiota is also thought to influence GC development. A recent review, looking at the gastric microbiome, concluded that whilst *H. pylori* infection was a risk factor for GC development, gastric colonization by other bacteria is also an important factor that needs consideration in relation to the subsequent risk of cancer development (Noto and Peek 2017).

Due to the acidic nature of the gastric environment, it has long been assumed that bacteria are unable to survive in the stomach. *H. pylori* has the appropriate enzymatic machinery to allow it to tolerate and colonize the harsh acidic conditions through urease activity, which allows it to neutralize the acid in its surrounding locality (Smith et al. 2006). This capability and the ensuing impact on gastric acid secretion directly impacts on the ability of other bacterial species to occupy the gastric niche and influence inflammatory processes that potentially facilitate carcinogenic progression (Sanduleanu et al. 2001). Whilst hugely informative studies to date fail to delineate the complex interplay between *H. pylori* and other components of the microbiota, they also only really inform us about the presence of various microbiota constituents but cannot provide us with information on the metabolic capabilities of the microbial consortium. One of the most fascinating factors when considering the role of the gastric microbiome in gastric disease remains that whilst *H. pylori* comprises the predominant constituent of the microbial community during the premalignant stage of gastric carcinogenesis, most likely facilitating the survival and colonization of other bacteria in the gastric environment, it then becomes only a minor component, or is entirely absent, in the microbiome signature of GC. In this case, the pioneer becomes the underdog and is essentially dethroned from its niche. Understanding the timing of this shift in power in the context of progression along the carcinogenesis process has not yet been explored, but could potentially provide a unique insight into the microbial drivers of tumour development. This evolution of the gastric microbiome through the various stages of tumorigenesis strongly supports the potential of other bacteria to be involved in driving carcinogenesis within the stomach.

Several studies have looked at the gastric microbiome in non-cancerous and premalignant states and have demonstrated that *H. pylori*-negative individuals have a highly diverse gastric microbiota, which is dominated by five major phyla: Proteobacteria, Firmicutes, Actinobacteria, Bacteroidetes and Fusobacteria (Bik et al. 2006; Andersson et al. 2008; Maldonado-Contreras et al. 2011). This unequivocally confirms that the stomach is not the sterile organ that we once assumed. There is also evidence that, much like in the faecal microbiome described above, microbial colonization of the stomach is present from soon after birth, whether *H. pylori* infection is present or absent (Moles et al. 2017; Brawner et al. 2017). The microbiome profile of *H. pylori*-negative children is similar to uninfected adult profiles (Brawner et al. 2017). Changes in the gastric microbiota composition have been noted in almost all studies in response to *H. pylori* infection or different pathological states. A reduced microbial richness is consistently seen in *H. pylori*-infected individuals. The microbial profile in *H. pylori*-positive individuals shifts dramatically with *H. pylori* accounting for between 70 and 97% of bacterial sequencing reads (Bik et al. 2006; Andersson et al. 2008). However, other than *H. pylori* infection, there is no consistently documented gastric microbiome signature based on pathology, reflecting the importance of environmental influences in disease pathogenesis (Bik et al. 2006; Maldonado-Contreras et al. 2011; Dicksved et al. 2009; Aviles-Jimenez et al. 2014). A recent animal study has also suggested that other members of the *Helicobacteraceae* can similarly influence gastric microbiome community structure (Nowroozilarki et al. 2017).

As high-throughput technology advances allowing us to ask more in-depth questions about the role of the microbiome, attempts have been made to assess the active gastric microbiome in order to understand its functional capacity, rather than merely documenting presence/absence. Based on RNA expression profiles, it is possible to look at the expression of microbial genes to correlate microbial metabolic capabilities with subsequent impact on the host. The use of RNA-based transcripts ensures that only viable microorganisms present in a sample will be interrogated, thus negating the potential confounder of dead cells which will still feature in DNA-based assessments. Assessing a cohort of patients undergoing endoscopy in Nicaragua, the gastric microbiome was detailed at different stages of *H. pylori*-induced pathology (Thorell et al. 2017). The findings further demonstrated that *H. pylori* influences the gastric microbiome composition and, for the first time, highlighted correlations between its presence and the presence of other bacterial genera—namely *Campylobacter, Sulfurospirillum* and *Deinococcus.* This study also highlighted increased expression of *H. pylori* genes involved in pH regulation and nickel transport, although this was not demonstrated on the basis of disease status. The application of metatranscriptomics is essential if we are to make sense of the complexity that surrounds multi-factorial disease processes. However, additional technological hurdles need to be overcome before this approach will become a mainstream option. Currently, challenges associated with ensuring that sufficient microbial signal can be derived from host-rich samples means that the required sequencing depth needed to generate sufficient microbial signal makes this approach costly, limiting sample size.

5.1 The Gastric Cancer Microbiome

Understanding that the GC microbiome is a unique "snapshot" in time, which does not necessarily allow us to derive information about the previous premalignant stages of the disease process, two recently published studies have provided the most in-depth insight into the microbiota composition of intestinal metaplasia and GC (Ferreira et al. 2018; Coker et al. 2018). Their findings demonstrate clear differences in the GC microbiome in comparison with gastritis alone. The GC microbiome is almost always associated with reduced bacterial richness and decreased abundance of *Helicobacter;* however, an increase in several oral and intestinal commensal species has also been documented in several studies (Castaño-Rodríguez et al. 2017a; Hsieh et al. 2018; Yu et al. 2017a, b; Sohn et al. 2017). The only study to indicate increased bacterial richness utilized RNA rather than DNA-based analysis, meaning findings cannot be directly compared (Castaño-Rodríguez et al. 2017a). The study also demonstrated an increased abundance of *Fusobacterium, Leptotrichia, Lactococcus* and *Veillonella* species and evidence of co-occurrence between bacterial taxa, with *Lactococcus* found to have strong co-occurrence with *Aneurinibacillus* and *Bacillus* species. An increased abundance of *Fusobacterium* alongside increased abundance of *Clostridium* and *Lactobacillus* species was also demonstrated in a DNA-based study of GC patients from Taiwan (Hsieh et al. 2018). Using network analysis to demonstrate co-occurrence/co-exclusion relationships between microbiome constituents, the study by Coker and co-workers identified five GC-enriched taxa—*Peptostreptococcus stomatis, Streptococcus anginosus, Parvimonas micra, Slackia exigua* and *Dialister pneumosintes* as being central to the GC occurrence network (Coker et al. 2017). A further 31 bacterial taxa (enrichment of 21 taxa and a reduction in 10 further taxa) were shown to be associated with disease progression (Coker et al. 2017). The co-occurrence of networks between gastric bacteria appears to be strongest in *H. pylori*-negative pre-cancerous patient samples.

These various studies further highlight that gastric microbiota community dynamics are heavily influenced by *H. pylori* presence/absence. However, defining a dysbiotic signature based purely on taxa presence/absence without considering the functional implications of the ensuing dysbiosis gives only a limited view with limited therapeutic scope.

6 *Helicobacter pylori* and the Colonic Microbiome

The microbial component of the colonic environment is many magnitudes higher than the gastric environment in terms of microbial abundance and also species richness (Louis et al. 2014). It has also been the subject of extensive characterization, both in terms of defining the healthy colonic microbiota across the colon (Eckburg et al. 2005) and attempting to define the microbial signatures of various colonic diseases and microbial therapeutic interventions (Marchesi et al. 2016; McIlroy et al. 2018).

Studies looking at the direct impact of *H. pylori* on the colonic microbiome are few and far between. A recent mouse model of *H. pylori* infection demonstrated that whilst the infection affected the local gastric environment, it also caused systemic effects, altering the colonic microbiota (Kienesberger et al. 2018). Another recent study showed that alterations in the faecal microbiota, especially the dominant phyla of Bacteroidetes, Firmicutes and Proteobacteria, may be involved in the process of *H. pylori*-related gastric lesion progression (Gao et al. 2018). One would assume that the most significant route of impact would be through *H. pylori*-induced physiological effects such as hypochlorhydria and hypergastrinaemia. It is then challenging to delineate whether changes were directly due to *H. pylori* infection or were due to the altered acidity per se. Studies aimed at de-linking altered acidity from *H. pylori* infection, through analysis of individuals on acid reducing medications, including proton pump inhibitors, will help to address this issue, which is currently unresolved. As discussed above, the impact of *H. pylori* on gut microbial colonization within childhood warrants specific and targeted study, not least because of the possibility of immunoregulatory effects from *H. pylori* on the host.

The other major *H. pylori*-induced influence on the colonic microbiome is through *H. pylori* eradication regimes. Antibiotic therapeutic regimens for eradication deliver a huge insult on the gastrointestinal microbiota, which have been demonstrated to significantly alter both oral and colonic microbiomes, dramatically decreasing bacterial diversity (Jakobsson et al. 2010; Oh et al. 2016). Probiotics are also sometimes used in the management of *H. pylori* infection to either improve eradication rates or to reduce the adverse effects of the antibiotic treatment (Homan and Orel 2015). The impact of probiotics on subsequent microbiome communities remains to be seen.

7 Concluding Remarks

Whilst the focus on *H. pylori* for a long time was rightly on its importance within the pathophysiology of PUD and GC, the net reduction in prevalence of the organism during the "The only good *H. pylori* is a dead *H. pylori*" era has led to new ways of thinking about our ancient cohabitant. Indeed, we are now entering an era of low *H. pylori* prevalence in the Western world, and since the organism is probably derived from parents or siblings, this reduction in prevalence will continue downwards in coming years. Given its World Health Organisation confirmed carcinogen status, there are unlikely any arguments in favour of countering this reduction in prevalence can be considered, so clinicians and researchers now need to view the gastrointestinal tract and gut microbiota through the prism of absent *H. pylori*. In the majority of individuals, *H. pylori* probably acts to reduce gastric acid, with PUD and GC being minority associations with infection/colonization. *H. pylori* also appears to have an immunoregulatory role and appears to reduce the incidence of IBD, mitigate against damage from gastrointestinal pathogens, and reduce the subsequent risk of asthma development when present in early childhood. Whether or not it has a fundamental role within sequential colonization of the gut in childhood remains to be seen, but it appears scientifically plausible at least.

The emergence of new diseases of the oesophagus, such as eosinophilic oesophagitis, and the rise of known diseases such as oesophageal adenocarcinoma and squamous cell carcinoma may be precipitated by the decline in *H. pylori* but require further study. Certainly, there are emerging microbial signatures to oesophageal cancer variants, though biological mechanisms remain to be identified. Some of these organisms, most prominently *Fusobacterium nucleatum*, are of potential prognostic importance. Understanding the role of *H. pylori* in the pathophysiology of or protection against these conditions would lend enormous insight into mechanisms and potentially suggest novel therapeutic options. Given the declining prevalence of *H. pylori*, it is imperative that studies of the microbiome within the upper gastrointestinal tract examine its role in the coming years, else this may become increasingly difficult, particularly in Western countries.

Finally, the gastrointestinal tract is a single system beginning at the mouth and ending at the anus, through which digestion and absorption of all human nutrition is facilitated. The major impact of *H. pylori* within the stomach in terms of acid secretion, immunoregulation and stimulation, and gastroduodenal pathologies warrants consideration within population-level studies exploring the microbiome at all gastrointestinal sites. Ideally, such studies would be longitudinal and commence in infancy, but in order to understand the impact of *H. pylori* on the colonic microbiome or associated diseases such as IBD, we first have to study it in these settings.

Whilst treating and eradicating *H. pylori* has been of enormous benefit within the sphere of PUD and GC, only now in its relative absence can we understand the "other side of the coin" and the benefits conferred to the host by the organism. Whilst it may be too late to prevent the decline of *H. pylori*, a greater understanding of its role might allow us to explore alternative strategies to achieve similar outcomes, for instance through acid suppression. The story of our ancient cohabitant is not yet fully told; however, our emphasis is starting to shift from considering it a simple pathogen to understanding a more complex relationship between the organism and its human host. A new paradigm of *H. pylori* may yet reveal that we were short-sighted in eradicating it whenever found. Whilst re-emergence of widespread colonization now seems unlikely, the organism still has lessons to teach us during its decline.

References

Andersson AF, Lindberg M, Jakobsson H, Backhed F, Nyren P, Engstrand L (2008) Comparative analysis of human gut microbiota by barcoded pyrosequencing. PLoS ONE 3:e2836

Arnold IC et al (2011) *Helicobacter pylori* infection prevents allergic asthma in mouse models through the induction of regulatory t cells. J Clin Invest 121:3088–3093

Arrieta MC et al (2015) Early infancy microbial and metabolic alterations affect risk of childhood asthma. Sci Transl Med 7:307ra152

Atherton JC, Blaser MJ (2009) Coadaptation of *Helicobacter pylori* and humans: ancient history, modern implications. J Clin Invest 119:2475–2487

Aviles-Jimenez F, Vazquez-Jimenez F, Medrano-Guzman R, Mantilla A, Torres J (2014) Stomach microbiota composition varies between patients with non-atrophic gastritis and patients with intestinal type of gastric cancer. Sci Rep 4:4202
Backhed F et al (2015) Dynamics and stabilization of the human gut microbiome during the first year of life. Cell Host Microbe 17:690–703
Benitez AJ et al (2015) Inflammation-associated microbiota in pediatric eosinophilic esophagitis. Microbiome 3:23
Bhat S et al (2011) Risk of malignant progression in barrett's esophagus patients: results from a large population-based study. J Natl Cancer Inst 103:1049–1057
Bik EM et al (2006) Molecular analysis of the bacterial microbiota in the human stomach. Proc Natl Acad Sci U S A 103:732–737
Blackett KL et al (2013) Oesophageal bacterial biofilm changes in gastro-oesophageal reflux disease, barrett's and oesophageal carcinoma: association or causality? Aliment Pharmacol Ther 37:1084–1092
Brawner KM et al (2017) *Helicobacter pylori* infection is associated with an altered gastric microbiota in children. Mucosal Immunol 10:1169–1177
Castaño-Rodríguez N, Goh KL, Fock KM, Mitchell HM, Kaakoush NO (2017a) Dysbiosis of the microbiome in gastric carcinogenesis. Sci Rep 7:15957
Castaño-Rodríguez N, Kaakoush NO, Lee WS, Mitchell HM (2017b) Dual role of *Helicobacter* and *Campylobacter* species in IBD: a systematic review and meta-analysis. Gut 66:235–249
Chen Y, Blaser MJ (2008) *Helicobacter pylori* colonization is inversely associated with childhood asthma. J Infect Dis 198:553–560
Chernikova DA et al (2018) The premature infant gut microbiome during the first 6 weeks of life differs based on gestational maturity at birth. Pediatr Res. https://doi.org/10.1038/s41390-018-0022-z
Coker OO et al (2018) Mucosal microbiome dysbiosis in gastric carcinogenesis. Gut 67:1024–1032
Cook MB, Chow WH, Devesa SS (2009) Oesophageal cancer incidence in the United States by race, sex, and histologic type, 1977–2005. Br J Cancer 101:855–859
Corley DA et al (2008) *Helicobacter pylori* infection and the risk of barrett's oesophagus: a community-based study. Gut 57:727–733
De Filippo C et al (2010) Impact of diet in shaping gut microbiota revealed by a comparative study in children from Europe and rural Africa. Proc Natl Acad Sci U S A 107:14691–14696
Dellon ES, Hirano (2018) Epidemiology and natural history of eosinophilic esophagitis. Gastroenterology 154:319–332
Den Hollander WJ et al (2015) Intergenerational reduction in *Helicobacter pylori* prevalence is similar between different ethnic groups living in a western city. Gut 64:1200–1208
Dicksved J, LindberG M, Rosenquist M, Enroth H, Jansson JK, Engstrand L (2009) Molecular characterization of the stomach microbiota in patients with gastric cancer and in controls. J Med Microbiol 58:509–516
Dominguez-Bello MG et al (2010) Delivery mode shapes the acquisition and structure of the initial microbiota across multiple body habitats in newborns. Proc Natl Acad Sci U S A 107:11971–11975
Eckburg PB et al (2005) Diversity of the human intestinal microbial flora. Science 308:1635–1638
Elliott DE, Urban JJ, Argo CK, Weinstock JV (2000) Does the failure to acquire helminthic parasites predispose to Crohn's disease? Faseb J 14:1848–1855
Elliott DRF, Walker AW, O'Donovan M, Parkhill J, Fitzgerald RC (2017) A non-endoscopic device to sample the oesophageal microbiota: a case-control study. Lancet Gastroenterol Hepatol 2:32–42
El-omar E et al (1994) Low prevalence of helicobacter pylori in inflammatory bowel disease: association with sulphasalazine. Gut 35:1385–1388
Ferreira RM et al (2018) Gastric microbial community profiling reveals a dysbiotic cancer-associated microbiota. Gut 67:226–236
Gao JJ et al (2018) Association between gut microbiota and *Helicobacter pylori*-related gastric lesions in a high-risk population of gastric cancer. Front Cell Infect Microbiol 8:202

Ge Z et al (2018) Helicobacter pylori-infected C57BL/6 mice with different gastrointestinal microbiota have contrasting gastric pathology, microbial and host immune responses. Sci Rep. 8(1):8014

Graham DY (1997) The only good *Helicobacter pylori* is a dead *Helicobacter pylori*. Lancet 350:70–71

Hansen R, Thomson JM, Fox JG, El-Omar EM, Hold GL (2011) Could *Helicobacter* organisms cause inflammatory bowel disease? FEMS Immunol Med Microbiol 61:1–14

Hansen R et al (2015) First-pass meconium samples from healthy term vaginally-delivered neonates: an analysis of the microbiota. PLoS ONE 10:e0133320

Harris JK et al (2015) Esophageal microbiome in eosinophilic esophagitis. PLoS ONE 10:e0128346

Higgins PD et al (2011) Prior *Helicobacter pylori* infection ameliorates Salmonella typhimurium-induced colitis: mucosal crosstalk between stomach and distal intestine. Inflamm Bowel Dis 17:1398–1408

HomaN M, Orel R (2015) Are probiotics useful in *Helicobacter pylori* eradication? World J Gastroenterol 21:10644–10653

Hsieh YY et al (2018) Increased abundance of *Clostridium* and *Fusobacterium* in gastric microbiota of patients with gastric cancer in Taiwan. Sci Rep 8:158

Hvid-Jensen F, Pedersen L, Drewes AM, Sorensen HT, Funch-Jensen P (2011) Incidence of adenocarcinoma among patients with barrett's esophagus. N Engl J Med 365:1375–1383

Islami F, Kamangar F (2008) *Helicobacter pylori* and esophageal cancer risk: a meta-analysis. Cancer Prev Res 1:329–338

Jakobsson HE, Jernberg C, Andersson AF, Sjolund-Karlsson M, Jansson JK, Engstrand L (2010) Short-term antibiotic treatment has differing long-term impacts on the human throat and gut microbiome. PLoS ONE 5:e9836

Kienesberger S et al (2018) Gastric *Helicobacter pylori* infection affects local and distant microbial populations and host responses. Cell Rep 14(6):1395–1407

Koenig JE et al (2011) Succession of microbial consortia in the developing infant gut microbiome. Proc Natl Acad Sci U S A 108(suppl 1):4578–4585

Lee JY et al (2016) *Helicobacter pylori* infection with atrophic gastritis is an independent risk factor for advanced colonic neoplasm. Gut Liver 10:902–909

Louis P, Hold GL, Flint HJ (2014) The gut microbiota, bacterial metabolites and colorectal cancer. Nat Rev Microbiol 12:661–672

Luther J, Dave M, Higgins PD, Kao JY (2010) Association between *Helicobacter pylori* infection and inflammatory bowel disease: a meta-analysis and systematic review of the literature. Inflamm Bowel Dis 16:1077–1084

Madan JC et al (2016) Association of cesarean delivery and formula supplementation with the intestinal microbiome of 6-week-old infants. Jama Pediatr 170:212–219

Malaty HM et al (2002) Age at acquisition of *Helicobacter pylori* infection: a follow-up study from infancy to adulthood. Lancet 359:931–935

Maldonado-contreras A et al (2011) Structure of the human gastric bacterial community in relation to *Helicobacter pylori* status. ISME J 5:574–579

Marchesi JR et al (2016) The gut microbiota and host health: a new clinical frontier. Gut 65:330–339

May M, Abrams JA (2018) Emerging insights into the esophageal microbiome. Curr Treat Options Gastroenterol 16:72–85

Mcilroy J, Ianiro G, Mukhopadhya I, Hansen R, Hold GL (2018) Review article: the gut microbiome in inflammatory bowel disease—avenues for microbial management. Aliment Pharmacol Ther 47:26–42

Mendall MA et al (1992) Childhood living conditions and *Helicobacter pylori* seropositivity in adult life. Lancet 339:896–897

Moles L et al (2017) Bacterial diversity of the gastric content of preterm infants during their first month of life at the hospital. Frontiers in Nutrition 4:12

Nam JH et al (2017) *Helicobacter pylori* infection is an independent risk factor for colonic adenomatous neoplasms. Cancer Causes Control 28:107–115

Noto JM, Peek RM (2017) The gastric microbiome, its interaction with *Helicobacter pylori,* and its potential role in the progression to stomach cancer. PLoS Pathog 13:e1006573

Nowroozilarki N, Jamshidi S, Zahraei Salehi T, Kolahian S (2017) Identification of *Helicobacter* and *Wolinella* spp. In oral cavity of toy breed dogs with periodontal disease. Topics in Companion Animal Medicine 32:96–99

Oh B et al (2016) The effect of probiotics on gut microbiota during the *Helicobacter pylori* eradication: randomized controlled trial. Helicobacter 21:165–174

O'Keeffe J, Moran AP (2008) Conventional, regulatory, and unconventional T cells in the immunologic response to *Helicobacter pylori.* Helicobacter 13:1–19

Palmer C, Bik EM, Digiulio DB, Relman DA, Brown PO (2007) Development of the human infant intestinal microbiota. PLoS Biol 5:e177

Pei Z, Bini EJ, Yang L, Zhou M, Francois F, Blaser MJ (2004) Bacterial biota in the human distal esophagus. Proc Natl Acad Sci U S A 101:4250–4255

Pei Z, Yang L, Peek RM, Levine SM, Pride DT, Blaser MJ (2005) Bacterial biota in reflux esophagitis and barrett's esophagus. World J Gastroenterol 11:7277–7283

Posselt G, Backert S, Wessler S (2013) The functional interplay of *Helicobacter pylori* factors with gastric epithelial cells induces a multi-step process in pathogenesis. Cell Commun Signal 11:77. https://doi.org/10.1186/1478-811X-11-77

Rad R et al (2006) CD25 +/FoxP3 + T cells regulate gastric inflammation and *Helicobacter pylori* colonization *in vivo.* Gastroenterology 131:525–537

Rautava S, Luoto R, Salminen S, Isolauri E (2012) Microbial contact during pregnancy, intestinal colonization and human disease. Nat Rev Gastroenterol Hepatol 9:565–576

Sanduleanu S, Jonkers D, De Bruine A, Hameeteman W, Stockbrugger RW (2001) Non-*Helicobacter pylori* bacterial flora during acid-suppressive therapy: differential findings in gastric juice and gastric mucosa. Aliment Pharmacol Ther 15:379–388

Smith MG, Hold GL, Tahara E, El-Omar EM (2006) Cellular and molecular aspects of gastric cancer. World J Gastroenterol 12:2979–2990

Smolka AJ, Backert S (2012) How Helicobacter pylori infection controls gastric acid secretion. J Gastroenterol 47(6):609–618

Smolka AJ, Schubert ML (2017) *Helicobacter pylori*-induced changes in gastric acid secretion and upper gastrointestinal disease. Curr Top Microbiol Immunol 400:227–252

Snider EJ, Freedberg DE, Abrams JA (2016) Potential role of the microbiome in barrett's esophagus and esophageal adenocarcinoma. Dig Dis Sci 61:2217–2225

Sohn SH et al (2017) Analysis of gastric body microbiota by pyrosequencing: possible role of bacteria other than *Helicobacter pylori* in the gastric carcinogenesis. J Cancer Prev 22:115–125

Sonnenberg A, Lash RH, Genta RM (2010) A national study of *Helicobacter pylori* infection in gastric biopsy specimens. Gastroenterology 139:1894–1901

Sonnenberg A, Turner KO, Spechler SJ, Genta RM (2017) The influence of *Helicobacter pylori* on the ethnic distribution of barrett's metaplasia. Aliment Pharmacol Ther 45:283–290

Thorell K et al (2017) *In vivo* analysis of the viable microbiota and *Helicobacter pylori* transcriptome in gastric infection and early stages of carcinogenesis. Infect Immun 85:e00031–17

Yamamura K et al (2016) Human microbiome *Fusobacterium nucleatum* in esophageal cancer tissue is associated with prognosis. Clin Cancer Res 22:5574–5581

Yang L, Lu X, Nossa CW, Francois F, Peek RM, Pei Z (2009) Inflammation and intestinal metaplasia of the distal esophagus are associated with alterations in the microbiome. Gastroenterology 137:588–597

Yassour M et al (2016) Natural history of the infant gut microbiome and impact of antibiotic treatment on bacterial strain diversity and stability. Sci Transl Med 8:343ra81

Yatsunenko T et al (2012) Human gut microbiome viewed across age and geography. Nature 486:222–227

Yu G et al (2017a) Gastric microbiota features associated with cancer risk factors and clinical outcomes: a pilot study in gastric cardia cancer patients from Shanxi, China. Int J Cancer 141:45–51

Yu G et al (2017b) Molecular characterization of the human stomach microbiota in gastric cancer patients. Front Cell Infect Microbiol 7:302

Zhang H et al (2018) *Helicobacter pylori* colonization protects against chronic experimental colitis by regulating Th17/Treg balance. Inflamm Bowel Dis 24:1481–1492

Resolution of Gastric Cancer-Promoting Inflammation: A Novel Strategy for Anti-cancer Therapy

M. Blanca Piazuelo, Rachel P. Riechelmann, Keith T. Wilson and Holly M. Scott Algood

Contents

M. B. Piazuelo
Department of Medicine, Vanderbilt Center for Mucosal Inflammation and Cancer, Vanderbilt University Medical Center, Nashville, TN, USA
e-mail: maria.b.piazuelo@vumc.org

R. P. Riechelmann
Department of Clinical Oncology, AC Camargo Cancer Center, Sao Paulo, SP, Brazil
e-mail: rachel.riechelmann@accamargo.org.br

K. T. Wilson · H. M. S. Algood (✉)
Tennessee Valley Healthcare System, Department of Veterans Affairs, Department of Medicine, Department of Pathology, Microbiology and Immunology, Vanderbilt Center for Mucosal Inflammation and Cancer, Vanderbilt Institute of Infection, Immunity and Inflammation (VI4), Vanderbilt Ingram Cancer Center, Vanderbilt University Medical Center, Nashville, TN, USA
e-mail: holly.m.algood@vumc.org

K. T. Wilson
e-mail: keith.wilson@vumc.org

S. Backert (ed.), *Molecular Mechanisms of Inflammation: Induction, Resolution and Escape by Helicobacter pylori*, Current Topics in Microbiology and Immunology 421, https://doi.org/10.1007/978-3-030-15138-6_13

Abstract The connection between inflammation and cancer was initially recognized by Rudolf Virchow in the nineteenth century. During the last decades, a large body of evidence has provided support to his hypothesis, and now inflammation is recognized as one of the hallmarks of cancer, both in etiopathogenesis and ongoing tumor growth. Infection with the pathogen *Helicobacter pylori* is the primary causal factor in 90% of gastric cancer (GC) cases. As we increase our understanding of how chronic inflammation develops in the stomach and contributes to carcinogenesis, there is increasing interest in targeting cancer-promoting inflammation as a strategy to treat GC. Moreover, once cancer develops and anti-cancer immune responses are suppressed, there is evidence of a substantial shift in the microenvironment and new targets for immune therapy emerge. In this chapter, we provide insight into inflammation-related factors, including T lymphocytes, macrophages, pro-inflammatory chemokines, and cytokines, which promote *H. pylori*-associated GC initiation and growth. While intervening with chronic inflammation is not a new practice in rheumatology or gastroenterology, this approach has not been fully explored for its potential to prevent carcinogenesis or to contribute to the treatment of GC. This review highlights current and possible strategies for therapeutic intervention including (i) targeting pro-inflammatory mediators, (ii) targeting growth factors and pathways involved in angiogenesis in the gastric tumor microenvironment, and (iii) enhancing anti-tumor immunity. In addition, we highlight a significant number of clinical trials and discuss the importance of individual tumor characterization toward offering personalized immune-related therapy.

1 Introduction

Rudolf Virchow initially described the connection between inflammation and cancer (Virchow 1863). During the last decades, a large body of evidence has provided support to his hypothesis, and now inflammation is recognized as one of the hallmarks of cancer (Colotta et al. 2009). Many types of cancer are preceded by a chronic inflammatory process, mostly initiated by infections or exposure to environmental factors. It is estimated that about 15% of new cancer cases worldwide in 2012 were attributable to carcinogenic infections, with *H. pylori* being the most important, accounting for about 770,000 cases of gastric cancer (GC) annually (Plummer et al. 2016).

GC is the fifth most common malignancy worldwide and the third leading cause of cancer-related mortality (Ferlay et al. 2015). Incidence rates of GC differ widely across geographic regions, with the highest rates observed in Asia, Eastern Europe, and some Latin American countries. Most GCs are adenocarcinomas but are highly heterogeneous with respect to histological architecture and molecular features (Cancer Genome Atlas Research 2014; Lee et al. 2016; Gullo et al. 2018). Histological classification systems (Lauren 1965; Lauwers et al. 2010) are clinically impractical to guide patient management. Due to differences in etiological and epidemiological factors, GCs are classified anatomically as cardia (proximal) and non-cardia (distal).

It is estimated that ~90% of cases of non-cardia cancer worldwide are caused by *H. pylori* infection (Plummer et al. 2015).

GC is a lethal disease, mainly due to the high rates of diagnosis at advanced stages. With the exception of Japan and Korea, where screening programs for early detection have been implemented, overall five-year survival rates after diagnosis are below 35% (Cancer Stat Facts: Stomach Cancer, SEER Cancer Statistics Review 1975–2015 2018; Zeng et al. 2018). Early GC is limited to the mucosa and submucosa, regardless of lymph node involvement, and surgical resection is the only curative treatment. Only a minority of patients with advanced disease respond to current modalities of treatment, which according to the stage, include a combination of adjuvant or neoadjuvant therapies with surgery (Van Cutsem et al. 2016). Recent advances in targeted therapy such as trastuzumab, an antibody against human epidermal growth factor receptor 2 (HER2), and ramucirumab, an antibody against VEGFR2 (Bang et al. 2010; Fuchs et al. 2014; Wilke et al. 2014), and immunotherapy modalities (Fuchs et al. 2018a) have produced encouraging results in the treatment of patients with certain subtypes of advanced GC.

Due to the wide heterogeneity of GCs, new strategies to treat this disease are a priority. Although most cancer research has focused on the molecular changes of the neoplastic cells, it is now recognized that non-tumoral cells in the tumor microenvironment, especially immune cells, proliferate with the tumor and provide essential support for its growth and invasion. The recognized protective effect of non-steroidal anti-inflammatory drugs against GC and other gastrointestinal tumors (Abnet et al. 2009; Epplein et al. 2008; Kim et al. 2018) supports the role of chronic inflammation in carcinogenesis.

In this chapter, we provide insight into inflammation-related factors that promote *H. pylori*-associated GC initiation and growth, focusing on current and potential strategies for therapeutic intervention (Fig. 1). A section is dedicated to novel immunotherapy modalities, especially promising in certain subtypes of GC, such as Epstein–Barr virus (EBV)-associated and microsatellite instable (MSI) tumors. Of note, *H. pylori* is also the main etiologic factor of mucosa-associated lymphoid tissue lymphoma and will be discussed in Chapter "MALT Lymphoma as a Model of Chronic Inflammation-Induced Gastric Tumor Development".

2 Inflammatory Mediators of Gastritis and the Tumor Microenvironment (TME)

2.1 *Chronic Inflammation in* H. pylori

Chronic inflammation of the gastric mucosa, termed gastritis, is a hallmark of *H. pylori* infection. *H. pylori* colonization leads to gastritis in all infected persons, but not all persons will develop symptoms. Both innate and adaptive immune cells are present and active during chronic inflammation (Posselt et al. 2013; Naumann

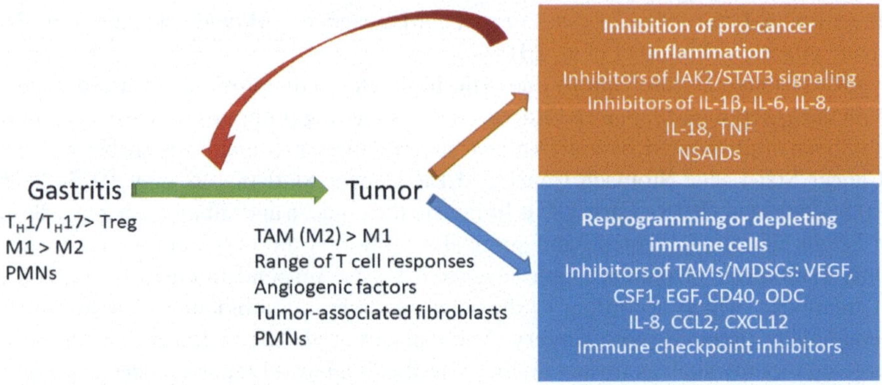

Fig. 1 Resolution of GC-promoting inflammation: a novel strategy for anti-cancer therapy. *H. pylori* infection leads to gastritis in infected persons, but only a subset will go on to develop GC. The microenvironment of the immune response during infection changes when a tumor develops. Immunotherapies could target several immune pathways. Some therapies could target the pro-inflammatory environment which drives the development of the tumor, while others would target the tumor microenvironment by reprogramming tumor infiltrating cells or inhibiting angiogenic factors

et al. 2017; see also Chapters "Role of NOD1 and ALPK1/TIFA Signalling in Innate Immunity Against *Helicobacter pylori* Infection" and "Carbohydrate-Dependent and Antimicrobial Peptide Defence Mechanisms Against *Helicobacter pylori* Infections" of this volume). Immune cell migration to the stomach and production of chemokines and cytokines culminate in ongoing activation of anti-microbial responses and the generation of reactive oxygen and nitrogen species (ROS and RNS). The chronic inflammatory response is believed to be required for the development of a sequence of epithelial transformations called the Correa cascade, which includes multifocal atrophic gastritis, intestinal metaplasia, dysplasia, and cancer (Correa et al. 1975, 2010; Mera et al. 2005, 2018).

Characteristic of mucosal surfaces, like the intestines, gastric epithelial cells (GECs) respond to microbes in the environment. GECs respond to *H. pylori* and produce a number of pro-inflammatory cytokines and chemokines, including interleukin (IL)-1β, IL-6, IL-8, and IL-18 (Brandt et al. 2005; Zhang et al. 2015; see also Chapter "Mechanisms of Inflammasome Signaling, microRNA Induction and Resolution of Inflammation by *Helicobacter pylori*" of this book). The local cytokine responses in human subjects indicate that there is increased tissue expression of IL-1β, IL-6, IL-8, and tumor necrosis factor (TNF, also referred to as TNF-α) (Lindholm et al. 1998). It has been suggested that elevated levels of many of these cytokines including IL-1β, IL-8, and TNF can also serve as biomarkers for GC (Macri et al. 2006). These cytokines and chemokines impact recruitment of immune cells—particularly polymorphonuclear cells (PMNs) and macrophages. As the infection persists, these cytokine responses also chronically persist. Many of these cytokines converge on signaling through Janus kinase/signal transducer and activator of transcription proteins (JAK/STATs) and activating nuclear factor-kappa B (NF-κB), which leads to transcriptional upregulation of anti-apoptotic proteins, pro-inflammatory cytokines and chemokines, adhesion molecules, and increased expression of inducible nitric oxide synthase (NOS2) or NADPH oxidase enzyme isoforms (Gobert and Wilson 2017). These changes in the microenvironment contribute to the development of carcinoma, because they can lead to increased DNA damage, dysfunctional DNA repair enzymes, and genetic instability (see also Chapter "Crosstalk Between DNA Damage and Inflammation in the Multiple Steps of Gastric Carcinogenesis"). Moreover, many of these inflammatory cytokines, such as IL-1β, IL-6, and IL-8, play a pivotal role in mediating the interaction between cancer stem cells and the microenvironment.

In addition to innate immune cell infiltration, cells of the adaptive arm of the immune system, including T lymphocytes and B lymphocytes, migrate into the gastric tissue in response to *H. pylori* infection. The T lymphocytes are predominantly $CD4^+$ T helper cells (Th) and exhibit pro-inflammatory phenotypes (Th1 and Th17) as they express interferon-gamma (IFNγ) and IL-17A, both pro-inflammatory cytokines associated with chronic inflammation. The differentiation of naïve T cells to activated Th1 or Th17 cells can be dictated by the innate cytokine environment. IL-1β, IL-6, IL-23, and transforming growth factor-beta (TGF-β) skew the T cell response toward IL-17A-producing cells, while the expression of IL-12 is likely to push naïve T cells toward IFNγ-producing T cells. In humans (but not in mice), IL-23 also plays a role in Th1 cell differentiation. Increased gene expression of IFNγ, IL-12p40, IL-17A, and IL-23 has been reported in stomach biopsies from *H. pylori*-infected adults and children (Bhuiyan et al. 2014; Staples et al. 2013), and the expression of IL-17A is associated with disease severity (Arachchi et al. 2017). Interestingly, not all studies have the same cytokine signature, in a study which evaluated *H. pylori*-positive gastritis patients versus *H. pylori*-negative gastritis patients, IL-12 expression was significantly elevated only in the *H. pylori*-positive patients, whereas many other cytokines were elevated in both groups, including TNF, IL-1β, IL-6, and IL-8 (Bauditz et al. 1999).

IL-17A and IFNγ subsequently further activate epithelial cells and macrophages in the tissue and can amplify the PMN response. Animal models have successfully defined roles for IFNγ and IL-17A in activating the proper signals required for the development of gastritis but also in activating anti-microbial responses against *H. pylori* (Algood et al. 2009; Sjokvist Ottsjo et al. 2015; Dixon et al. 2016). In *H. pylori*-infected individuals, the frequencies of IFNγ and IL-$17A^+$ cells were increased in the antrum (Luzza et al. 2000), particularly in patients with *H. pylori*-induced gastric ulcers (Adamsson et al. 2017). Other T cell cytokines are also involved in the chronic inflammatory response including IL-21. Levels of IL-21 are increased in *H. pylori*-infected samples from patients with gastritis (Caruso et al. 2007) and in *H. pylori*-infected mice (Carbo et al. 2014). In human subjects, IL-21 expression correlates with the activation of STAT3 and more severe gastritis (Bagheri et al. 2015). Mice deficient in IL-21 infected with *H. pylori* do not develop gastritis, but are colonized with a higher level of *H. pylori* than wild-type controls (Carbo et al. 2014). These data suggest again that IL-21 can drive inflammation, but also that inflammation is necessary to bring anti-microbial responses to the stomach control *H. pylori* colonization.

Additional insight related to the role of macrophages in *H. pylori* immunopathogenesis and inflammation-associated cancer risk stems from studies related to polyamines. Polyamines are pleiotropic polycations that have many cellular functions, including the regulation of gene transcription, protein translation, cell growth, proliferation, and differentiation (Hardbower et al. 2017; Pegg 2006, 2009). The production of the three major polyamines—putrescine, spermidine, and spermine—is tightly regulated and centers on the rate-limiting enzyme, ornithine decarboxylase (ODC1, hereafter referred to as ODC) (Asim et al. 2010). ODC uses the substrate L-ornithine to produce putrescine via a decarboxylation reaction (Asim et al. 2010). ODC has been implicated in several malignancies, including breast, colorectal, and gastric cancers. Most of the studies related to ODC have been focused on its role in epithelial cell function. However, ODC expression is upregulated in macrophages by *H. pylori* in vitro (Asim et al. 2010) and in infected gastritis tissues of mice and humans (Hardbower et al. 2017). Importantly, macrophage ODC is immunosuppressive, impairing M1-dependent host defense against *H. pylori*; mice with myeloid-specific deletion of *Odc* exhibited marked upregulation of M1 responses, including NOS2 expression/NO production; M1 gene signatures (*Nos2*, *Il1b*, *Tnf*, *Il6*, *Il12a*, *Il12b*) and M1 protein responses (IL-1β and TNF-α) as well as pro-inflammatory chemokines and IL-17A (Hardbower et al. 2017). In parallel, there was increased gastric inflammation (both acute and chronic) but a clear benefit of reduced *H. pylori* bacterial colonization levels. Another crucial observation was that the immunosuppressive effects of ODC activity were linked to the ability of putrescine to cause histone modifications (specific acetylation and methylation events) favoring the suppression of gene transcription, thus blocking M1 response (Hardbower et al. 2017), suggesting that polyamines have a deleterious effect of restricting mucosal immune responses. These findings lead to the question of the potential role of ODC/polyamines in GC development induced by *H. pylori*; this is discussed in Sect. 4.2 below.

Many of these cytokines and signaling pathways likely contribute to the development of cancer. However, the microenvironment in the stomach of a patient with chronic gastritis likely differs from the tumor microenvironment.

2.2 *The TME in Gastric Cancer*

The TME is a complex network of tumor cells and numerous types of non-tumor cells, including lymphocytes, myeloid cells, endothelial cells, and fibroblasts. As a dynamic environment, the TME involves a large variety of molecules such as growth factors, cytokines, chemokines, antibodies, proteases, and metabolites as well as the extracellular matrix. Non-resolving inflammation derived from chronic infection with *H. pylori* is one of the characteristics of the TME in GC and is considered to play an essential role in tumor initiation and growth. In a long-term failed attempt to promote healing, this complex network of mediators in the gastric mucosa leads to the upregulation of pathways that increase cell survival, activate stem cells, and promote epithelial proliferation.

Defined almost two decades ago, cancer immunoediting is the process by which the immune system can either restrain or promote cancer development, ultimately favoring the outgrowth of tumor cells with reduced immunogenicity (Shankaran et al. 2001; Schreiber et al. 2011). Mechanisms that lead to inhibition of anti-tumor immune responses involve multiple components within the TME. Tumor cells secrete cytokines and chemokines to recruit myeloid-derived suppressor cells (MDSCs), regulatory T cells (Tregs), and tumor-associated macrophages (TAMs). These cells directly suppress the functions of natural killer (NK) cells and $CD8^+$ T cells through the production and expression of various factors, ultimately favoring tumor growth and invasion (Kitamura et al. 2015).

The macrophage is a key player in the innate immune response that then modulates chronic inflammatory responses, therefore it plays an important role during *H. pylori* infection and carcinogenesis. Circulating monocytes are recruited across the vasculature into tumors by tumor-derived chemoattractants such as colony-stimulating factor 1 (CSF1), CC ligand 2 (CCL2), vascular endothelial growth factor A (VEGFA, commonly referred to as VEGF), or CXCL12 (Murdoch et al. 2008; Noy and Pollard 2014). In the tissues, macrophages adjust to the particular conditions of the environment, adopting either pro-inflammatory (M1) or anti-inflammatory (M2) phenotypes (Mills 2012). M1 macrophages are activated by bacterial constituents and Th1 cytokines (e.g., IFNγ) and show anti-tumor activity through high antigen-presenting capacity, phagocytosis, and upregulation of pro-inflammatory Th1 responses. In contrast, M2 macrophages are activated by Th2 cytokines (e.g., TGF-β, IL-4, IL-10, and IL-13), leading to the suppression of adaptive immunity and promotion of tissue remodeling, angiogenesis, and tumor growth (Fig. 2) (Tiemessen et al. 2007; Mills 2012). Multiple characteristics of tumors, including hypoxia and abundant cell death, help direct macrophage function toward attempting a "homeostatic" restoration (Ruffell et al. 2012). Thus, it has been argued that most TAMs exhibit a predominantly M2 phenotype (Mantovani et al. 2002). However, the large variety of functions, in which TAMs are engaged, suggests that extreme forms of M1/M2 polarization may not exist in the TME (Qian and Pollard 2010). In any case, high densities of TAMs or M2 macrophages have been associated with worse overall survival in several malignancies, including GC (Zhang et al. 2012; Jiang et al. 2017).

TAMs promote the suppression of effective anti-tumor immunity via different pathways including the production of anti-inflammatory cytokines (including IL-10 and TGF-β), prostaglandin E2 (PGE2), and expression of programmed death ligand 1 (PD-L1) (Mantovani et al. 2014). In addition, TAMs regulate vascular programming of tumors by the production of VEGF and other pro-angiogenic factors (Ruffell et al. 2012). Macrophage infiltration correlates significantly with tumor vascularity in human GC (Ohta et al. 2003). TAMs can also contribute to the invasiveness of tumor cells by remodeling the extracellular matrix and by opening the way to exit the tumor and colonize the surrounding tissues (Guiet et al. 2011). As the tumor progresses, hypoxic regions develop, caused by high metabolic and proliferative rates. Hypoxia is a potent inducer of VEGF, and this is mediated by the transcription factor hypoxia-inducible factor-1 (HIF-1) (Semenza 2003, 2012). In addition to the role in angiogenesis, VEGF that is secreted by tumor cells can function in an autocrine

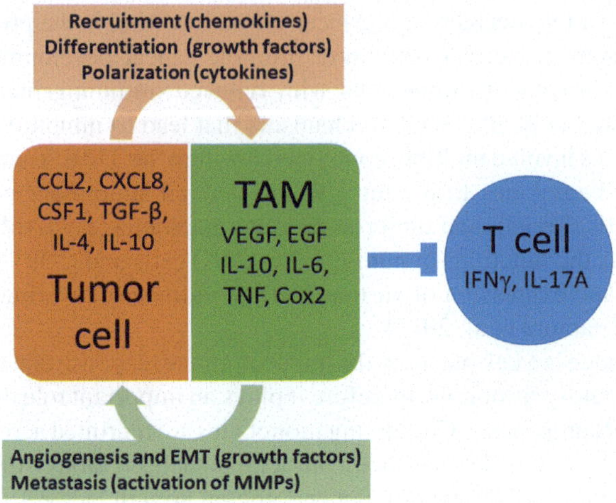

Fig. 2 Tumor-associated macrophages and their role in the tumor microenvironment. Tumor cells can produce chemokines, cytokines and growth factors which drive recruitment, differentiation and polarization of TAMs. When the tumor-associated macrophages enter the TME they can suppress anti-tumor T cell activity, contribute to angiogenesis through production of growth factors, and promote metastasis through activation of matrix metalloproteases

manner promoting proliferation, dedifferentiation, and transition from an epithelial to mesenchymal phenotype, enhancing stromal invasion and tumor growth. This autocrine signaling, which is mediated by VEGFR2 and by neuropilins, could be necessary for the function of cancer stem cells because it seems to maintain the stem cell reserve and to sustain self-renewal (Goel and Mercurio 2013).

Recently, next-generation sequencing and large-scale genomics have led to new molecular classifications of GC (Cancer Genome Atlas Research 2014; Cristescu et al. 2015). The Cancer Genome Atlas (TCGA) project classified GC into four subtypes: EBV-positive, MSI, genomically stable, and chromosomally unstable tumors (Cancer Genome Atlas Research 2014). At the same time, advances in cancer immunotherapy have opened new frontiers for patient care across a variety of tumors. Among the GC subtypes, EBV and MSI are the most promising in this regard. Amplification of *CD274* and *PDCD1LG2*, which encode PD-L1 and PDCD1LG2, respectively, is often found in EBV-positive tumors with high levels of PD-L1 protein expression detected on tumor cells (Derks et al. 2016). The interaction of PD-L1 with its receptor PD-1 (commonly found on T cells) inhibits T lymphocyte migration, proliferation, and secretion of cytotoxic mediators, ultimately favoring tumor escape from the immune response.

Tumor-immune interactions are increasingly recognized as drivers of the clinical outcome and as potential targets for therapy. Recently, Thorsson and co-workers (2018) characterized the immune TME of 33 cancer types (more than 10,000 tumors) into six "immune subtypes": (1) wound healing, (2) IFNγ, (3) inflammatory, (4) lymphocyte depleted, (5) immunologically quiet, and (6) TGF-β dominant. About 80% of gastric adenocarcinomas were grouped within immune subtypes 1 (wound healing, characterized by elevated expression of angiogenic genes, a Th2 cell bias and high proliferating rate) and 2 (IFNγ dominant, with the highest M1/M2 polarization, a strong CD8 signal, but also a high proliferating rate). The best overall survival (combining all cancer sites) was observed in the immune subtype 3 (inflammatory, defined by elevated Th17 and Th1 genes and low/moderate tumor proliferation), while types 1 and 2 had less favorable outcomes despite having a substantial immune component (Thorsson et al. 2018). This study highlights the importance of the immune interactions within the TME on prognosis and the need for individual tumor characterization for an effective personalized choice of immune-related therapies. Although the study by Thorsson and colleagues found that the immune subtype 3 TME (determined by immunogenetics) had the best overall survival of all cancer types, the situation may be different in the GC TME. A recent publication investigating the expression of several cytokines and their relationship with clinicopathological characteristics in GCs revealed that IL-17 expression (measured by immunohistochemistry) was associated with decreased survival (Kim et al. 2017). Taken together, these results indicate a need for a deeper understanding of T cell cytokines in the TME in GC.

3 Anti-cancer Strategies Targeting the Pro-inflammatory Mediators in the TME

3.1 Cell Signaling Inhibitors

One way to target several cytokines/inflammatory mediators is to inhibit the transcription factor NF-κB. Several drugs are available to modulate the NF-κB pathway and thereby reduce expression of IL-8 and other pro-inflammatory cytokines. Many of these drugs target NF-κB indirectly by reducing ROS production including resveratrol, anthocyanin, apigenin, and RK-1-123. Because resveratrol is a member of the polyphenol flavonoids class of antioxidants produced by a restricted number of plants, it has received significant attention as a potential treatment/adjunct therapy for GC patients and several reviews have addressed it directly (Zulueta et al. 2015). Resveratrol has been shown to inhibit the proliferation of a number of cancer cell lines, and it behaves as a chemo-preventive agent in assays that measure the three stages of carcinogenesis (Holian et al. 2002). While human pilot studies in patients with colorectal liver metastases have demonstrated that preoperative resveratrol reduced cancer cell proliferation (Patel et al. 2010) and increased apoptosis in resected tumor tissues (Howells et al. 2011), there have been no clinical trials with resveratrol in GC.

3.2 Cytokine Antagonists (IL-1, IL-6, IL-18, TNF)

While many of these cytokines have been discussed in detail in other areas of this book (Chaps. 7–11), here a review will be provided for cytokines that may be candidates for targeting during cancer therapy.

IL-1β is expressed by a number of different cell types in vitro in response to *H. pylori,* and all of these cells may participate in the inflammatory response in vivo, including dendritic cells (DCs), monocytes/macrophages, and GECs (Kim et al. 2013; Semper et al. 2014; Pachathundikandi et al. 2016). IL-1β is a cytokine that has received significant attention because in several studies, polymorphisms associated with increased expression of IL-1β are significantly associated with the development of GC (El-Omar et al. 2000; Camargo et al. 2006). Expression is notably correlated with clinical and pathological features of GC (Yin et al. 2016). Solid tumors in which IL-1β has been shown to be up-regulated include breast, colon, lung, head, and neck cancers, and melanomas, and patients with IL-1β-producing tumors have a generally poor prognosis. IL-1β inhibits acid secretion by downregulating H^+/K^+ATPase expression and gastrin expression (Smolka and Backert 2012). Moreover, transgenic expression of IL-1β in the stomach causes gastritis-associated GC with the recruitment of MDSCs (Tu et al. 2008) and in human xenograph models, elevated levels of IL-1β are correlated with advanced metastatic disease (Lewis et al. 2006). There are several possibilities for targeting IL-1β, including approved treatments already utilized to treat patients with rheumatoid arthritis (Nikfar et al. 2018). IL-1 receptor antagonist (IL-1RA, anakinra) is a naturally occurring protein that has been shown to decrease tumor growth, angiogenesis, and metastases in murine xenograft models (Weinreich et al. 2003). This is the IL-1β-blocking therapy that has received the most attention. Anakinra is well absorbed in humans, and its safety is well documented with few adverse reactions (Sota et al. 2018), making it a candidate to be tested in combination with standard chemotherapy in GC. Currently, there is only one clinical trial in the NCI database utilizing anakinra for the treatment of cancer, and it is specifically focused on early-stage multiple myeloma (clinicaltrials.gov ID# NCT02492750). In addition to the IL-1RA, there several other agents available to inhibit the inflammatory and tumor-promoting effects of IL-1β including anti-interleukin-1 monoclonal antibodies, the soluble IL-1 receptor type II, IL-1β-converting enzyme inhibitors, and IL-1β cytokine traps.

IL-6 is another pleiotropic cytokine which impacts inflammatory T cell biology as well as tissue regeneration and carcinogenesis. While the findings on associations between IL-6 polymorphisms and risk of GC are controversial, there is increasing evidence that levels of IL-6 may be a prognostic marker for spread (Ashizawa et al. 2005). In inflammatory cells, IL-6 is well known for activating STAT3 signaling, which can induce a pro-carcinogenic, tumorigenic microenvironment. STAT3 signaling leads to the activation of NF-κB in inflammatory cells and drives a positive feedback loop between immune cells and tumor cells that further stimulates the cancer stem cell components and may contribute to metastasis and resistance to cancer therapies. GC cell lines also express high levels of IL-6 receptor, and IL-6 activation of GC cells leads to STAT3 activation and VEGF production (Huang et al. 2004; Lee et al. 2010; Wang et al. 2013); moreover, in ex vivo assays this leads to human umbilical vein endothelial cell proliferation, tube formation, and vascularization in a Matrigel plug assay (Huang et al. 2004). IL-6 also inhibits H_2O_2-induced apoptosis and blocks repair of oxidative DNA lesions in human GC cells through upregulation of anti-apoptotic gene, *MCL-1* (Lin et al. 2001). Again, methods targeting IL-6 have been developed for the treatment of IL-6-associated diseases, such as rheumatoid arthritis and Castleman disease, but not for cancers. These therapeutics include anti-IL-6 antibodies (siltuximab and sirukumab), anti-IL-6 receptor antibodies (cilizumab and tocilizumab), soluble gp130 (also a receptor of IL-6, designed to inhibit IL-6 binding to IL-6R), and some selective small molecules which inhibit JAK/STAT signaling as described above. While the concept to inhibit IL-6 and/or IL-6 signaling is not new (Jones et al. 2011; Sansone and Bromberg 2012), few clinical trials have been performed with these therapeutics in solid tumors (Ruffell and Coussens 2015), and at the time of writing this chapter, no clinical trials were published utilizing these biological treatments in GC. Another way to block IL-6 indirectly is through the inhibition of STAT3. Napabucasin, an oral inhibitor of cancer stem cells through STAT3 signaling blocking, is being tested in many gastrointestinal tumors, either alone or in combination with standard chemotherapy. Currently, there is one clinical trial evaluating the association between napabucasin and weekly paclitaxel (a first-line cytotoxic agent) as second-line therapy for patients with GC (NCT01278956). An interesting strategy in GC would be to investigate the efficacy and safety of napabucasin combined with an anti-IL-6 antagonist.

TNF is a pro-inflammatory cytokine, which was first recognized for its inhibitory effect in some tumors when present at high concentrations, but it is also key cytokine for orchestrating inflammation and the host immune responses. TNF-activated chemokine gradients recruit immune cells to the sites of infection/inflammation. TNF is induced during *H. pylori* infection (Lindholm et al. 1998). The level of TNF may dictate its functional consequences; for it is thought to be pro-angiogenic in tumors, but a potent anti-vascular cytokine at higher doses and can be used clinically to destroy tumor vasculature. Anti-TNF therapy is currently used for rheumatoid arthritis, Crohn's disease, and other inflammatory diseases (reviewed by Udalova et al. 2016), but targeting TNF as an anti-cancer therapy has led to some scrutiny. One can appreciate why there are two very different hypotheses as to the effect of anti-TNF therapy on cancer. On the one hand, anti-TNF therapy could inhibit cancer development by reducing chronic inflammation, but on the other hand, if TNF induces apoptosis or has suppressive effects on gene expression, anti-TNF therapies may enhance the development of certain tumors. There have been many studies to assess the possibility that existing anti-TNF treatments increase the risk for cancer, but because of heterogeneity within these studies, there has been no consensus (reviewed by Solomon et al. 2012). Interestingly, Fan and colleagues (Chen et al. 2009) investigated the opposing approach of delivering the TNF protein (not anti-TNF) to gastric tumors. They fused TNF with a peptide (GX-1) known to target the human GC vasculature and injected the construct into the circulation of *nude* mice containing tumors of human GC cells. This targeted approach of delivering TNF delayed tumor growth and was less toxic than TNF alone. But, administering TNF has not been tested in clinical trials.

As mentioned earlier, some chronic inflammation is driven by T cell-derived cytokines including IL-17A and IFNγ. Targeting IFNγ to reduce inflammation or treat cancer has not been strongly considered, because it has such an important role for control of infections—both viral and bacterial. The level of immunosuppression created by inhibiting this pro-inflammatory pathway would be unacceptable. On the other hand, targeting IL-17A has been considered both in chronic inflammatory disease and cancer. There is evidence of increased IL-17A expression in GC compared to normal gastric tissue (Yamada et al. 2012), and this may contribute to an imbalance of Th17/Treg cells (Li et al. 2013). On the other hand, there is conflicting evidence as to whether Th17 cells are increased in the peripheral blood of GC patients and a good marker for tumor progression (Liu et al. 2012a; Zhang et al. 2008; Yamada et al. 2012). IL-17 inhibitors approved by the FDA include secukinumab and ixekizumab. These were designed to treat inflammatory disorders including psoriasis (Hueber et al. 2010; Sanford and McKeage 2015; Markham 2016), but they have not been used in a cancer setting and could be tested in patients with GC.

3.3 Non-steroidal Anti-inflammatory Drugs

Substantial evidence from epidemiological studies suggests that the use of aspirin and other non-steroidal anti-inflammatory drugs (NSAIDs) is protective against GC as well as other gastrointestinal tumors (Abnet et al. 2009; Epplein et al. 2009; Algra and Rothwell 2012; Huang et al. 2017; Kong et al. 2016; Rothwell et al. 2012; Zhang et al. 2014). Although the mechanisms by which NSAIDs protect GC are not completely understood, it is recognized that NSAIDs primarily reduce the production of prostaglandins (PGs) by inhibiting the activity of cyclooxygenase enzymes (Cox1 and/or Cox2). Cox1, encoded by *PTGS1,* is constitutively expressed and responsible for the production of prostanoids during basal conditions in the gastrointestinal tract and other tissues. Cox2, encoded by *PTGS2,* is an inducible isoform upregulated at the sites of inflammation and in some cancers, including GC (Ristimaki et al. 1997). Cox enzymes participate in the conversion of arachidonic acid into prostanoids, including PGs and thromboxane A2 (TxA_2) (Wang and DuBois 2018). Besides promoting inflammation, prostanoids may facilitate tumor progression by several mechanisms, including stimulation of proliferation and inhibition of apoptosis of cancer cells, stimulation of tumor invasion and angiogenesis, and suppression of immune responses. A comprehensive review on the role of prostanoids in gastrointestinal cancer was recently published (Wang and DuBois 2018). Among prostanoids, PGE2 is the most abundant in human GC (Uefuji et al. 2000), and the measurement of its metabolite (PGE-M) levels in urine has shown that PGE-M could be used as a biomarker for predicting GC risk and prognosis (Wang et al. 2017).

In the context of colorectal cancer, a Cox2 selective inhibitor, celecoxib, was the first approved agent for patients with familial adenomatous polyposis. However, due to the cardiovascular side effects (Bresalier et al. 2005), long-term use of Cox2-selective inhibitors for cancer chemoprevention is no longer recommended. In contrast, long-term regular use of aspirin has proven beneficial for the prevention of both cancer and cardiovascular diseases. Based on the evidence suggesting that aspirin therapy reduces the incidence of colorectal cancer after 5–10 years of use, the U. S. Preventive Services Task Force now recommends low-dose aspirin use for the primary prevention of colorectal cancer in adults aged 50–59 who meet certain criteria (Bibbins-Domingo and Force 2016; Chubak et al. 2016). Regarding GC, the evidence on the role of aspirin on prevention has been more limited, and there is no current recommendation as a chemopreventive agent. Consistent with previous evidence, however, a recent meta-analysis (Huang et al. 2017) and a longitudinal study covering the whole population of South Korea concluded that long-term aspirin use was associated with a reduction in GC risk (Kim et al. 2018). In this high GC risk population, the protective effect was significant in those individuals with cumulative aspirin daily dose use for at least three years. The evidence on the use of non-aspirin NSAIDs has shown less consistent results in protection against GC across studies (Kim et al. 2018; Abnet et al. 2009; Epplein et al. 2009; Huang et al. 2017). A recent study aimed to evaluate the protective effects of low-dose aspirin use after GC diagnosis found no association with GC-specific mortality after one year of follow-up (Spence et al. 2018). Currently, a phase III clinical trial assessing the long-term effects of regular aspirin use on recurrence and survival in various types of cancer is ongoing (Add-aspirin, NCT02804815). For the use of aspirin on GC prevention, the well-known risks, including renal and platelet dysfunction, gastric ulceration, and gastrointestinal bleeding (Lanas et al. 2011), should be considered. Nevertheless, the strength of the associations consistently seen in observational studies, along with the high GC mortality rate, supports the need for further research on the potential of NSAID chemoprevention trials, especially aspirin, among select high-risk populations.

Overall, blocking the inflammatory response in GC through the inhibition of pro-inflammatory cytokines comprises an attractive approach for clinical trials. A list of selected ongoing studies assessing therapeutic agents targeting cancer-promoting inflammation in patients with GC (excluding antiangiogenic agents) is presented in Table 1. However, it is unlikely that the blockade of a single cytokine would result in dramatic clinical effect. Rather, there is a good rationale to combine targeted agents directed to different cytokines in GC, including systemic chemotherapy. Yet, like most of the new clinical trials of advanced cancer, trials in GC should enroll moleculari selected patients. In the context of cytokines, well-conducted studies should be undertaken to identify predictive biomarkers; with that information, patients would have their GC tissues tested for the biomarker in order to enroll into a specific "anti-inflammatory" trial.

Table 1 Selected ongoing studies assessing therapeutic agents targeting cancer-promoting inflammation in patients with GC (excluding antiangiogenic agents)

Strategy	Agent	Target	Mode of action	Combination or comparator	Tumor types	Phase	ClinicalTrials.gov identifier
Blocking pro-inflammatory mediators	Napabucasin (BBI608)	STAT3	Inhibitor of cancer cell stemness; indirectly inhibits IL-6 through STAT3 inhibition	Napabucasin + paclitaxel	Metastatic or locally recurrent GC or GEJC	III	NCT01278956
	Aspirin	COX1/COX2	COX enzymes inhibition	Aspirin monotherapy	Non-metastatic solid tumors (adjuvant setting)	III	NCT02804815
Inhibiting recruitment of immune cells	Emactuzumab (RG7155)	CSFR1	Monoclonal antibody	Emactuzumab + atezolizumab (anti-PD-L1 mAB)	Locally advanced or metastatic solid tumors	IB	NCT02323191
				Emactuzumab + selicrelumab (CD40 agonist)[a]	Locally advanced or metastatic solid tumors	IA/IB	NCT02760797
	Pexidartinib (PLX3397)	CSFR1	Tyrosine kinase inhibitor	Pexidartinib + paclitaxel	Advanced, incurable solid tumors	IB	NCT01525602
	HuMax-IL8 (BMS-986253)	IL-8	Monoclonal antibody	HuMax-IL8 + nivolumab	Advanced, incurable solid tumors	I/II	NCT03400332
	DKN-01	DKK-1	Monoclonal antibody	DKN-01 + paclitaxel or pembrolizumab	Metastatic or locally recurrent GC or GEJC	I	NCT02013154

(continued)

Table 1 (continued)

Strategy	Agent	Target	Mode of action	Combination or comparator	Tumor types	Phase	ClinicalTrials.gov identifier
Reprogramming TAMs from M2 to M1	CDX-1140	CD40	Monoclonal antibody (CD40 agonist)	CDX-1140 + standard therapy	Advanced solid tumors	I	NCT03329950
	Selicrelumab (RO7009789)	CD40	Monoclonal antibody (CD40 agonist)	Selicrelumab + atezolizumab (anti-PD-L1 mAB)	Locally advanced or metastatic solid tumors	IB	NCT02304393
Immuno-conjugates	CEA-TCB antibody (RO6958688)	CEA expressed in tumor tissues	Bispecific anti-CEA anti-CD3 monoclonal antibody	Monotherapy	Advanced CEA-expressing solid tumors	I	NCT02324257

[a]This trial combines an inhibitor of immune cell recruitment and an agent to reprogram macrophages to M1 phenotype. CD3: cluster of differentiation 3; CD40: cluster of differentiation 40; CEA: carcinoembryonic antigen; CEA-TCB: carcinoembryonic antigen T-cell bispecific; COX-1/COX-2: cyclooxygenases 1 and 2; CSF1R: colony-stimulating factor 1 receptor; DKK-1: dickkopf-related protein 1; GC: gastric adenocarcinoma; GEJC: gastroesophageal adenocarcinoma; mAb: monoclonal antibody; NCT: clinicalTrials.gov identifier; PD-L1: programmed death-ligand 1; STAT3: signal transducer and activator of transcription 3; TAMs: tumor-associated macrophages

4 Strategies Targeting Growth Factors Involved in Angiogenesis

4.1 Growth Factors as Target (VEGF/VEGFR2, CSF1/CSF1R, EGF/EGFR)

Vascular endothelial growth factor A (VEGFA, often referred to as VEGF) is a vascular permeability factor and the main regulator of tumor angiogenesis (Senger et al. 1983; Ferrara 2002). VEGF is a member of a family of growth factors and primarily binds to tyrosine kinase receptors VEGFR2 and VEGFR1 (Chung et al. 2010). VEGF is secreted by both tumor and non-tumor cells, such as macrophages, endothelial cells, and fibroblasts (Goel and Mercurio 2013). VEGFR2 is mainly expressed by endothelial cells, but also by a variety of cells, including tumor cells. The binding of VEGF to VEGFR2 is considered critical for the regulation of tumor angiogenesis, by promoting the proliferation and migration of endothelial cells, as well as the degradation and remodeling of the extracellular matrix. Independent of the role of VEGF in tumor development, a large body of evidence supports a role for VEGF in the pathogenesis and maintenance of chronic inflammatory disorders. Notably, immune cells can express VEGF receptors, and the functions of these cells can be regulated by VEGF signaling. VEGF promotes the adherence of leukocytes to the vascular endothelium and the release of pro-inflammatory cytokines, such as IL-6 and TNF, which support tumor development in chronic inflammatory diseases (Waldner and Neurath 2012). The role of VEGF signaling in cancer-associated inflammation was demonstrated by Waldner and co-workers (Waldner et al. 2010). Patients with inflammatory bowel disease or with colitis-associated cancer showed increased expression of VEGFR2 on intestinal epithelial cells. Results from in vivo and in vitro experiments demonstrated that chronic inflammation induces VEGFR2 expression on intestinal epithelial cells and that VEGFR2 signaling is necessary for tumor growth (Waldner et al. 2010). There is also significant evidence that VEGF/VEGFR2 signaling has an important role in GC pathogenesis (Lieto et al. 2008; Murukesh et al. 2010; Suzuki et al. 2010), and the inhibition of this interaction is the main target of anti-VEGF therapeutics (Ferrara 2009).

The monoclonal anti-VEGF antibody bevacizumab was the first agent developed targeting the VEGF pathway. Bevacizumab is now approved for first- and/or second-line treatment of a variety of tumors including colorectal cancer, but clinical trials in GC have not obtained encouraging results (Ohtsu et al. 2011; Shen et al. 2015). Ramucirumab, a human monoclonal antibody that targets VEGFR2 (Spratlin et al. 2010), is the first drug targeting angiogenesis that showed to prolong survival in patients with previously treated advanced GC or gastroesophageal junction carcinomas (GEJC) in phase III clinical trials (Fuchs et al. 2014; Wilke et al. 2014). Currently, ramucirumab is indicated as a single agent or in combination with paclitaxel, for the treatment of patients with advanced or metastatic, GC or GEJC with disease progression on or after fluoropyrimidine- or platinum-containing chemotherapy (Ajani et al. 2016).

Apatinib, a selective VEGFR2 tyrosine kinase inhibitor, was tested in Asian patients with previously treated, advanced GC and showed prolonged overall and progression-free survival (Li et al. 2016). Apatinib was approved by the FDA in 2017 as third-line therapy for refractory GC or GEJC. Experimental studies have suggested that apatinib not only has anti-angiogenesis effects, but also possesses substantial angiogenesis-independent effects, inhibiting cell proliferation in vitro and delaying xenograph tumor growth in vivo (Lin et al. 2017). A large number of clinical trials that include apatinib treatment for patients with GC are currently underway, including phase III and IV trials (NCT03042611 and NCT02426034). Regorafenib, a multikinase inhibitor that targets VEGFR2 and is used for refractory colorectal cancer (Wilhelm et al. 2011; Riechelmann and Grothey 2017), is currently being tested in clinical trials for GC and has provided promising results (Pavlakis et al. 2016).

Colony-stimulating factor 1 (CSF1 or also known as MCSF) is a hematopoietic growth factor constitutively expressed by many cell types (Hamilton et al. 2016). CSF1 is the major lineage regulator of most populations of macrophages, but it is also a chemotactic factor for macrophages. CSF1 exerts its effects through a tyrosine kinase receptor (CSF1R), which is expressed on monocytes and macrophages, but also on other myeloid cells within the TME (Cannarile et al. 2017). In human GC, elevated expression of CSF1 or CSF1R significantly correlated with disease progression and also with poor overall survival and disease-free survival (Okugawa et al. 2018). Targeting TAMs by the inhibition of CSF1/CSF1R has shown encouraging results in preclinical cancer models in a variety of tumors, not only decreasing the number of TAMs, but also reprogramming remaining TAMs to support antigen presentation and bolster T cell activation (Ries et al. 2014; Zhu et al. 2014; Pyonteck et al. 2013; Quail et al. 2016; DeNardo et al. 2011). Experimental evidence has shown that macrophages can mediate chemotherapy resistance by providing survival factors or activating anti-apoptotic pathways in cancer cells. In a mouse model of breast cancer, cytotoxic therapy showed to induce CSF1-dependent macrophage recruitment (DeNardo et al. 2011). In this model, blockade of macrophage recruitment with CSF1R-signaling antagonists, in combination with paclitaxel, showed the promotion of Th1 responses and improved mouse survival by a reduction in primary and metastatic tumors.

In human tumors, the most promising evidence targeting the CSF1/CSF1R axis has been documented in patients with the diffuse type of tenosynovial giant cell tumor (TGCT). TGCT is a rare neoplasm associated with inflammation and joint destruction, in part due to infiltration of CSFR1-bearing macrophages (Gelhorn et al. 2016). Because TGCT is associated with overexpression of CSF1, therapies targeting the CSF1/CSF1R axis have been tested in patients with locally advanced or relapsed TGCT (Brahmi et al. 2016). Studies have shown significant clinical improvement with emactuzumab (RG-7155), a recombinant humanized monoclonal antibody targeting CSF1R (Ries et al. 2014; Cassier et al. 2015), and with the small molecule CSF1R inhibitor pexidartinib (PLX3397) (Tap et al. 2015). In the context of GC, a large number of clinical trials assessing the potential of various CSFR1 inhibitors (emactuzumab, pexidartinib, DCC-3014, and others) either as monotherapies or in combination with other therapeutic modalities are underway (see Table 1).

Additional recent studies have implicated several other potential master regulators of macrophage function and polarization in the context of *H. pylori* infection and gastric carcinogenesis. We recently reported (Hardbower et al. 2016) that epidermal growth factor receptor (EGFR) signaling is a crucial component of the response of macrophages to bacterial infections, with *H. pylori* a prototypical example. While EGFR signaling in GECs has been documented and related to both ligand-dependent (EGF) and independent (TNF) responses (Yan et al. 2009), the response of mouse and human macrophage cell lines and mouse bone marrow-derived primary macrophages was ligand-independent and involved both tyrosine 1068 and serine 1046/7 phosphorylation sites (Hardbower et al. 2016). Importantly, human gastric tissues exhibited marked phosphorylation of EGFR in gastric macrophages along the entire cascade from gastritis to gastric adenocarcinoma, and mice with myeloid-specific deletion of *Egfr*, exhibited attenuated gastric inflammation scores, increased *H. pylori* colonization, and reduction in M1 macrophage and Th1/Th17 responses.

It should be noted that these studies were in a model of chronic infection, not cancer. However, in the azoxymethane-dextran sulfate sodium model of colitis-associated colon carcinogenesis, mice with myeloid deletion of *Egfr* showed a marked reduction in tumor development (Hardbower et al. 2017). Moreover, these findings were associated with attenuation of M2 responses and angiogenesis and associated signaling. Surprisingly, mice with epithelial deletion of *Egfr* did not show protection from colon tumorigenesis. There are a series of studies from other groups similarly showing that deletion of *Egfr* in myeloid cells reduces liver cancer (Lanaya et al. 2014) and other colon cancer models (Srivatsa et al. 2017). There are also studies related to pancreatic cancer implicating macrophage EGFR signaling in M2 macrophage polarization (Ma et al. 2016).

The addition of the EGFR inhibitor, gefitinib, was very effective in rodent models of GC (Sierra et al. 2018). Specifically, when added to the diet, this agent significantly reduced the development of dysplasia and intramucosal carcinoma in *H. pylori*-infected INS-GAS mice and dysplasia and invasive GCs in infected gerbils. Gefitinib treatment reduced PMN infiltration and chemokine expression, as well as epithelial DNA damage in both rodent models. Gefitinib was effective if given as a pretreatment before the infection or if administered after infection and inflammation was already established, and still had a benefit if given to animals after antibiotic eradication of the *H. pylori* (Sierra et al. 2018). It should be noted that the use of a pharmacologic approach does not distinguish the offending cellular source of EGFR signaling. To this end, the effect of GEC-specific deletion of *Egfr* was investigated; this resulted in less gastric inflammation, DNA damage, and chemokine expression. Thus, the Wilson laboratory is generating mice with myeloid- and epithelial-specific deletion of *Egfr* on the cancer-prone FVB/N INS-GAS mouse and will be determining the effect on GC during *H. pylori* infection.

Despite the strong scientific rationale of antiangiogenic agents in GC and other solid tumors, the overall results of clinical trials with these agents have been quite modest, with survival improvements generally measured in weeks. Likewise, despite the promising preclinical data with anti-EGFR agents in GC, randomized trials with these agents, with the exception of trastuzumab for HER2-positive GC (Song et al. 2016), were not effective in patients with molecularly unselected metastatic GC. Therefore, their future in the drug development process in GC is likely to be undermined by more innovative agents, such as immunotherapy.

4.2 Strategies Inhibiting M1 to M2 Transitions and Promoting M1 Phenotype

Besides the mentioned effects of targeting the CSF1/CSF1R axis on macrophage polarization, reprogramming TAMs into M1-phenotype macrophages can be achieved through a variety of other therapeutic modalities including chemotherapy, immunotherapy, and radiotherapy (Ruffell et al. 2012; Genard et al. 2017). One of the modalities under investigation is based on the use of CD40 agonists. CD40 is a member of the TNF receptor superfamily that is present on a variety of immune cell types. CD40 activation plays a critical role in triggering T and B cell immunity, by activation of antigen-presenting cells, resulting in an enhanced anti-tumor immune response (Vonderheide 2018). By stimulating CD40, monoclonal antibodies against CD40 similarly have shown to reprogram TAMs from M2 phenotype to M1 macrophages. In a mouse model of pancreatic cancer, CD40-activated macrophages rapidly infiltrated tumors, showed anti-tumor properties, and facilitated the depletion of tumor stroma (Beatty et al. 2011). In addition, the activation of CD40 present on the surfaces of some solid tumor cells leads to direct tumor cell apoptosis and decreased tumor growth. In the context of GC, several clinical trials involving patients with a variety of advanced solid tumors are underway (Table 1). Some evidence has indicated that it is also possible to re-educate TAMs by exposure to specific immunological mediators which may promote M1 macrophage development, such as IFNγ (Duluc et al. 2009; De Palma et al. 2008).

Having evidenced that loss of ODC enhances host defense against *H. pylori* (Hardbower et al. 2017), a related question is the potential to inhibit ODC in vivo. Efficacy of the pharmacologic agent α-difluoromethylornithine (DFMO), which blocks ODC activity, has been demonstrated in clinical trials related to the prevention of colon polyps (Meyskens et al. 2008; Zell et al. 2010) and has also been used for the treatment of neuroblastoma (Bassiri et al. 2015; Evageliou et al. 2016; Saulnier Sholler et al. 2015). The mechanism of action is unproven, but has been conceptually related to reduction of epithelial cell growth and DNA replication. The downstream enzyme, spermine oxidase (SMOX), which generates oxidative stress during the back conversion of spermine to spermidine, is upregulated by *H. pylori* in both macrophages (Chaturvedi et al. 2004) and GECs (Chaturvedi et al. 2014; Chaturvedi et al. 2011). Moreover, this generation of ROS leads to oxidative DNA damage in the gastric epithelium (Chaturvedi et al. 2014; Chaturvedi et al. 2011), which is especially notable in the presence of gastric intestinal metaplasia (Chaturvedi et al. 2015). Inhibition of ODC with DFMO or inhibition of SMOX reduces DNA damage and GC development in gerbils infected with a gerbil-adapted *H. pylori* strain derived from a patient with gastric dysplasia from the Andean mountain region of Colombia, where GC risk is high (Chaturvedi et al. 2015).

Based on these studies, we are currently conducting a phase II clinical trial using DFMO (eflornithine, 500 mg po per day) versus placebo in Latin American sites (Honduras and Puerto Rico, NCT02794428) in patients with precancerous gastric lesions (i.e., intestinal metaplasia). The primary end point is DNA damage at 6 months of treatment, with secondary endpoints of DNA damage at 18 months and histopathology at 18 months. The trial is ongoing, but final results will not be expected until 2020. An additional goal of the study is to determine the effects of DFMO on immune cells versus epithelial cells.

4.3 *Inhibition of Myeloid-derived Suppressor Cells (MDSCs)*

MDSCs, a heterogeneous group of immature myeloid cells that inhibit the anti-tumor activity of T lymphocytes and NK cells and which are absent in physiologic conditions, make an important component of TME. Studies in mice have demonstrated that the depletion of MDSCs is associated with delayed tumor growth (Schroder et al. 2017; Davis et al. 2017). Based on such findings, MDSCs have been investigated as therapeutic targets in GC. The Dickkopf-related protein 1 (DKK-1), a Wnt regulator, promotes immunosuppression in TME through the stimulation of MDSCs (Moehler et al. 2018). Two DKK-1-directed antibodies (BHQ880 and DKN-01) are currently being tested in several tumor types; an ongoing phase I trial is testing a DKK-1 monoclonal antibody in monotherapy or in combination with paclitaxel or pembrolizumab for patients with heavily pre-treated GC or GEJC (NCT02013154). Another strategy to inhibit the functionality of MDSCs is through the blockade of the glucocorticoid-induced TNFR-related protein (GITR). This protein is expressed in normal monocytes, MDSCs, and macrophages, and its suppression has led to reduced tumor progression and increased T-cell infiltration in an animal model of pancreatic cancer (Moehler et al. 2018). Interestingly, the inhibition of GITR was associated with upregulation of cytotoxic T lymphocyte-associated antigen 4 (CTLA4) on T cells and PD-L1 on tumor cells, proposing that a combination with checkpoint inhibitors may be required for this treatment to be effective.

4.4 *Targeting Chemokines and Their Receptors*

The TME can also be targeted through the disruption of chemokine networks. Chemokines, small proteins normally involved in immune cell migration and lymphoid tissue expansion, are implicated in the TME immunosuppression through complex mechanisms, including the stimulation of MDSCs (Nagarsheth et al. 2017).

Because chemokines and their receptors participate in major roles in inflammation and related inflammatory diseases, research has been conducted to explore the modulation of certain chemokines as a form of cancer- (and TME-) directed therapy. Chemokine inhibitors, mostly in combination with immune checkpoint inhibitors, are undergoing clinical testing in different tumor types.

The overexpression of chemokine CCL2, the monocyte chemoattractant protein (MCP1), induces angiogenesis and tumorigenesis of GC in nude mice via macrophage recruitment (Kuroda et al. 2005). An elevated level of CCL2 has been reported in patients with GC, was correlated with lymph node metastasis, and was associated with lower overall survival rate (Tonouchi et al. 2002; Futagami et al. 2008; Tao et al. 2014). However, while the inhibition of the CCL2-CCR2 (CCL2 main receptor) signaling pathway represented an attractive approach in GC, clinical trials have shown disappointing results. Carlumab, also known as CNTO888, is a human $IgG1_k$ monoclonal antibody that binds CCL2 with high affinity and which has been tested in two different phase I trials in patients with solid tumors. In the first trial, carlumab was administered in monotherapy; it offered modest anti-tumor activity, with the best responses observed being tumor stabilization in ovarian cancer, ocular melanoma, and neuroendocrine tumor (Sandhu et al. 2013). In the second phase I trial, where carlumab was combined with different standard chemotherapeutic agents, only one out of 19 patients experienced a partial response (Brana et al. 2015). Another approach to CCL2/CCR2 interference is to inhibit the CCR2 receptor. A humanized IgG1 antibody, MLN1202, has been successful in several inflammation-related diseases such as multiple sclerosis and atherosclerosis. However, a phase II trial with this agent in patients with bone metastases did not show efficacy (Vela et al. 2015). Given the above results, further trials with CCL2/CCR2 inhibitors are on hold.

CXCR4 and CXCL12, its ligand, are immunohistochemically overexpressed in GC by comparison with normal gastric tissue and associated with survival, proliferation, angiogenesis, and migration of cancer cells. Studies have demonstrated that such chemokine expression patterns are prognostic factors for survival and metastases (Xue et al. 2017). Also, overexpression of CXCL12 in gastric mucosa contributes to carcinogenesis in the presence of inflammatory stimuli such as *H. pylori* (Shibata et al. 2013). CXCL12 also activates the PI3K/mTOR and MAPK/ERK signaling pathways (Rubie et al. 2016). The blockade of the CXCL12-CXCR4 axis is the target of drugs used to treat the human immunodeficiency virus (HIV) infection or to help with the mobilization and collection of CD34-positive hematopoietic stem cells for transplantation in patients with certain hematological malignancies. Recently several pre-clinical studies in different tumor types, including GC, report encouraging anti-tumor effects from this biological class (Xue et al. 2017). However, clinical trials with CXCL12-CXCR4 inhibitors have just begun. A phase I trial of LY2510924, a peptide antagonist, which blocks stromal cell-derived factor-1 from CXCR4 binding, was conducted in patients with advanced and refractory solid tumors (although there were not any patients with GC in the study); while the safety profile was good, the best response observed was tumor stabilization in 20% of patients and the median duration of treatment was only 1.9 months (Galsky et al. 2014). Both randomized phase II trials in non-small cell lung cancer and renal cell carcinoma were negative for their primary endpoints of progression-free survival (Salgia et al. 2017; Hainsworth et al. 2016). Despite these preliminary negative results, there are a number of ongoing clinical trials in multiple tumor types, but at the time of this publication, there are no trials in GC. An interesting approach in GC is to combine CXCL12-CXCR4 inhibitors with other targeted agents that block the same pathways (e.g., mTOR inhibitors and MEK inhibitors) or with other chemokine inhibitors.

Interleukin-8 (IL-8) is a potent neutrophil chemotactic cytokine with potential impact on the tumor microenvironment. In chronic *H. pylori*-associated gastritis, GECs express increased levels of IL-8 within the lamina propria (Crabtree et al. 1994b). IL-8 is the single most up-regulated gene in whole-genome profiling of *H. pylori* exposed GECs (Eftang et al. 2012). Moreover, IL-8 expression is correlated with the functional *cag*PAI status of *H. pylori* strains (Crabtree et al. 1994a, 1995), but IL-8 can also be strongly induced by IL-17 responses. The result of IL-8 expression is the recruitment of PMNs to the tissue for higher degrees of PMN infiltration, which is correlated with an increase in the secretion of TNF, IL-1β, IL-6, and IL-8 (Bauditz et al. 1999). In a study out of northern India, circulating levels of IL-8 were higher among patients with GC compared to healthy controls, but levels were comparable to patients with functional dyspepsia (Kumar et al. 2015). The expression of IL-8 directly correlates with a poor prognosis in GC (Yamada et al. 2013; Lee et al. 2004).

In addition to its potent neutrophilic chemotactic activity, IL-8 can induce proliferation and migration of cancer cells (Wilson et al. 1999; Brew et al. 2000). For this reason, there has been interest in targeting it as a cancer therapy. IL-8 increases the proliferation, migration, and survival of endothelial cells, and enhances epithelial-mesenchymal transition (EMT) and survival of cancer cells (Fernando et al. 2011). Moreover, there is evidence that IL-8 may enhance macrophage activity in tumors through activating VEGF expression, a potent angiogenic factor (Martin et al. 2009). IL-8 may modulate invasiveness and/or extracellular matrix remodeling through enhancement of matrix metalloprotease expression (MMP2/MMP9) (Inoue et al. 2000; Kim et al. 2001; Li et al. 2005). In addition to general mechanisms to block NF-κB signaling and IL-8 induction discussed above, small molecule inhibitors targeting IL-8 receptors (CXCR1 and CXCR2) have been developed. G31P and SCH-527123 were initially developed for use in prostate and colon cancers, respectively, with the intention of reducing cell migration and increasing apoptosis of cancer cells (Liu et al. 2012b; Ning et al. 2012). Repertaxin, another inhibitor of CXCR1 and CXCR2, has shown to decrease tumor proliferation in the GC cell line MKN45 (Wang et al. 2016). These inhibitors have only been tested in vitro on GC cells and no clinical trials have been performed with them in GC.

Meanwhile, there is also interest in directly blocking IL-8, especially since the findings that high serum IL-8 levels correlate with poor prognosis in various tumors (Sanmamed et al. 2014). Moreover, IL-8 was found to stimulate recruitment of MDSCs and promote EMT in tumors conferring resistance to immune-mediated killing (reviewed by David et al. 2016). An IL-8-specific monoclonal antibody, known as HuMax-IL8, has been shown to reduce mesenchymal features in cancer cells leading to enhanced susceptibility to NK and T cell-mediated lysis and to decrease the frequency of granulocytic MDSCs in xenograft models (Collins et al. 2018). The HuMax-IL8 drug (BMS-986253) is also designed to inhibit IL-8 and thus far the only clinical trial utilizing this drug is to test it in combination with nivolumab in patients with advanced solid tumors, but no results have been published yet (NCT03400332).

5 Enhancing Anti-tumor Immunity (Therapy-Induced Inflammation)

Another form of targeting inflammation as an anti-cancer strategy is to modulate the immunosuppressive TME through therapies that stimulate the anti-tumor immune response. The immune checkpoint inhibitors, currently the monoclonal antibodies against PD-1/PD-L1 and CTLA4, represent a successful class and have been approved to treat numerous solid tumors and hematological malignancies. The use of immunotherapy in GC has just started and will be discussed below.

5.1 Checkpoint Inhibitors (Anti-PD-1/L1, Anti-CTLA4)

Currently, two immune checkpoint inhibitors (ICPi), pembrolizumab and nivolumab, have been approved in many countries for advanced/metastatic GC and both are anti-PD-1 monoclonal antibodies. The landmark phase II trial by Le and coworkers (2017) demonstrated dramatic tumor shrinkage and significantly prolonged progression-free survival with pembrolizumab for patients with heavily pre-treated MSI solid tumors, mostly Lynch syndrome-associated colorectal cancer. These results led to the FDA's first-ever site-agnostic approval. Given that tumor lymphocytic infiltration is one of the hallmarks of MSI tumors, it was conceived that ICPi would work mostly in "inflamed" tumors. This has proven to be true; larger magnitudes of benefit have been observed in inflammation-associated tumors with higher mutation burden, such as lung cancer, melanoma, and Merkel cell carcinoma. In GC, the overall benefit of immunotherapy has been modest. However, in specific subgroups where cancer-associated inflammation is present, i.e., MSI, higher response rates and prolonged disease control have been reported. While EBV-positive nasopharyngeal carcinoma patients are more likely to benefit from ICPi (Kao et al. 2015), clinical data in GC have not been published, but are eagerly awaited. Currently, it is unknown whether concurrent or prior *H. pylori* infection predicts response and survival in GC.

In the KEYNOTE-059 multi-cohort trial, pembrolizumab was administered in different settings of patients with advanced GC or GEJC (Wainberg et al. 2017). Cohort 1 enrolled 259 patients with at least two prior lines of therapy and 57% of the patients had PD-L1 positive tumors (defined as positive immunohistochemistry expression in at least 1% in tumor cells); in this cohort, pembrolizumab monotherapy offered an overall objective response rate of 12% (16% in PD-L1 positive and 6% in PD-L1 negative, respectively), with 3% of patients achieving complete response, regardless of PD-L1 status. As expected, among the 4% of patients with MSI tumors, the objective response rate was 57.1% (Fuchs et al. 2017). This uncontrolled phase II trial led to the approval of this drug in the third line setting of PD-L1-positive GC. Recently, a phase III trial of pembrolizumab versus paclitaxel in second line was negative for its primary endpoint, overall survival, for patients with combined positive score ≥1 tumors—the number of PD-L1 positive staining tumor, lymphocytes, or macrophages divided by the total number of viable tumor cells and multiplied by 100 (Fuchs et al. 2018b). Interestingly, the study showed that the higher the combined positive score, the higher the benefit of pembrolizumab when compared to paclitaxel, reinforcing that the more intense the tumor inflammation, the higher the likelihood of anti-tumor activity from treatment with an ICPi.

In the phase III trial named ATTRACTION-2, 493 Asian patients with advanced GC or GEJC previously treated with chemotherapy were randomized to receive nivolumab or placebo until disease progression (Kang et al. 2017). Similar to the overall clinically irrelevant, albeit statistically significant, results with pembrolizumab, this trial demonstrated a modest gain (measured in days) in the overall survival. In contrast to the pembrolizumab trial, PD-L1 expression was not predictive of a survival benefit with nivolumab. Based on these results, nivolumab was approved in Japan for GC.

Overall clinical trials in GC with anti-CTLA4 monoclonal antibodies have shown discouraging results, with limited efficacy signals (Bang et al. 2017; Ralph et al. 2010). Similarly, the results with the combination of anti-PD-L1 and anti-CTLA4 agents were disappointing. In two cohorts of the CheckMate 032 phase I/phase II trial where one-quarter to one-third of patients had PD-L1-positive tumors, nivolumab in combination with ipilimumab was delivered on two different dose schedules (Janjigian et al. 2017). These treatments led to an overall objective response in 8% and 24% of patients, respectively, but with median overall survival times of only 6.9 and 4.8 months. These results clearly demonstrate that although immunotherapy is one of the greatest advances in modern oncology, it does not work in molecularly unselected GC cancer patients. The molecularly distinct subtypes of GC (Cancer Genome Atlas Research 2014; Cristescu et al. 2015) should be explored for predictive biomarkers to ICPi responsiveness.

5.2 *Other Immunotherapies—Immunoconjugates*

A new class of immunotherapeutic agents, immunoconjugates, is being tested and demonstrated promising results in certain tumor types. An carcinoembryonic antigen (CEA)-anti-CD3 bispecific antibody (CEA-TCB) has shown impressive results in phase I trials in colorectal cancer (Parkhurst et al. 2011). The CEA-TCB RO6958688 was given as monotherapy or in combination with atezolizumab, an anti-PD-L1 monoclonal antibody, to patients with advanced solid tumors, the majority with colorectal cancer, with positive immunohistochemistry expression of CEA (Tabernero et al. 2017); tumor inflammation was evidenced in computerized tomographies (CT) scans within a few days of administration in the higher dose levels cohorts, and after 4 weeks, metabolic response by 18-FDG PET scan was observed in 28% of patients in the monotherapy group and in 60% of those in the combination arm. Interestingly, the most common adverse events were inflammation-related, such as pyrexia (56.3%), and dose-limiting toxicities were associated with inflammation in metastatic lesions. Although preliminary, the CEA-TCB RO6958688 could be tested in a great variety of CEA + tumors (NCT02324257, NCT02650713). Indeed, phase III trials are planned in advanced colorectal and GC.

6 Concluding Remarks

In this chapter, we have provided insights into inflammation-related factors that promote *H. pylori*-associated GC initiation and growth. Strategies for immune therapy could fall into three major classifications including targeting pro-cancer inflammation and reprogramming or depleting immune cells in the TAM to increase anti-cancer immune responses. Some of these strategies will likely be most successful when therapy is provided as an adjunct to current chemotherapy, whereas others may prove to be more effective as monotherapy. For example, anti-VEGF agents provide some benefit when combined with chemotherapy, but offer limited efficacy as monotherapy. On the other hand, immunoconjugates and immune checkpoint inhibitors seem to offer anti-tumor activity alone or combined among them, while the addition of chemotherapy does not appear to improve outcomes in GC. Moreover, because of the diversity of the microenvironments (gastritis to tumor), which have been reported and the array of potential targets, it is clear that immune subtyping and tumor characterization are crucial for the field to move to more successful immune therapies to treat persons at risk of GC and/or to treat existing cancer. And lastly, the drug development process, including pharma and academia, should concentrate efforts to identify predictive biomarkers in GC and to design clinical trials with enriched populations so we can offer GC patients treatments that make a significant difference in their lives.

Acknowledgements This work was supported by U. S. National Institutes of Health (NIH) grants R01AT004821, R01CA190612, P01CA116087, P01CA028842; Office of Medical Research, Veterans Affairs Merit Review grants IBX000915A(to H.M.S.A.) and I01BX001453(to K.T.W.); and Vanderbilt University Digestive Disease Research Center supported by NIH grant P30DK058404.

References

Abnet CC, Freedman ND, Kamangar F, Leitzmann MF, Hollenbeck AR, Schatzkin A (2009) Non-steroidal anti-inflammatory drugs and risk of gastric and oesophageal adenocarcinomas: results from a cohort study and a meta-analysis. Br J Cancer 100(3):551–557. https://doi.org/10.1038/sj.bjc.6604880

Adamsson J, Ottsjo LS, Lundin SB, Svennerholm AM, Raghavan S (2017) Gastric expression of IL-17A and IFNgamma in *Helicobacter pylori* infected individuals is related to symptoms. Cytokine 99:30–34. https://doi.org/10.1016/j.cyto.2017.06.013

Ajani JA, D'Amico TA, Almhanna K, Bentrem DJ, Chao J, Das P, Denlinger CS, Fanta P, Farjah F, Fuchs CS, Gerdes H, Gibson M, Glasgow RE, Hayman JA, Hochwald S, Hofstetter WL, Ilson DH, Jaroszewski D, Johung KL, Keswani RN, Kleinberg LR, Korn WM, Leong S, Linn C, Lockhart AC, Ly QP, Mulcahy MF, Orringer MB, Perry KA, Poultsides GA, Scott WJ, Strong VE, Washington MK, Weksler B, Willett CG, Wright CD, Zelman D, McMillian N, Sundar H (2016) Gastric Cancer, Version 3.2016, NCCN Clinical Practice Guidelines in Oncology. Journal of the National Comprehensive Cancer Network: JNCCN 14 (10):1286–1312

Algood HM, Allen SS, Washington MK, Peek RM Jr, Miller GG, Cover TL (2009) Regulation of gastric B cell recruitment is dependent on IL-17 receptor A signaling in a model of chronic bacterial infection. J Immunol 183(9):5837–5846. https://doi.org/10.4049/jimmunol.0901206

Algra AM, Rothwell PM (2012) Effects of regular aspirin on long-term cancer incidence and metastasis: a systematic comparison of evidence from observational studies versus randomised trials. Lancet Oncol 13(5):518–527. https://doi.org/10.1016/S1470-2045(12)70112-2

Arachchi PS, Fernando N, Weerasekera MM, Senevirathna B, Weerasekera DD, Gunasekara CP (2017) Proinflammatory Cytokine IL-17 Shows a Significant Association with *Helicobacter pylori* Infection and Disease Severity. Gastroenterol Res Pract 2017:6265150. https://doi.org/10.1155/2017/6265150

Ashizawa T, Okada R, Suzuki Y, Takagi M, Yamazaki T, Sumi T, Aoki T, Ohnuma S, Aoki T (2005) Clinical significance of interleukin-6 (IL-6) in the spread of gastric cancer: role of IL-6 as a prognostic factor. Gastric Cancer: Off. J. Int. Gastric Cancer Assoc. Jpn. Gastric Cancer Assoc. 8(2):124–131. https://doi.org/10.1007/s10120-005-0315-x

Asim M, Chaturvedi R, Hoge S, Lewis ND, Singh K, Barry DP, Algood HS, de Sablet T, Gobert AP, Wilson KT (2010) *Helicobacter pylori* induces ERK-dependent formation of a phospho-c-Fos c-Jun activator protein-1 complex that causes apoptosis in macrophages. J Biol Chem 285(26):20343–20357. https://doi.org/10.1074/jbc.M110.116988

Bagheri N, Azadegan-Dehkordi F, Shirzad M, Zamanzad B, Rahimian G, Taghikhani A, Rafieian-Kopaei M, Shirzad H (2015) Mucosal interleukin-21 mRNA expression level is high in patients with *Helicobacter pylori* and is associated with the severity of gastritis. Cent-Eur J Immunol 40(1):61–67. https://doi.org/10.5114/ceji.2015.50835

Bang YJ, Cho JY, Kim YH, Kim JW, Di Bartolomeo M, Ajani JA, Yamaguchi K, Balogh A, Sanchez T, Moehler M (2017) Efficacy of sequential ipilimumab monotherapy versus best supportive care for unresectable locally advanced/metastatic gastric or gastroesophageal junction cancer. Clin. Cancer Res.: Off. J. Am. Assoc. Cancer Res. 23(19):5671–5678. https://doi.org/10.1158/1078-0432.CCR-17-0025

Bang YJ, Van Cutsem E, Feyereislova A, Chung HC, Shen L, Sawaki A, Lordick F, Ohtsu A, Omuro Y, Satoh T, Aprile G, Kulikov E, Hill J, Lehle M, Ruschoff J, Kang YK, To GATI (2010) Trastuzumab in combination with chemotherapy versus chemotherapy alone for treatment of HER2-positive advanced gastric or gastro-oesophageal junction cancer (ToGA): a phase 3, open-label, randomised controlled trial. Lancet 376(9742):687–697. https://doi.org/10.1016/S0140-6736(10)61121-X

Bassiri H, Benavides A, Haber M, Gilmour SK, Norris MD, Hogarty MD (2015) Translational development of difluoromethylornithine (DFMO) for the treatment of neuroblastoma. Transl Pediatr 4(3):226–238. https://doi.org/10.3978/j.issn.2224-4336.2015.04.06

Bauditz J, Ortner M, Bierbaum M, Niedobitek G, Lochs H, Schreiber S (1999) Production of IL-12 in gastritis relates to infection with *Helicobacter pylori*. Clin Exp Immunol 117(2):316–323

Beatty GL, Chiorean EG, Fishman MP, Saboury B, Teitelbaum UR, Sun W, Huhn RD, Song W, Li D, Sharp LL, Torigian DA, O'Dwyer PJ, Vonderheide RH (2011) CD40 agonists alter tumor stroma and show efficacy against pancreatic carcinoma in mice and humans. Science 331(6024):1612–1616. https://doi.org/10.1126/science.1198443

Bhuiyan TR, Islam MM, Uddin T, Chowdhury MI, Janzon A, Adamsson J, Lundin SB, Qadri F, Lundgren A (2014) Th1 and Th17 responses to *Helicobacter pylori* in Bangladeshi infants, children and adults. PLoS ONE 9(4):e93943. https://doi.org/10.1371/journal.pone.0093943

Bibbins-Domingo K, Force USPST (2016) Aspirin use for the primary prevention of cardiovascular disease and colorectal cancer: U.S. Preventive services task force recommendation statement. Ann Intern Med 164(12):836–845. https://doi.org/10.7326/m16-0577

Brahmi M, Vinceneux A, Cassier PA (2016) Current systemic treatment options for tenosynovial giant cell tumor/pigmented villonodular synovitis: targeting the CSF1/CSF1R Axis. Curr Treat Options Oncol 17(2):10. https://doi.org/10.1007/s11864-015-0385-x

Brana I, Calles A, LoRusso PM, Yee LK, Puchalski TA, Seetharam S, Zhong B, de Boer CJ, Tabernero J, Calvo E (2015) Carlumab, an anti-C-C chemokine ligand 2 monoclonal antibody, in combination with four chemotherapy regimens for the treatment of patients with solid tumors: an open-label, multicenter phase 1b study. Target Oncol 10(1):111–123. https://doi.org/10.1007/s11523-014-0320-2

Brandt S, Kwok T, Hartig R, König W, Backert S (2005) NF- B activation and potentiation of proinflammatory responses by the *Helicobacter pylori* CagA protein. Proc Natl Acad Sci 102:9300–9305. https://doi.org/10.1073/pnas.0409873102

Bresalier RS, Sandler RS, Quan H, Bolognese JA, Oxenius B, Horgan K, Lines C, Riddell R, Morton D, Lanas A, Konstam MA, Baron JA, Adenomatous Polyp Prevention on Vioxx Trial I (2005) Cardiovascular events associated with rofecoxib in a colorectal adenoma chemoprevention trial. N Engl J Med. 352(11):1092–1102. https://doi.org/10.1056/nejmoa050493

Brew R, Erikson JS, West DC, Kinsella AR, Slavin J, Christmas SE (2000) Interleukin-8 as an autocrine growth factor for human colon carcinoma cells in vitro. Cytokine 12(1):78–85. https://doi.org/10.1006/cyto.1999.0518

Camargo MC, Mera R, Correa P, Peek RM Jr, Fontham ET, Goodman KJ, Piazuelo MB, Sicinschi L, Zabaleta J, Schneider BG (2006) Interleukin-1beta and interleukin-1 receptor antagonist gene polymorphisms and gastric cancer: a meta-analysis. Cancer Epidemiol Biomark Prev: Publ Am Assoc Cancer Res, Cosponsored Am Soc Prev Oncol 15(9):1674–1687. https://doi.org/10.1158/1055-9965.EPI-06-0189

Cancer Genome Atlas Research N (2014) Comprehensive molecular characterization of gastric adenocarcinoma. Nature 513(7517):202–209. https://doi.org/10.1038/nature13480

Cancer Stat Facts: Stomach Cancer, SEER Cancer Statistics Review, 1975–2015. (2018) National Cancer Institute. https://seer.cancer.gov/csr/1975_2015/. 2018

Cannarile MA, Weisser M, Jacob W, Jegg AM, Ries CH, Ruttinger D (2017) Colony-stimulating factor 1 receptor (CSF1R) inhibitors in cancer therapy. J Immunother Cancer 5(1):53. https://doi.org/10.1186/s40425-017-0257-y

Carbo A, Olivares-Villagomez D, Hontecillas R, Bassaganya-Riera J, Chaturvedi R, Piazuelo MB, Delgado A, Washington MK, Wilson KT, Algood HM (2014) Systems modeling of the role of interleukin-21 in the maintenance of effector CD4 + T cell responses during chronic *Helicobacter pylori* infection. mBio **5**(4):e01243–01214. https://doi.org/10.1128/mbio.01243-14

Caruso R, Fina D, Peluso I, Fantini MC, Tosti C, Del Vecchio Blanco G, Paoluzi OA, Caprioli F, Andrei F, Stolfi C, Romano M, Ricci V, MacDonald TT, Pallone F, Monteleone G (2007) IL-21 is highly produced in *Helicobacter pylori*-infected gastric mucosa and promotes gelatinases synthesis. J Immunol 178(9):5957–5965

Cassier PA, Italiano A, Gomez-Roca CA, Le Tourneau C, Toulmonde M, Cannarile MA, Ries C, Brillouet A, Muller C, Jegg AM, Broske AM, Dembowski M, Bray-French K, Freilinger C, Meneses-Lorente G, Baehner M, Harding R, Ratnayake J, Abiraj K, Gass N, Noh K, Christen

RD, Ukarma L, Bompas E, Delord JP, Blay JY, Ruttinger D (2015) CSF1R inhibition with emactuzumab in locally advanced diffuse-type tenosynovial giant cell tumours of the soft tissue: a dose-escalation and dose-expansion phase 1 study. Lancet Oncol 16(8):949–956. https://doi.org/10.1016/S1470-2045(15)00132-1

Chaturvedi R, Asim M, Piazuelo MB, Yan F, Barry DP, Sierra JC, Delgado AG, Hill S, Casero RA, Jr., Bravo LE, Dominguez RL, Correa P, Polk DB, Washington MK, Rose KL, Schey KL, Morgan DR, Peek RM, Jr., Wilson KT (2014) Activation of EGFR and ERBB2 by *Helicobacter pylori* results in survival of gastric epithelial cells with DNA damage. Gastroenterology **146**(7):1739–1751 e1714. https://doi.org/10.1053/j.gastro.2014.02.005

Chaturvedi R, Asim M, Romero-Gallo J, Barry DP, Hoge S, de Sablet T, Delgado AG, Wroblewski LE, Piazuelo MB, Yan F, Israel DA, Casero RA, Jr., Correa P, Gobert AP, Polk DB, Peek RM, Jr., Wilson KT (2011) Spermine oxidase mediates the gastric cancer risk associated with *Helicobacter pylori* CagA. Gastroenterology **141**(5):1696–1708 e1691–1692. https://doi.org/10.1053/j.gastro.2011.07.045

Chaturvedi R, Cheng Y, Asim M, Bussiere FI, Xu H, Gobert AP, Hacker A, Casero RA Jr, Wilson KT (2004) Induction of polyamine oxidase 1 by *Helicobacter pylori* causes macrophage apoptosis by hydrogen peroxide release and mitochondrial membrane depolarization. J Biol Chem 279(38):40161–40173. https://doi.org/10.1074/jbc.M401370200

Chaturvedi R, de Sablet T, Asim M, Piazuelo MB, Barry DP, Verriere TG, Sierra JC, Hardbower DM, Delgado AG, Schneider BG, Israel DA, Romero-Gallo J, Nagy TA, Morgan DR, Murray-Stewart T, Bravo LE, Peek RM Jr, Fox JG, Woster PM, Casero RA Jr, Correa P, Wilson KT (2015) Increased *Helicobacter pylori*-associated gastric cancer risk in the Andean region of Colombia is mediated by spermine oxidase. Oncogene 34(26):3429–3440. https://doi.org/10.1038/onc.2014.273

Chen B, Cao S, Zhang Y, Wang X, Liu J, Hui X, Wan Y, Du W, Wang L, Wu K, Fan D (2009) A novel peptide (GX1) homing to gastric cancer vasculature inhibits angiogenesis and cooperates with TNF alpha in anti-tumor therapy. BMC Cell Biol 10:63. https://doi.org/10.1186/1471-2121-10-63

Chubak J, Whitlock EP, Williams SB, Kamineni A, Burda BU, Buist DS, Anderson ML (2016) Aspirin for the Prevention of Cancer Incidence and Mortality: Systematic Evidence Reviews for the U.S. Preventive Services Task Force. Ann Intern Med **164**(12):814–825. https://doi.org/10.7326/m15-2117

Chung AS, Lee J, Ferrara N (2010) Targeting the tumour vasculature: insights from physiological angiogenesis. Nat Rev Cancer 10(7):505–514. https://doi.org/10.1038/nrc2868

Collins JM, Heery CR, Donahue RN, Palena C, Mada RA, Strauss J, Gatti-Mays ME, Schlom J, Gulley JL, Bilusic M (2018) Phase I trial of BMS-986253, an anti-IL-8 monoclonal antibody, in patients with metastatic or unresectable solid tumors. Paper presented at the 2018 ACOS annual meeting, Chicago, Illinois, June 4, 2018

Colotta F, Allavena P, Sica A, Garlanda C, Mantovani A (2009) Cancer-related inflammation, the seventh hallmark of cancer: links to genetic instability. Carcinogenesis 30(7):1073–1081. https://doi.org/10.1093/carcin/bgp127

Correa P, Haenszel W, Cuello C, Tannenbaum S, Archer M (1975) A model for gastric cancer epidemiology. Lancet 2(7924):58–60

Correa P, Piazuelo MB, Wilson KT (2010) Pathology of gastric intestinal metaplasia: clinical implications. Am J Gastroenterol 105(3):493–498. https://doi.org/10.1038/ajg.2009.728

Crabtree JE, Covacci A, Farmery SM, Xiang Z, Tompkins DS, Perry S, Lindley IJ, Rappuoli R (1995) *Helicobacter pylori* induced interleukin-8 expression in gastric epithelial cells is associated with CagA positive phenotype. J Clin Pathol 48(1):41–45

Crabtree JE, Farmery SM, Lindley IJ, Figura N, Peichl P, Tompkins DS (1994a) CagA/cytotoxic strains of *Helicobacter pylori* and interleukin-8 in gastric epithelial cell lines. J Clin Pathol 47(10):945–950

Crabtree JE, Wyatt JI, Trejdosiewicz LK, Peichl P, Nichols PH, Ramsay N, Primrose JN, Lindley IJ (1994b) Interleukin-8 expression in *Helicobacter pylori* infected, normal, and neoplastic gastroduodenal mucosa. J Clin Pathol 47(1):61–66

Cristescu R, Lee J, Nebozhyn M, Kim KM, Ting JC, Wong SS, Liu J, Yue YG, Wang J, Yu K, Ye XS, Do IG, Liu S, Gong L, Fu J, Jin JG, Choi MG, Sohn TS, Lee JH, Bae JM, Kim ST, Park SH, Sohn I, Jung SH, Tan P, Chen R, Hardwick J, Kang WK, Ayers M, Hongyue D, Reinhard C, Loboda A, Kim S, Aggarwal A (2015) Molecular analysis of gastric cancer identifies subtypes associated with distinct clinical outcomes. Nat Med 21(5):449–456. https://doi.org/10.1038/nm.3850

David JM, Dominguez C, Hamilton DH, Palena C (2016) The IL-8/IL-8R Axis: a double agent in tumor immune resistance. Vaccines 4(3). https://doi.org/10.3390/vaccines4030022

Davis RJ, Moore EC, Clavijo PE, Friedman J, Cash H, Chen Z, Silvin C, Van Waes C, Allen C (2017) Anti-PD-L1 Efficacy Can Be Enhanced by Inhibition of Myeloid-Derived Suppressor Cells with a Selective Inhibitor of PI3Kdelta/gamma. Can Res 77(10):2607–2619. https://doi.org/10.1158/0008-5472.CAN-16-2534

De Palma M, Mazzieri R, Politi LS, Pucci F, Zonari E, Sitia G, Mazzoleni S, Moi D, Venneri MA, Indraccolo S, Falini A, Guidotti LG, Galli R, Naldini L (2008) Tumor-targeted interferon-alpha delivery by Tie2-expressing monocytes inhibits tumor growth and metastasis. Cancer Cell 14(4):299–311. https://doi.org/10.1016/j.ccr.2008.09.004

DeNardo DG, Brennan DJ, Rexhepaj E, Ruffell B, Shiao SL, Madden SF, Gallagher WM, Wadhwani N, Keil SD, Junaid SA, Rugo HS, Hwang ES, Jirstrom K, West BL, Coussens LM (2011) Leukocyte complexity predicts breast cancer survival and functionally regulates response to chemotherapy. Cancer Discov 1(1):54–67. https://doi.org/10.1158/2159-8274.CD-10-0028

Derks S, Liao X, Chiaravalli AM, Xu X, Camargo MC, Solcia E, Sessa F, Fleitas T, Freeman GJ, Rodig SJ, Rabkin CS, Bass AJ (2016) Abundant PD-L1 expression in Epstein-Barr Virus-infected gastric cancers. Oncotarget 7(22):32925–32932. https://doi.org/10.18632/oncotarget.9076

Dixon BR, Radin JN, Piazuelo MB, Contreras DC, Algood HM (2016) IL-17a and IL-22 induce expression of antimicrobials in gastrointestinal epithelial cells and may contribute to epithelial cell defense against *Helicobacter pylori*. PLoS ONE 11(2):e0148514. https://doi.org/10.1371/journal.pone.0148514

Duluc D, Corvaisier M, Blanchard S, Catala L, Descamps P, Gamelin E, Ponsoda S, Delneste Y, Hebbar M, Jeannin P (2009) Interferon-gamma reverses the immunosuppressive and protumoral properties and prevents the generation of human tumor-associated macrophages. Int J Cancer 125(2):367–373. https://doi.org/10.1002/ijc.24401

Eftang LL, Esbensen Y, Tannaes TM, Bukholm IR, Bukholm G (2012) Interleukin-8 is the single most up-regulated gene in whole genome profiling of *H. pylori* exposed gastric epithelial cells. BMC Microbiol 12:9. https://doi.org/10.1186/1471-2180-12-9

El-Omar EM, Carrington M, Chow WH, McColl KE, Bream JH, Young HA, Herrera J, Lissowska J, Yuan CC, Rothman N, Lanyon G, Martin M, Fraumeni JF Jr, Rabkin CS (2000) Interleukin-1 polymorphisms associated with increased risk of gastric cancer. Nature 404(6776):398–402. https://doi.org/10.1038/35006081

Epplein M, Nomura AM, Hankin JH, Blaser MJ, Perez-Perez G, Stemmermann GN, Wilkens LR, Kolonel LN (2008) Association of *Helicobacter pylori* infection and diet on the risk of gastric cancer: a case-control study in Hawaii. Cancer Sauses Control: CCC 19(8):869–877. https://doi.org/10.1007/s10552-008-9149-2

Epplein M, Nomura AM, Wilkens LR, Henderson BE, Kolonel LN (2009) Nonsteroidal antiinflammatory drugs and risk of gastric adenocarcinoma: the multiethnic cohort study. Am J Epidemiol 170(4):507–514. https://doi.org/10.1093/aje/kwp162

Evageliou NF, Haber M, Vu A, Laetsch TW, Murray J, Gamble LD, Cheng NC, Liu K, Reese M, Corrigan KA, Ziegler DS, Webber H, Hayes CS, Pawel B, Marshall GM, Zhao H, Gilmour SK, Norris MD, Hogarty MD (2016) Polyamine antagonist therapies inhibit neuroblastoma initiation and progression. Clin Cancer Res: Off J Am Assoc Cancer Res 22(17):4391–4404. https://doi.org/10.1158/1078-0432.CCR-15-2539

Ferlay J, Soerjomataram I, Dikshit R, Eser S, Mathers C, Rebelo M, Parkin DM, Forman D, Bray F (2015) Cancer incidence and mortality worldwide: sources, methods and major patterns in GLOBOCAN 2012. Int J Cancer 136(5):E359–386. https://doi.org/10.1002/ijc.29210

Fernando RI, Castillo MD, Litzinger M, Hamilton DH, Palena C (2011) IL-8 signaling plays a critical role in the epithelial-mesenchymal transition of human carcinoma cells. Can Res 71(15):5296–5306. https://doi.org/10.1158/0008-5472.CAN-11-0156

Ferrara N (2002) VEGF and the quest for tumour angiogenesis factors. Nat Rev Cancer 2(10):795–803. https://doi.org/10.1038/nrc909

Ferrara N (2009) VEGF-A: a critical regulator of blood vessel growth. Eur Cytokine Netw 20(4):158–163. https://doi.org/10.1684/ecn.2009.0170

Fuchs C, Doi T, Jang RW-J, Muro K, Satoh T, Machado M, Sun W, Jalal SI, Shah MA, Metges J-P, Garrido M, Golan T, Mandala M, Wainberg ZA, Catenacci DVT, Bang Y, Wang J, Koshiji M, Dalal R, Yoon HH (2017) KEYNOTE-059 cohort 1: efficacy and safety of pembrolizumab (pembro) monotherapy in patients with previously treated advanced gastric cancer. Paper presented at the American Society of Clinical Oncology, Chicago, Illinois, 2017

Fuchs CS, Doi T, Jang RW, Muro K, Satoh T, Machado M, Sun W, Jalal SI, Shah MA, Metges JP, Garrido M, Golan T, Mandala M, Wainberg ZA, Catenacci DV, Ohtsu A, Shitara K, Geva R, Bleeker J, Ko AH, Ku G, Philip P, Enzinger PC, Bang YJ, Levitan D, Wang J, Rosales M, Dalal RP, Yoon HH (2018) Safety and efficacy of pembrolizumab monotherapy in patients with previously treated advanced gastric and gastroesophageal junction cancer: phase 2 clinical KEYNOTE-059 trial. JAMA oncology 4(5):e180013. https://doi.org/10.1001/jamaoncol.2018.0013

Fuchs CS, Ozguoglu M, Bang Y, Bartolomeo MD, Mandala M, Ryu M-H, Fornaro L, Olesinski T, Caglevic C, Chung HC, Muro K, Goekkurt E, WMansoor W, McDermott RS, Shacham-Shmueli E, Chen X, Kang SP, Mayo CA, Ohtsu A, Shitara K Pembrolizumab (pembro) vs paclitaxel (PTX) for previously treated advanced gastric or gastroesophageal junction (G/GEJ) cancer: Phase 3 KEYNOTE-061 trial. In: ASCO Annual Meeting, Chicago, Illinois, June 4, 2018 2018b. J Clin Oncol

Fuchs CS, Tomasek J, Yong CJ, Dumitru F, Passalacqua R, Goswami C, Safran H, Dos Santos LV, Aprile G, Ferry DR, Melichar B, Tehfe M, Topuzov E, Zalcberg JR, Chau I, Campbell W, Sivanandan C, Pikiel J, Koshiji M, Hsu Y, Liepa AM, Gao L, Schwartz JD, Tabernero J, Investigators RT (2014) Ramucirumab monotherapy for previously treated advanced gastric or gastro-oesophageal junction adenocarcinoma (REGARD): an international, randomised, multicentre, placebo-controlled, phase 3 trial. Lancet 383(9911):31–39. https://doi.org/10.1016/S0140-6736(13)61719-5

Futagami S, Tatsuguchi A, Hiratsuka T, Shindo T, Horie A, Hamamoto T, Ueki N, Kusunoki M, Miyake K, Gudis K, Tsukui T, Sakamoto C (2008) Monocyte chemoattractant protein 1 and CD40 ligation have a synergistic effect on vascular endothelial growth factor production through cyclooxygenase 2 upregulation in gastric cancer. J Gastroenterol 43(3):216–224. https://doi.org/10.1007/s00535-007-2151-8

Galsky MD, Vogelzang NJ, Conkling P, Raddad E, Polzer J, Roberson S, Stille JR, Saleh M, Thornton D (2014) A phase I trial of LY2510924, a CXCR4 peptide antagonist, in patients with advanced cancer. Clin Cancer Res: Off J Am Assoc Cancer Res 20(13):3581–3588. https://doi.org/10.1158/1078-0432.CCR-13-2686

Gelhorn HL, Tong S, McQuarrie K, Vernon C, Hanlon J, Maclaine G, Lenderking W, Ye X, Speck RM, Lackman RD, Bukata SV, Healey JH, Keedy VL, Anthony SP, Wagner AJ, Von Hoff DD, Singh AS, Becerra CR, Hsu HH, Lin PS, Tap WD (2016) Patient-reported Symptoms of Tenosynovial Giant Cell Tumors. Clin Ther 38(4):778–793. https://doi.org/10.1016/j.clinthera.2016.03.008

Genard G, Lucas S, Michiels C (2017) Reprogramming of tumor-associated macrophages with anticancer therapies: radiotherapy versus chemo- and immunotherapies. Front Immunol 8:828. https://doi.org/10.3389/fimmu.2017.00828

Gobert AP, Wilson KT (2017) Human and *Helicobacter pylori* Interactions Determine the Outcome of Gastric Diseases. Curr Top Microbiol Immunol 400:27–52. https://doi.org/10.1007/978-3-319-50520-6_2

Goel HL, Mercurio AM (2013) VEGF targets the tumour cell. Nat Rev Cancer 13(12):871–882. https://doi.org/10.1038/nrc3627

Guiet R, Van Goethem E, Cougoule C, Balor S, Valette A, Al Saati T, Lowell CA, Le Cabec V, Maridonneau-Parini I (2011) The process of macrophage migration promotes matrix metalloproteinase-independent invasion by tumor cells. J Immunol 187(7):3806–3814. https://doi.org/10.4049/jimmunol.1101245

Gullo I, Carneiro F, Oliveira C, Almeida GM (2018) Heterogeneity in gastric cancer: from pure morphology to molecular classifications. Pathobiol: J Immunopathol, Mol Cell Biol 85(1–2):50–63. https://doi.org/10.1159/000473881

Hainsworth JD, Reeves JA, Mace JR, Crane EJ, Hamid O, Stille JR, Flynt A, Roberson S, Polzer J, Arrowsmith ER (2016) A Randomized, open-label phase 2 study of the CXCR4 Inhibitor LY2510924 in combination with sunitinib versus sunitinib alone in patients with metastatic renal cell carcinoma (RCC). Target Oncol 11(5):643–653. https://doi.org/10.1007/s11523-016-0434-9

Hamilton JA, Cook AD, Tak PP (2016) Anti-colony-stimulating factor therapies for inflammatory and autoimmune diseases. Nat Rev Drug Discovery 16(1):53–70. https://doi.org/10.1038/nrd.2016.231

Hardbower DM, Coburn LA, Asim M, Singh K, Sierra JC, Barry DP, Gobert AP, Piazuelo MB, Washington MK, Wilson KT (2017) EGFR-mediated macrophage activation promotes colitis-associated tumorigenesis. Oncogene 36(27):3807–3819. https://doi.org/10.1038/onc.2017.23

Hardbower DM, Singh K, Asim M, Verriere TG, Olivares-Villagomez D, Barry DP, Allaman MM, Washington MK, Peek RM Jr, Piazuelo MB, Wilson KT (2016) EGFR regulates macrophage activation and function in bacterial infection. J Clin Investig 126(9):3296–3312. https://doi.org/10.1172/JCI83585

Holian O, Wahid S, Atten MJ, Attar BM (2002) Inhibition of gastric cancer cell proliferation by resveratrol: role of nitric oxide. Am J Physiol Gastrointest Liver Physiol 282(5):G809–816. https://doi.org/10.1152/ajpgi.00193.2001

Howells LM, Berry DP, Elliott PJ, Jacobson EW, Hoffmann E, Hegarty B, Brown K, Steward WP, Gescher AJ (2011) Phase I randomized, double-blind pilot study of micronized resveratrol (SRT501) in patients with hepatic metastases–safety, pharmacokinetics, and pharmacodynamics. Cancer Prev Res 4(9):1419–1425. https://doi.org/10.1158/1940-6207.CAPR-11-0148

Huang SP, Wu MS, Shun CT, Wang HP, Lin MT, Kuo ML, Lin JT (2004) Interleukin-6 increases vascular endothelial growth factor and angiogenesis in gastric carcinoma. J Biomed Sci 11(4):517–527. https://doi.org/10.1159/000077902

Huang XZ, Chen Y, Wu J, Zhang X, Wu CC, Zhang CY, Sun SS, Chen WJ (2017) Aspirin and non-steroidal anti-inflammatory drugs use reduce gastric cancer risk: a dose-response meta-analysis. Oncotarget 8(3):4781–4795. https://doi.org/10.18632/oncotarget.13591

Hueber W, Patel DD, Dryja T, Wright AM, Koroleva I, Bruin G, Antoni C, Draelos Z, Gold MH, Psoriasis Study G, Durez P, Tak PP, Gomez-Reino JJ, Rheumatoid Arthritis Study G, Foster CS, Kim RY, Samson CM, Falk NS, Chu DS, Callanan D, Nguyen QD, Uveitis Study G, Rose K, Haider A, Di Padova F (2010) Effects of AIN457, a fully human antibody to interleukin-17A, on psoriasis, rheumatoid arthritis, and uveitis. Sci Transl Med 2(52):52ra72. https://doi.org/10.1126/scitranslmed.3001107

Inoue K, Slaton JW, Eve BY, Kim SJ, Perrotte P, Balbay MD, Yano S, Bar-Eli M, Radinsky R, Pettaway CA, Dinney CP (2000) Interleukin 8 expression regulates tumorigenicity and metastases in androgen-independent prostate cancer. Clin Cancer Res: Off J Am Assoc Cancer Res 6(5):2104–2119

Janjigian YY, Ott PA, Calvo E, Kim JW, Ascierto PA, Sharma P (2017) Nivolumab ± ipilimumab in pts with advanced (adv)/metastatic chemotherapy-refractory (CTx-R) gastric (G), esophageal (E), or gastroesophageal junction (GEJ) cancer: CheckMate 032 study. J Clin Oncol 35(15):4014–4014. https://doi.org/10.1200/JCO.2017.35.15_suppl.4014

Jiang W, Liu K, Guo Q, Cheng J, Shen L, Cao Y, Wu J, Shi J, Cao H, Liu B, Tao K, Wang G, Cai K (2017) Tumor-infiltrating immune cells and prognosis in gastric cancer: a systematic review and meta-analysis. Oncotarget 8(37):62312–62329. https://doi.org/10.18632/oncotarget.17602

Jones SA, Scheller J, Rose-John S (2011) Therapeutic strategies for the clinical blockade of IL-6/gp130 signaling. J Clin Investig 121(9):3375–3383. https://doi.org/10.1172/JCI57158

Kang YK, Boku N, Satoh T, Ryu MH, Chao Y, Kato K, Chung HC, Chen JS, Muro K, Kang WK, Yeh KH, Yoshikawa T, Oh SC, Bai LY, Tamura T, Lee KW, Hamamoto Y, Kim JG, Chin K, Oh DY, Minashi K, Cho JY, Tsuda M, Chen LT (2017) Nivolumab in patients with advanced gastric or gastro-oesophageal junction cancer refractory to, or intolerant of, at least two previous chemotherapy regimens (ONO-4538-12, ATTRACTION-2): a randomised, double-blind, placebo-controlled, phase 3 trial. Lancet 390(10111):2461–2471. https://doi.org/10.1016/S0140-6736(17)31827-5

Kao HF, Hsu C, Huang HC, Wang CW, Cheng JD, Hong RL (2015) Correlation between plasma Epstein-Barr virus DNA and clinical response to pembrolizumab in patients with advanced or metastatic nasopharyngeal carcinoma. Paper presented at the European Cancer Congress, Vienna, Austria

Kim DJ, Park JH, Franchi L, Backert S, Nunez G (2013) The Cag pathogenicity island and interaction between TLR2/NOD2 and NLRP3 regulate IL-1beta production in *Helicobacter pylori* infected dendritic cells. Eur J Immunol 43(10):2650–2658. https://doi.org/10.1002/eji.201243281

Kim JY, Bae BN, Kang G, Kim HJ, Park K (2017) Cytokine expression associated with *Helicobacter pylori* and Epstein-Barr virus infection in gastric carcinogenesis. APMIS: Acta Pathologica, Microbiologica, et Immunologica Scandinavica 125(9):808–815. https://doi.org/10.1111/apm.12725

Kim MH, Chang J, Kim WJ, Banerjee S, Park SM (2018) Cumulative dose threshold for the chemopreventive effect of aspirin against gastric cancer. Am J Gastroenterol 113(6):845–854. https://doi.org/10.1038/s41395-018-0097-5

Kim SJ, Uehara H, Karashima T, McCarty M, Shih N, Fidler IJ (2001) Expression of interleukin-8 correlates with angiogenesis, tumorigenicity, and metastasis of human prostate cancer cells implanted orthotopically in nude mice. Neoplasia 3(1):33–42. https://doi.org/10.1038/sj/neo/7900124

Kitamura T, Qian BZ, Pollard JW (2015) Immune cell promotion of metastasis. Nat Rev Immunol 15(2):73–86. https://doi.org/10.1038/nri3789

Kong P, Wu R, Liu X, Liu J, Chen S, Ye M, Yang C, Song Z, He W, Yin C, Yang Q, Jiang C, Liao F, Peng R, Zhou Z, Xu D, Xia L (2016) The effects of anti-inflammatory drug treatment in gastric cancer prevention: an update of a meta-analysis. J Cancer 7(15):2247–2257. https://doi.org/10.7150/jca.16524

Kumar S, Kumari N, Mittal RD, Mohindra S, Ghoshal UC (2015) Association between pro-(IL-8) and anti-inflammatory (IL-10) cytokine variants and their serum levels and *H. pylori*-related gastric carcinogenesis in northern India. Meta Gene 6:9–16. https://doi.org/10.1016/j.mgene.2015.07.008

Kuroda T, Kitadai Y, Tanaka S, Yang X, Mukaida N, Yoshihara M, Chayama K (2005) Monocyte chemoattractant protein-1 transfection induces angiogenesis and tumorigenesis of gastric carcinoma in nude mice via macrophage recruitment. Clin Cancer Res: Off J Am Assoc Cancer Res 11(21):7629–7636. https://doi.org/10.1158/1078-0432.CCR-05-0798

Lanas A, Wu P, Medin J, Mills EJ (2011) Low doses of acetylsalicylic acid increase risk of gastrointestinal bleeding in a meta-analysis. Clin Gastroenterol Hepatol: Off Clin Pract J Am Gastroenterol Assoc 9(9):762–768 e766. https://doi.org/10.1016/j.cgh.2011.05.020

Lanaya H, Natarajan A, Komposch K, Li L, Amberg N, Chen L, Wculek SK, Hammer M, Zenz R, Peck-Radosavljevic M, Sieghart W, Trauner M, Wang H, Sibilia M (2014) EGFR has a tumour-promoting role in liver macrophages during hepatocellular carcinoma formation. Nat Cell Biol 16(10):972–977. https://doi.org/10.1038/ncb3031

Lauren P (1965) The two histological main types of gastric carcinoma: diffuse and so-called intestinal-type carcinoma. an attempt at a histo-clinical classification. Acta Pathologica et Microbiologica Scandinavica 64:31–49

Lauwers G, Carneiro F, Graham D (2010) Gastric carcinoma. WHO classification of tumours of the digestive system, 4th edn. International Agency for Research on Cancer, Lyon, France

Le DT, Durham JN, Smith KN, Wang H, Bartlett BR, Aulakh LK, Lu S, Kemberling H, Wilt C, Luber BS, Wong F, Azad NS, Rucki AA, Laheru D, Donehower R, Zaheer A, Fisher GA, Crocenzi TS, Lee JJ, Greten TF, Duffy AG, Ciombor KK, Eyring AD, Lam BH, Joe A, Kang SP, Holdhoff M, Danilova L, Cope L, Meyer C, Zhou S, Goldberg RM, Armstrong DK, Bever KM, Fader AN, Taube J, Housseau F, Spetzler D, Xiao N, Pardoll DM, Papadopoulos N, Kinzler KW, Eshleman JR, Vogelstein B, Anders RA, Diaz LA Jr (2017) Mismatch repair deficiency predicts response of solid tumors to PD-1 blockade. Science 357(6349):409–413. https://doi.org/10.1126/science.aan6733

Lee J, Bass AJ, Ajani JA (2016) Gastric adenocarcinoma: an update on genomics, immune system modulations, and targeted therapy. Am Soc Clin Oncol Educ Book Am Soc Clin Oncol Meet 35:104–111. https://doi.org/10.14694/EDBK_159091

Lee KH, Bae SH, Lee JL, Hyun MS, Kim SH, Song SK, Kim HS (2004) Relationship between urokinase-type plasminogen receptor, interleukin-8 gene expression and clinicopathological features in gastric cancer. Oncology 66(3):210–217. https://doi.org/10.1159/000077997

Lee SA, Choi SR, Jang JS, Lee JH, Roh MH, Kim SO, Kim MC, Kim SJ, Jeong JS (2010) Expression of VEGF, EGFR, and IL-6 in gastric adenomas and adenocarcinomas by endoscopic submucosal dissection. Dig Dis Sci 55(7):1955–1963. https://doi.org/10.1007/s10620-009-0967-1

Lewis AM, Varghese S, Xu H, Alexander HR (2006) Interleukin-1 and cancer progression: the emerging role of interleukin-1 receptor antagonist as a novel therapeutic agent in cancer treatment. J Transl Med 4:48. https://doi.org/10.1186/1479-5876-4-48

Li A, Varney ML, Valasek J, Godfrey M, Dave BJ, Singh RK (2005) Autocrine role of interleukin-8 in induction of endothelial cell proliferation, survival, migration and MMP-2 production and angiogenesis. Angiogenesis 8(1):63–71. https://doi.org/10.1007/s10456-005-5208-4

Li J, Qin S, Xu J, Xiong J, Wu C, Bai Y, Liu W, Tong J, Liu Y, Xu R, Wang Z, Wang Q, Ouyang X, Yang Y, Ba Y, Liang J, Lin X, Luo D, Zheng R, Wang X, Sun G, Wang L, Zheng L, Guo H, Wu J, Xu N, Yang J, Zhang H, Cheng Y, Wang N, Chen L, Fan Z, Sun P, Yu H (2016) Randomized, double-blind, placebo-controlled phase iii trial of apatinib in patients with chemotherapy-refractory advanced or metastatic adenocarcinoma of the stomach or gastroesophageal junction. J Clin Oncol: Off J Am Soc Clin Oncol 34(13):1448–1454. https://doi.org/10.1200/JCO.2015.63.5995

Li Q, Li Q, Chen J, Liu Y, Zhao X, Tan B, Ai J, Zhang Z, Song J, Shan B (2013) Prevalence of Th17 and Treg cells in gastric cancer patients and its correlation with clinical parameters. Oncol Rep 30(3):1215–1222. https://doi.org/10.3892/or.2013.2570

Lieto E, Ferraraccio F, Orditura M, Castellano P, Mura AL, Pinto M, Zamboli A, De Vita F, Galizia G (2008) Expression of vascular endothelial growth factor (VEGF) and epidermal growth factor receptor (EGFR) is an independent prognostic indicator of worse outcome in gastric cancer patients. Ann Surg Oncol 15(1):69–79. https://doi.org/10.1245/s10434-007-9596-0

Lin MT, Juan CY, Chang KJ, Chen WJ, Kuo ML (2001) IL-6 inhibits apoptosis and retains oxidative DNA lesions in human gastric cancer AGS cells through up-regulation of anti-apoptotic gene mcl-1. Carcinogenesis 22(12):1947–1953

Lin Y, Zhai E, Liao B, Xu L, Zhang X, Peng S, He Y, Cai S, Zeng Z, Chen M (2017) Autocrine VEGF signaling promotes cell proliferation through a PLC-dependent pathway and modulates Apatinib treatment efficacy in gastric cancer. Oncotarget 8(7):11990–12002. https://doi.org/10.18632/oncotarget.14467

Lindholm C, Quiding-Jarbrink M, Lonroth H, Hamlet A, Svennerholm AM (1998) Local cytokine response in *Helicobacter pylori*-infected subjects. Infect Immun 66(12):5964–5971

Liu T, Peng L, Yu P, Zhao Y, Shi Y, Mao X, Chen W, Cheng P, Wang T, Chen N, Zhang J, Liu X, Li N, Guo G, Tong W, Zhuang Y, Zou Q (2012a) Increased circulating Th22 and Th17 cells are associated with tumor progression and patient survival in human gastric cancer. J Clin Immunol 32(6):1332–1339. https://doi.org/10.1007/s10875-012-9718-8

Liu X, Peng J, Sun W, Yang S, Deng G, Li F, Cheng JW, Gordon JR (2012b) G31P, an antagonist against CXC chemokine receptors 1 and 2, inhibits growth of human prostate cancer cells in nude mice. Tohoku J Exp Med 228(2):147–156

Luzza F, Parrello T, Monteleone G, Sebkova L, Romano M, Zarrilli R, Imeneo M, Pallone F (2000) Up-regulation of IL-17 is associated with bioactive IL-8 expression in *Helicobacter pylori*-infected human gastric mucosa. J Immunol 165(9):5332–5337

Ma X, Wu D, Zhou S, Wan F, Liu H, Xu X, Xu X, Zhao Y, Tang M (2016) The pancreatic cancer secreted REG4 promotes macrophage polarization to M2 through EGFR/AKT/CREB pathway. Oncol Rep 35(1):189–196. https://doi.org/10.3892/or.2015.4357

Macri A, Versaci A, Loddo S, Scuderi G, Travagliante M, Trimarchi G, Teti D, Famulari C (2006) Serum levels of interleukin 1beta, interleukin 8 and tumour necrosis factor alpha as markers of gastric cancer. Biomark: Biochem Indic Expo, Response Susceptibility Chem 11(2):184–193. https://doi.org/10.1080/13547500600565677

Mantovani A, Allavena P, Sica A, Balkwill F (2008) Cancer-related inflammation. Nature 454(7203):436–444. https://doi.org/10.1038/nature07205

Mantovani A, Sozzani S, Locati M, Allavena P, Sica A (2002) Macrophage polarization: tumor-associated macrophages as a paradigm for polarized M2 mononuclear phagocytes. Trends Immunol 23(11):549–555

Mantovani A, Vecchi A, Allavena P (2014) Pharmacological modulation of monocytes and macrophages. Curr Opin Pharmacol 17:38–44. https://doi.org/10.1016/j.coph.2014.07.004

Markham A (2016) Ixekizumab: First Global Approval. Drugs 76(8):901–905. https://doi.org/10.1007/s40265-016-0579-y

Martin D, Galisteo R, Gutkind JS (2009) CXCL8/IL8 stimulates vascular endothelial growth factor (VEGF) expression and the autocrine activation of VEGFR2 in endothelial cells by activating NFkappaB through the CBM (Carma3/Bcl10/Malt1) complex. J Biol Chem 284(10):6038–6042. https://doi.org/10.1074/jbc.C800207200

Mera R, Fontham ET, Bravo LE, Bravo JC, Piazuelo MB, Camargo MC, Correa P (2005) Long term follow up of patients treated for *Helicobacter pylori* infection. Gut 54(11):1536–1540. https://doi.org/10.1136/gut.2005.072009

Mera RM, Bravo LE, Camargo MC, Bravo JC, Delgado AG, Romero-Gallo J, Yepez MC, Realpe JL, Schneider BG, Morgan DR, Peek RM Jr, Correa P, Wilson KT, Piazuelo MB (2018) Dynamics of *Helicobacter pylori* infection as a determinant of progression of gastric precancerous lesions: 16-year follow-up of an eradication trial. Gut 67(7):1239–1246. https://doi.org/10.1136/gutjnl-2016-311685

Meyskens FL Jr, McLaren CE, Pelot D, Fujikawa-Brooks S, Carpenter PM, Hawk E, Kelloff G, Lawson MJ, Kidao J, McCracken J, Albers CG, Ahnen DJ, Turgeon DK, Goldschmid S, Lance P, Hagedorn CH, Gillen DL, Gerner EW (2008) Difluoromethylornithine plus sulindac for the prevention of sporadic colorectal adenomas: a randomized placebo-controlled, double-blind trial. Cancer prevention research 1(1):32–38. https://doi.org/10.1158/1940-6207.CAPR-08-0042

Mills CD (2012) M1 and M2 macrophages: oracles of health and disease. Crit Rev Immunol 32(6):463–488

Moehler M, Gopfert K, Lenz HJ (2018) Outlook: immunotherapy in gastrointestinal carcinoma—innovative strategies. Oncol Res Treat 41(5):313–315. https://doi.org/10.1159/000489047

Murdoch C, Muthana M, Coffelt SB, Lewis CE (2008) Role Myeloid Cells Promot Tumour Angiogenesis. Nat Rev Cancer 8(8):618–631. https://doi.org/10.1038/nrc2444

Murukesh N, Dive C, Jayson GC (2010) Biomarkers of angiogenesis and their role in the development of VEGF inhibitors. Br J Cancer 102(1):8–18. https://doi.org/10.1038/sj.bjc.6605483

Nagarsheth N, Wicha MS, Zou W (2017) Chemokines in the cancer microenvironment and their relevance in cancer immunotherapy. Nat Rev Immunol 17(9):559–572. https://doi.org/10.1038/nri.2017.49

Naumann M, Sokolova O, Tegtmeyer N, Backert S (2017) *Helicobacter pylori*: a paradigm pathogen for subverting host cell signal transmission. Trends Microbiol 25(4):316–328. https://doi.org/10.1016/j.tim.2016.12.004

Nikfar S, Saiyarsarai P, Tigabu BM, Abdollahi M (2018) Efficacy and safety of interleukin-1 antagonists in rheumatoid arthritis: a systematic review and meta-analysis. Rheumatol. Int. https://doi.org/10.1007/s00296-018-4041-1

Ning Y, Labonte MJ, Zhang W, Bohanes PO, Gerger A, Yang D, Benhaim L, Paez D, Rosenberg DO, Nagulapalli Venkata KC, Louie SG, Petasis NA, Ladner RD, Lenz HJ (2012) The CXCR2 antagonist, SCH-527123, shows anti-tumor activity and sensitizes cells to oxaliplatin in preclinical colon cancer models. Mol Cancer Ther 11(6):1353–1364. https://doi.org/10.1158/1535-7163.MCT-11-0915

Noy R, Pollard JW (2014) Tumor-associated macrophages: from mechanisms to therapy. Immunity 41(1):49–61. https://doi.org/10.1016/j.immuni.2014.06.010

Ohta M, Kitadai Y, Tanaka S, Yoshihara M, Yasui W, Mukaida N, Haruma K, Chayama K (2003) Monocyte chemoattractant protein-1 expression correlates with macrophage infiltration and tumor vascularity in human gastric carcinomas. Int J Oncol 22(4):773–778

Ohtsu A, Shah MA, Van Cutsem E, Rha SY, Sawaki A, Park SR, Lim HY, Yamada Y, Wu J, Langer B, Starnawski M, Kang YK (2011) Bevacizumab in combination with chemotherapy as first-line therapy in advanced gastric cancer: a randomized, double-blind, placebo-controlled phase III study. J Clin Oncol: Off J Am Soc Clin Oncol 29(30):3968–3976. https://doi.org/10.1200/JCO.2011.36.2236

Okugawa Y, Toiyama Y, Ichikawa T, Kawamura M, Yasuda H, Fujikawa H, Saigusa S, Ohi M, Araki T, Tanaka K, Inoue Y, Tanaka M, Miki C, Kusunoki M (2018) Colony-stimulating factor-1 and colony-stimulating factor-1 receptor co-expression is associated with disease progression in gastric cancer. Int J Oncol 53(2):737–749. https://doi.org/10.3892/ijo.2018.4406

Pachathundikandi SK, Müller A, Backert S (2016) Inflammasome activation by *Helicobacter pylori* and Its implications for persistence and immunity. Curr Top Microbiol Immunol 397:117–131. https://doi.org/10.1007/978-3-319-41171-2_6

Parkhurst MR, Yang JC, Langan RC, Dudley ME, Nathan DA, Feldman SA, Davis JL, Morgan RA, Merino MJ, Sherry RM, Hughes MS, Kammula US, Phan GQ, Lim RM, Wank SA, Restifo NP, Robbins PF, Laurencot CM, Rosenberg SA (2011) T cells targeting carcinoembryonic antigen can mediate regression of metastatic colorectal cancer but induce severe transient colitis. Mol Ther: J Am Soc Gene Ther 19(3):620–626. https://doi.org/10.1038/mt.2010.272

Patel KR, Brown VA, Jones DJ, Britton RG, Hemingway D, Miller AS, West KP, Booth TD, Perloff M, Crowell JA, Brenner DE, Steward WP, Gescher AJ, Brown K (2010) Clinical pharmacology of resveratrol and its metabolites in colorectal cancer patients. Can Res 70(19):7392–7399. https://doi.org/10.1158/0008-5472.CAN-10-2027

Pavlakis N, Sjoquist KM, Martin AJ, Tsobanis E, Yip S, Kang YK, Bang YJ, Alcindor T, O'Callaghan CJ, Burnell MJ, Tebbutt NC, Rha SY, Lee J, Cho JY, Lipton LR, Wong M, Strickland A, Kim JW, Zalcberg JR, Simes J, Goldstein D (2016) Regorafenib for the treatment of advanced gastric cancer (INTEGRATE): a multinational placebo-controlled phase II trial. J Clin Oncol: Off J Am Soc Clin Oncol 34(23):2728–2735. https://doi.org/10.1200/JCO.2015.65.1901

Pegg AE (2006) Regulation of ornithine decarboxylase. J Biol Chem 281(21):14529–14532. https://doi.org/10.1074/jbc.R500031200

Pegg AE (2009) Mammalian polyamine metabolism and function. IUBMB Life 61(9):880–894. https://doi.org/10.1002/iub.230

Plummer M, de Martel C, Vignat J, Ferlay J, Bray F, Franceschi S (2016) Global burden of cancers attributable to infections in 2012: a synthetic analysis. Lancet Glob Health 4(9):e609–616. https://doi.org/10.1016/S2214-109X(16)30143-7

Plummer M, Franceschi S, Vignat J, Forman D, de Martel C (2015) Global burden of gastric cancer attributable to *Helicobacter pylori*. Int J Cancer 136(2):487–490. https://doi.org/10.1002/ijc.28999

Posselt G, Backert S, Wessler S (2013) The functional interplay of *Helicobacter pylori* factors with gastric epithelial cells induces a multi-step process in pathogenesis. Cell Commun Signal 11:77. https://doi.org/10.1186/1478-811X-11-77

Pyonteck SM, Akkari L, Schuhmacher AJ, Bowman RL, Sevenich L, Quail DF, Olson OC, Quick ML, Huse JT, Teijeiro V, Setty M, Leslie CS, Oei Y, Pedraza A, Zhang J, Brennan CW, Sutton JC, Holland EC, Daniel D, Joyce JA (2013) CSF-1R inhibition alters macrophage polarization and blocks glioma progression. Nat Med 19(10):1264–1272. https://doi.org/10.1038/nm.3337

Qian BZ, Pollard JW (2010) Macrophage diversity enhances tumor progression and metastasis. Cell 141(1):39–51. https://doi.org/10.1016/j.cell.2010.03.014

Quail DF, Bowman RL, Akkari L, Quick ML, Schuhmacher AJ, Huse JT, Holland EC, Sutton JC, Joyce JA (2016) The tumor microenvironment underlies acquired resistance to CSF-1R inhibition in gliomas. Science 352 (6288):aad3018. https://doi.org/10.1126/science.aad3018

Ralph C, Elkord E, Burt DJ, O'Dwyer JF, Austin EB, Stern PL, Hawkins RE, Thistlethwaite FC (2010) Modulation of lymphocyte regulation for cancer therapy: a phase II trial of tremelimumab in advanced gastric and esophageal adenocarcinoma. Clin Cancer Res: Off J Am Assoc Cancer Res 16(5):1662–1672. https://doi.org/10.1158/1078-0432.CCR-09-2870

Riechelmann R, Grothey A (2017) Antiangiogenic therapy for refractory colorectal cancer: current options and future strategies. Ther Adv Med Oncol 9(2):106–126. https://doi.org/10.1177/1758834016676703

Ries CH, Cannarile MA, Hoves S, Benz J, Wartha K, Runza V, Rey-Giraud F, Pradel LP, Feuerhake F, Klaman I, Jones T, Jucknischke U, Scheiblich S, Kaluza K, Gorr IH, Walz A, Abiraj K, Cassier PA, Sica A, Gomez-Roca C, de Visser KE, Italiano A, Le Tourneau C, Delord JP, Levitsky H, Blay JY, Ruttinger D (2014) Targeting tumor-associated macrophages with anti-CSF-1R antibody reveals a strategy for cancer therapy. Cancer Cell 25(6):846–859. https://doi.org/10.1016/j.ccr.2014.05.016

Ristimaki A, Honkanen N, Jankala H, Sipponen P, Harkonen M (1997) Expression of cyclooxygenase-2 in human gastric carcinoma. Can Res 57(7):1276–1280

Rothwell PM, Price JF, Fowkes FG, Zanchetti A, Roncaglioni MC, Tognoni G, Lee R, Belch JF, Wilson M, Mehta Z, Meade TW (2012) Short-term effects of daily aspirin on cancer incidence, mortality, and non-vascular death: analysis of the time course of risks and benefits in 51 randomised controlled trials. Lancet 379(9826):1602–1612. https://doi.org/10.1016/S0140-6736(11)61720-0

Rubie C, Kauffels A, Kolsch K, Glanemann M, Justinger C (2016) CXCL12/CXCR4 display an inverse mRNA expression profile in gastric carcinoma that correlates with tumor progression. Oncology letters 11(1):360–364. https://doi.org/10.3892/ol.2015.3850

Ruffell B, Affara NI, Coussens LM (2012) Differential macrophage programming in the tumor microenvironment. Trends Immunol 33(3):119–126. https://doi.org/10.1016/j.it.2011.12.001

Ruffell B, Coussens LM (2015) Macrophages and therapeutic resistance in cancer. Cancer Cell 27(4):462–472. https://doi.org/10.1016/j.ccell.2015.02.015

Salgia R, Stille JR, Weaver RW, McCleod M, Hamid O, Polzer J, Roberson S, Flynt A, Spigel DR (2017) A randomized phase II study of LY2510924 and carboplatin/etoposide versus carboplatin/etoposide in extensive-disease small cell lung cancer. Lung Cancer 105:7–13. https://doi.org/10.1016/j.lungcan.2016.12.020

Sandhu SK, Papadopoulos K, Fong PC, Patnaik A, Messiou C, Olmos D, Wang G, Tromp BJ, Puchalski TA, Balkwill F, Berns B, Seetharam S, de Bono JS, Tolcher AW (2013) A first-in-human, first-in-class, phase I study of carlumab (CNTO 888), a human monoclonal antibody against CC-chemokine ligand 2 in patients with solid tumors. Cancer Chemother Pharmacol 71(4):1041–1050. https://doi.org/10.1007/s00280-013-2099-8

Sanford M, McKeage K (2015) Secukinumab: first global approval. Drugs 75(3):329–338. https://doi.org/10.1007/s40265-015-0359-0

Sanmamed MF, Carranza-Rua O, Alfaro C, Onate C, Martin-Algarra S, Perez G, Landazuri SF, Gonzalez A, Gross S, Rodriguez I, Munoz-Calleja C, Rodriguez-Ruiz M, Sangro B, Lopez-Picazo JM, Rizzo M, Mazzolini G, Pascual JI, Andueza MP, Perez-Gracia JL, Melero I (2014)

Serum interleukin-8 reflects tumor burden and treatment response across malignancies of multiple tissue origins. Clin Cancer Res: Off J Am Assoc Cancer Res 20(22):5697–5707. https://doi.org/10.1158/1078-0432.CCR-13-3203
Sansone P, Bromberg J (2012) Targeting the interleukin-6/Jak/stat pathway in human malignancies. J Clin Oncol: Off J Am Soc Clin Oncol 30(9):1005–1014. https://doi.org/10.1200/JCO.2010.31.8907
Saulnier Sholler GL, Gerner EW, Bergendahl G, MacArthur RB, VanderWerff A, Ashikaga T, Bond JP, Ferguson W, Roberts W, Wada RK, Eslin D, Kraveka JM, Kaplan J, Mitchell D, Parikh NS, Neville K, Sender L, Higgins T, Kawakita M, Hiramatsu K, Moriya SS, Bachmann AS (2015) A phase i trial of DFMO targeting polyamine addiction in patients with relapsed/refractory neuroblastoma. PLoS ONE 10(5):e0127246. https://doi.org/10.1371/journal.pone.0127246
Schreiber RD, Old LJ, Smyth MJ (2011) Cancer immunoediting: integrating immunity's roles in cancer suppression and promotion. Science 331(6024):1565–1570. https://doi.org/10.1126/science.1203486
Schroder M, Loos S, Naumann SK, Bachran C, Krotschel M, Umansky V, Helming L, Swee LK (2017) Identification of inhibitors of myeloid-derived suppressor cells activity through phenotypic chemical screening. Oncoimmunology 6(1):e1258503. https://doi.org/10.1080/2162402X.2016.1258503
Semenza GL (2003) Targeting HIF-1 for cancer therapy. Nat Rev Cancer 3(10):721–732. https://doi.org/10.1038/nrc1187
Semenza GL (2012) Hypoxia-inducible factors in physiology and medicine. Cell 148(3):399–408. https://doi.org/10.1016/j.cell.2012.01.021
Semper RP, Mejias-Luque R, Gross C, Anderl F, Muller A, Vieth M, Busch DH, Prazeres da Costa C, Ruland J, Gross O, Gerhard M (2014) *Helicobacter pylori*-induced IL-1beta secretion in innate immune cells is regulated by the NLRP3 inflammasome and requires the cag pathogenicity island. J Immunol 193(7):3566–3576. https://doi.org/10.4049/jimmunol.1400362
Senger DR, Galli SJ, Dvorak AM, Perruzzi CA, Harvey VS, Dvorak HF (1983) Tumor cells secrete a vascular permeability factor that promotes accumulation of ascites fluid. Science 219(4587):983–985
Shankaran V, Ikeda H, Bruce AT, White JM, Swanson PE, Old LJ, Schreiber RD (2001) IFNgamma and lymphocytes prevent primary tumour development and shape tumour immunogenicity. Nature 410(6832):1107–1111. https://doi.org/10.1038/35074122
Shen L, Li J, Xu J, Pan H, Dai G, Qin S, Wang L, Wang J, Yang Z, Shu Y, Xu R, Chen L, Liu Y, Yu S, Bu L, Piao Y (2015) Bevacizumab plus capecitabine and cisplatin in Chinese patients with inoperable locally advanced or metastatic gastric or gastroesophageal junction cancer: randomized, double-blind, phase III study (AVATAR study). Gastric Cancer: Off J Int Gastric Cancer Assoc Jpn. Gastric Cancer Assoc 18(1):168–176. https://doi.org/10.1007/s10120-014-0351-5
Shibata W, Ariyama H, Westphalen CB, Worthley DL, Muthupalani S, Asfaha S, Dubeykovskaya Z, Quante M, Fox JG, Wang TC (2013) Stromal cell-derived factor-1 overexpression induces gastric dysplasia through expansion of stromal myofibroblasts and epithelial progenitors. Gut 62(2):192–200. https://doi.org/10.1136/gutjnl-2011-301824
Sierra JC, Asim M, Verriere TG, Piazuelo MB, Suarez G, Romero-Gallo J, Delgado AG, Wroblewski LE, Barry DP, Peek RM Jr, Gobert AP, Wilson KT (2018) Epidermal growth factor receptor inhibition downregulates *Helicobacter pylori*-induced epithelial inflammatory responses. DNA Damage Gastric Carcinogenesis. Gut 67(7):1247–1260. https://doi.org/10.1136/gutjnl-2016-312888
Sjokvist Ottsjo L, Flach CF, Nilsson S, Malefyt Rde W, Walduck AK, Raghavan S (2015) Defining the Roles of IFN-gamma and IL-17A in Inflammation and Protection against *Helicobacter pylori* Infection. PLoS ONE 10(7):e0131444. https://doi.org/10.1371/journal.pone.0131444
Smolka AJ, Backert S (2012) How *Helicobacter pylori* infection controls gastric acid secretion. J Gastroenterol 47(6):609–618. https://doi.org/10.1007/s00535-012-0592-1
Solomon DH, Mercer E, Kavanaugh A (2012) Observational studies on the risk of cancer associated with tumor necrosis factor inhibitors in rheumatoid arthritis: a review of their methodologies and results. Arthritis Rheum 64(1):21–32. https://doi.org/10.1002/art.30653

Song H, Zhu J, Lu D (2016) Molecular-targeted first-line therapy for advanced gastric cancer. Cochrane Database Syst Rev 7:CD011461. https://doi.org/10.1002/14651858.cd011461.pub2

Sota J, Vitale A, Insalaco A, Sfriso P, Lopalco G, Emmi G, Cattalini M, Manna R, Cimaz R, Priori R, Talarico R, de Marchi G, Frassi M, Gallizzi R, Soriano A, Alessio M, Cammelli D, Maggio MC, Gentileschi S, Marcolongo R, La Torre F, Fabiani C, Colafrancesco S, Ricci F, Galozzi P, Viapiana O, Verrecchia E, Pardeo M, Cerrito L, Cavallaro E, Olivieri AN, Paolazzi G, Vitiello G, Maier A, Silvestri E, Stagnaro C, Valesini G, Mosca M, de Vita S, Tincani A, Lapadula G, Frediani B, De Benedetti F, Iannone F, Punzi L, Salvarani C, Galeazzi M, Angotti R, Messina M, Tosi GM, Rigante D, Cantarini L, Working Group" of Systemic Autoinflammatory Diseases of SIR (2018) Safety profile of the interleukin-1 inhibitors anakinra and canakinumab in real-life clinical practice: a nationwide multicenter retrospective observational study. Clinical rheumatology. https://doi.org/10.1007/s10067-018-4119-x

Spence AD, Busby J, Johnston BT, Baron JA, Hughes CM, Coleman HG, Cardwell CR (2018) Low-dose aspirin use does not increase survival in 2 independent population-based cohorts of patients with esophageal or gastric cancer. Gastroenterology 154(4):849–860 e841. https://doi.org/10.1053/j.gastro.2017.10.044

Spratlin JL, Cohen RB, Eadens M, Gore L, Camidge DR, Diab S, Leong S, O'Bryant C, Chow LQ, Serkova NJ, Meropol NJ, Lewis NL, Chiorean EG, Fox F, Youssoufian H, Rowinsky EK, Eckhardt SG (2010) Phase I pharmacologic and biologic study of ramucirumab (IMC-1121B), a fully human immunoglobulin G1 monoclonal antibody targeting the vascular endothelial growth factor receptor-2. J Clin Oncol: Off J Am Soc Clin Oncol 28(5):780–787. https://doi.org/10.1200/JCO.2009.23.7537

Srivatsa S, Paul MC, Cardone C, Holcmann M, Amberg N, Pathria P, Diamanti MA, Linder M, Timelthaler G, Dienes HP, Kenner L, Wrba F, Prager GW, Rose-John S, Eferl R, Liguori G, Botti G, Martinelli E, Greten FR, Ciardiello F, Sibilia M (2017) EGFR in tumor-associated myeloid cells promotes development of colorectal cancer in mice and associates with outcomes of patients. Gastroenterology 153(1):178–190 e110. https://doi.org/10.1053/j.gastro.2017.03.053

Staples E, Ingram RJ, Atherton JC, Robinson K (2013) Optimising the quantification of cytokines present at low concentrations in small human mucosal tissue samples using Luminex assays. J Immunol Methods 394(1–2):1–9. https://doi.org/10.1016/j.jim.2013.04.009

Suzuki S, Dobashi Y, Hatakeyama Y, Tajiri R, Fujimura T, Heldin CH, Ooi A (2010) Clinicopathological significance of platelet-derived growth factor (PDGF)-B and vascular endothelial growth factor-a expression, PDGF receptor-beta phosphorylation, and microvessel density in gastric cancer. BMC Cancer 10:659. https://doi.org/10.1186/1471-2407-10-659

Tabernero J, Melero I, Ros W, Argiles G, Marabelle A, Rodriguez-Ruiz M (2017) Phase Ia and Ib studies of the novel carcinoembryonic antigen (CEA) T-cell bispecific (CEA CD3 TCB) antibody as a single agent and in combination with atezolizumab: Preliminary efficacy and safety in patients with metastatic colorectal cancer (mCRC). J Clin Oncol, 35. ACOS. https://doi.org/10.1200/jco.2017.35.15_suppl.3002

Tao LL, Shi SJ, Chen LB, Huang GC (2014) Expression of monocyte chemotactic protein-1/CCL2 in gastric cancer and its relationship with tumor hypoxia. World J Gastroenterol 20(15):4421–4427. https://doi.org/10.3748/wjg.v20.i15.4421

Tap WD, Wainberg ZA, Anthony SP, Ibrahim PN, Zhang C, Healey JH, Chmielowski B, Staddon AP, Cohn AL, Shapiro GI, Keedy VL, Singh AS, Puzanov I, Kwak EL, Wagner AJ, Von Hoff DD, Weiss GJ, Ramanathan RK, Zhang J, Habets G, Zhang Y, Burton EA, Visor G, Sanftner L, Severson P, Nguyen H, Kim MJ, Marimuthu A, Tsang G, Shellooe R, Gee C, West BL, Hirth P, Nolop K, van de Rijn M, Hsu HH, Peterfy C, Lin PS, Tong-Starksen S, Bollag G (2015) Structure-Guided Blockade of CSF1R Kinase in Tenosynovial Giant-Cell Tumor. N Engl J Med 373(5):428–437. https://doi.org/10.1056/NEJMoa1411366

Thorsson V, Gibbs DL, Brown SD, Wolf D, Bortone DS, Ou Yang TH, Porta-Pardo E, Gao GF, Plaisier CL, Eddy JA, Ziv E, Culhane AC, Paull EO, Sivakumar IKA, Gentles AJ, Malhotra R, Farshidfar F, Colaprico A, Parker JS, Mose LE, Vo NS, Liu J, Liu Y, Rader J, Dhankani V, Reynolds SM, Bowlby R, Califano A, Cherniack AD, Anastassiou D, Bedognetti D, Rao A,

Chen K, Krasnitz A, Hu H, Malta TM, Noushmehr H, Pedamallu CS, Bullman S, Ojesina AI, Lamb A, Zhou W, Shen H, Choueiri TK, Weinstein JN, Guinney J, Saltz J, Holt RA, Rabkin CE, Cancer Genome Atlas Research N, Lazar AJ, Serody JS, Demicco EG, Disis ML, Vincent BG, Shmulevich L (2018) The immune landscape of cancer. Immunity 48(4):812–830 e814. https://doi.org/10.1016/j.immuni.2018.03.023

Tiemessen MM, Jagger AL, Evans HG, van Herwijnen MJ, John S, Taams LS (2007) CD4 + CD25 + Foxp3 + regulatory T cells induce alternative activation of human monocytes/macrophages. Proc Natl Acad Sci U S A 104(49):19446–19451. https://doi.org/10.1073/pnas.0706832104

Tonouchi H, Miki C, Tanaka K, Kusunoki M (2002) Profile of monocyte chemoattractant protein-1 circulating levels in gastric cancer patients. Scand J Gastroenterol 37(7):830–833

Tu S, Bhagat G, Cui G, Takaishi S, Kurt-Jones EA, Rickman B, Betz KS, Penz-Oesterreicher M, Bjorkdahl O, Fox JG, Wang TC (2008) Overexpression of interleukin-1beta induces gastric inflammation and cancer and mobilizes myeloid-derived suppressor cells in mice. Cancer Cell 14(5):408–419. https://doi.org/10.1016/j.ccr.2008.10.011

Udalova I, Monaco C, Nanchahal J, Feldmann M (2016) Anti-TNF therapy. microbiology spectrum 4(4). https://doi.org/10.1128/microbiolspec.mchd-0022-2015

Uefuji K, Ichikura T, Mochizuki H (2000) Cyclooxygenase-2 expression is related to prostaglandin biosynthesis and angiogenesis in human gastric cancer. Clin Cancer Res: Off J Am Assoc Cancer Res 6(1):135–138

Van Cutsem E, Sagaert X, Topal B, Haustermans K, Prenen H (2016) Gastric cancer. Lancet 388(10060):2654–2664. https://doi.org/10.1016/S0140-6736(16)30354-3

Vela M, Aris M, Llorente M, Garcia-Sanz JA, Kremer L (2015) Chemokine receptor-specific antibodies in cancer immunotherapy: achievements and challenges. Front Immunol 6:12. https://doi.org/10.3389/fimmu.2015.00012

Virchow R (1863) Die Krankhaften Geschwülste (translation, Diseased Tumors). Hirschwald, Berlin, Germany

Vonderheide RH (2018) The immune revolution: a case for priming. Not Checkpoint. Cancer cell 33(4):563–569. https://doi.org/10.1016/j.ccell.2018.03.008

Wainberg ZA, Jalal S, Muro K, Yoon HH, Garrido M, Golan T, Doi T, DCatenacci DV, Geva R, Ku G, Bleeker J, Bang Y, Hara H, Chung H, Savage M, Want J, Koshiji M, Dalal R, Fuchs CS (2017) KEYNOTE-059 Update: efficacy and safety of pembrolizumab alone or in combination with chemotherapy in patients with advanced gastric or gastroesophageal (G/GEJ) cancer. Annu Oncol 28(suppl 5):v616–v617

Waldner MJ, Neurath MF (2012) Targeting the VEGF signaling pathway in cancer therapy. Expert Opin Ther Targets 16(1):5–13. https://doi.org/10.1517/14728222.2011.641951

Waldner MJ, Wirtz S, Jefremow A, Warntjen M, Neufert C, Atreya R, Becker C, Weigmann B, Vieth M, Rose-John S, Neurath MF (2010) VEGF receptor signaling links inflammation and tumorigenesis in colitis-associated cancer. J Exp Med 207(13):2855–2868. https://doi.org/10.1084/jem.20100438

Wang D, DuBois RN (2018) Role of prostanoids in gastrointestinal cancer. J Clin Investig https://doi.org/10.1172/jci97953

Wang J, Hu W, Wang K, Yu J, Luo B, Luo G, Wang W, Wang H, Li J, Wen J (2016) Repertaxin, an inhibitor of the chemokine receptors CXCR1 and CXCR2, inhibits malignant behavior of human gastric cancer MKN45 cells in vitro and in vivo and enhances efficacy of 5-fluorouracil. Int J Oncol 48(4):1341–1352. https://doi.org/10.3892/ijo.2016.3371

Wang T, Cai H, Zheng W, Michel A, Pawlita M, Milne G, Xiang YB, Gao YT, Li HL, Rothman N, Lan Q, Shu XO, Epplein M (2017) A Prospective Study of Urinary Prostaglandin E2 Metabolite, *Helicobacter pylori* Antibodies, and Gastric Cancer Risk. Clin Infect Dis: Off Publ Infect Dis Soc Am 64(10):1380–1386. https://doi.org/10.1093/cid/cix106

Wang Z, Si X, Xu A, Meng X, Gao S, Qi Y, Zhu L, Li T, Li W, Dong L (2013) Activation of STAT3 in human gastric cancer cells via interleukin (IL)-6-type cytokine signaling correlates with clinical implications. PLoS ONE 8(10):e75788. https://doi.org/10.1371/journal.pone.0075788

Weinreich DM, Elaraj DM, Puhlmann M, Hewitt SM, Carroll NM, Feldman ED, Turner EM, Spiess PJ, Alexander HR (2003) Effect of interleukin 1 receptor antagonist gene transduction on human melanoma xenografts in nude mice. Can Res 63(18):5957–5961

Wilhelm SM, Dumas J, Adnane L, Lynch M, Carter CA, Schutz G, Thierauch KH, Zopf D (2011) Regorafenib (BAY 73-4506): a new oral multikinase inhibitor of angiogenic, stromal and oncogenic receptor tyrosine kinases with potent preclinical anti-tumor activity. Int J Cancer 129(1):245–255. https://doi.org/10.1002/ijc.25864

Wilke H, Muro K, Van Cutsem E, Oh SC, Bodoky G, Shimada Y, Hironaka S, Sugimoto N, Lipatov O, Kim TY, Cunningham D, Rougier P, Komatsu Y, Ajani J, Emig M, Carlesi R, Ferry D, Chandrawansa K, Schwartz JD, Ohtsu A, Group RS (2014) Ramucirumab plus paclitaxel versus placebo plus paclitaxel in patients with previously treated advanced gastric or gastro-oesophageal junction adenocarcinoma (RAINBOW): a double-blind, randomised phase 3 trial. Lancet Oncol 15(11):1224–1235. https://doi.org/10.1016/s1470-2045(14)70420-6

Wilson AJ, Byron K, Gibson PR (1999) Interleukin-8 stimulates the migration of human colonic epithelial cells in vitro. Clin Sci 97(3):385–390

Xue LJ, Mao XB, Ren LL, Chu XY (2017) Inhibition of CXCL12/CXCR4 axis as a potential targeted therapy of advanced gastric carcinoma. Cancer Med 6(6):1424–1436. https://doi.org/10.1002/cam4.1085

Yamada S, Kato S, Matsuhisa T, Makonkawkeyoon L, Yoshida M, Chakrabandhu T, Lertprasertsuk N, Suttharat P, Chakrabandhu B, Nishiumi S, Chongraksut W, Azuma T (2013) Predominant mucosal IL-8 mRNA expression in non-cagA Thais is risk for gastric cancer. World J Gastroenterol 19(19):2941–2949. https://doi.org/10.3748/wjg.v19.i19.2941

Yamada Y, Saito H, Ikeguchi M (2012) Prevalence and clinical relevance of Th17 cells in patients with gastric cancer. J Surg Res 178(2):685–691. https://doi.org/10.1016/j.jss.2012.07.055

Yan F, Cao H, Chaturvedi R, Krishna U, Hobbs SS, Dempsey PJ, Peek RM, Jr., Cover TL, Washington MK, Wilson KT, Polk DB (2009) Epidermal growth factor receptor activation protects gastric epithelial cells from *Helicobacter pylori*-induced apoptosis. Gastroenterology **136**(4):1297–1307, e1291–1293. https://doi.org/10.1053/j.gastro.2008.12.059

Yin S, Lan C, Pei H, Zhu Z (2016) Expression of interleukin 1beta in gastric cancer tissue and its effects on gastric cancer. OncoTargets Therapy 9:31–35. https://doi.org/10.2147/OTT.S94277

Zell JA, McLaren CE, Chen WP, Thompson PA, Gerner EW, Meyskens FL (2010) Ornithine decarboxylase-1 polymorphism, chemoprevention with eflornithine and sulindac, and outcomes among colorectal adenoma patients. J Natl Cancer Inst 102(19):1513–1516. https://doi.org/10.1093/jnci/djq325

Zeng H, Chen W, Zheng R, Zhang S, Ji JS, Zou X, Xia C, Sun K, Yang Z, Li H, Wang N, Han R, Liu S, Li H, Mu H, He Y, Xu Y, Fu Z, Zhou Y, Jiang J, Yang Y, Chen J, Wei K, Fan D, Wang J, Fu F, Zhao D, Song G, Chen J, Jiang C, Zhou X, Gu X, Jin F, Li Q, Li Y, Wu T, Yan C, Dong J, Hua Z, Baade P, Bray F, Jemal A, Yu XQ, He J (2018) Changing cancer survival in China during 2003–15: a pooled analysis of 17 population-based cancer registries. The Lancet Glob Health 6(5):e555–e567. https://doi.org/10.1016/S2214-109X(18)30127-X

Zhang B, Rong G, Wei H, Zhang M, Bi J, Ma L, Xue X, Wei G, Liu X, Fang G (2008) The prevalence of Th17 cells in patients with gastric cancer. Biochem Biophys Res Commun 374(3):533–537. https://doi.org/10.1016/j.bbrc.2008.07.060

Zhang QW, Liu L, Gong CY, Shi HS, Zeng YH, Wang XZ, Zhao YW, Wei YQ (2012) Prognostic significance of tumor-associated macrophages in solid tumor: a meta-analysis of the literature. PLoS ONE 7(12):e50946. https://doi.org/10.1371/journal.pone.0050946

Zhang S, Zhang XQ, Ding XW, Yang RK, Huang SL, Kastelein F, Bruno M, Yu XJ, Zhou D, Zou XP (2014) Cyclooxygenase inhibitors use is associated with reduced risk of esophageal adenocarcinoma in patients with Barrett's esophagus: a meta-analysis. Br J Cancer 110(9):2378–2388. https://doi.org/10.1038/bjc.2014.127

Zhang XS, Tegtmeyer N, Traube L, Jindal S, Perez-Perez G, Sticht H, Backert S, Blaser MJ (2015) A specific A/T polymorphism in Western tyrosine phosphorylation B-motifs regulates *Helicobacter pylori* CagA epithelial cell interaction. PloS Pathog 11:e1004621. https://doi.org/10.1371/journal.ppat.1004621

Zhu Y, Knolhoff BL, Meyer MA, Nywening TM, West BL, Luo J, Wang-Gillam A, Goedegebuure SP, Linehan DC, DeNardo DG (2014) CSF1/CSF1R blockade reprograms tumor-infiltrating macrophages and improves response to T-cell checkpoint immunotherapy in pancreatic cancer models. Can Res 74(18):5057–5069. https://doi.org/10.1158/0008-5472.CAN-13-3723

Zulueta A, Caretti A, Signorelli P, Ghidoni R (2015) Resveratrol: a potential challenger against gastric cancer. World J Gastroenterol 21(37):10636–10643. https://doi.org/10.3748/wjg.v21.i37.10636

The manufacturer's authorised representative in the EU is Springer Nature Customer Service Centre GmbH, Europaplatz 3, 69115 Heidelberg, Germany. If you have any concerns regarding our products, please contact ProductSafety@springernature.com

Printed and bound by CPI Group (UK) Ltd, Croydon, CR0 4YY
05/10/2026
02228130-0012